Two Manuals Combined into One Book



# NBC Decontamination

# And

# Ranger Skills Handbook

Department of Defense

Pentagon
Publishing

***FM 3-5**
***MCWP 3-37.3**

Field Manual
No. 3-5

Headquarters,
Department of the Army

Marine Corps Warfighting Publication (MCWP)
3-37.3

Commandant,
US Marine Corps

Washington, DC, 28 July 2000

# NBC DECONTAMINATION

## Contents





* This publication supersedes Field Manual (FM) 305/Fleet Marine Force Manual (FMFM) 11-10, 17 November 1993.

Page

# Preface

FM 3-5 integrates the nuclear, biological, and chemical (NBC) fundamentals published in FM 3-100 and implements Standardized North Atlantic Treaty Organization Agreement (STANAG) 2426.

The extent and timing of decon depends on the tactical situation, the mission, the area of contamination, and the decon resources available. Survivability and reducing the effect of any chemical threat are the ultimate goals of decon. This manual provides detailed guidance on conducting decon operations performed by chemical and nonchemical personnel. Individual soldiers and leaders must be familiar with the basic procedures and concepts in this manual.

Appendix A contains an English to metric measurement conversion chart.

The proponent of this publication is Headquarters TRADOC. To submit changes for improving this publication use Department of the Army (DA) Form 2028 (Recommended Changes to Publications and Blank Forms) and forward to Commandant, US Army Chemical School, ATTN: ATSN-CMZ, Fort Leonard Wood, Missouri 65473-8926.

*Unless this publication states otherwise, nouns and pronouns do not refer exclusively to men.*

# Chapter 1

# Introduction

The use of NBC weapons creates unique residual hazards that may require decon. In addition to the deliberate use of these weapons, collateral damage, natural disasters, and industrial emitters may require decon. The presence of contamination generally reduces the effectiveness of our combat power. Contamination forces us into protective equipment that degrades our ability to perform individual and collective tasks.

## CONTAMINATION FORMS

1-1. The following are the different forms of contamination:

- Solids – radioactive particles, biological spores, or dusty agents that could appear as a fine dust.
- Liquids – liquid droplets that fall like rain. Droplets can range from thick and sticky to the consistency of water.
- Vapors or gases – created by bursting munitions or generators. These clouds are affected by the weather and can cover large areas.
- Aerosols – fine liquids or solid particles suspended in the air. They behave much like vapors.

## CONTAMINATION-HAZARDS TRANSMISSION

1-2. Contamination hazards can be transmitted in the following manner:

- **Transfer**. Anything that touches a surface covered with liquid or solid contamination will tend to pick up that contamination and move it from one surface to another.
- **Spread**. Touching a surface covered with liquid chemical agent can spread contamination on the same surface, thereby, increasing the size of the contaminated area.
- **Vapor**. Vapors can be carried through the air in the form of a dust, atomized liquids (aerosols), or true gases. Vapors in an open/outdoor area disperse rapidly, so there is no need to decon.
- **Desorption**. Liquid-chemical contamination absorbs into porous surfaces. Once absorbed, it begins to desorb or give off gas; that is, low levels of vapor pass out of the contaminated surface into the air and can be transferred to any surface that contacts it, including bare skin.
- **Radiation**. Radiation is given off by radioactive dust or dirt, most of which appears as fallout. For decon purposes, radiation can be thought of as a solid.

## AGENT CLASSIFICATION

1-3. Depending on the length of time agents will be a hazard, they are classified as—

- Nonpersistent – an immediate threat that lasts a few minutes. They rarely require decon.
- Persistent – takes a longer time to act and may last for days. In a protected environment, these agents can last for long periods of time. All agents are affected to some extent by the weather. Even radiological particles can "rain out" of the air and form hot spots on the ground. Biological organisms are greatly affected by sunlight. Chemical agents can be decontaminated by the weather. The duration of a hazard is a complex estimation that is based on numerous factors which include the—
  - Type of contamination.
  - Contamination density and droplet size.
  - Temperature.
  - Wind speed.
  - Sunlight.
  - Humidity and rain.
  - Composition of the contaminated surface.
  - Type of soil and terrain.

## DECONTAMINANTS

1-4. Decontaminants are described as follows:

- Natural – weather (rain, wind, and humidity).
- Standard – supertropical bleach (STB) and decontaminating solution number 2 (DS2).
- Nonstandard – soaps and detergents.

1-5. For more information on the types of decontaminants, see Appendix B.

## DECON CONCEPTS

1-6. Whenever soldiers are unable to avoid contamination and have to use protective measures, decon is necessary to allow them to remove their protective gear and resume normal operations. Weathering is the most desirable means of decon; however, time and operational needs may not permit that option.

1-7. This chapter describes when, where, and how much to decon. Protective clothing (mission-oriented protective posture [MOPP] gear), protective equipment, and collective protective shelters (CPSs) offer only a temporary solution. Decon is the removal, destruction, or naturalization of contamination. If you become contaminated, some decon must occur as soon as possible.

## DECON DECISIONS

1-8. The decision to decon is a risk assessment and is made within the context of the mission, enemy, terrain, troops, time available, and civilian consideration (METT-TC) and the resources available. Decon must be considered if the contamination levels exceed the negligible risk levels as follows:

- Chemical and biological contamination causes mild incapacitation in 5 percent or less of unprotected soldiers operating for 12 continuous hours within 1 meter of contamination. For the chemical-agent monitor (CAM), this equates to a one-bar reading at a distance of 1 inch from the surface.
- Radiological contamination can cause mild incapacitation in 2.5 percent or less of unprotected, nonpreviously exposed soldiers. This equates to a reading of no more than 0.33 centigray per hour (cGy/hr).

1-9. MOPP-gear exchange provides excellent protection against field concentrations of agents; however, wearing the gear causes performance degradation. Decon is performed to restore the normal operating tempo, but the logistical support that is required to keep soldiers in MOPP gear impacts operations. Therefore, decon should be conducted as soon as practical. Table 1-1 provides comparison data for decon levels/techniques.

**Table 1-1. Comparison Data for Decon Levels/Techniques**

| Levels | Techniques[1] | Best Start Time | Performed by | Advantages |
|---|---|---|---|---|
| Immediate | Skin decon<br>Personal wipe down<br>Operator's spray down | Before 1 minute<br>Within 15 minutes | Individual<br>Individual or crew | Stops agent from penetrating. |
| Operational | MOPP-gear exchange[2]<br>Vehicle washdown[3] | Within 6 hours | Unit<br>Battalion crew or decon platoon (-) | Provides possible temporary relief from MOPP4. Limits liquid agent spread. |
| Thorough | DED and DAD<br>DTD | When mission allows reconstitution | Decon platoon<br>Unit | Provides probable long-term MOPP reduction with minimum risk. |

[1]The techniques become less effective the longer they are delayed.

[2]Performance degradation and risk assessment must be considered when exceeding 6 hours. See FM 3-4 for battle-dress overgarment (BDO) risk assessment.

[3]Vehicle washdown is most effective if started within 1 hour.

## DECON PRINCIPLES

1-10. Decon immediately for a chemical agent on the skin. Perform higher levels of decon as a result of the risk assessment. Personnel should—

- Decon as soon as possible. The sooner the contamination is removed, the sooner MOPP levels can be reduced and combat power can be restored.
- Decon only what is necessary. Weathering is the least costly method of decon. Expend resources where they count.
- Decon as far forward as possible. While METT-TC dependent, performing decon as close to the point as possible reduces the spread of contamination and minimizes any transfer hazard.
- Decon by priority. The commander identifies which items are most critical to the mission. Restoring combat power and reestablishing operating tempo are key.

## DECON LEVELS

1-11. The three levels of decon operations are immediate, operational, and thorough (see Table 1-1, page 1-3).

### IMMEDIATE

1-12. Skin decon is a soldier's basic survival skill and should be performed within 1 minute of being contaminated using the skin decontaminating kit (SDK).

1-13. Personal wipe down should be performed within 15 minutes. This is done to remove contamination from individual equipment by using an individual equipment decon kit (IEDK). Use detector paper or a CAM to locate the agent. For radiological contamination, use a radiac set to locate it and then brush, wipe, or shake it off.

1-14. Operators' spray down should be done within 15 minutes. Operators use the on-board decon apparatus to decon surfaces that they must touch or contact to operate the equipment. Radiological contamination in the form of dust particles may be wiped, scraped, or brushed off.

### OPERATIONAL

1-15. A MOPP-gear exchange should be performed within 6 hours of being contaminated when thorough decon cannot be done. Soldiers will continue to wear MOPP gear and the operating tempo will be reduced.

1-16. Vehicle washdown should be performed within 6 hour of being contaminated when the mission does not permit a thorough decon. This process removes gross contamination and limits the spread of it.

### THOROUGH

1-17. Detailed equipment decon (DED) and detailed aircraft decon (DAD) restore items so that they can be used without MOPP gear. Normally, the DED and the DAD are conducted as part of a reconstitution or during breaks in combat operations. These operations require support from a chemical decon unit.

1-18. Detailed troop decon (DTD) normally takes place in conjunction with DED/DAD. The contaminated unit conducts this process and supports the DED/DAD operations.

## CHEMICAL-AGENT-RESISTANT COATING (CARC)

1-19. Army equipment is painted with CARC, which precludes the absorption of chemical agents and, thus, facilitates decon. By preventing penetration of an agent, contamination is easier to remove. Thickened agents are still a problem, but detergent and water should remove them.

1-20. The proper CARC paint colors should be used to paint bumper numbers, spot painting, and so forth. Do not use other paints in lieu of CARC.

## WEATHER EFFECTS

1-21. If METT-TC allows, leaving the contamination alone is the easiest method of decon. However, marking, reporting, and periodically rechecking the contamination are required.

### TEMPERATURE

1-22. The higher the temperature, the faster the rate of evaporation. Contamination persistency decreases as the temperature rises. It has no effect on radiological contamination.

### WIND

1-23. Aeration aids in decon. Agents are dispersed by the wind, thus reducing the concentration.

### HUMIDITY AND RAIN

1-24. Moisture tends to break down chemical agents but does so slowly. Heavy rain physically removes contamination; however, contaminated runoff may build up in drainage areas. Biological agents dehydrate (dry out) in low humidity, thus reducing their persistency. Rain may prevent the desorption of aerosols and leach contamination into the soil.

### SUNLIGHT

1-25. Sunlight hastens the evaporation and decomposition of agents. Ultraviolet and infrared radiation in sunlight rapidly kills most biological agents.

### TIME

1-26. Only time will neutralize and destroy radiological contamination. Other techniques merely shield or remove the problem.

## DECON IN COMBAT

1-27. Combat effectiveness is degraded by MOPP. Decon can remove the contamination and restore combat power. The detection and warning network is vital to ensure that soldiers and equipment are protected when contamination is encountered. Considerations of decon in combat are—

- Immediate decon. It saves lives and permits the use of individual equipment and key systems.
- Operational decon. It reduces the spread and the level of contamination. In some cases, when combined with weathering, MOPP levels may be reduced without further decon.
- Thorough decon. It removes the unit from the fight but allows it to return with restored effectiveness.

1-28. Figure 1-1 depicts the drop in effectiveness as the affected unit and soldiers react.

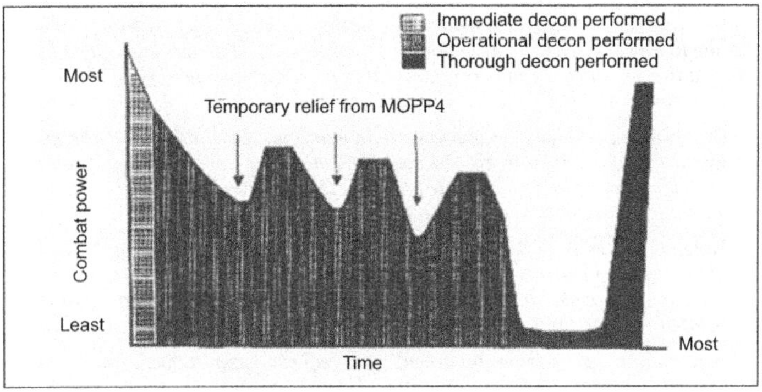


**Figure 1-1. How Decon Affects Combat Effectiveness**

## Decon-Operation Scenario

*You are the commander of the 155-millimeter Self-Propelled Howitzer Battery. Your unit is hit by a persistent nerve agent. Soldiers react by taking cover and going to MOPP4. The warning system provided sufficient alert and only a few have to conduct skin decon; most contamination is on the outside of your howitzers and vehicles.*

*The ongoing fire mission continues with minimal disruption, and the enemy is being detected. Your well-trained soldiers complete a personal wipe down and an operator's spray down as soon as possible. During this period, you are capable of providing fire support and moving, as necessary.*

*The temperature is 55°F, and humidity is low. While the work rate is strenuous, your troops drink water and suffer no heat casualties. However, after several hours, your response times to the fire mission are getting longer and soldiers have not eaten in more than 12 hours.*

*Upon making a decision to decon, you request support from your battalion. Coordination is made for an operational decon en route to an alternate position. You have 2 ½ hours to complete the movement and be in a firing position. The alternate position is 20 minutes away, with the decon site about halfway.*

*You send the advance party to meet the supporting power-driven decon-equipment (PDDE) crew and organize the site. The crew sets up and runs the vehicle washdown while the advance party, with battery supply personnel, prepares the MOPP-gear exchange.*

*The battalion chemical noncommissioned officer (NCO) directs the vehicle washdown as you rotate one platoon at a time through the site. The firing platoons go quickly as it is critical to have the artillery firing tables available. You complete the operation and are in position to fire at the prescribed time. Your soldiers are in MOPP4 because a vapor hazard still exists from the residual contamination. The advance party found the new area clear of contamination, and ammo resupply commences.*

*As sections begin improving their positions, gun crews one and two leave one soldier on each gun and move the rest of the soldiers 65 feet upwind. After a check on contamination proves negative, the battery NBC NCO directs unmasking procedures using two crew members. When no symptoms show up, the crews eat and get relief from the masks. The wind direction is monitored to ensure that the soldiers stay upwind of any vapors desorbing from their equipment.*

*Soldiers on the guns rotate into the clean area for rest and relief. Those returning to the guns assume MOPP4.*

# Chapter 2

# Immediate Decon

Once a soldier is aware of chemical or biological contamination on his bare skin, he initiates immediate decon techniques, without command, by using his personal SDK. He decontaminates his hood, mask, gloves, and weapon using the IEDK or an additional SDK. To remove radiological contamination from equipment and personnel, brush it off and/or use soap and water.

**NOTE: Throughout this manual, SDK refers to the M291 kit and IEDK refers to the M295 kit.**

## SKIN DECON

2-1. Start the skin-decon techniques within 1 minute of becoming contaminated. Some toxic chemical agents, especially nerve agents, kill in minutes.

### CHEMICAL

2-2. Use the SDKs within 1 minute of contamination (see Figure 2-1). Instructions for use are listed on the outside of the kit itself, on the individual packet within the kit, and in STP 21-1 SMCT.

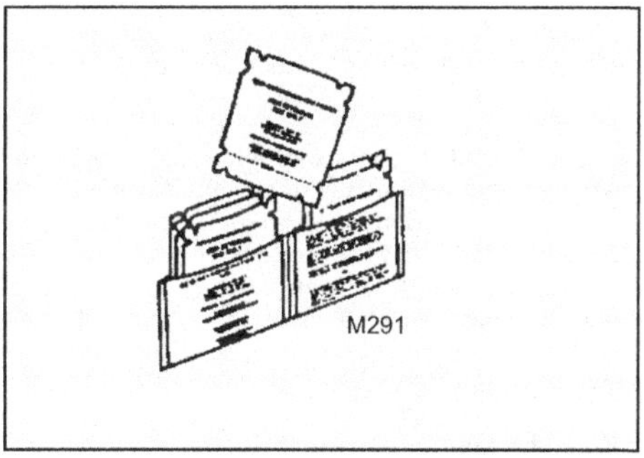


**Figure 2-1. M291 SDK**

2-3. If an SDK is not available, chemical contamination may be pinch-blotted from the skin with a cloth and flushed with water from a canteen. Soap, if available, can also be used to wash the agent from the skin. Washing with soap and water (or hot water) is the next best method for toxic-agent decon if SDKs are not available, but this method is not as effective as using the decon kits.

### BIOLOGICAL

2-4. Washing with soap and water removes nearly all biological agents from the skin. A 0.5 percent chlorine (calcium hypochlorite [HTH] or household bleach) solution is an effective biological decontaminant. See Appendix B for details on how to make this solution.

### RADIOLOGICAL

2-5. To remove radiological dust particles, brush, wash, or wipe them off.

## PERSONAL WIPE DOWN

2-6. The personal wipe-down techniques are most effective when done within 15 minutes of being contaminated. The CAM and/or M8/M9 detector paper is used to detect and monitor equipment.

**NOTE: Very high concentrations of DS2 can cause a CAM response in the G mode.**

2-7. Wipe down your mask, hood, gloves, and other essential gear. Do not attempt to remove chemical contamination from your protective overgarment unless there are obvious clumps of the agent. In this case, scrape off the material. Brush off radiological or frozen chemical-agent contamination. See FM 3-4 for detailed information on your overgarment's protective qualities.

### CHEMICAL

2-8. Decon individual equipment using the IEDKs. Wearing your Kevlar® helmet protective cover will prevent or reduce the adsorption of any liquid chemical agent. Washing with soap and water and bleach solutions is partially effective (see Appendix B).

### BIOLOGICAL

2-9. Wash with soap and water. If water is not available, use SDKs in the same manner as described for chemical-agent decon.

### RADIOLOGICAL

2-10. Locate radiological contamination with monitoring equipment and remove by brushing and shaking it off. Wash exposed areas of your skin and pay particular attention to your hair and fingernails. Avoid breathing the dust particles you shake off by wearing your protective mask or a piece of cloth over your nose and mouth. If wet, conduct a MOPP-gear exchange as soon as possible because brushing or shaking will not remove the contamination. Wipe off your equipment with warm, soapy water using rags or damp paper towels.

## OPERATOR'S SPRAY DOWN

2-11. Decon other mission-essential portions of your equipment before continuing your mission. The CAM and/or M8/M9 detector paper is used to determine what surfaces require decon.

2-12. Operator's spray down is most effective when done within 15 minutes of contamination.

## CHEMICAL

2-13. Decon those surfaces that you must touch on the exterior of the vehicle or equipment you must use to do your job with the on-board portable decon apparatus (such as the M11 or M13) (see Figure 2-2). The IEDK may be used on equipment that DS2 may cause damage to by corrosive action (see Figure 2-3).

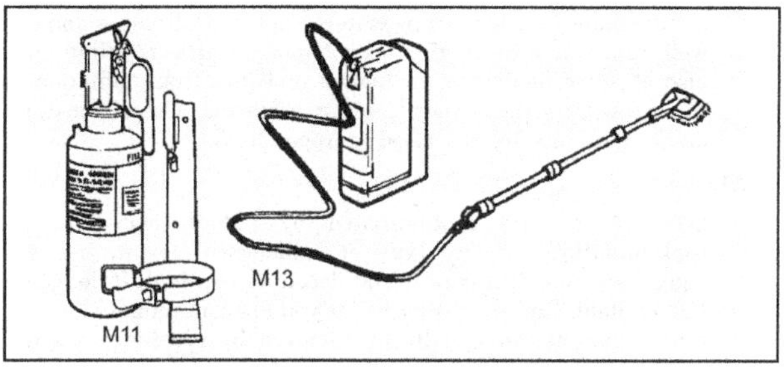


**Figure 2-2. Portable Decon Apparatuses**

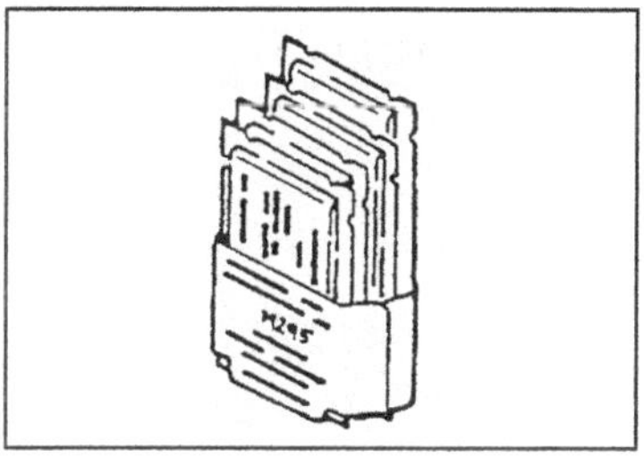


**Figure 2-3. M295 IEDK**

**WARNING**

**Use extreme caution at all times when handling DS2. Do not mix DS2 and STB because it will cause a fire. Do not allow DS2 to be sprayed on personnel or protective clothing. DS2 is a combustible solution. Do not spray DS2 on hot engine blocks. Severe chemical burns can result if personnel fail to observe all safety precautions. DS2 can severely injure eyes and skin and, if inhaled, can cause illness. DS2 can damage the NBC protective overgarment. Long-term contact with DS2 can damage the NBC protective gloves, hood, and overboots.**

2-14. Scrub the DS2 into the exterior surface with brushes. Wait 30 minutes, then wash off. If a decon apparatus is not available, use the field-expedient resources that are available to apply DS2 or STB from bulk containers. If necessary, use the nonstandard decontaminants that are discussed in Appendix B.

## BIOLOGICAL

2-15. Use a bleach solution for biological contamination; however, if bleach is not available, use hot, soapy water. Apply with brushes and scrub the surface well. Rinse the surface after scrubbing (no wait is required). DS2 and STB are also effective against most known biological contamination, but because of their caustic nature, they are not preferred. Other nonstandard biological decontaminants are described in Appendix B.

## RADIOLOGICAL

2-16. If you are contaminated by fallout, rain out, neutron-induced contamination, or any type of radiological agent, use your monitoring equipment to help locate it and decon as required. If detection equipment is not available and you suspect that you are contaminated, decon. Radiological contamination can usually be removed by brushing or scraping. Water is effective for flushing away radiological contamination; however, the runoff should be controlled by using drainage ditches that flow into a sump. Remember, you have not destroyed the contamination, it has just been moved. The runoff will still be hazardous. If you have time, brush or scoop away the top inch of soil from your fighting position to lower the amount of radiological contamination affecting you.

# Chapter 3

# Operational Decon

An operational decon limits the spread of contamination, allows temporary relief from MOPP4, and facilitates additional decon requirements. By speeding up the weathering process, the need for a thorough decon may be eliminated. This process requires about a 120-square-yard area and, depending on the number of contaminated vehicles, may not require a nearby water source. An operational decon consists of a vehicle washdown and a MOPP-gear exchange. The techniques that may be used to conduct an operational decon are decentralized company control, centralized battalion control, or centralized brigade control.

## TECHNIQUES

3-1. Decentralized company control is provided by the unit commander. The commander requests PDDE support, selects the decon site, links up with the PDDE crew, and executes the operational decon.

3-2. Centralized battalion control is similar to the above except that the battalion chemical NCO controls the decon. The site is selected by the battalion commander, and the battalion chemical NCO directs site setup and security and provides command and control ($C^2$). The battalion chemical NCO travels with the PDDE and communicates via the radio.

3-3. Centralized brigade control is done when decon assets within the brigade have been consolidated. In this technique, the brigade chemical NCO performs those functions described for the battalion chemical NCO. The advantages and disadvantages of operational-decon techniques are listed in Table 3-1, page 3-2.

## PHASES

3-4. The three phases for an operational decon are preparation, execution, and site clearance.

### PREPARATION

3-5. The preparation phase starts with the decision to conduct an operational decon and ends with a site that is set up and ready for operation. (The decision to decon was described in Chapter 1.) The CAM is used to identify which vehicles need decontaminating, and PDDE support is requested. Coordination with higher headquarters (HQ) includes identifying the linkup point and the unit that will provide the PDDE support. Signal operating instructions (SOI) information is exchanged and the technique (decentralized company control, centralized battalion control, or centralized brigade control) is selected. The number and type of vehicles and the number of personnel for a MOPP-gear exchange are provided.

**Table 3-1. Advantages and Disadvantages of Operational-Decon Techniques**

| | Decentralized Company Control | Centralized Battalion Control | Centralized Brigade Control |
|---|---|---|---|
| Advantages | Flexibility<br>Dispersion | $C^2$<br>Flexibility<br>Dispersion<br>Planning<br>Sustained operations | $C^2$<br>Synchronized support assets<br>Rehearsal<br>Time<br>Planning<br>Sustained operations |
| Disadvantages | $C^2$<br>Synchronized support assets<br>Rehearsal<br>Time<br>Planning<br>Sustained operations | Synchronized support assets<br>Rehearsal<br>Time | Flexibility<br>Dispersion |

3-6. Site selection is done by the controlling HQ. The site may have been designated in the operation order (OPORD) or selected based on the current METT-TC. For site characteristics, see Table 3-2.

3-7. Linkup at the designated point includes establishing security, positioning any necessary marshalling areas, camouflaging removal points, designating the direction of flow into and out of the site, and ensuring that unit supply has sufficient replacement MOPP gear.

3-8. Site setup requires the PDDE crew to position itself upwind from the entrance. Set up the MOPP-gear exchange about 164 feet upwind from the vehicle washdown at a 45° angle. Notify the contaminated unit when the site is ready.

## EXECUTION

3-9. The execution phase starts with establishing positive control over the site and ends with processing the last contaminated vehicle and soldier through the site. Vehicles move from the predecon area to the site either on order or by watching the vehicle ahead. Contaminated soldiers leave the vehicle and proceed to the MOPP-gear-exchange point. The contaminated unit provides security.

3-10. Each vehicle receives a 2- to 3-minute washdown with hot, soapy water and moves to a point near the MOPP-gear exchange. Any remaining soldiers requiring MOPP-gear exchange will do so then. When the squad or crew has finished, it remounts its vehicles and moves into an after-decon assembly area (AA) to await further instructions or moves to its next battle position.

## SITE CLEARANCE

3-11. The site-clearance phase includes cleanup, marking, and reporting. METT-TC will dictate the cleanup requirements. The PDDE crew buries or double bags and removes contaminated MOPP gear and other expendables. Standard NBC contamination markers are posted, and the PDDE crew submits an NBC 5 report.

**Table 3-2. Operational-Decon Phases**

| | |
|---|---|
| **Preparation** | |
| Decon assessment | Identify soldiers and equipment to be decontaminated. |
| Coordination | Request PDDE support. The chemical section conducts coordination with the contaminated unit on the linkup point. Decon operations should be done between 1 and 6 hours after becoming contaminated. |
| Site selection | Is selected by the controlling HQ.<br>Is off the main route but has easy access.<br>Is a large enough area (120 square yards per site for a squad-size element).<br>Has good overhead concealment.<br>Has a good water source (plan for 100 gallons per vehicle).<br>Has good drainage. |
| Linkup | Ensure that the noncommissioned officer in charge (NCOIC) knows where to link up with the contaminated unit and the location for site setup. Radio communication is essential for operations. |
| Site setup | Ensure that the PDDE is positioned properly and ready to dispense hot, soapy water.<br>Ensure that the contaminated unit sets up and operates the MOPP-gear exchange at the same time as the vehicle washdown. Consider contamination runoff when positioning the PDDE. |
| **Execution** | |
| Site control and security | Ensure that the drivers of the contaminated vehicles know when to move into position at the washdown location.<br>Ensure that the contaminated unit has provided site security. |
| Processing | Ensure that the decon site NCOIC is processing vehicles at a rate of 2 to 3 minutes per vehicle. Also, ensure that soldiers are going through the MOPP-gear exchange at the rate of 30 minutes per squad/crew. |
| **Site Clearance** | |
| Cleanup | Decon site NCOIC ensures that the MOPP-gear-exchange area is cleaned up. |
| Marking and reporting | Decon site NCOIC ensures that his team properly marks the decon site and sends the NBC 5 report forward. |

## VEHICLE WASHDOWN

3-12. A vehicle washdown may be conducted with or without PDDE and in either a one- or two-lane configuration. An unsupported washdown requires the contaminated unit to have washing equipment that can produce 60 to 120 pounds per square inch (psi) of water pressure. The capacity to heat water and inject soap increases effectiveness. A supported washdown requires PDDE organic to the unit or from a supporting chemical decon or dual-purpose company. A two-lane washdown is simply a one-lane washdown parallel with another (see Figure 3-1, page 3-4).

3-13. Other configurations are limited only by the constraints of the METT-TC. Figure 3-2, page 3-5, illustrates a dispersed operational-decon setup. Table 3-3, page 3-5, describes the vehicle-washdown process.

## MOPP-GEAR EXCHANGE

3-14. A MOPP-gear exchange can be done with a buddy team, a triple buddy team, or an individual (emergency). For planning purposes, estimate about 60 minutes for a squad or a platoon to complete the exchange. Since individual performance is severely degraded after 6 hours at MOPP4, the MOPP-gear exchange should be scheduled before that time.

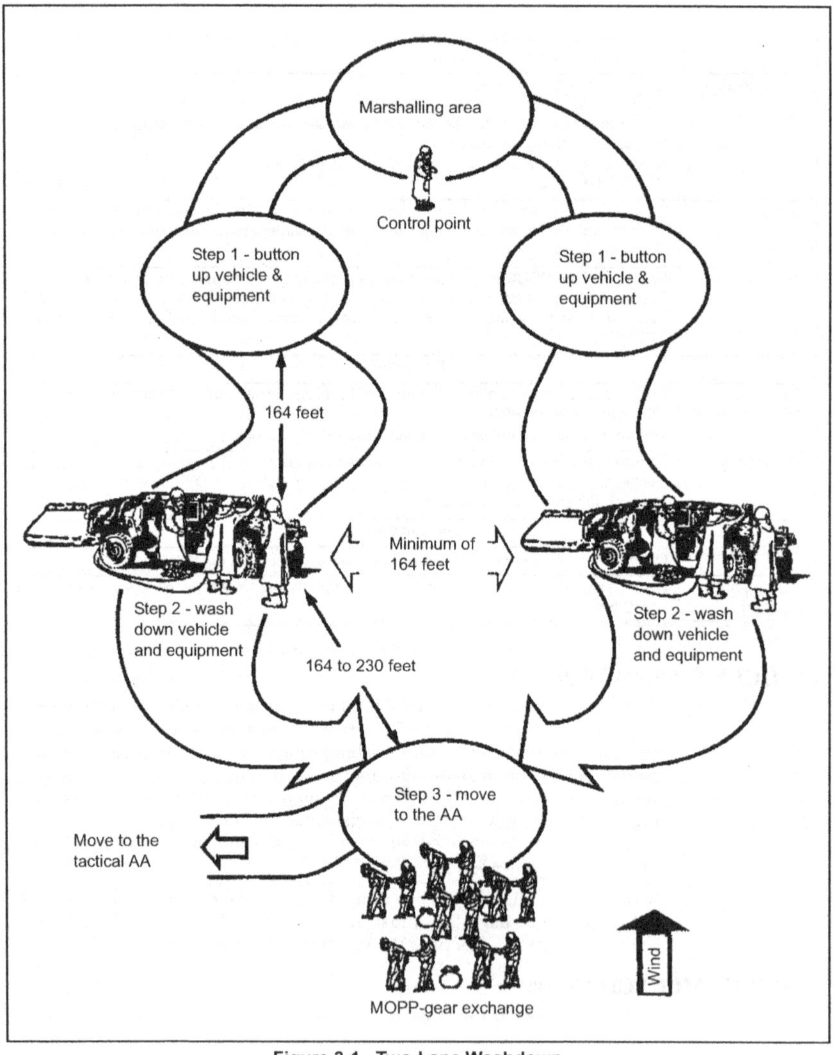


**Figure 3-1. Two-Lane Washdown**

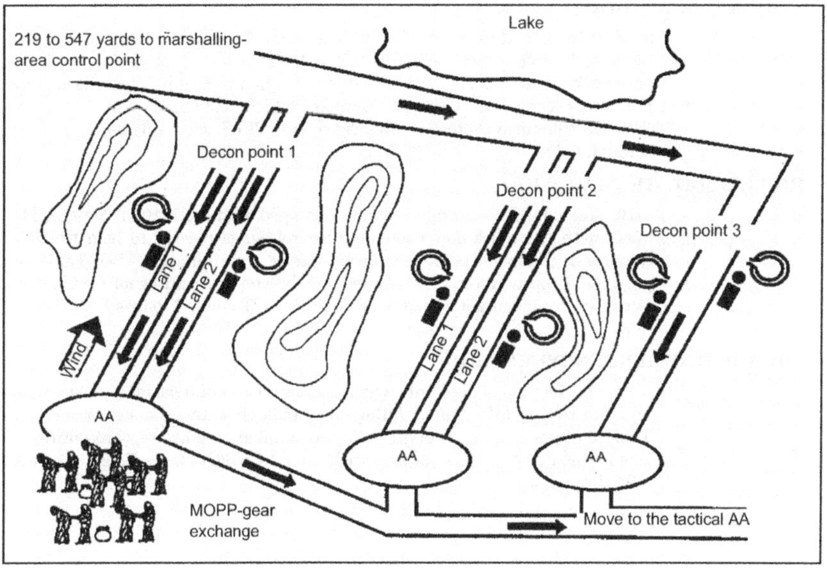


**Figure 3-2. Dispersed Operational-Decon Setup**

**Table 3-3. Vehicle-Washdown Process**

| Steps | Equipment | Procedures |
|---|---|---|
| 1. Assemble vehicles. Unit is tactically dispersed. Personnel at the control point direct movement. | Watch | Personnel at the control point supervise preparing vehicles and directing movement out of the AA. |
| 2. Prepare vehicles. | None | Individual/crew closes all access doors, hatches, windows, and other openings. Remove camouflage and cover muzzles. If required, crews (less drivers) move to the MOPP-gear-exchange area. Move to the wash area on order. |
| 3. Perform washdown. | Delivery system (such as the M12 and M17, 65-gallon-per-minute [GPM] pump, fire-fighting equipment, and so forth) that delivers hot, soapy water at 60 to 120 psi. Also sufficient water, fuel, and detergent for vehicles. | Soldiers/personnel wash equipment from top to bottom. Decon crew wears a toxicological apron, protective (TAP) or wet-weather gear over MOPP gear. |
| 4. Move out to the next position. | None | Vehicles move to the MOPP-gear-exchange area (if required) or the next battle position. |

## BUDDY-TEAM METHOD

3-15. This method uses pairs of soldiers under the supervision of their squad/team leader to conduct the buddy-team MOPP-gear exchange. See Table 3-4, pages 3-7 through 3-17, for detailed step-by-step procedures for this method. For units equipped with the combat vehicle crewman uniform system (CVCUS)/chemical protective undergarment (CPU), see Table 3-5, pages 3-18 through 3-21.

## TRIPLE BUDDY-TEAM METHOD

3-16. This method is used by soldiers equipped with the M40A1, M42, or M43 mask with the quick-doff hood. A third soldier is needed to hold the filter canister and hose to prevent the transfer of contamination. See Table 3-6, pages 3-22 through 3-27, for detailed step-by-step procedures for this method. For units equipped with the CVCUS/CPU, see Table 3-7, pages 3-28 through 3-30.

## INDIVIDUAL (EMERGENCY) METHOD

3-17. This method is used only when a soldier does not have buddy to help him and the risk of MOPP-gear failure demands that an exchange occur. See Table 3-8, pages 3-31 through 3-33, for detailed step-by-step procedures for this method. For units equipped with the CVCUS/CPU, see Table 3-9, pages 3-34 through 3-37.

**Table 3-4. Steps for the Buddy-Team Method MOPP-Gear Exchange**

| Steps | Contamination Type | Required Equipment | Procedures |
|---|---|---|---|
| 1. Decon gear—removes gross contamination from individual gear (weapons, helmet, load-bearing equipment, and mask carrier). | All contamination | Four long-handled brushes<br>A large piece of plastic (poncho or similar material) | The soldier mixes three parts earth to two parts STB. |
| | Chemical/ biological | One 30-gallon container, STB dry mix, and one IEDK | The soldier removes and discards the chemical protective helmet cover if worn. He brushes or rubs STB onto his individual gear (helmet and mask carrier) and the hose of the M42 or M43 mask if worn. He gently shakes off any excess STB and sets aside his gear on an uncontaminated surface. |
| | Radiological | Hot, soapy water | The soldier brushes or wipes radiological contamination from his individual gear. He washes it with hot, soapy water (if available) then sets it aside to dry on an uncontaminated surface (plastic, poncho, or similar material). |
| 2. Prepare to decon—facilitates later removal of overgarment trousers and overboots. | All contamination | Cutting tool | Buddy #1 unfastens the shoulder straps on Buddy #2's hood and pulls them over his shoulder and reattaches them to the Velcro® fasteners. He loosens the drawcord on Buddy #2's hood (see Figure 3-3, page 3-8). He removes the M9 detector paper from Buddy #2's overgarment. He unties the drawcords on the trouser legs of Buddy #2's overgarment. He unzips Buddy #2's trouser legs and rolls a cuff in each trouser leg, ensuring that the cuffs do not come above the tops of his overboots (see Figure 3-4, page 3-8). He unfastens or cuts the fasteners on Buddy #2's overboots. **NOTE: A soldier can do this step by himself or with the help of his buddy. When wearing overboots, unsnap both quick releases on the overboots. The M40 voice amplifier (M7) and the M42A2 detachable microphone cannot be decontaminated and will be disposed of as contaminated waste; however, ensure that these items are contaminated before disposing of them.** |
| 3. Decon mask and hood—removes gross contamination. | Chemical/ biological | Two IEDKs per person | Buddy #1 uses an IEDK to wipe Buddy #2's eye-lens outserts from the top down. **(NOTE: Do not press so hard that you break Buddy #2's face-mask seal.)** He then wipes the rest of Buddy #2's hood from the top of the head to the bottom of the hood (see Figure 3-5, page 3-9). After he has finished wiping Buddy #2's mask, he must wipe his own gloves in preparation for rolling Buddy #2's hood. He starts from the rear and rolls Buddy #2's hood, using 2-inch tucks, until it reaches the center of his head (see Figure 3-6, page 3-9). He rolls the front of Buddy #2's hood tightly under the outlet valve and filter. He ensures that the hood is off of Buddy #2's garment (see Figure 3-7, page 3-10). |

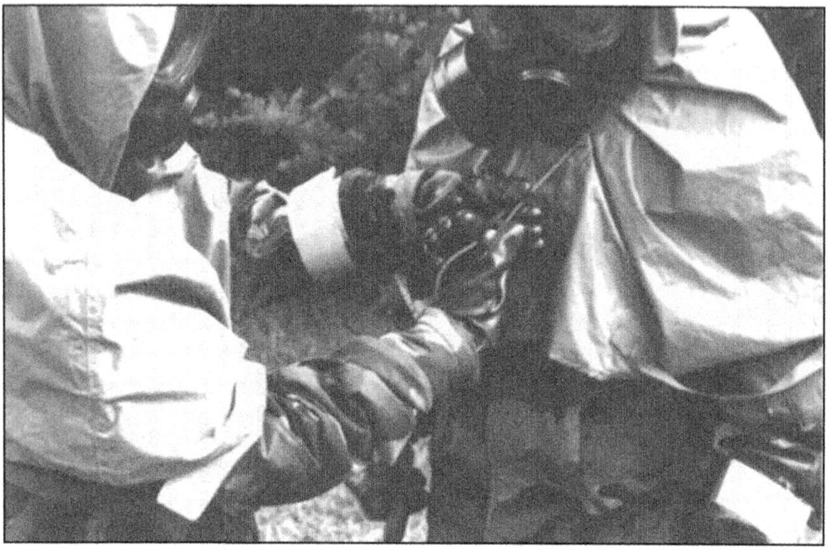

Figure 3-3. Loosening the Drawcord on the Hood

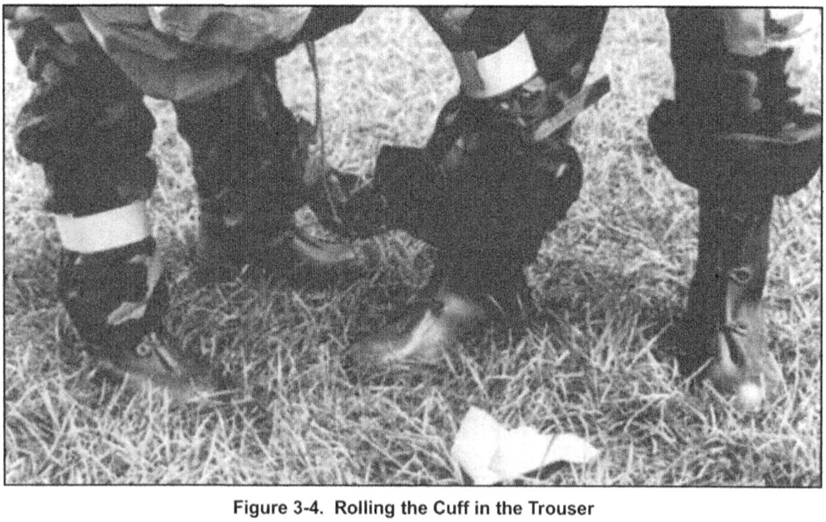

Figure 3-4. Rolling the Cuff in the Trouser

Figure 3-5. Wiping From the Head to the Bottom of the Hood

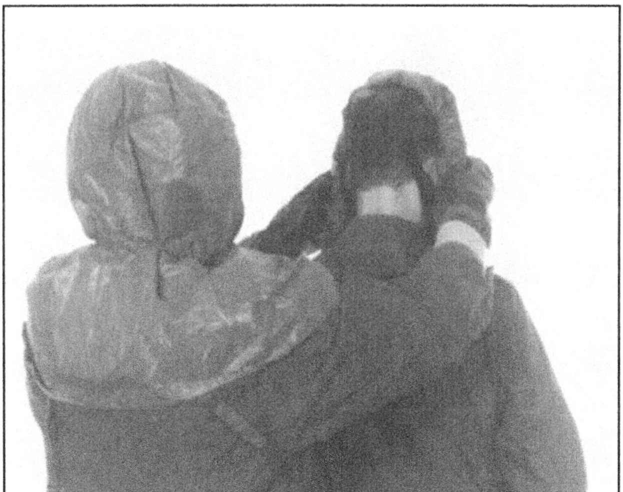

Figure 3-6. Rolling the Hood to the Center of the Head

Figure 3-7. Ensuring That the Hood is off the Garment

**Table 3-4. Steps for the Buddy-Team Method MOPP-Gear Exchange (Continued)**

| Steps | Contamination Type | Required Equipment | Procedures |
|---|---|---|---|
| 3. Decon mask and hood (continued). | Radiological | Three containers (about 3-gallon capacity)<br>Two sponges<br>Soapy water<br>Rinse water<br>Paper towels or similar drying material | Buddy #1 wipes Buddy #2's mask and hood with a sponge dipped in hot, soapy water and rinses them with a sponge dipped in clean water. He dries Buddy #2's mask and hood with paper towels or rags. Buddy #2 wipes his own gloves. **NOTE: Cool, soapy water is not as effective for removing contamination, but it can be used if you scrub longer.** If the water supply is limited, use drinking water from your canteen with a wet sponge or cloth. If water is not available, brush and dust off the radioactive dust particles. ***Do not reverse roles.*** Only Buddy #2's hood will be decontaminated and rolled at this time. |
| 4. Remove overgarments and overboots—limits the spread of agents and helps prevent agents from penetrating through to the undergarments or the skin. | All contamination | Two discard containers (for example, plastic bags) | Buddy #1 grasps Buddy #2's overgarment jacket, unsnaps the snaps individually, and unties the drawcord at the bottom of the jacket (see Figure 3-8, page 3-12). He unfastens the Velcro at the wrist of Buddy #2's jacket and then refastens it. He unfastens the Velcro closure over the zippered front of Buddy #2's jacket and unzips the jacket. He grasps Buddy #2's jacket at the shoulders and instructs him to make a fist. He then pulls Buddy #2's jacket down and away from him, ensuring that the black part of the jacket is not touched (see Figure 3-9, page 3-12). He lays Buddy #2's overgarment jacket on the ground, black side up. (It will be used to stand on later.) He carefully unfastens and unzips Buddy #2's trousers. **NOTE: Do not loosen Buddy #2's waist tabs.** He instructs Buddy #2 to loosen his overboots by alternately stepping on each heel and pulling up on his foot (see Figure 3-10, page 3-13). He grasps Buddy #2's trousers and pulls them down to his knees. He instructs Buddy #2 to walk out of his trousers/overboots simultaneously and step onto the black side of the jacket (see Figure 3-11, page 3-13). Buddy #2 should step onto the jacket wearing his mask, battle-dress uniform (BDU), combat boots, and gloves. |
| 5. Remove gloves. | All contamination | Two discard containers (from step 4) | The soldier holds the fingertips of his gloves and partially slides his hand out. When the fingers of both hands are free, he holds his arms away from his body and lets the gloves drop (see Figure 3-12, page 3-14). |

Figure 3-8. Unsnapping the Trousers From the Jacket

Figure 3-9. Pulling the Jacket Down and Away

Figure 3-10. Loosening the Overboots

Figure 3-11. Stepping on the Black Lining of the Jacket

Figure 3-12. Removing the Gloves

**Table 3-4. Steps for the Buddy-Team Method MOPP-Gear Exchange (Continued)**

| Steps | Contamination Type | Required Equipment | Procedures |
|---|---|---|---|
| 6. Put on overgarment. | All contamination | One set of chemical protective overgarments per person | Buddy #1 opens the package containing the new overgarment without touching the inside of the package. Buddy #2 pulls out the overgarment without touching the outside of the package. He puts on the overgarment and fastens it, leaving the trouser legs open (see Figure 3-13, page 3-16). ***Do not reverse roles.*** Only Buddy #2 will put on the clean overgarment at this time. |
| 7. Put on overboots and gloves. | All contamination | One set of chemical protective overboots per person<br>One set of chemical protective gloves per person (correct size)<br>M9 detector paper | Buddy #1 opens the package of clean overboots without touching the inside of the package. Buddy #2 removes the overboots from the package without touching the outside of the package, puts them on, and fastens his trouser legs. Buddy #1 opens the package of clean gloves without touching the inside of the package. Buddy #2 removes the gloves from the package without touching the outside of the package and puts them on. Buddy #2 puts on the M9 detector paper (see Figure 3-14, page 3-16). ***Do not reverse roles.*** Only Buddy #2 will put on the clean overboots and gloves at this time. |

Figure 3-13. Putting on the New Overgarment

Figure 3-14. Putting on the M9 Detector Paper

Table 3-4. Steps for the Buddy-Team Method MOPP-Gear Exchange (Continued)

| Steps | Contamination Type | Required Equipment | Procedures |
|---|---|---|---|
| 8. Secure hood. | All contamination | One IEDK | Buddy #1 uses an IEDK to wipe Buddy #2's gloves. He unrolls Buddy #2's hood and attaches the straps and tightens the neck cord. Buddy #1 and Buddy #2 reverse roles and repeat steps 2 through 8. |
| 9. Secure gear. | All contamination | One chemical protective helmet cover per person | The soldier secures his individual gear and puts it back on (see Figure 3-15). He puts on a new chemical protective helmet cover and moves to the AA. He uses the buddy system to check the fit of all secured gear. |

Figure 3-15. Securing the Hood

**Table 3-5. Steps for the CVCUS/CPU (Buddy-Team Method) MOPP-Gear Exchange**

| Steps | Contamination Type | Required Equipment | Procedures |
|---|---|---|---|
| 1. Decon gear—removes gross contamination from individual gear (weapons, helmet, load-bearing equipment, and mask carrier). | All contamination | Four long-handled brushes<br>A large piece of plastic (poncho or similar material) | The soldier mixes three parts earth to two parts STB. |
| | Chemical/ biological | One 30-gallon container, STB dry mix, and one IEDK | The soldier removes and discards the chemical protective helmet cover if worn. He brushes or rubs the STB onto his individual gear and the hose of the M42 mask if worn. He gently shakes off any excess STB and sets aside his gear on an uncontaminated surface. |
| | Radiological | Hot, soapy water | The soldier brushes or wipes radiological contamination from his individual gear. He washes it with hot, soapy water (if available) then sets it aside to dry on an uncontaminated surface (plastic, poncho, or similar material). |
| 2. Prepare for decon—facilitates MOPP-gear removal. | All contamination | Cutting tool | Buddy #1 unfastens the shoulder straps on Buddy #2's hood and pulls them over his shoulder and reattaches them to the Velcro fastener. He loosens the drawcord on Buddy #2's hood. He removes the M9 detector paper from Buddy #2's CVCUS. He unzips the ankles on both legs of Buddy #2's CVCUS. He unfastens or cuts the fasteners on Buddy #2's overboots. **NOTE: A soldier can do this step by himself or with the help of his buddy.** |
| 3. Decon mask and hood—removes gross contamination. | Chemical/ biological | Two IEDKs per person | Buddy #1 uses an IEDK to wipe Buddy #2's eye-lens outserts from the top down. **(NOTE: Do not press so hard that you break Buddy #2's face-mask seal.)** He then wipes the rest of Buddy #2's hood from the top of the head to the bottom of the hood (see Figure 3-5, page 3-9). After he has finished wiping Buddy #2's mask, he must wipe his own gloves in preparation for rolling Buddy #2's hood. He starts from the rear and rolls Buddy #2's hood, using 2-inch tucks, until it reaches the center of his head (see Figure 3-6, page 3-9). He rolls the front of Buddy #2's hood tightly under the outlet valve and filter. He ensures that the hood is off of Buddy #2's garment (see Figure 3-7, page 3-10). |

**Table 3-5. Steps for the CVCUS/CPU (Buddy-Team Method) MOPP-Gear Exchange (Continued)**

| Steps | Contamination Type | Required Equipment | Procedures |
|---|---|---|---|
| 3. Decon mask and hood (continued). | Radiological | Three containers (about 3-gallon capacity)<br>Two sponges<br>Soapy water<br>Rinse water<br>Paper towels or similar drying material | Buddy #1 wipes Buddy #2's mask and hood with a sponge dipped in hot soapy water and rinses them with a sponge dipped in clean water. He dries Buddy #2's mask and hood with paper towels or rags. Buddy #2 wipes down his own gloves. **NOTE: Cool, soapy water is not as effective for removing contamination, but it can be used if you scrub longer.** If the water supply is limited, use drinking water from your canteen with a wet sponge or cloth. Only Buddy #2's mask and hood will be decontaminated and rolled at this time. |
| 4. Remove CVCUS/ overboots—limits the spread of contamination. | All contamination | Two discard containers (for example, plastic bags) | Buddy #1 opens all the zippers on Buddy #2's CVCUS in the following order: wrist, ankles, and front. While standing behind Buddy #2 and adjacent to the clean area, Buddy #1 grasps the shoulders of Buddy #2's CVCUS and instructs him to make a fist to ensure that his gloves do not come off. He then pulls Buddy #2's CVCUS down below his knees, turning it inside out. He has Buddy #2 to partially remove his overboots by alternately stepping on each heel and pulling up on his foot. While kneeling, Buddy #1 grasps one leg of Buddy #2's CVCUS and his overboot and instructs him to remove his leg and foot simultaneously (see Figure 3-16, page 3-20). As Buddy #2 removes each leg and foot, he steps directly onto the clean area. |
| 5. Remove CPU jacket/gloves. | All contamination | Two discard containers | Buddy #2 holds the fingertips of his gloves and partially slides his hand out. When the fingers of both hands are free, he holds his arms away from his body and lets the gloves drop. Buddy #1 unzips and removes Buddy #2's CPU jacket by grasping it at the shoulders and pulling down, turning it inside out (see Figure 3-17, page 3-20). He places it on the clean area with the clean side up. He removes Buddy #2's protective glove inserts. |

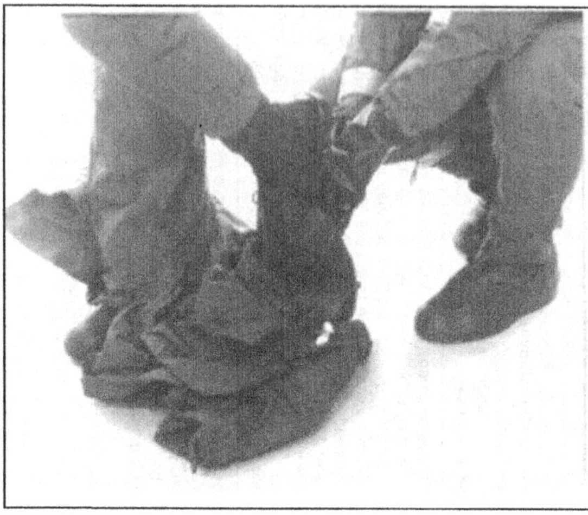

Figure 3-16. Removing Leg and Foot From the Garment

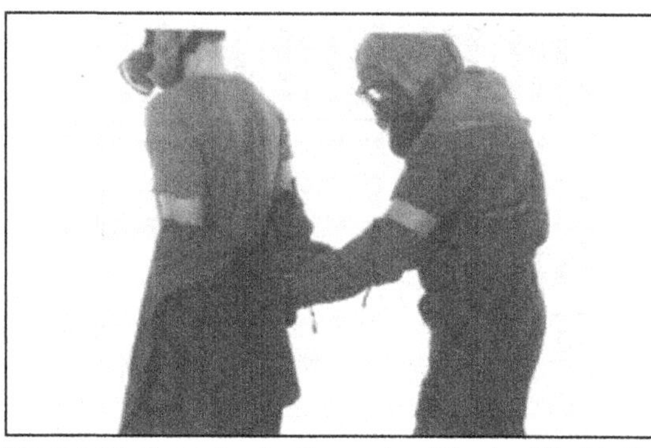

Figure 3-17. Removing the Garment and Turning it Inside Out

Table 3-5. Steps for the CVCUS/CPU (Buddy-Team Method) MOPP-Gear Exchange (Continued)

| Steps | Contamination Type | Required Equipment | Procedures |
|---|---|---|---|
| 6. Remove combat boots/ CPU pants. | All contamination | Two discard containers | Buddy #1 instructs Buddy #2 to stand on the clean area (but not on the CPU jacket) and unlace his combat boots. Buddy #2 partially removes his boots by alternately stepping on each heel and pulling up on his foot. Buddy #1 grasps one of Buddy #2's boots and instructs him to remove his foot. As Buddy #2 removes his foot, he will step directly onto the clean side of the CPU jacket. Repeat the procedure for the other foot. Next, Buddy #1 removes Buddy #2's CPU drawers by pulling down on the sides and turning them inside out. Buddy #2 is now stripped to his underwear and mask and is ready to put on new clothing. **NOTE: If Buddy #2's feet come in contact with the contaminated pants of the CPU, decon them.** |
| 7. Put on CPU/ CVCUS. | All contamination | One CPU/CVCUS per soldier | Buddy #1 opens the package containing the new undergarment without touching the inside of the package. Buddy #2 pulls out the undergarment without touching the outside of the package. He puts on the new clothing in the following order: CPU drawers, CPU jacket, combat boots, and CVCUS (over the CPU drawers). |
| 8. Put on overboots and gloves. | All contamination | One set of chemical protective overboots per soldier<br>One set of chemical protective gloves per soldier<br>M9 detector paper | Buddy #1 opens the package of clean overboots without touching the inside of the package. Buddy #2 removes the overboots without touching the outside of the package and puts them on. Buddy #1 opens the package of clean gloves without touching the inside. Buddy #2 removes the gloves without touching the outside of the package and puts them on. Buddy #2 puts on the M9 detector paper (see Figure 3-18, page 3-23). |
| 9. Secure hood. | Chemical/ biological | One IEDK | Buddy #1 wipes his rubber gloves with an IEDK. He unrolls Buddy #2's hood and attaches the straps and tightens the neck cord. Buddy #1 and Buddy #2 reverse rolls and repeat steps 2 through 9. |
| 10. Secure gear. | All contamination | One chemical protective helmet cover per person | The soldier secures his individual gear and puts it back on. He puts on a new chemical protective helmet cover and moves to the AA. He uses the buddy system to check the fit of all secured gear. |

**Table 3-6. Steps for the Triple-Buddy Method MOPP-Gear Exchange With the M40A1 and M42**

| Steps | Contamination Type | Required Equipment | Procedures |
|---|---|---|---|
| 1. Decon gear—removes gross contamination from individual gear (weapons, helmet, load-bearing equipment, and mask carrier). | All contamination | Four long-handled brushes<br>A large piece of plastic (poncho or similar material) | The soldier mixes three parts earth to two parts STB. |
| | Chemical/ biological | One 30-gallon container, STB dry mix, and one IEDK | The soldier removes and discards the chemical protective helmet cover if worn. He brushes or rubs STB onto his individual gear (helmet and mask carrier) and the hose of the M42 or M43 mask if worn. He gently shakes off any excess STB and sets aside his gear on an uncontaminated surface. |
| | Radiological | Hot, soapy water | The soldier brushes or wipes radiological contamination from his individual gear. He washes it with hot, soapy water (if available) then sets it aside to dry on an uncontaminated surface (plastic, poncho, or similar material). |
| 2. Prepare to decon—facilitates later removal of overgarment trousers and overboots. | All contamination | Cutting tool | Buddy #1 unfastens the shoulder straps on Buddy #2's hood and pulls them over his shoulder and reattaches them to the Velcro fastener. He loosens the drawcord on Buddy #2's hood and ties off the microphone cord to the hose of his mask (see Figure 3-19). He removes the M9 detector paper from Buddy #2's overgarment. He unzips Buddy #2's trouser legs and rolls a cuff in each trouser leg, ensuring that the cuffs do not come above the tops of his overboots. He unfastens or cuts the fasteners on Buddy #2's overboots. **NOTE: A soldier can do this step by himself or with the help of his buddy.** |

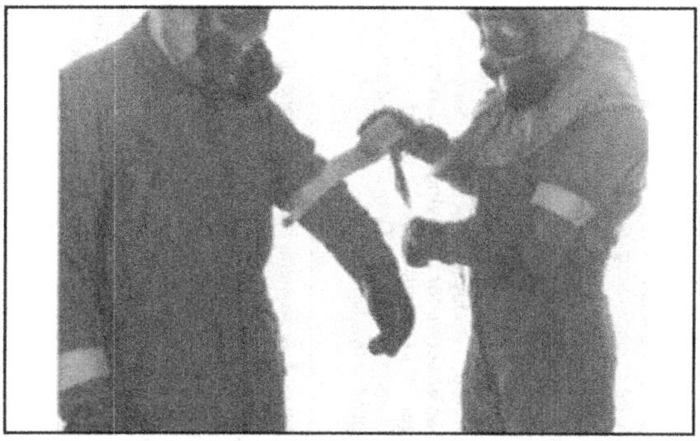

Figure 3-18. Reapplying the M9 Detector Paper

Figure 3-19. Tying the Microphone Cord to the Hose of the Mask

Table 3-6. Steps for the Triple-Buddy Method MOPP-Gear Exchange With the M40A1 and M42 (Continued)

| Steps | Contamination Type | Required Equipment | Procedures |
|---|---|---|---|
| 3. Decon mask and hood—removes gross contamination. | Chemical/ biological | Two IEDKs per person | Buddy #1 uses an IEDK to wipe Buddy #2's eye-lens outserts from the top down. **(NOTE: Do not press so hard that you break Buddy #2's face-mask seal.)** He then wipes the rest of Buddy #2's hood from the top of the head to the bottom of the hood (see Figure 3-5, page 3-9). After he has finished wiping Buddy #2's mask, he must wipe his own gloves in preparation for rolling Buddy #2's hood. He starts from the rear and rolls Buddy #2's hood, using 2-inch tucks, until it reaches the center of his head (see Figure 3-6, page 3-9). He rolls the front of Buddy #2's hood tightly under the outlet valve and filter. He ensures that the hood is off of Buddy #2's garment (see Figure 3-7, page 3-10). |
| | Radiological | Three containers (about 3-gallon capacity)<br>Two sponges<br>Soapy water<br>Rinse water<br>Paper towels or similar drying material | Buddy #1 wipes Buddy #2's mask and hood with a sponge dipped in hot, soapy water and rinses them with a sponge dipped in clean water. He dries Buddy #2's mask and hood with paper towels or rags. Buddy #2 wipes down his own gloves. **NOTE: Cool, soapy water is not as effective for removing contamination, but it can be used if you scrub longer.** If the water supply is limited, use drinking water from your canteen with a wet sponge or cloth. Only Buddy #2's mask and hood will be decontaminated and rolled at this time. |
| 4. Remove overgarment overboots—limits the spread of agents and helps prevent agents from penetrating through to the undergarments or the skin. | All contamination | Two discard containers (for example, plastic bags) | Buddy #1 grasps Buddy #2's outside overgarment jacket, unsnaps the snaps individually, and unties the drawcord at the bottom of the jacket. He unfastens the Velcro at the wrist of Buddy #2's jacket and then refastens it. He unfastens the Velcro closure over the zippered front of Buddy #2's jacket and unzips the jacket. Buddy #1 grasps Buddy #2's jacket by the shoulders and instructs him to make a fist. He then pulls the jacket down and away from Buddy #2, ensuring that the black part of the jacket is not touched (see Figure 3-20, page 3-26). He lays Buddy #2's overgarment jacket on the ground, black side up. (It will be used to stand on later.) Buddy #1 carefully unfastens and unzips Buddy #2's trousers. **NOTE: Do not loosen Buddy #2's waist tabs.** He instructs Buddy #2 to break the seals on his overboots by alternately stepping on each heel and pulling up on his foot. He grasps Buddy #2's trousers and pulls them down to his knees. Buddy #1 instructs Buddy #2 to walk out of his trousers/overboots, taking care not to step on the contaminated side of the overgarment. If Buddy #2 is wearing overboots, Buddy #1 removes Buddy #2's trousers first then helps him step out of his overboots onto the black side of the jacket. Buddy #2 should step onto the jacket wearing his mask, BDU, combat boots, and gloves. |

**Table 3-6. Steps for the Triple-Buddy Method MOPP-Gear Exchange With the M40A1 and M42 (Continued)**

| Steps | Contamination Type | Required Equipment | Procedures |
|---|---|---|---|
| 5. Remove gloves. | All contamination | Discard container (from Step 4) | The soldier holds the fingertips of his gloves and partially slides his hand out. When the fingers of both hands are free, he holds his arms away from his body and lets the gloves drop. |
| 6. Put on overgarment. | All contamination | One set of chemical protective overgarments per person | Buddy #1 opens the package containing the new overgarment without touching the inside of the package. Buddy #2 pulls out the overgarment without touching the outside of the package. He puts on the overgarment and fastens it, leaving the trouser legs open. ***Do not reverse roles.*** Only Buddy #2 will put on the clean overgarments at this time. |

Figure 3-20. Removing the Outer Garment

**Table 3-6. Steps for the Triple-Buddy Method MOPP-Gear Exchange With the M40A1 and M42 (Continued)**

| Steps | Contamination Type | Required Equipment | Procedures |
|---|---|---|---|
| 7. Put on overboots and gloves. | All contamination | One set of chemical protective overboots per person<br>One set of chemical protective gloves per person<br>M9 detector paper | Buddy #1 opens the package of clean overboots without touching the inside of the package. Buddy #2 removes the overboots without touching the outside of the package, puts them on, and fastens his trouser legs. Buddy #1 opens the package of clean gloves without touching the inside of the package. Buddy #2 removes the gloves without touching the outside of the package and puts them on. Buddy #2 puts on the M9 detector paper. ***Do not reverse roles.*** Only Buddy #2 will put on clean overboots and gloves at this time. **NOTE: The gloves and overboots will have a light powdery coating. This is normal. It is not necessary to remove it. It will not affect the protective qualities.** |
| 8. Secure hood. | All contamination | One IEDK | Buddy #1 wipes his rubber gloves with an IEDK. He unrolls Buddy #2's hood and attaches the straps and tightens the neck cord. He checks the clips and neck cord on Buddy #2's hood. Buddy #1 and Buddy #2 reverse roles and repeat steps 2 through 8. |
| 9. Secure gear. | All contamination | One chemical protective helmet cover per person | The soldier secures his individual gear and puts it back on. He puts on a new chemical protective helmet cover and moves to the AA. He uses the buddy system to check the fit of all secured gear. |

**Table 3-7. Steps for the CVCUS/CPU (Triple-Buddy Method) MOPP-Gear Exchange With the M40A1, M42, and M43 (Aviation and Armor)**

| Steps | Contamination Type | Required Equipment | Procedures |
|---|---|---|---|
| 1. Decon gear—removes gross contamination from individual gear (weapons, helmet, load-bearing equipment, and mask carrier). | All contamination | Four long-handled brushes<br>A large piece of plastic (poncho or similar material) | The soldier mixes three parts earth to two parts STB. |
| | Chemical/ biological | One 30-gallon container, STB dry mix, and one IEDK | The soldier removes and discards the chemical protective helmet cover if worn. He brushes or rubs STB onto his individual gear (helmet and mask carrier) and the hose of the M43 mask if worn. He gently shakes off any excess STB and sets aside his gear on an uncontaminated surface. |
| | Radiological | Hot, soapy water | The soldier brushes or wipes radiological contamination from his individual gear. He washes it with hot, soapy water (if available) and sets it aside to dry on an uncontaminated surface (plastic, poncho, or similar material). |
| 2. Prepare to decon—facilitates later removal of overgarment trousers and overboots. | All contamination | Cutting tool | Buddy #1 unfastens the shoulder straps on Buddy #2's hood, pulls them over his shoulder, and reattaches them to the Velcro fastener. He loosens the drawcord on Buddy #2's hood and ties off the microphone cord to the hose of his mask. He removes the M9 detector paper from Buddy #2's overgarment. He unzips Buddy #2's trouser legs and rolls a cuff in each trouser leg, ensuring that the cuffs do not come above the top of his overboots. He unfastens or cuts the fasteners on Buddy #2's overboots. **NOTE: A soldier can do this step by himself or with the help of his buddy.** |

**Table 3-7. Steps for the CVCUS/CPU (Triple-Buddy Method) MOPP-Gear Exchange With the M40A1, M42, and M43 (Aviation and Armor) (Continued)**

| Steps | Contamination Type | Required Equipment | Procedures |
|---|---|---|---|
| 3. Decon mask and hood—removes gross contamination. | Chemical/ biological | Two IEDKs per person | If wearing the M42 or M43 mask, use the buddy-team method (three buddies). Buddy #1 uses an IEDK to wipe Buddy #2's hood, mask, canister, and hose of the M42 or M43 mask. **NOTE: Do not press so hard that you break Buddy #2's face-mask seal.** He uses as many wipes as necessary. Buddy #3 wipes off his gloves and holds Buddy #1's hose and canister away from his contaminated overgarment. After Buddy #1 wipes Buddy #2's mask and then his own gloves, he rolls Buddy #2's hood. He starts from the rear and rolls Buddy #2's hood, using 2-inch tucks, until it reaches the center of his head (see Figure 3-6, page 3-9). He rolls the front of Buddy #2's hood tightly under the outlet valve and filter. He ensures that the hood is off of Buddy #2's garment (see Figure 3-7, page 3-10). |
| | Radiological | Three containers (about 3-gallon capacity)<br>Two sponges<br>Soapy water<br>Rinse water<br>Paper towels or similar drying material | Buddy #1 wipes Buddy #2's mask and hood with a sponge dipped in hot, soapy water and rinses them with a sponge dipped in clean water. He dries Buddy #2's mask and hood with paper towels or rags. Buddy #2 wipes his own gloves. **NOTE: Cool, soapy water is not as effective for removing contamination, but it can be used if you scrub longer.** If the water supply is limited, use drinking water from you canteen with a wet sponge or cloth. If water is not available, brush and dust off the radioactive dust particles. ***Do not reverse roles***. Only Buddy #2's hood will be decontaminated and rolled at this time. |
| 4. Remove CVCUS overboots—limits the spread of contamination. | All contamination | Two discard containers (for example, plastic bags) | Buddy #1 opens all the zippers on Buddy #2's CVCUS in the following order: wrist, ankles, and front. While standing behind Buddy #2 and adjacent to the clean area, Buddy #1 grasps the shoulders of Buddy #2's CVCUS and instructs him to make a fist to ensure that his gloves do not come off. He then pulls Buddy #2's CVCUS down below his knees, turning it inside out. He has Buddy #2 to partially remove his overboots by alternately stepping on each heel and pulling up on his foot. While kneeling, Buddy #1 grasps one leg of Buddy #2's CVCUS and his overboot and instructs him to remove his leg and foot simultaneously. As Buddy #2 removes each leg and foot, he steps directly onto the clean area. Buddy #1 decons his gloves again. |

**Table 3-7. Steps for the CVCUS/CPU (Triple-Buddy Method) MOPP-Gear Exchange With the M40A1, M42, and M43 (Aviation and Armor) (Continued)**

| Steps | Contamination Type | Required Equipment | Procedures |
|---|---|---|---|
| 5. Remove CPU jacket/ gloves. | All contamination | Two discard containers | Buddy #2 holds the fingertips of his gloves and partially slides his hand out. When the fingers of both hands are free, he holds his arms away from his body and lets the gloves drop. Buddy #1 unzips and removes Buddy #2's CPU jacket by grasping it at the shoulders and pulling down, turning it inside out. He places it on the clean area with the clean side up. He removes Buddy #2's protective glove inserts. |
| 6. Remove combat boots/ CPU pants. | All contamination | Two discard containers | Buddy #1 instructs Buddy #2 to stand on the clean area (but not on the CPU jacket) and unlace his combat boots. Buddy #2 partially removes his boots by alternately stepping on each heel and pulling up on his foot. Buddy #1 grasps one of Buddy #2's boots and instructs him to remove his foot. As Buddy #2 removes his foot, he steps directly onto the clean side of the CPU jacket. Repeat the procedure on the other foot. Next, Buddy #1 removes Buddy #2's CPU drawers by pulling down on the sides and turning them inside out. Buddy #2 is now stripped to his underwear and mask and is ready to put on new clothing. **NOTE: If Buddy #2's feet come in contact with the contaminated pants of the CPU, decon them.** |
| 7. Put on CPU/ CVCUS. | All contamination | One CPU/CVCUS per person | Buddy #1 opens the package containing the new undergarment without touching the inside. Buddy #2 pulls out the undergarment without touching the outside of the package. He puts on the new clothing in the following order: CPU drawers, CPU jacket, combat boots, and CVCUS (over the CPU drawers). |
| 8. Put on overboots and gloves. | All contamination | One set of chemical protective over-boots per person<br>One set of chemical protective gloves per person<br>M9 detector paper | Buddy #1 opens the package of clean overboots without touching the inside of the package. Buddy #2 removes the overboots without touching the outside of the package and puts them on. Buddy #1 opens the package of clean gloves without touching the inside of the package. Buddy #2 removes the gloves without touching the outside of the package and puts them on. Buddy #2 puts on the M9 detector paper. |
| 9. Secure hood. | Chemical/ biological | One IEDK | Buddy #1 wipes his rubber gloves with an IEDK. He unrolls Buddy #2 hood and attaches the straps and tightens the neck cord. He checks the snaps and neck cord on Buddy #2's hood to ensure that they are closed. Buddy #1 and Buddy #2 reverse rolls and repeat steps 2 through 9. |
| 10. Secure gear. | All contamination | One chemical protective helmet cover per person | The soldier secures his individual gear and puts it back on. He puts on a new chemical protective helmet cover and moves to the AA. He uses the buddy system to check the fit of all secured gear. |

**Table 3-8. Steps for the Individual (Emergency Method) MOPP-Gear Exchange**

| Steps | Contamination Type | Required Equipment | Procedures |
|---|---|---|---|
| 1. Decon gear—removes gross contamination from individual gear (weapon, helmet, load-bearing equipment, and mask carrier). | Chemical/ biological | One IEDK | The soldier uses M8 detector paper to determine the areas of gross contamination and field-expedient absorbents (sand, dirt, or rags) to remove gross liquid contamination. He should take special care to avoid touching these areas during overgarment removal. He uses an IEDK to decon his individual gear (helmet, load-carrying equipment (LCE), weapon, and mask carrier). |
| | Radiological | Hot, soapy water | The soldier brushes or wipes radiological contamination from his individual gear. He washes it with hot, soapy water (if available) then sets it aside to dry on an uncontaminated surface (plastic, poncho, or similar material). |
| 2. Prepare to decon—facilities removal of overgarment trousers and overboots. | All contamination | Cutting tool | The soldier unfastens the shoulder straps on his hood, pulls them over his shoulders, and reattaches them. He loosens the drawcord on the hood of his protective mask. **NOTE: If wearing the M43 protective mask, tie off the microphone cord to the hose of the mask.** He removes the M9 detector paper from his overgarment and unties/cuts the drawcords on the trouser legs of his overgarment. He unzips the trouser legs and rolls a cuff in each trouser leg, ensuring that the cuffs do not come above the top of his overboots. He unfastens or cuts the fasteners on the overboots. |
| 3. Decon mask and hood—removes gross contamination from the mask and hood. | Chemical/ biological | One IEDK | The soldier wipes the eye-lens outserts on his mask from the top down. He wipes his mask and gloves and rolls his hood. He grasps the straps of the hood and lifts the hood off his shoulders and partially over his head until most of the back of his head is exposed. He rolls the hood, starting at the chin, and works around the entire mask until the rolled hood will stay up and off of his shoulders. He tucks the straps and neck cord into the roll. He rolls the hood tightly against his mask without pulling the hood off the back of his head. He tucks the tail between the upper part of the canister and the mask. (Tie the tail over and under the hose for the M42 mask.)<br><br>The soldier removes the applicator mitt from the package with his nondominant hand. Making a "V", he wipes down his hand, paying particular attention to the areas between his fingers. Once he has thoroughly wiped down his dominant hand, he must insert it into the applicator mitt and thoroughly wipe down his other hand. He gently pats the voicemitter with black powder until it is covered. |

Table 3-8. Steps for the Individual (Emergency Method) MOPP-Gear Exchange (Continued)

| Steps | Contamination Type | Required Equipment | Procedures |
|---|---|---|---|
| 3. Decon mask and hood (continued). | | | The soldier starts at the top of his hood and wipes down and away, patting until the surface of the hood is covered by the black powder. He rewipes his gloves, starting with his nondominant hand.<br><br>The soldier lifts the hood off of his shoulders by grasping the shoulder straps in one hand and placing the other hand on top of the head and pulling the hood over his head until the elastic band is over his knuckles and most of the back of his head is exposed. He should not expose his ears or pull the hood completely over his face or mask. He tucks his shoulder straps, underarm straps, and the rolled portion of his hood under the elastic band.<br><br>Follow these procedures when using the M40A1 mask with the quick-doff hood: The soldier removes the underarm straps from the front of his hood and places them over his shoulders. He refastens them on the front of his hood. **NOTE: Anytime you decon your mask or roll your hood, place two fingers on the voicemitter of the mask to prevent accidental breakage of the seal.** |
| | Radiological | Three containers (about 3-gallon capacity)<br>Two sponges<br>Soapy water<br>Rinse water<br>Paper towels or similar drying material | The soldier wipes his mask and hood with a sponge dipped in hot, soapy water and rinses them with a sponge dipped in clean water. He dries the mask and hood with paper towels or rags. **NOTE: Cool, soapy water is not as effective for removing contamination, but it can be used if you scrub longer.** If water is not available, soldier should brush and dust off the radioactive dust particles. |
| 4. Remove overgarment/ overboots—prevents agent from penetrating through to the undergarments or the skin. | All contamination | Two discard containers (recommend plastic bags).<br>Chemical protective suit with protective gloves and overboots | The soldier grasps his overgarment jacket and unsnaps the snaps individually. He unties the drawcord at the bottom of the jacket. He unfastens the Velcro at the waist and then refastens it. He unfastens the Velcro closure over the zippered front of the jacket and unzips the jacket. He grasps the front of the jacket and pulls the jacket back until it is off his shoulders. He puts his arms behind his back and works his arms out of the sleeves. He should not let the outside of the jacket touch his body. When the jacket is off, he lays it on the ground with the black side up. He unfastens and unzips his trousers. He should not loosen his waist tabs. He loosens his overboots by alternately stepping on each heel and pulling up on his foot. He grasps his trousers and pushes them down to his knees. He walks out of his trousers/ overboots simultaneously and steps onto the black side of the jacket. He steps onto the jacket wearing his mask, BDU, combat boots, and gloves. |

**Table 3-8. Steps for the Individual (Emergency Method) MOPP-Gear Exchange (Continued)**

| Steps | Contamination Type | Required Equipment | Procedures |
|---|---|---|---|
| 5. Remove gloves. | All contamination | Two discard containers (from step 4)<br>One IEDK | The soldier wipes around the edges of the packages containing the new items (gloves, overgarments, and overboots) with an IEDK. He opens the new packages. He holds the fingertips of his gloves and partially slides his hands out. When the fingers of both hands are free, he holds his arms away from his body and lets his gloves drop. |
| 6. Put on overgarment. | All contamination | One set of chemical protective overgar-ments per person<br>One IEDK | The soldier removes the overgarment from its package without touching the outside. He puts on the overgarment and fastens it, leaving the trouser legs open until he puts on the new overboots. |
| 7. Put on overboots and gloves. | All contamination | One set of chemical protective overboots per person<br>One set of chemical protective gloves per person | The soldier removes the overboots (one at a time) from their package without touching the outside and puts them on. He removes the gloves from their package without touching the outside, puts them on, and fastens his trouser legs. **NOTE: The gloves and overboots will have a light powdery coating. This is normal. Do not remove it. It will not affect the protective qualities.** |
| 8. Secure hood. | All contamination | One chemical protective helmet cover per person | The soldier secures his individual gear and puts it back on. He puts on a new chemical protective helmet cover and moves to the AA. |

**Table 3-9. Steps for the CVCUS/CPU, Individual (Emergency Method) MOPP-Gear Exchange**

| Steps | Contamination Type | Required Equipment | Procedures |
|---|---|---|---|
| 1. Decon gear—removes gross contamination from individual gear (weapon, helmet, load-bearing equipment, and mask carrier). | Chemical/ biological | One IEDK | The soldier uses M8 detector paper to determine the areas of gross contamination and field-expedient absorbents (sand, dirt, or rags) to remove gross liquid contamination. He should take special care to avoid touching these areas during overgarment removal. He uses an IEDK to decon his individual gear (helmet LCE, weapon, and mask carrier). |
| | Radiological | Hot, soapy water | The soldier brushes or wipes radiological contamination from his individual gear. He washes it with hot, soapy water (if available) then sets it aside to dry on an uncontaminated surface (plastic, poncho, or similar material). |
| 2. Prepare to decon—facilities removal of overgarment trousers and overboots. | All contamination | Cutting tool | The soldier unfastens the shoulder straps on his hood, pulls them over his shoulders, and reattaches them to the front of the hood. He loosens the drawcord on the hood of his protective mask. **NOTE: If wearing the M43 protective mask, tie off the microphone cord to the hose of the mask.** He removes the M9 detector paper from his overgarment and unties/cuts the drawcords on the trouser legs of his overgarment. He unzips the trouser legs and rolls a cuff in each trouser leg, ensuring that the cuffs do not come above the top of his overboots. He unfastens or cuts the fasteners on the overboots. |
| 3. Decon mask and hood—removes gross contamination from the mask and hood. | Chemical/ biological | One IEDK | The soldier wipes the eye-lens outserts on his mask from the top down. He wipes his mask and gloves and rolls his hood. He grasps the straps of the hood and lifts the hood off his shoulders and partially over his head until most of the back of his head is exposed. He rolls the hood, starting at the chin, and works around the entire mask until the rolled hood will stay up and off of his shoulders. He tucks the straps and neck cord into the roll. He rolls the hood tightly against his mask without pulling the hood off the back of his head. He tucks the tail between the upper part of the canister and the mask. (Tie the tail over and under the hose for the M42 mask.)<br><br>The soldier removes the applicator mitt from the package with his nondominant hand. Making a "V", he wipes down his hand, paying particular attention to the areas between his fingers. Once he has thoroughly wiped down his dominant hand, he must insert it into the applicator mitt and thoroughly wipe down his other hand. |

**Table 3-9. Steps for the CVCUS/CPU, Individual (Emergency Method) MOPP-Gear Exchange (Continued)**

| Steps | Contamination Type | Required Equipment | Procedures |
|---|---|---|---|
| 3. Decon mask and hood (continued). | | | The soldier gently pats the voicemitter with black powder until it has been covered. He starts at the top of his hood and wipes down and away, patting until the surface of the hood is covered by the black powder. He rewipes his gloves, starting with his nondominant hand.<br><br>The soldier lifts the hood off of his shoulders by grasping the shoulder straps in one hand and placing the other hand on top of the head and pulling the hood over his head until the elastic band is over his knuckles and most of the back of his head is exposed. He should not expose his ears or pull the hood completely over his face or mask. He tucks his shoulder straps, underarm straps, and the rolled portion of his hood under the elastic band.<br><br>Follow these procedures when using the M40A1 mask with the quick-doff hood: The soldier removes the underarm straps from the front of his hood and places them over his shoulders. He refastens them on the front of his hood. **NOTE: Anytime you decon your mask or roll your hood, place two fingers on the voicemitter of the mask to prevent accidental breakage of the seal.** |
| | Radiological | Three containers (about 3-gallon capacity)<br>Two sponges<br>Soapy water<br>Rinse water<br>Paper towels or similar drying material | The soldier wipes his mask and hood with a sponge dipped in hot, soapy water and rinses them with a sponge dipped in clean water. He dries the mask and hood with paper towels or rags. **NOTE: Cool, soapy water is not as effective for removing contamination, but it can be used if you scrub longer.** If water is not available, the soldier should brush and dust off the radioactive dust particles. |
| 4. Remove CVCUS/ overboots. | All contamination | Two discard containers (plastic bags) | The soldier opens all the zippers on his CVCUS in the following order: wrists, ankles, and front. While standing adjacent to the clean area, he grasps the front of his CVCUS and pulls it off of his shoulders. He makes a fist to ensure that his gloves do not come off when removing his CVCUS. He puts his arms behind his back and works his arms out of the sleeves. He loosens his overboots by alternately stepping on each heel and pulling up on his foot. He pulls his CVCUS down below his knees, turning it inside out. He should not let the outside of his CVCUS touch his body. He walks out of his CVCUS/overboots simultaneously and steps onto the clean area. |

**Table 3-9. Steps for the CVCUS/CPU, Individual (Emergency Method) MOPP-Gear Exchange (Continued)**

| Steps | Contamination Type | Required Equipment | Procedures |
|---|---|---|---|
| 5. Remove jacket/ gloves. | All contamination | Two discard containers | The soldier wipes his gloves and around the edges of the packages containing the new items (gloves, undergarments, and overboots) with an IEDK. He opens the new packages. He holds the fingertips of the gloves and partially slides his hands out. When the fingers of both hands are free, he holds his arms away from his body and lets the gloves drop.<br><br>The soldier opens all Velcro wrist fasteners and unzips his CPU jacket. He removes the CPU jacket by grasping it at the shoulders and pulling down, turning it inside out. He places the jacket on a clean area with the clean side up. He removes the protective glove inserts. |
| 6. Remove combat boot/CPU pants. | All contamination | Two discard containers | The soldier unlaces his combat boots, removes them, and steps directly onto his CPU jacket. He takes off his CPU drawers by placing his hands beneath the waistband and removing them, turning them inside out. The soldier is now stripped to his underwear and mask and is ready to put on his new clothing.<br><br>The soldier follows these procedures if wearing the BDU. He stands on the clean area (but not on the CPU jacket) and unlaces his combat boots. He partially removes his boots by alternately stepping on each heel and pulling up on his foot. He removes each trouser leg and combat boot simultaneously and steps directly onto the clean side of the CPU jacket. |
| 7. Put on CPU. | All contamination | One set of chemical protective undergarments per soldier | The soldier removes the undergarment from the package without touching the outside. He puts on the undergarment in the following order: CPU drawers, CPU jacket, combat boots, and CVCUS (over the CPU drawers). |
| 8. Put on overboots and gloves. | Chemical/ biological | One set of chemical protective overboots per soldier<br>One set of chemical protective gloves per soldier<br>M9 detector paper | The soldier removes the overboots from their package without touching the outside and puts them on. He removes the gloves from their package without touching the outside and puts them on. He puts on the M9 detector paper. |

Table 3-9. Steps for the CVCUS/CPU, Individual (Emergency Method) MOPP-Gear Exchange (Continued)

| Steps | Contamination Type | Required Equipment | Procedures |
|---|---|---|---|
| 9. Secure hood. | Chemical/ biological | One IEDK | The soldier wipes his rubber gloves with an IEDK. He unrolls his hood, attaches the straps, and tightens the neck cord. |
| 10. Secure gear. | All contamination | One chemical protective helmet cover per soldier | The soldier secures his individual gear and puts it back on. He puts on a new chemical protective helmet cover and moves to the AA. |

# Chapter 4

# Thorough Decon

Thorough-decon operations reduce and sometimes eliminate contamination. They restore combat power by removing nearly all contamination from unit and individual equipment so that troops can operate it safely for extended periods at reduced MOPP levels. Since a small risk of contamination remains, periodic checks with the CAM, the M8/M9 detector paper, or the M256A1 detector kit must be performed after each operation. A thorough decon can be done as part of reconstitution, or it can be done to support operations throughout the battlefield. After a thorough decon, the unit moves out of the decon site into a tactical AA. The unit, while in this tactical assembly, may undergo reconstitution or prepare for future operations. Combat-service-support (CSS) elements replenish combat stocks, refit equipment, and replace personnel and equipment, as required. The contaminated unit, with some assistance from a chemical unit, does the DTD. A supporting chemical unit does the DED or DAD. The planning considerations that are required to conduct thorough-decon operations and the methods that various chemical units (decon and smoke/decon) use to conduct DTD/DED are discussed in this chapter. The exact layout of a thorough-decon site is determined by METT-TC.

## PLANNING

4-1. A thorough decon is the most effective type of decon but is the most resource intensive. A combination of weather conditions, agent used, mission requirements, time, troops, and supplies available may make weathering the decon option of choice. Commanders decide which type of decon is required based on the recommendation of the chemical personnel.

4-2. Generally, thorough-decon operations are conducted beyond the range of enemy direct-fire systems. If a contaminated unit requires a thorough decon as part of reconstitution operations, the decon site is established near the reconstitution area or the brigade's rear area. Usually, company-sized units are reconstituted in their brigade's rear area while battalion-level units are reconstituted in their division's rear area. Organizations larger than a battalion will be reconstituted in the corps's rear area. Further information about reconstitution can be found in FMs 100-5 and 100-9.

4-3. All echelons prepare for thorough-decon operations as part of the overall planning process. From the commander's general guidance, the chemical staff can begin to develop the decon plan. Coordination with higher HQ is required to determine the availability of engineer support for site preparation and closure. Engineers can provide support for sump preparation, drainage ditches, and sumps. Coordinate with the civil-affairs office for host-nation support (personnel, equipment, and supplies) and for environmental requirements and restrictions.

4-4. The chemical staff selects possible decon sites throughout the unit's area of operation (AO) based on such factors as decon type, terrain, mission, threat, road network, and availability of water. After the decon sites are selected, linkup points are chosen to support each site. A site may have more than one linkup point.

4-5. Since decon assets are limited, the commander must establish a priority of decon support. The priority of support lists the units in the order they will receive decon support. This can change from phase to phase during an operation. The chemical staff develops the priority of support based on an understanding of the commander's intent.

4-6. To give priority of support to the lead task force during the assault phase may not be the best choice since any contaminated elements will most likely not stop for decon until after the assault is completed. The commander should establish a priority of work that specifies the order in which equipment will be decontaminated. For example, a priority of work may be in this order: engineer equipment, artillery pieces, main battle tank, and long-haul vehicles. A limiting factor is the availability of water. A typical vehicle requires 500 gallons of water during the DED. The actual amount of water required varies by the vehicle and its contamination level. The supported unit's chemical staff must develop a water resupply plan for thorough-decon operations.

4-7. A water resupply plan can be as simple as selecting a series of linkup points along a route where the chemical unit can link up with a bulk water truck. More complex water resupply plans include caching water throughout the AO, coordinating for the movement of water bladders by aircraft, and identifying water sources in the unit's AO. The use of nonpotable, salt, and brackish waters should be considered.

# PREPARATION

4-8. Subordinate units review their higher HQ decon plan. If additional decon sites or assets are required, request them from the higher HQ. The supporting decon-platoon leader conducts a leader's recon of the designated decon sites. Since the decon sites are initially selected by map recon, the actual site may not be suitable. If the selected site cannot support decon operations, the chemical unit leader attempts to find another site close to the original site and notifies the supported unit and his higher HQ of the change.

4-9. The decon-platoon unit leader determines the support requirements that are needed to operate the decon site. Engineers are required to dig sumps, improve access to the site, and dig ditches for runoff. Air defense (AD) and military police (MP) are required to provide security. Medical support is also required. Depending on the operational situation and the threat, some DED sites may be prepared in advance. Routes must be established and published in the OPORD, leading the contaminated unit to the linkup point.

**NOTE: Units must be informed not to proceed directly to the decon site.**

## EXECUTION

4-10. The actual decon operation begins once a unit requests decon support. The unit leader must assess his situation and accurately report it to his HQ. This situation report should include the type and extent of contamination, the current location, his assessment of the unit's ability to perform the current mission while contaminated, and the time he expects the unit to become combat ineffective. The contaminated unit should perform immediate-decon techniques to increase its survivability and to limit the spread of contamination. An operational decon should also be considered and performed, as appropriate. The request for decon support must contain several essential elements of information to assist the chemical staff and the decon-platoon leader in coordinating the decon operation. These essential elements of information are the—

- Designation of the contaminated unit.
- Location of the contaminated unit.
- Time the unit became contaminated.
- Earliest possible time the unit can move/begin decon.
- Type of contamination.
- Number and type of vehicles contaminated.
- Special requirements (patient-decon station, recovery assets, unit decon team, and so forth).
- Supported unit's frequency and call sign to facilitate initial linkup.

4-11. The supporting chemical unit is given a warning order (WO) to conduct a decon operation. Subsequent WOs provide more detailed information. After requesting decon support, the supported unit issues an OPORD/fragmentary order (FRAGO) to the chemical unit.

4-12. The supported unit must coordinate the movement of the contaminated unit to the linkup point. If the contaminated unit is out of the supported unit's AO, the move must be coordinated with the higher HQ.

4-13. Once the chemical unit is set up, the decon-platoon leader moves to the linkup point. Other elements may provide assets to support the decon operations, such as medical, engineer, AD, and MP. These elements linkup with the chemical unit before the arrival of the contaminated unit (see Table 4-1, page 4-4).

4-14. The supported unit is responsible for site security and overall control. The decon-platoon leader is responsible for operating the predecon area and processing vehicles.

4-15. After linkup is achieved and all support assets are in position, the actual decon operation begins. The decon-platoon leader, with help from the commander of the contaminated unit or the supported unit's command and control cell, supervises the decon operation. All equipment not necessary for decon operations should be located in a clear area to limit the spread of contamination.

**Table 4-1. Thorough-Decon Support Matrix**

| Tasks | Units | | | | | | | |
|---|---|---|---|---|---|---|---|---|
| | Division CMO | Division | Brigade | DISCOM | FSB | Chemical Unit | Contaminated Unit | Battalion |
| **Preparation-Phase Tasks** | | | | | | | | |
| Request | S | | | | | | P | S |
| Coordination | S | | S | | | | | P |
| Site selection | | S OR S | | | | P | | |
| Advance-party linkup | | | | | | S | P | S |
| Site setup | | | | S | S | P | | |
| **Execution-Phase Tasks** | | | | | | | | |
| Site control/security | | | | | | S | P | |
| Predecon actions | | | | | | P | S | |
| Processing | | | | | | P | S | |
| **Site-Clearance-Phase Tasks** | | | | | | | | |
| Cleanup | | | | | | P | P | |
| Marking and reporting | | | | | | P | | |

S - supporting responsibility
P - primary responsibility

4-16. A thorough-decon site consists of the following areas: predecon staging area, DED and DTD areas, and postdecon AA (see Figure 4-1). The chemical unit leader selects these areas based on operational guidance, road network, available cover and concealment, and water supply. The contaminated unit uses the predecon staging area to ready itself. The postdecon AA is the location where the vehicles and soldiers exiting the DED and DTD areas are linked up before moving from the decon site.

## PREDECON STAGING AREA

4-17. The contaminated unit moves to the predecon staging area about 274 to 547 yards downwind of the thorough-decon site. The contaminated unit performs predecon actions to include segregating vehicles by checking for contamination.

4-18. For chemical contamination, use the CAM and the M8/M9 detector paper. The contamination level can be estimated by comparing the M8/M9 detector paper to Figure 4-2. When monitoring vehicles for contamination, there should be about 33 to 49 feet between each vehicle to prevent false positive readings with the CAM. For radiological contamination, use the AN/PDR-77 or AN/VDR-2 radiac detector. If the vehicle only has isolated areas of contamination, use the M11 or M13 to decon those areas.

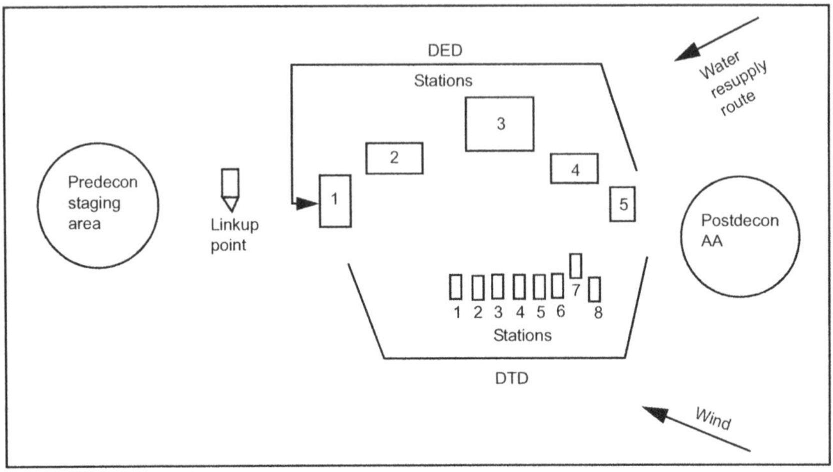


**Figure 4-1. Thorough-Decon Site**

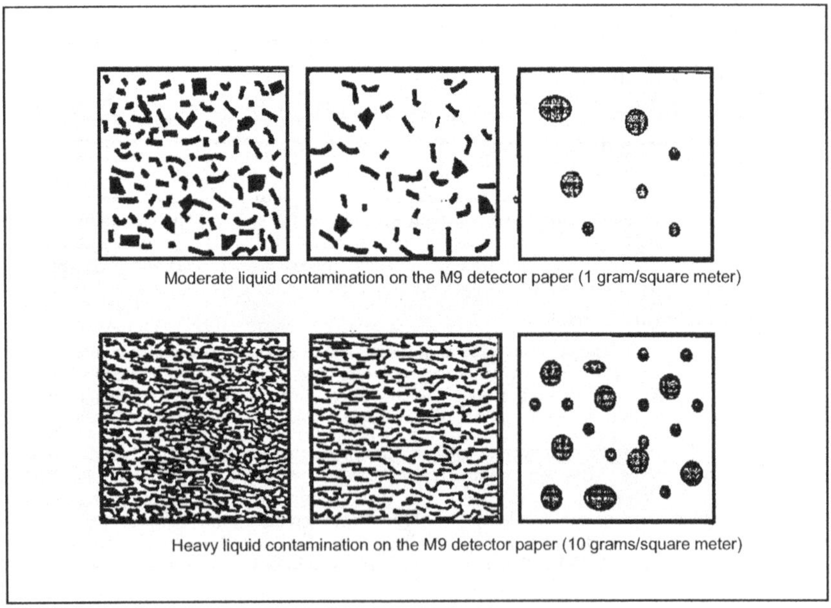


**Figure 4-2. Contamination Levels**

### Vehicle Crews

4-19. The vehicle crews, except for the drivers, dismount. As the crews dismount, they remove all contaminated equipment, to include electro-optical systems from the top and sides of the vehicles. Once the crews have exited the vehicle, they will not reenter. This prevents contamination from being spread into the vehicle interior.

### Vehicle Preparation

4-20. Using the pioneer tools from the contaminated vehicles, the crews remove all heavy mud and debris. They concentrate on the vehicle's undercarriage, which is the most likely place for contamination to collect and the hardest place to decon. Once the crews are finished with the pioneer tools, they are placed back on the vehicle. Initial removal of the mud and debris makes it more likely that the decon wash will remove any remaining contamination.

4-21. Seat covers (when applicable), canvas items, camouflage netting, and any other materials that can absorb liquid contamination are removed. These items create a potential transfer hazard and are not easily decontaminated. Left untreated, absorbed chemical agents will desorb after being decontaminated and will create a vapor hazard. Remove items that cannot be decontaminated by the standard methods used in the DTD, and place them at the collection point. Chemical-unit personnel provide advice concerning the decon or disposal of these items.

4-22. Design vehicle-loading plans to minimize the amount of equipment carried on the outside of the vehicle that cannot be readily decontaminated. Whenever possible, NBC covers should be used when a chemical attack is expected (see FM 3-4). All NBC covers are removed and disposed of as contaminated waste during the predecon actions.

## DED AND DTD AREAS

4-23. In coordination with the chemical unit leader operating the decon site, the contaminated unit begins sending contaminated vehicles by priority for processing. Communication is maintained between the predecon staging area and the chemical unit leaders. All assistant vehicle drivers are the first individuals sent through the DTD to ensure that there is a driver exchange at station 3 (see paragraph 4-94).

## POSTDECON AA

4-24. The chemical unit leader selects the general location for the postdecon AA. It must be big enough to hold the entire unit undergoing the thorough decon with proper cover and concealment. The postdecon AA is located about 0.62-mile upwind from the DED and DTD areas. The unit assembles in the postdecon AA after completing DTD and DED operations. The unit occupies the postdecon AA before moving to a reconstitution location or a tactical AA to prepare for future operations.

# DETAILED TROOP DECON

4-25. The contaminated unit or its higher HQ is responsible for setting up, operating, manning, and closing the DTD area at the thorough-decon site. The chemical unit leader determines the general location of the DTD within the decon site and provides technical advice on setting up, operating, and closing the DTD area. The supervisor of the DTD must establish a work/rest cycle.

4-26. There are eight stations for a DTD. Spacing between the stations is 10 to 16 feet (see Figure 4-3, page 4-8). Recapitulation of personnel and equipment for a DTD is provided at Table 4-2, page 4-9.

## STATION 1 - INDIVIDUAL-GEAR DECON

4-27. At this station, contamination is removed from individual gear (LCE, mask carrier, helmet, and weapon) to a negligible risk level (see Chapter 1 of this manual).

### Preparation

4-28. At this station, the following equipment and supplies are needed:

- Four 30-gallon containers for each two decon lanes used.
- 2 long-handled brushes for each two decon lanes used.
- Sufficient STB slurry mix for chemical contamination.
- Hot, soapy water and a sponge for radiological contamination.
- Rinse water.
- 2 ponchos or plastic sheets.
- Engineer tape.

4-29. Dig a sump that is 6 feet long, 6 feet wide, and 4 feet deep (minimum). Place three 30-gallon containers near the sump for ease of changing. Fill two containers with an STB slurry mix. Fill the other two containers with clean water for rinsing and place it about 3 feet forward of the STB cans. Place two long-handled scrub brushes at each can of STB slurry.

4-30. To prepare the slurry, mix 100 pounds of STB with 20 gallons of hot water. The chemical unit provides the hot water. Change the mixture after 20 soldiers have decontaminated their gear. The rinse water should be changed after every 10 soldiers or when it appears dirty (place the waste mixture in the sump).

4-31. Place a poncho or a plastic sheet on the ground at the checkpoint. Divide the poncho or the sheet in half using engineer tape. This is the contamination control line. The checkpoint will be a minimum of 10 feet from all other stations in order to get a true reading on the detection equipment. Place the following chemical detection/identification equipment at the checkpoint for an average company-sized unit:

- 1 CAM.
- 8 books of M8 detector paper.
- 4 M256A1 detector kits.
- 100 plastic trash bags.
- 1 AN/VDR-2 for radiological contamination.

4-32. Three soldiers are required to operate this station. One soldier supervises the decon of the individual gear and takes the decontaminated equipment to the checkpoint. He also prepares a new slurry mixture as necessary. One soldier remains at the checkpoint and checks all the gear using the detection equipment to ensure that it is decontaminated. One soldier transports the decontaminated gear to the reissue point.

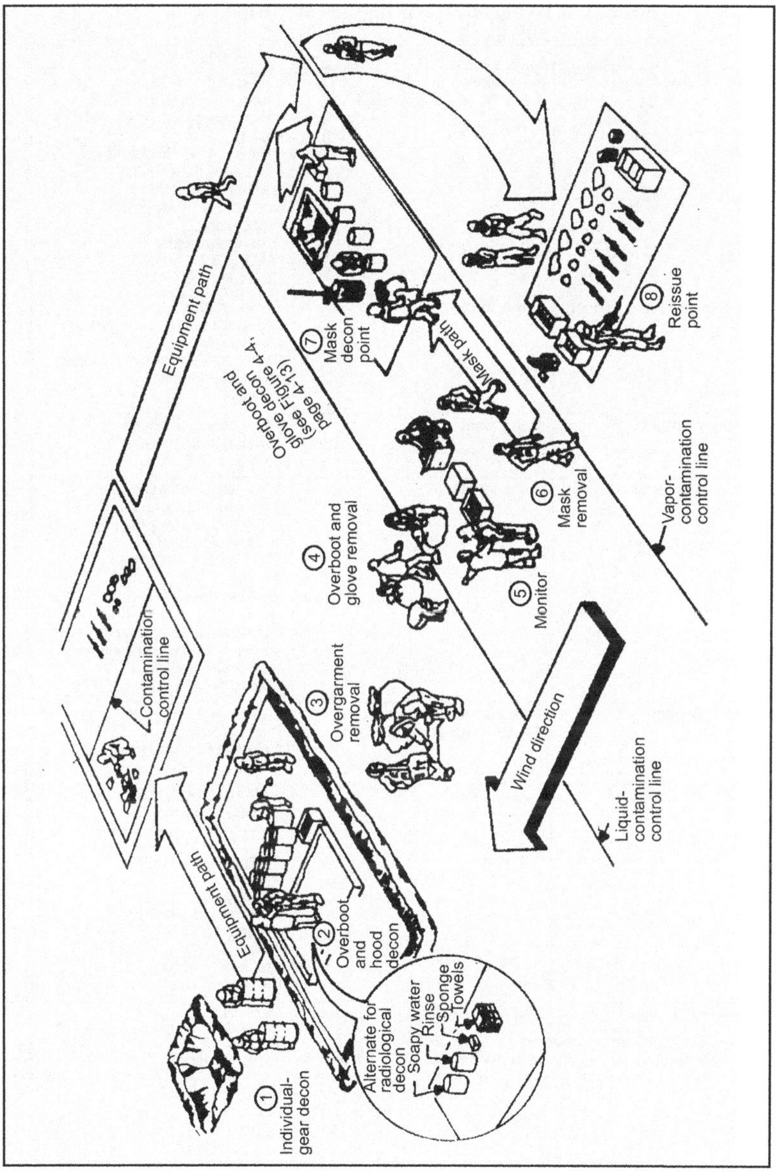


**Figure 4-3. DTD Layout**

**Table 4-2. DTD Personnel and Equipment Recapitulation**

| Stations | Personnel | Equipment/Supplies |
|---|---|---|
| Station 1 - individual-gear decon | 2 attendants<br>1 monitor (CAM operator) | 4 30-gallon containers<br>2 long-handled brushes<br>2 ponchos or plastic sheets<br>1 CAM<br>8 books of M8 detector paper<br>4 M256A1 detector kits<br>100 plastic trash bags<br>Sufficient STB slurry mix<br>Rinse water<br>*Hot, soapy water<br>*AN/VDR-2<br>*Sponge |
| Station 2 - overboots and hood decon | 1 attendant | Cutting tool<br>One SKD or IEDK per person<br>Large plastic sheet<br>Plastic trash bags (as required)<br>10 drums of STB<br>1 shovel |
| Station 3 - overgarment removal | 1 attendant | 2 30-gallon containers<br>100 plastic trash bags<br>10 boxes of SDKs |
| Station 4 - overboot and glove removal | 3 attendants | 8 30-gallon containers<br>2 long-handled brushes<br>M8 detector paper (as required)<br>1 immersion heater<br>100 plastic trash bags<br>1 CAM<br>Engineer tape<br>1 cutting tool<br>2 ponchos or large plastic sheets<br>10 percent STB/HTH solution<br>Hot, soapy water<br>Cold rinse water |
| Station 5 - monitor | 1 attendant (CAM operator)<br>1 aidman (or combat lifesaver) | 1 CAM<br>First aid supplies<br>5 books of M8 detector papers<br>1 case of SDKs |
| Station 6 - mask removal | 2 attendants | 1 M8A1 or M22 ACAA<br>Engineer tape |
| Station 7 - mask decon point | 2 attendants<br>1 monitor | 1 30-gallon container<br>4 3-gallon containers<br>1 CAM<br>2 sponges<br>One case of paper towels<br>1 immersion heater<br>Mask sanitizing solution<br>Hot, soapy water<br>Rinse water |
| Station 8 - reissue point | Unit NBC NCO/supply sergeant | Mask PLL |
| * Denotes the equipment needed for radiological contamination.<br>**NOTE: Assume that an operational decon was done before the thorough decon.** | | |

### Execution

4-33. The soldier decontaminates his gear by washing and scrubbing it for 6 minutes in a decontaminant container with hot, soapy water or an STB slurry mix. If he is wearing the M42 mask, he should use hot, soapy water and a sponge or an STB slurry mix to decon the hose and canister.

4-34. The soldier dips his gear into the clean water and rinses it for 4 minutes then hands it to the attendant and proceeds to the next station. The attendant takes the gear to the equipment checkpoint and places the decontaminated gear on the "dirty" side of the contamination control line and returns to the containers to pick up more gear. The monitor at the checkpoint checks the gear using the appropriate detection device and the procedures associated with that device. If the residual contamination exceeds negligible risks, recycle the gear and decon it again (see Chapter 1 of this manual). If the gear passes the check, place it on the clean side of the contamination control line. The attendant will carry the equipment to the reissue point.

4-35. Depending on the time available, more extensive washing and checking procedures may be used. The longer the gear is washed or left out in the air after washing, the lower the contamination level. The gear may be put in closed areas or plastic bags and checked for hazardous vapors with the M256A1 detector kit or the CAM. The CAM only detects the G-series nerve-agent vapors and the H-series blister-agent vapors.

### Risk

4-36. If this step is not done properly, contamination may remain on the equipment. The resulting vapor hazard could cause casualties to unmasked personnel, particularly in closed areas (vehicle interiors) or heavily wooded areas where air circulation is poor.

## STATION 2 - OVERBOOT AND HOOD DECON

4-37. At this station, gross contamination on overboots, trouser legs, mask, and hood is neutralized. If ample hoods are available at the reissue point, the hood should be cut away.

### Preparation

4-38. Prepare a shuffle pit by digging a shallow pit about 3 feet long, 3 feet wide, and 6 inches deep. Fill the shuffle pit with an STB dry mix or an STB slurry, depending on water availability. Prepare the STB dry mix by mixing three parts of earth to two parts of STB. Prepare the STB slurry in the same manner as at station 1 (see paragraph 4-30). Add more STB to the mix after 10 soldiers have processed through the shuffle pit. The chemical unit will provide 10 drums of STB for every company-sized unit that goes through the station.

4-39. At this station, the following equipment and supplies are needed for the mask and the shuffle pit: 1 cutting tool for each decon lane used, 1 SDK or IEDK per person, plastic trash bags (as required), 10 drums of STB, and 1 shovel (to mix).

4-40. One soldier is required to operate this station. The attendant directs and "observes" the soldiers as they decon their overboots and hoods.

### Execution

4-41. The soldier walks into the shuffle pit and spreads his legs apart (double-shoulder width), bends at the waist, and uses his hands to thoroughly rub the STB dry mix or the STB slurry on his overboots and lower trousers legs. He should take special care to rub the rear of his overboots. He should also remove any excess decontaminant from his gloves.

4-42. If a replacement hood is available, remove the hood as follows. Buddy #1 cuts the shoulder straps and drawcord on Buddy #2's hood. Buddy #1 pulls Buddy #2's hood inside out over the front of the mask, being careful not to touch the exposed neck or head. Buddy #1 gathers Buddy #2's hood in one hand and using a cutting tool, cuts away the hood as close as possible to the eye-lens outsert, voicemitter, and inlet-valve covers. Make sure nothing is left dangling below the bottom of the mask.

4-43. If a replacement hood is not available, Buddy #1 decontaminates and rolls Buddy #2's hood in the same manner as for a MOPP-gear exchange. When the task is completed, Buddy #1 and Buddy #2 reverse roles.

**NOTE: Soldiers should check their overboots, rubber gloves, and overgarment for damage. Any rips, tears, or punctures in these items should be reported to the monitor at station 5. This allows the monitor at station 5 to check the soldiers for chemical-agent symptoms and their clothing for possible contamination.**

### Risk

4-44. If this step is not done properly, contamination can be transferred to the combat boots and the head and neck area from the hood.

## STATION 3 - OVERGARMENT REMOVAL

4-45. At this station, contaminated overgarments are removed before the agent penetrates the overgarment material and touches the undergarments or the skin.

### Preparation

4-46. At this station, the following equipment and supplies are needed:

- Two 30-gallon containers for each two decon lanes used.
- 100 plastic trash bags (or about one per soldier).
- 10 boxes of SDKs.

4-47. One soldier is required to operate this station. He directs and "monitors" the soldiers as they remove their overgarments in the same manner as a MOPP-gear exchange.

### Execution

4-48. The attendant assists the soldier in removing his overgarment. The attendant cuts and removes the M9 detector paper from around the soldier's wrist. He unfastens the Velcro closure over the jacket zipper, waist cord, and wrist Velcro straps on the soldier's jacket. He unfastens the back snaps and instructs the soldier to make a fist. He then pulls the soldier's jacket down and away from him.

4-49. The attendant cuts and removes the M9 detector paper from the soldier's trousers. He unfastens the Velcro straps and zippers on the cuffs of the soldier's trousers. He also unfastens the front waist snaps and unzips the front zipper. He has the soldier lift one leg and point that foot down and bend slightly at the knees for stability. The attendant grasps the cuff of the elevated foot with a hand on each side and pulls the cuff in an alternating, jerking motion until the soldier can step out of the trouser leg. The process is repeated on the other leg.

**NOTE: The attendant ensures that the soldier steps wide enough so as not to rub his clean leg against the contaminated boot and overgarment.**

### Risk

4-50. If this step is not done properly, the agent may be transferred to the undergarment or the skin.

## STATION 4 - OVERBOOT AND GLOVE REMOVAL

4-51. At this station, contaminated overboots and gloves are removed to limit the spread of contamination. The overboots and gloves may also be decontaminated for reissue (if serviceable).

### Preparation

4-52. At this station, the following equipment and supplies are needed for every two lanes used:

- Engineer tape.
- 1 cutting tool for each lane used.
- 100 plastic trash bags (or about one per soldier).

4-53. If the overboots and gloves are not being decontaminated, two 30-gallon containers are needed.

4-54. If the overboots and gloves are being decontaminated, the following are needed:

- Eight 30-gallon containers.
- 2 long-handled brushes.
- 2 ponchos or large plastic sheets.
- 1 CAM.
- 1 AN/VDR-2 for radiological contamination.
- 4 M256A1 detection kits.
- 1 immersion heater.
- 100 plastic trash bags.
- M8 detector paper, as required.
- 10 percent STB/HTH solution.
- Hot, soapy water.
- Cold rinse water.

4-55. If replacement overboots and gloves are available, set up the station as follows (see Figure 4-4). Establish a liquid-contamination control line and set

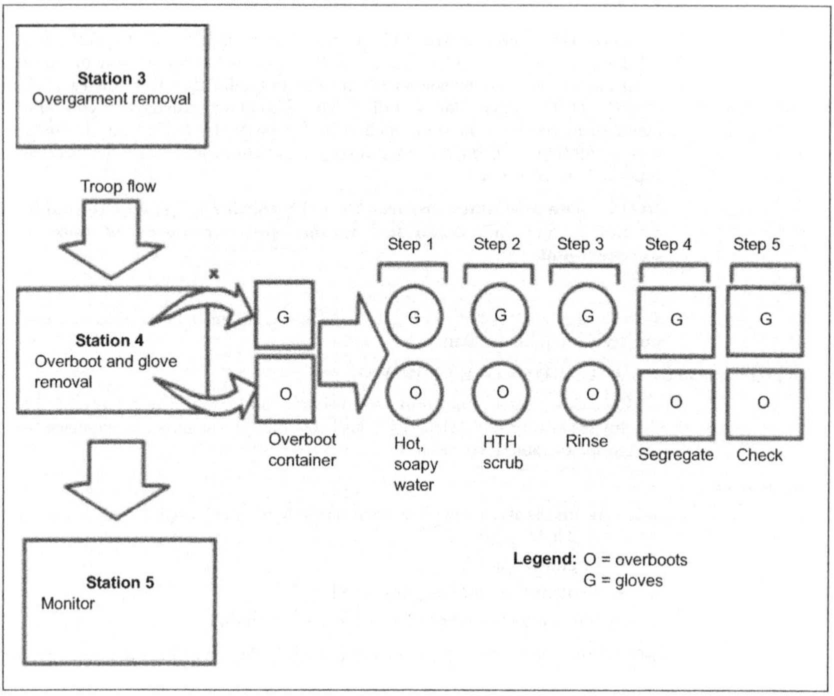


Figure 4-4. Station 4 Layout

two 30-gallon containers 1 foot back from the line. Soldiers support themselves using the containers and discard their overgarments into the containers. An attendant directs and monitors the soldiers as they remove their overboots and gloves in the same manner as a MOPP-gear exchange; however, the soldiers step over the control line instead of onto a jacket.

4-56. If replacement overboots and gloves are not available, set up the station as follows: Fill two 30-gallon containers with hot, soapy water and two 30-gallon containers with a 10 percent STB/HTH solution, placing two scrub brushes near the containers. Then fill two 30-gallon containers with cold rinse water. One attendant supervises and assists the soldiers wearing the M42 mask. The other two attendants decon the overboots and the gloves; one processes the overboots, while the other processes the gloves.

**NOTE: Replace the water in the container once 20 items have been processed. When available, the decon platoon will assist with the water requirements.**

4-57. For both situations, use engineer tape to mark the liquid-contamination control line on the ground. Place the cutting tools, two containers, and plastic bags on the "dirty" side of the liquid-contamination control line. The liquid-contamination control line separates the "dirty" and "clean" areas. No liquid agent should be tracked on the ground beyond the liquid-contamination control line.

### Execution

4-58. The liquid-contamination control line separates the "dirty" and "clean" areas. The attendant unfastens or cuts the elastic closures on the soldier's overboots. The soldier faces the liquid-contamination control line and steps back from it about 12 inches. The attendant steps on the back of the soldier's overboot and instructs him to lift his heel and work his foot out of the overboot and step across the liquid-contamination control line. Repeat the process on the other foot. If the overboot cannot be removed by this process, the attendant cuts it off and discards it in the designated container. The soldier holds the fingertips of his gloves and partially slides his hands out. The attendant remove the soldier's gloves.

4-59. If the soldier is wearing the M42 protective mask, the attendant from station 6 carries the soldier's filter canister until it is removed.

**NOTE: The attendant at station 4 performs his duty from the "dirty" side of the liquid-contamination control line.**

4-60. The overboots and gloves are decontaminated using the following steps:

**NOTE: Check all items for holes, tears, and punctures and discard any item with this defect. Do not decon any item that is unserviceable.**

- Step 1. Submerge the gloves and overboots in their respective container of hot, soapy water. Some of the contamination is removed during this step. When the overboots and gloves are removed from the container, ensure that no water remains inside them.
- Step 2. Submerge the gloves and overboots in their respective containers of STB/HTH solution. Thoroughly scrub the items until no visible contamination remains. After scrubbing, submerge each item once more before moving to the rinse container.
- Step 3. Thoroughly rinse the scrubbed items, making sure that they are rinsed inside as well as out.
- Step 4. Place usable items on a poncho or a plastic sheet to air dry and weather.
- Step 5. Place usable items in plastic trash bags along with an M256A1 detector kit. If the detector kit shows contamination remaining, the attendants can recycle the items or discard them. However, if the kit show no contamination, the items can be reused.

### Risk

4-61. If the overboots are not properly removed, the combat boots may become contaminated and contamination may be spread to clean areas. If the gloves are not properly removed, the undergarments and skin may become contaminated.

## STATION 5 - MONITOR

4-62. At this station, contamination on personnel is identified, spot-decon capabilities are provided, and medical aid is provided, as required.

### Preparation

4-63. At this station, the following equipment and supplies are needed:

- First aid supplies.
- 1 CAM.
- 1 AN/VDR-2 for radiological contamination.
- 5 books of M8 detector paper per 100 soldiers.
- 1 case of SDKs.

4-64. An aidman should be present to treat any soldiers suffering from chemical-agent symptoms. If an aidman is unavailable, a combat lifesaver should be present.

### Execution

4-65. The attendant checks the soldiers for contamination using the CAM. Liquid agents can be detected with M8 detector paper. Small quantities of agent vapor can be detected with the CAM. Symptoms of agent poisoning are the most obvious indication of skin contamination. At this station, the medic checks each soldier for symptoms of agent poisoning and treats, as required. Soldiers should report any damage to their MOPP gear that was identified at stations 2, 3, and 4. The attendant can decon any areas identified as contaminated with an SDK. Soldiers are remonitored after decon. It is possible that all liquid-chemical contamination is absorbed into the clothing. If so, M8/M9 detector paper will indicate negative, even though there is a hazard.

### Risk

4-66. If this station is omitted, the soldier could become a casualty. After this station, the soldier is not checked for contamination or decontamined again. Commanders can choose to conduct more extensive contamination checks here if medical assistance and time are available. This decreases the risk of casualties.

## STATION 6 - MASK REMOVAL

4-67. At this station, the mask is removed without contaminating the soldier. The mask is taken to a mask decon point, limiting agent transfer at the station.

### Preparation

4-68. At this station, the M8A1 or M22 automatic chemical-agent alarm (ACAA) and engineer tape are needed. Two soldiers are needed to operate this station. They remove and carry the masks to station 7.

### Execution

4-69. If the hood is still attached to the mask, the attendant pulls the hood over the front of the mask, grabs the mask by the voicemitter cover, and pulls the mask off the soldier. The soldier holds his breath as the mask is removed. If the mask has optical inserts, the attendant holds the mask open so that the soldier can remove the inserts without touching the outside of the mask. The soldier walks upwind 16 feet, crosses the vapor-contamination control line, and then resumes breathing. The attendant brings the mask to station 7.

4-70. If the wind direction remains constant, no chemical vapor hazard is expected beyond the vapor-contamination control line. Position the M8A1 or the M22 ACAA upwind of the station to warn of vapor hazards. The soldier getting decontaminated moves straight ahead while his mask, which may still give off vapors, is held on the vapor dirty side of the line and taken to station 7 where it is decontaminated.

### Risk

4-71. If the step is not done properly, the soldier could breathe toxic vapors. There is a high probability that the vapor hazard is still present on the mask and the hood. The soldier must not touch the outside of the mask because it could contaminate his bare hands. The soldier proceeds to station 8.

## STATION 7 - MASK DECON POINT

4-72. At this station, all the contamination is removed from the mask.

**NOTE: Once 20 items have been processed in the wash containers, replace the water. Once 10 item have been processed in the rinse water, replace it. Place the contaminated waste into the sump.**

### Preparation

4-73. At this station, the following materials and equipment are needed:

- 4 containers (about a 3-gallon capacity).
- 1 CAM (for chemical only).
- Hot, soapy water.
- Rinse water.
- Mask sanitizing solution.
- Immersion heater.
- One 30-gallon container.
- 2 sponges.
- 1 case of paper towels per company.

4-74. Dig a sump that is 4 feet long, 4 feet wide, and 4 feet deep in which to discard used filters and canisters. Place the equipment and materials as shown in Figure 4-3, page 4-8. Three soldiers are needed to operate this station. Two soldiers strip, wash, rinse, sanitize, and dry masks. The other soldier checks the masks and carriers them to the reissue point.

### Execution

4-75. Remove the eye-lens outserts and the hood if the hood was not cut off at station 2. Remove and discard the filters or canisters. Put the items into the properly marked containers. Wash the mask, hood, eye-lens outserts, and the hoses on the M42 and M43 masks in hot, soapy water. Rinse these items in clean water, dip them into the sanitizing solution, agitate them for 5 minutes, and then rinse them again in clean water. Add one tube of mask sanitizing solution (calcium hypochlorite) to each quart of water. Wipe the masks with rags until they are almost dry. Discard each container of mask sanitizing solution into a sump after every 10 masks. The attendant checks the masks for contamination with a CAM. If the masks are still contaminated, the attendant recycles them for more decon and then decons his rubber gloves. If the masks are not contaminated, the attendant takes the masks to the reissue point. The attendant must take care not to contaminate the reissue point or himself.

### Risk

4-76. If this step is not done properly, the soldiers may become contaminated when the masks are reissued to them at the reissue point. Even though the step is done correctly, there is still a possible danger when many masks are stacked together. Small amounts of residual vapor from each mask can become potentially dangerous.

**NOTE: Runners between stations 7 and 8 are in MOPP2 and are prepared to go into MOPP4.**

## STATION 8 - REISSUE POINT

4-77. At this station, the mask with its components is provided to the soldier for reassembly.

### Preparation

4-78. At this station, the protective-mask prescribed load-list (PLL) parts are needed. The unit NBC NCO/supply sergeant sets up the reissue point to provide the soldiers with replacement parts for all types of protective masks and assist in mask maintenance.

### Execution

4-79. The mask with its components is reissued to the soldier who assembles it in the AA. The unit chemical NCO affixes canisters to the cleaned M42 and M43 hoses. The soldier picks up individual gear and moves to the postdecon AA.

### Risk

4-80. If this step is not done properly, the soldier may be inadequately equipped for future operations.

**NOTE: If time is not available, the NBC NCO will have replacement chemical suits, overboots, and gloves at this station for reissue. If time is available, personnel will receive this equipment at the postdecon AA.**

# RESUPPLY SUPPORT RESPONSIBILITIES

4-81. The supporting chemical unit (decon or smoke/decon) must supply the majority of the equipment and supplies that are required to operate the DTD. This does not relieve the supported unit of the requirement to maintain adequate supplies to conduct DTD procedures.

4-82. A chemical unit's operational control continues to receive support from its parent organization. The supported unit provides the chemical unit with replacements, supplies, and material at the end of the DED. This shortens the time the chemical unit is nonmission capable following the DED. The supported unit must coordinate for the supplies and equipment necessary to operate the DTD.

4-83. Typical nonchemical units do not have more than two CAMs per company-sized unit. However, the DTD requires a minimum of three CAMs. The supported HQ may need to provide additional CAMs. The division support command (DISCOM) sustains the division's chemical units. A chemical unit attached to a maneuver force identifies the resources that are needed to accomplish the mission and passes these requirements to the maneuver staff's Supply Officer (S4)/Logistics Staff at Corps and Division (G4). The S4/G4 then coordinates with the supporting CSS element to fill the requirement. The supply units that operate the Class II points issue the protective masks and overgarments. They normally package protective clothing as complete sets of MOPP equipment to support a predetermined number of soldiers. Decon supplies may be issued in preconfigured "push" packages.

## DETAILED EQUIPMENT DECON

4-84. Chemical units (decon and smoke/decon platoons) are responsible for setting up, operating, and closing the DED portion of the thorough-decon operation. The DED for chemical and biological contamination is comprised of five stations. For radiological contamination, the DED uses all but station 2. Stations are normally 98 to 164 feet apart; however, spacing is METT-TC dependent.

### DECON STATIONS

4-85. The actions that are taken at each of the stations are as follows:

#### Station 1 - Primary Wash

4-86. At this station, the gross contamination and dirt are removed from the vehicle. The vehicle is sprayed for 2 to 3 minutes with hot, soapy water. The vehicle is then scrubbed to help remove caked-on dirt. The mechanical action of scrubbing also helps remove thickened chemical agents. Although the undersurfaces are difficult to reach, try to remove as much dirt as possible. After scrubbing the vehicle, spray it again for 2 to 3 minutes to remove loosened dirt and contamination. This station uses about 250 gallons of water per vehicle. Larger vehicles with large quantities of dirt use more water. The runoff from this station is contaminated and must be treated as hazardous. The contaminated runoff must be controlled; if available, engineer support may be used. This station requires high water-pressure systems (M12A1 power-driven decon apparatus [PDDA] and M17 Lightweight Decontaminating System [LDS]) rather than low water-volume systems (65-GPM pumps).

**NOTE: Thirty-five cubic feet of space per 250 gallons of liquid runoff should be used when calculating the size for the drainage pump.**

4-87. The effectiveness of the wash depends on the type of wash (hot, soapy water; hot water; cold water; or steam). The relative effectiveness rankings for selected surfaces are listed in Table 4-3, page 4-20. Hot, soapy water is water heated to about 120$^{0}$ to 140$^{0}$ F to which a detergent has been added to reduce its surface tension. The detergent removes the agent by emulsification, which is followed by the mechanical displacement of the suspension. Hot water alone is less effective than hot, soapy water. Because of the high temperature, some agents are best removed by steam through vaporization. Finally, for some chemical agents, cold water exhibits better solvent characteristics.

### Station 2 - DS2 Application

4-88. At this station, a decontaminant is applied to the entire vehicle. The vehicle is divided into four parts, and a member of the scrubbing team is assigned that part of the vehicle. This limits the workload of each member of the scrubbing team and avoids duplication of work. DS2 is applied starting at the top of the vehicle and working towards the undercarriage. Every effort is made to apply DS2 to the undercarriage, especially if the vehicle has crossed a contaminated area. The mop is the least tiring method of applying DS2. Using a mop to apply DS2 creates a large amount of spillage. However, continual use of the M13 decontaminating apparatus, portable (DAP) requires the scrubbing team to exert more energy than with using the mop. The M13 DAP can be used to apply DS2 to hard-to-reach areas. Planning factors for DS2 application are provided at Table 4-4, page 4-20.

4-89. Before starting the decon operation, the scrubbing team pours 5-gallon cans of DS2 into 30-gallon containers if mops are going to be used instead of M13 DAPs. Each member of the scrubbing team wears a TAP apron or wet-weather gear to protect himself from the DS2.

4-90. Water adversely affects the DS2's ability to react with chemical agents. If there is excess water remaining on the vehicle from station 1, the scrubbing team can—

- Wait for the majority of the water to evaporate.
- Remove the excess water.
- Increase the amount of DS2 applied.

4-91. There must be sufficient DS2 on the item being decontaminated for complete neutralization to occur. The DS2-to-agent ratio needs to be 55 to 1 for H agents and 25 to 1 for G agents. For a vehicle the size of an M1A1 tank, this corresponds to 15 and 7 gallons, respectively.

### Station 3 - Contact Time/Interior Decon

4-92. At this station, the DS2 is allowed to completely neutralize the chemical agent and the interior of the vehicle is decontaminated. Vehicles are moved to a concealed position. Vehicles will remain in station 3 for no less than 30 minutes. DS2 reacts with most chemical agents within 5 minutes. However, by allowing the DS2 to remain on the contaminated surface for 30 minutes, the amount of agent that will later desorb (off gas) will be significantly reduced. When there is a 30-minute contact time, there will be no desorption after decon operations for most chemical agents. However, studies indicate that distilled-mustard (HD) vapors desorb after decon even if DS2 is allowed to remain for 30 minutes.

4-93. While the vehicle is held at this station for the DS2 to completely react, the driver inspects the interior of the vehicle for liquid contamination. He is given M8 detector paper to check for chemical contamination. If he identifies chemical contamination, he is given decon supplies to decon the interior of the vehicle. The best decon solution for use in the interior of vehicles is a 5 percent solution of HTH or STB. The driver wipes all reasonably accessible

**Table 4-3. Effectiveness of Types of Wash**

| Agents/Surfaces | Relative Effectiveness[1] |
|---|---|
| TGD on alkyd-painted metal | Hot water and/or steam; hot, soapy water; cold, soapy water |
| TGD on CARC-painted metal | Hot, soapy water; hot water and/or steam; cold water |
| THD on alkyd-painted metal | Hot water and/or steam; hot, soapy water; cold water |
| THD on CARC-painted metal | Hot, soapy water; hot water and/or steam; cold water |
| HD on alkyd-painted metal | Hot, soapy water; hot water and or steam; cold water |
| HD on CARC-painted metal | Hot, soapy water and/or steam; cold water; hot water |
| VX on alkyd-painted metal | Steam; cold water; hot water and/or hot, soapy water |
| TGD on canvas/webbing | Steam; hot water and/or hot, soapy water; cold water |
| GD on canvas/webbing | Steam; hot, soapy water and/or hot water; cold water |
| THD on canvas/webbing | Steam; hot, soapy water; cold water; hot water |
| HD on canvas/webbing | Steam; hot water; hot, soapy water; cold water |
| VX on canvas/webbing | Steam; hot water and/or hot, soapy water; cold water |
| Overall | Steam; hot water and/or hot, soapy water; cold water |

[1]The types of wash are listed in the order of effectiveness, starting with the most effective to the least effective.

**Table 4-4. Planning Factors for DS2 Application**

| Planning Categories | Tank | Truck | APC |
|---|---|---|---|
| Gallons of DS2 applied | 15 | 8 | 7 |
| Minutes to apply solution | 35 | 29 | 18 |

surfaces with a rag or sponge soaked in the HTH or STB solution. He should not attempt to decon areas where there is little likelihood of contamination (electrical assemblies, area beneath the turret floor, and so forth).

**CAUTION**

Do not mix HTH or STB with DS2. If mixed, a violent reaction will occur.

4-94. Once the interior decon is completed, the driver places a cover over the seat and floor of the vehicle. Then he dismounts from the vehicle and proceeds to the start of the DTD. The assistant driver, having completed the DTD, mounts the vehicle and moves it to the next station. Drivers must exercise caution when entering or exiting the vehicle. A DS2-coated surface is slippery, and the DS2 may react with chemical protective footwear. Personnel should avoid stepping in DS2 and tracking it into the vehicle.

4-95. For radiological contamination, use an AN/PDR-77 or AN/VDR-2 radiac detector to determine the extent and location of contamination inside the vehicle. If there is contamination, determine the intensity of the contamination inside the vehicle. If the contamination has an intensity greater than 0.33cGy, the interior of the vehicle must be decontaminated. Use a sponge to wipe the interior of the vehicle.

### Station 4 - Rinse

4-96. At this station, the DS2 is removed from the vehicle. The vehicle is sprayed with water from top to bottom. Take care to rinse the undercarriage. This station uses about 200 gallons of water per vehicle. Failure to remove all DS2 from the vehicle may cause a false-positive reading at station 5. If high water-pressure systems (M12A1 PDDA and M17 LDS) are not available, large-volume water pumps (65-GPM pumps) should be used at this station. The driver removes plastic or other material (if present) covering the seats and floor and disposes of it as hazardous waste. See Table 4-5 for the planning factors for the rinse station.

**Table 4-5. Planning Factors for the Rinse Station**

| Planning Categories | M1 Tank | M2 BFV | M113 APC | M109A PALADIN | HEMTT | 5-Ton Truck | HMMWV |
|---|---|---|---|---|---|---|---|
| M12A1 PDDA rinse (gallons applied) | 325 | 325 | 203 | 325 | 180 | 158 | 90 |
| M12A1 PDDA rinse (minutes applied) | 12 | 12 | 9 | 12 | 8 | 7 | 4 |
| M17 LDS rinse[1] (gallons applied) | 57 | 57 | 38 | 57 | 30 | 42 | 23 |
| M17 LDS rinse[1] (minutes applied) | 14 | 14 | 10 | 14 | 12 | 11 | 6 |
| [1] Rinse is done with a spray wand. | | | | | | | |

### Station 5 - Check

4-97. At this station, the vehicle is checked to see if it has a negligible contamination level or if it still has significant contamination remaining. Detection procedures will vary depending on the type of contamination. If significant contamination is found on the vehicle, the vehicle will be recycled to station 2 for chemical contamination or station 1 for radiological contamination. An assistant driver takes the vehicle to the AA.

4-98. **Chemical.** The CAM is used to check for the presence of vapor from residual liquid contamination. A one bar or lower reading on the CAM indicates a negligible contamination level. Once the CAM indicates the presence of vapor contamination, M8 detector paper is used to verify the presence of liquid contamination. If it is suspected that both the CAM and M8 detector paper are producing a false positive, use an M256A1 detector kit to confirm or deny the presence of contamination. See Table 4-6, page 4-22, for a list of common interferences that can cause false-positive readings on the CAM. If the vehicle has significant contamination remaining, recycle it. The commander may modify the recycle criteria based on mission requirements.

4-99. There will be desorption of chemical agents from the surfaces after decon. The desorption of vapors on surfaces painted with the CARC will stop sooner then those surfaces painted with alkyd. Consider this when checking decontaminated items for overall decon effectiveness.

Table 4-6. Common Interferences of the CAM

| Interferences | G-Bar Response | H-Bar Response |
|---|---|---|
| DS2 | Low | |
| Insect repellent | Low to very high | |
| Brake fluid | High to very high | Very high |
| General-purpose cleaner | High | |
| Burning kerosene | | High |
| Breath mints | High | |
| Gasoline vapors | Low | Low |
| Burning grass | Low to high | Low |
| Burning gas | Low | |
| Green smoke | Low | Low to high |
| Break-free oil | Low | |
| Ammonia | Very high | |

4-100. **Radiological.** Use the AN/PDR-77 or AN/VDR-2 to determine if any contamination remains. If there is contamination remaining, determine the intensity of the contamination inside and outside of the vehicle. If the contamination has an intensity greater than 0.33cGy/hr, the vehicle is recycled to station 1.

## RECYCLE CRITERIA

4-101. The commander, with the chemical unit leader's help, establishes the recycle criteria before starting decon operations. The recycle criteria determines which vehicles will return to station 1 after contamination is detected at station 5. If the unit has sufficient time and resources, any vehicle having more contamination than the acceptable level should be recycled. However, time and resources are usually limited, and not all vehicles can be recycled. The recycle criteria is based on the weathering effects.

## RECONSTITUTION CRITERIA

4-102. The operational- and thorough-decon operations that are performed on vehicles or major equipment will be recorded on DA Form 2404 (Army) or NAVMC 10245 (Marine). This information will become a permanent record for these vehicles or major equipment. The entries that are required on these forms are the type of decon performed, decontaminant used, date/time group completed, location of decon site, and type of monitoring equipment used to verify decon completeness. These forms will remain a permanent part of the unit's, vehicle's, and equipment's records. This information will assist with reconstitution operations once the conflict or war has ended.

## DED LAYOUTS

4-103. Dual-purpose and decon platoons establish thorough decon sites differently because of organization and equipment differences. The optimum layout for each type of platoon is described in the following paragraphs (see

Figures 4-5 through 4-8, pages 4-24 through 4-27). This layout provides the maximum output for platoons at 100 percent personnel and equipment. The equipment and personnel requirements for the optimum layout are identified for both the chemical unit (decon platoon) and supported unit (augmenter) (see Tables 4-7 through 4-10, pages 4-28 through 4-31).

4-104. Since it is likely that all platoons will not be at 100 percent strength, alternate layouts are discussed. For each alternate layout, chemical-unit equipment requirements are identified. However, personnel requirements are not identified by chemical unit and supported unit. The total number of personnel that are needed is identified, with a minimum number of chemical unit personnel.

## ALTERNATE-LAYOUT PLANNING CONSIDERATIONS

4-105. Chemical unit leaders use METT-TC to determine the best possible DED layout to execute their mission. When determining alternate DED layouts, use the following guidelines:

- The ability to spray hot, soapy water or steam under pressure must be retained at station 1.
- The largest number of people is required at station 2.
- Experienced and qualified CAM operators are required at station 5.
- Water does not have to be hot to rinse off the DS2; however, the lower the water pressure the more water that is required for the rinse.

## DECON PLATOON

4-106. The optimum DED layout for an M12A1 PDDA-equipped platoon requires the use of all authorized equipment and personnel (see Figure 4-5, page 4-24). While this DED layout is manpower and equipment intensive, it provides for the rapid decon of vehicles and equipment (eight vehicles processed per hour). This layout uses dual lanes at stations 1, 4, and 5 to process two vehicles at one time. Since the most time- and labor-intensive work takes place at station 3, this station is designed to process three vehicles at a time. The processing rate of this layout will be affected by any work/rest cycle (see Table 4-11, page 4-32).

## SMOKE/DECON PLATOON

4-107. Smoke/decon platoons equipped with the M17 LDS will set up the DED differently then the decon units equipped with the M12A1 PDDA. While the optimum DED layout is manpower and equipment intensive, it provides for the rapid decon of vehicles and equipment (eight vehicles processed per hour). The layout uses a dual lane at stations 1, 4, and 5 to process two vehicles at a time. Since the most time- and labor-intensive work takes place at station 3, this station is designed to process three vehicles at a time. The processing rate of this layout will be affected by any work/rest cycle (see Table 4-11, page 4-32).

4-108. It may not be possible, for a variety of reasons, for an M17 LDS-equipped decon platoon to use the optimum DED layout. Figure 4-7, page 4-26, shows an example of an optimum DED layout. Limited personnel or equipment will affect the DED layout. Figure 4-8, page 4-27, shows an example of an alternate DED layout.

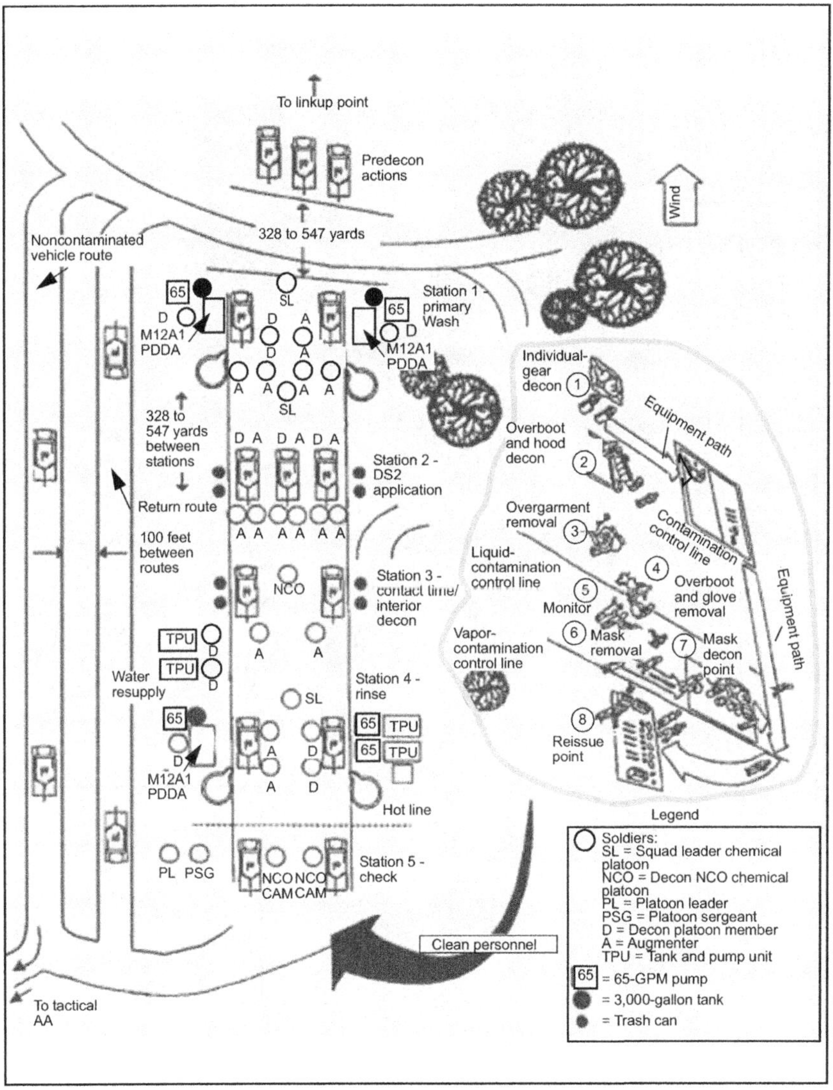


**Figure 4-5. Optimum DED Layout for the M12A1 PDDA-Equipped Unit**

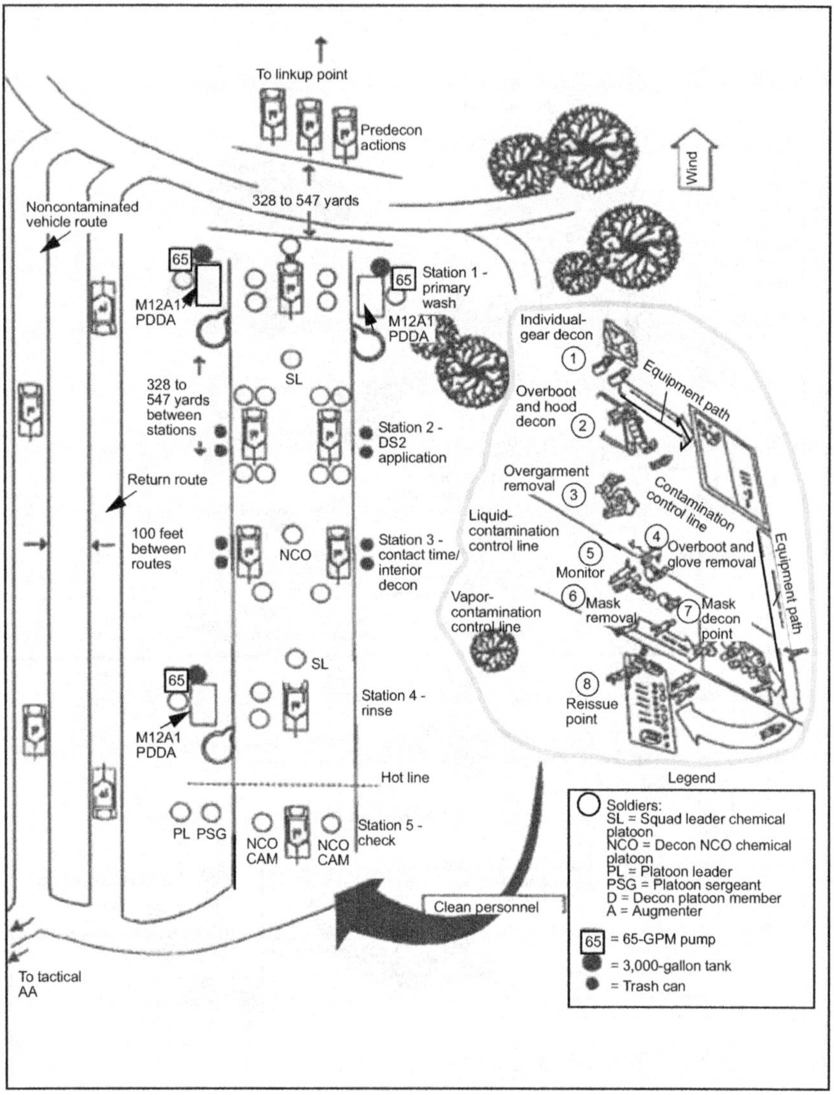


**Figure 4-6. Alternate DED Layout for the M12A1 PDDA-Equipped Unit**

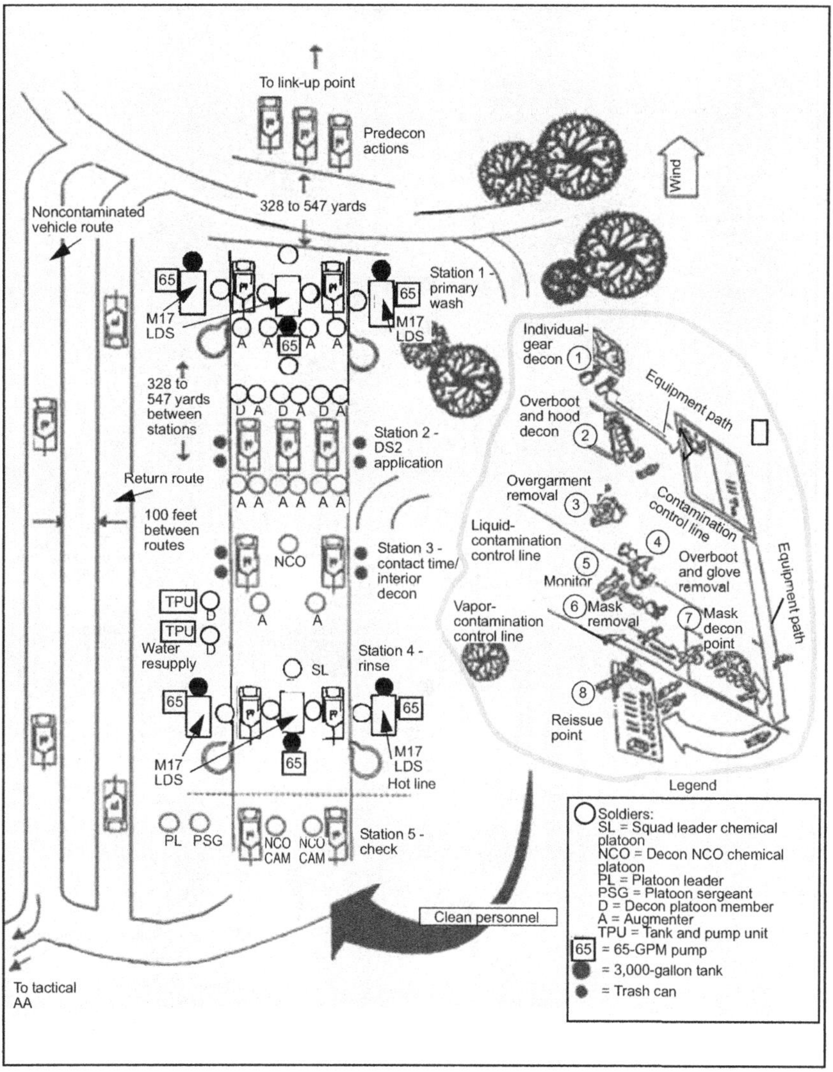


**Figure 4-7. Optimum DED Layout for the M17 LDS-Equipped Unit**

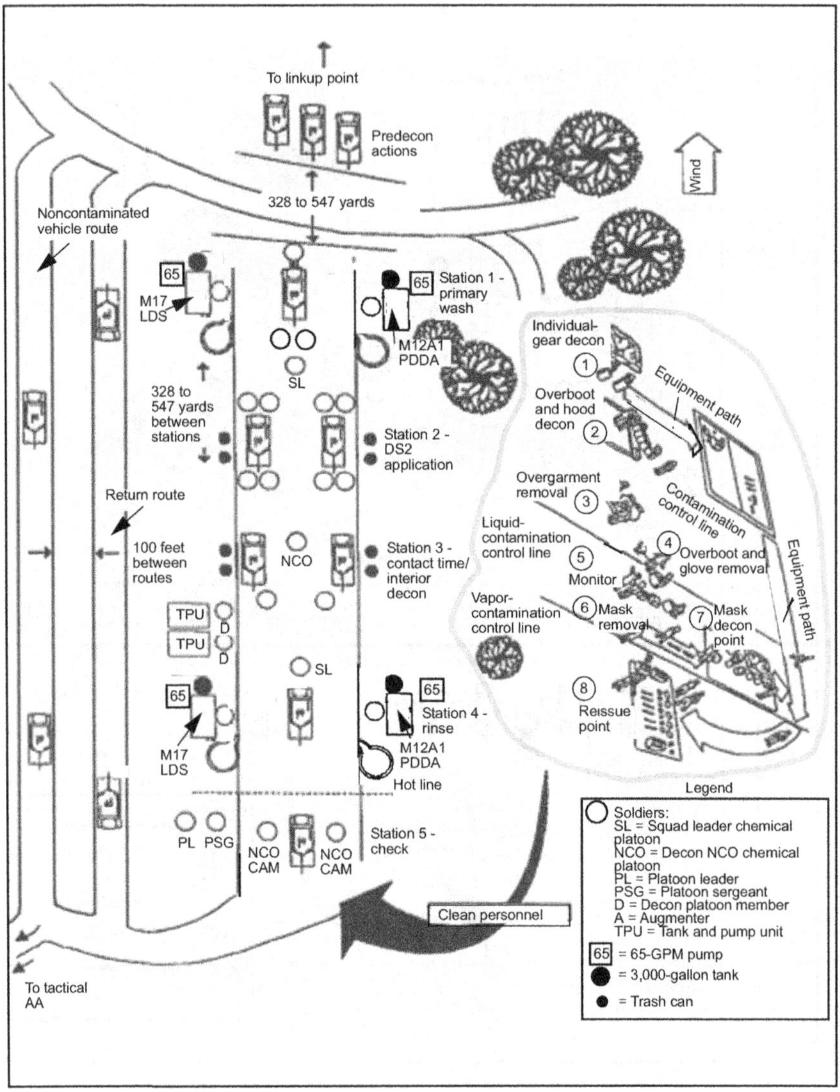


**Figure 4-8. Alternate DED Layout for the M17 LDS-Equipped Unit**

**Table 4-7. Personnel and Equipment Requirements for the Optimum DED Layout of the M12A1 PDDA-Equipped Unit**

| Stations | Personnel | | Equipment/Supplies |
|---|---|---|---|
| | Decon Platoon | Augmenter | |
| Station 1 - primary wash | 1 squad leader<br>2 PDDE operators<br>4 sprayers | 4 scrubbers | 2 M12A1 PDDAs<br>2 3,000-gallon tanks<br>2 65-GPM pumps<br>6 long-handled brushes<br>8 TAP aprons<br>Liquid detergent |
| Station 2 - DS2 application | 1 squad leader<br>3 appliers | 9 appliers | 18 long-handled brushes<br>9 mops with extra mop heads<br>3 30-gallon containers<br>9 M13 DAPs<br>Sufficient DS2 |
| Station 3 - contact time/ interior decon | 1 NCO | 2 interior decon assis-tants | 2 AN/VDR-2<br>3 TAP aprons<br>6 30-gallon containers<br>10 books of M8 detector paper<br>30 sponges<br>8 M256A1 detector kits<br>50 trash bags<br>1 clipboard with a pen<br>1 stopwatch |
| Station 4 - rinse | 1 squad leader<br>1 PDDE operator<br>2 pump operators | 2 sprayers | 1 M12A1 PDDA<br>1 3,000-gallon tank<br>3 65-GPM pumps<br>2 TPUs<br>2 TAP aprons |
| Station 5 - check | 2 NCO/CAM opera-tors | | 2 CAMs<br>10 M256A1 detector kits<br>20 books of M8 detector paper<br>2 AN/VDR-2<br>2 M8A1 or M22 ACAAs |
| $C^2$ | 1 PL<br>1 PSG | | 1 HMMWV with a radio<br>3 NBC marking kits |
| Total personnel | 20 | 17 | |

**WARNING**

**Chemical-unit leaders must consider the impact of the work/rest cycle on their unit's ability to process vehicles through the DED. Failure to initiate a work/rest cycle could result in heat casualties and mission failure.**

4-109. It may not be possible, for a variety of reasons, for an M12A1 PDDA-equipped decon platoon to use the optimum DED layout. Limited personnel or equipment will affect the DED layout. Figure 4-6, page 4-25, shows an example of an alternate DED layout.

**Table 4-8. Personnel and Equipment Requirements for the Alternate DED Layout of the M12A1 PDDA-Equipped Unit**

| Stations | Personnel | Equipment/Supplies |
|---|---|---|
| Station 1 -<br>primary wash | 1 NCOIC*<br>2 PDDE operators*<br>4 sprayers<br>2 scrubbers | 2 M12A1 PDDAs<br>2 3,000-gallon tanks<br>2 65-GPM pumps<br>4 long-handled brushes<br>6 TAP aprons<br>Liquid detergent |
| Station 2 -<br>DS2 application | 1 NCOIC*<br>8 appliers | 14 long-handled brushes<br>8 mops with extra mop heads<br>3 30-gallon containers<br>8 M13 DAPs<br>Sufficient DS2 |
| Station 3 -<br>contact time/<br>interior decon | 1 NCO*<br>2 interior decon assistants | 2 AN/VDR-2<br>3 TAP aprons<br>6 30-gallon containers<br>10 books of M8 detector paper<br>30 sponges<br>8 M256A1 detector kits<br>50 trash bags<br>1 clipboard with a pen<br>1 stopwatch |
| Station 4 -<br>rinse | 1 NCOIC*<br>1 PDDE operator*<br>2 sprayers | 1 M12A1 PDDA<br>1 3,000-gallon tank<br>3 65-GPM pumps<br>2 TPUs<br>2 TAP aprons |
| Station 5 -<br>check | 2 NCO/CAM operators* | 2 CAMs<br>10 M256A1 detector kits<br>20 books of M8 detector paper<br>2 AN/VDR-2<br>2 M8A1 or M22 ACAAs |
| $C^2$ | 1 PL*<br>1 PSG* | 1 HMMWV with a radio<br>3 NBC marking kits |
| Total personnel | 29 (Minimum of 11 soldiers from the decon platoon) | |
| *** These individuals should be from the decon platoon.** | | |

## THOROUGH-DECON-SITE CLOSURE

4-110. Once all vehicles and soldiers from the contaminated unit have processed through the thorough-decon site, the site can be closed. Coordination with the supported unit is necessary before actually closing the site to ensure that all contaminated elements have been processed.

4-111. The chemical unit (decon or smoke/decon platoons) closes the DED first. Once the DED is closed, the chemical unit processes through the DTD. After the chemical unit has processed through the DTD, the DTD is closed. At that point, the chemical unit marks the area as a contaminated area and reports its exact location to the supported unit, using an NBC-5 report. The exact procedures for closing the DED and DTD are described as follows:

**Table 4-9. Personnel and Equipment Requirements for the Optimum DED Layout of the M17 LDS-Equipped Unit**

| Stations | Personnel | | Equipment/Supplies |
|---|---|---|---|
| | Decon Platoon | Augmenter | |
| Station 1 - primary wash | 1 squad leader<br>4 sprayers<br>2 scrubbers | 2 scrubbers | 3 M12A1 PDDAs<br>3 1,500-gallon tanks<br>3 65-GPM pumps<br>6 long-handled brushes<br>8 TAP aprons<br>Liquid detergent |
| Station 2 - DS2 application | 1 squad leader<br>3 appliers | 9 appliers | 18 long-handled brushes<br>9 mops with extra mop heads<br>3 30-gallon containers<br>9 M13 DAPs<br>Sufficient DS2 |
| Station 3 - contact time/ interior decon | 1 NCO<br>2 interior decon assistants | | 2 AN/VDR-2<br>3 TAP aprons<br>6 30-gallon containers<br>10 books of M8 detector paper<br>30 sponges<br>8 M256A1 detector kits<br>50 trash bags<br>1 clipboard with a pen<br>1 stopwatch |
| Station 4 - rinse | 1 squad leader<br>4 sprayers | | 3 M17 LDSs<br>3 3,000-gallon tanks<br>3 65-GPM pumps<br>2 TPUs<br>4 TAP aprons |
| Station 5 - check | 2 NCO/CAM operators | | 2 CAMs<br>10 M256A1 detector kits<br>20 books of M8 detector paper<br>2 AN/VDR-2<br>2 M8A1 or M22 ACAAs |
| $C^2$ | 1 PL<br>1 PSG | | 1 HMMWV with a radio<br>3 NBC marking kits |
| Total personnel | 23 | 11 | |

## CLOSING THE DED AREA

4-112. The DED is closed in sequence, starting at station 1. All vehicles, equipment, and nonexpendable supplies are inspected for contamination. If contamination is found on an item, it is decontaminated. The actions at each station are described as follows:

### Station 1 - Primary Wash

4-113. Spray all vehicles and equipment with hot, soapy water to remove any contamination that could have been transferred during the primary wash operations. Drain the water billets or fabric tanks of water. Inspect all equipment and vehicles for contamination using the appropriate detection equipment.

**Table 4-10. Personnel and Equipment Requirements for the Alternate DED Layout of the M17 LDS-Equipped Unit**

| Stations | Personnel | Equipment/Supplies |
|---|---|---|
| Station 1 -<br>primary wash | 1 NCOIC*<br>4 sprayers<br>4 scrubbers | 3 M17 LDSs<br>3 3,000-gallon tanks<br>3 65-GPM pumps<br>4 long-handled brushes<br>8 TAP aprons<br>Liquid detergent |
| Station 2 -<br>DS2 application | 1 NCOIC*<br>8 appliers | 14 long-handled brushes<br>8 mops with extra mop heads<br>3 30-gallon containers<br>8 M13 DAPs<br>Sufficient DS2 |
| Station 3 -<br>contact time/<br>interior decon | 1 NCO*<br>2 interior decon assistants | 2 AN/VDR-2<br>3 TAP aprons<br>6 30-gallon containers<br>10 books of M8 detector paper<br>30 sponges<br>8 M256A1 detector kits<br>50 trash bags<br>1 clipboard with a pen<br>1 stopwatch |
| Station 4 -<br>rinse | 1 NCOIC*<br>4 sprayers | 2 M17 LDS<br>2 3,000-gallon tanks<br>2 65-GPM pumps<br>4 TAP aprons |
| Station 5 -<br>check | 2 NCO/CAM operators* | 2 CAMs<br>10 M256A1 detector kits<br>20 books of M8 detector paper<br>2 AN/VDR-2<br>2 M8A1 or M22 ACAAs |
| $C^2$ | 1 PL*<br>1 PSG* | 1 HMMWV with a radio<br>3 NBC marking kits |
| Water resupply | 2 drivers* | 2 TPUs |
| Total personnel | 32 (Minimum of 10 soldiers from the smoke/decon platoon) | |
| *** These individuals should be from the smoke/decon platoon.** | | |

### Station 2 - DS2 Application

4-114. Throw the mops and brushes that were used in applying the DS2 into a sump or bury them. Load the unused cans of the DS2 and the M13 DAPs on a vehicle.

### Station 3 - Contact Time/Interior Decon

4-115. Inspect any unused supplies and equipment for contamination. If no contamination is detected, load the equipment and supplies on a vehicle. Throw all contaminated supplies into the nearest sump. In calculating sump requirements, 35 cubic feet of space is required for 250 gallons of liquid drain off. Once a sump is 2/3 full, prepare a new sump.

Table 4-11. Work/Rest Table

| Temperatures | Stations and Workload | | | | |
|---|---|---|---|---|---|
| | Station 1 – Primary Wash | Station 2- DS2 Application | Station 3 – Contact Time/Interior Decon | Station 4 - Rinse | Station 5 - Check |
| | Physical Exertion | | | | |
| | Moderate | Heavy | Moderate | Moderate | Light |
| Cool (<68°F) | 60 work 15 rest | 30 work 30 rest | 60 work 15 rest | 60 work 15 rest | 60 work 15 rest |
| Warm (68° - 74°F) | 45 work 45 rest | 20 work 20 rest | 45 work 45 rest | 45 work 45 rest | 50 work 50 rest |
| Hot (74° - 84°F) | 30 work 60 rest | 15 work 30 rest | 30 work 60 rest | 30 work 60 rest | 40 work 80 rest |
| Very Hot (>84°F) | 20 work 60 rest | 10 work 30 rest | 20 work 60 rest | 20 work 60 rest | 25 work 75 rest |

**NOTE: When operating in temperatures above 75°F, consider the ability of the soldiers to accomplish the mission. Once the soldiers have reached their maximum workload for heat stress, they cannot recover quickly enough to accomplish the decon mission. A viable option is to postpone the decon operation until a cooler part of the day or evening. This will reduce the heat-stress load on the soldiers and increase the probability of mission success. For further information, see FM 3-4.**

### Station 4 – Rinse

4-116. Spray all vehicles and equipment with hot, soapy water to remove any contamination that could have been transferred during the rinse operations. Drain the water billets or fabric tanks of water. Inspect all equipment and vehicles for contamination using the appropriate detection equipment. If no contamination is detected, load the equipment on the vehicles. Spread one can of STB dry mix into each of the sumps and then cover them. Ensure that DS2 is not present in the sumps. Post NBC hazard markers near the covered sumps. **(NOTE: DS2 destroys chemical agents; however, some of the by-products created from it are toxic.)**

### Station 5 – Check

4-117. Ensure that all equipment has been checked for contamination. If it is not contaminated, load it on a vehicle. If it is contaminated, decon it according to the appropriate technical manual (TM). Throw any contaminated supplies into the nearest sump. Move all vehicles upwind of station 5 and inspect them again for contamination. If any contamination is detected, the crew will use either the M11 or M13 DAP to decon the identified areas. Once the vehicles are staged, all the soldiers will proceed to the DTD.

## CLOSING THE DTD AREA

4-118. Once all the soldiers from the DED have processed through the DTD, it may be closed. After the last person has exited the DTD—

- Pick up all the used supplies from station 7 and put them in station 7's sump. Take up the contamination control line. If engineer tape was used, dispose of it in station 7's sump.
- Move all usable supplies and equipment from all the stations to station 1. Discard all unusable supplies from stations 3, 4, and 5 in station 1's sump.
- Decon all supplies and equipment collected at station 1 using the decon and rinse water at station 1. Empty the decon and rinse water from the station into the sump and decontaminant the containers.
- Mark the entire decon area. Remove your overgarment using the MOPP-gear-exchange technique and dispose of it in the sump at station 1.
- Move any equipment used to fill the sump upwind of the decon area. Decon your rubber gloves and move all equipment and supplies in station 1 upwind of the decon area. Keep this equipment and supplies separate from that used to fill the sump.

## THOROUGH DECON UNDER UNUSUAL CONDITIONS

4-119. Decon operations are not restricted to daylight hours. The enemy may employ chemical and biological weapons at night, since weather conditions are usually more favorable for their employment at that time. Consequently, NBC contamination encountered at night may require decon before daylight.

4-120. Conducting decon operations at night is a challenge. "White light" cannot be used without possibly revealing your location. However, decon personnel must have illumination to perform such essential decon tasks as spraying water, applying decontaminants, using detection equipment, and taking off the MOPP gear.

4-121. Use M8/M9 detector paper to check for chemical contamination and to see how well the chemical decon was performed. If contamination is present, the paper color changes to red and cannot be seen if red-filtered light is used. The CAM and M256A1 detector kit can be used to identify and detect chemical agents but will require a light source in order to obtain a reading.

4-122. Several vehicles awaiting or undergoing decon present a significant infrared "signature" due to the hot engines. To enter, exit, and move within the decon site is difficult under blackout conditions.

4-123. Night-decon operations are difficult but can be accomplished. Whenever possible, conduct thorough-decon operations in built-up areas. Use a building for DTD operations and a warehouse for DED operations. This allows for the use of white light. Soldiers can see what they are decontaminating, see where they are going, and read the color changes of the M8/M9 detector paper.

4-124. There will be a need for additional soldiers to act as ground guides in the DED area. The NCOICs of stations 1, 2, and 4 will inspect each vehicle with a flashlight or chemical light before the vehicle proceeds to the next station. There should be at least two flashlights at each station in both the DED and DTD areas.

## Chapter 5

# Special Decon Operations

The decon of terrain or structures allows personnel to increase their stay time in an area and facilitates their passage through an area. Large-scale terrain decon requires extensive amounts of equipment, material, and time. Thus, terrain decon is limited to areas of critical importance. Contaminated areas and structures that can be avoided do not need to be decontaminated except by weathering. If they must be used, decon may be necessary. When decon is not practical, use protection to deal with the hazard. Most terrain and structural surfaces (excluding unpainted metal and glass) absorb the agent. Therefore, the agent may be transferred to clothing by physical contact. The agent may also be transferred through vaporization and subsequent inhalation or diffusion into the eyes and the clothing. The decon of equipment containing vulnerable components presents certain challenges. Electronics and optics are especially vulnerable to damage if not carefully decontaminated. Some materials, such as canvas, absorb chemical agents and may not be decontaminated and reused.

## TERRAIN DECON

5-1. A terrain decon provides more long-term rather than short-term benefits. Generally, you cannot decon occupied terrain well enough to allow you to reduce your MOPP level right away. At best, all you can do is speed up the weathering process. This may allow you to reduce your MOPP level a little sooner, for example, in 2 days instead of 4 or 5. Terrain decon can be so expensive and so ineffective that you should consider relocating.

5-2. If you decide on a terrain decon, limit the amount you do. Build a causeway or breach a path through the contamination for entry/exit into a major facility. Restrict your efforts to areas where you must work or move. No matter what you do, you will probably have to remain in MOPP; but you can reduce the danger of picking up liquid contamination and spreading it to other areas.

5-3. The most effective techniques for a terrain decon are the natural methods (weathering or aging). Using standard decontaminants will quickly exhaust your resources (see Appendix B). Usually the most practical method is to scrap off contaminated surfaces. Fire is a terrain-decon method; however, consideration must be given to the potential downwind vapor hazard. If possible, a contaminated area should be evacuated until time and weather remove the hazard. An STB slurry can be used for vital areas. To determine the persistency of chemicals at various temperatures, refer to FM 3-9.

5-4. A terrain decon may be accomplished using several methods. These methods are listed in the order of decreasing effectiveness. Although they reduce the contact hazard, none will prevent hazardous vapors desorbing from contaminated surfaces. Therefore, appropriate protective measures must still be taken.

## WEATHERING

5-5. Weathering is the simplest and easiest form of decon.

### Chemical and Biological Contamination

5-6. Warm, windy weather can significantly reduce terrain contamination. In some cases, this occurs in a few hours or it may take a few days. Many variables affect the persistency of chemical and biological hazards, so it is impossible to accurately predict how long it takes such contamination to weather. Sunlight is especially effective against most biological agents.

### Radiological Contamination

5-7. Although the term "weathering" is used to describe the decon process, weather has little effect on radiological hazards. Heavy rain and wind may remove some contamination, but only time will reduce the radiation emanating from the contamination.

## REMOVING/COVERING

5-8. Removing or covering contamination does not destroy it, but it does keep the hazards away from you temporarily.

### Chemical and Biological Contamination

5-9. Contamination can be removed from paths and unpaved roadways by scraping off a layer of earth using heavy earthmoving equipment (bulldozers and road graders). Contaminated paths and paved surfaces can be covered with a 4-inch layer of earth, roofing paper, plastic sheets, and wood mats. This is a temporary measure because the agents may penetrate the covering. When the contamination penetrates the covering or when the covering is removed, the hazards will reappear. The coverings may also extend the life of the contamination hazard by reducing its exposure to air and sunlight.

### Radiological Contamination

5-10. Radiological contamination must be covered by thick layers of dense material, like earth. Three inches of earth will decrease radiation dose rates by half because of the shielding provided by the soil. However, 12 inches is more effective. The job will be easier with earthmoving equipment, but the equipment and operators probably will have to undergo decon.

## NEUTRALIZING

5-11. The STB dry mix or STB slurry may be used effectively against chemical or biological contamination, but not against radiological contamination. Trained chemical troops and equipment are required when using the STB slurry, which is costly both in time and material.

### Chemical and Biological Contamination

5-12. The STB dry mix is spread on solid surfaces and raked into soft surfaces like sand or earth.

5-13. The STB slurry is applied to terrain by using a spray hose that is attached to an M12A1 decon apparatus. Operators wear TAP aprons over their MOPP gear and sit on the front fenders of a vehicle to spray the STB slurry. The speed of the vehicle should be adjusted according to the terrain; this is normally 2 to 4 miles per hour. The operators spray a fan pattern 3 to 5 feet ahead of the vehicle. The pattern overlaps on the center and extends past both sides of the vehicle. The STB slurry forms a seal over the surface and must be renewed periodically due to deterioration from traffic or weathering. One load of STB slurry will decon a concrete surface about 328 feet long by 33 feet wide. Other surfaces may require more decontaminants. It takes about an hour to load and mix one load of SBT slurry.

### Radiological Contamination

5-14. The STB dry mix and STB slurry have no effect on radiological contamination or its hazards.

## BURNING

5-15. Burning works well against chemically or biologically contaminated vegetation; but it is of no value against radiological contamination.

### Chemical and Biological Contamination

5-16. Fuel may be used to burn grass or short undergrowth. Burning also works on dirt surfaces. Soak the area with diesel fuel, kerosene, or fuel oil and ignite remotely. Do not use gasoline; it burns too quickly. Burning will cause vapor hazards downwind, and protective measures will have to be used by downwind units. Area commanders must warn these units of the vapor hazards.

### Radiological Contamination

5-17. Radiologically contaminated surfaces should not be burned. Burning will not destroy radiological contamination or its hazards. It may spread contamination if radioactive particles become suspended in smoke spread by the wind.

## EXPLODING

5-18. Explosives can be used to blow up areas that are chemically and biologically contaminated; however they are of no value against radiological contamination.

### Chemical and Biological Contamination

5-19. Detonating cord, bangalore torpedoes, or mine-clearing line charges may be used with the STB dry mix to clear small paths through a contaminated area. The two methods that are used follow:

- First method. Remove the holding band and loosen the lid of the bleach drum. Set the drum with the lid in place upside down over about 3 feet of looped detonation cord. Remove the drum and leave the bleach piled on the lid as shown in Figure 5-1, A, page 5-4. Lids may be fired in a series with a single strand of detonation cord connecting the loops.

- Second method. Remove the holding band and loosen the lid of the bleach drum. Set the drum in a shallow hole as shown in Figure 5-1, B. Make sure the lid is loose. Use 5 feet of looped detonating cord (7 feet in soft ground) for each drum. The drums may be fired individually or in a series. For gross contamination, additional firings may be necessary.

5-20. For either emplacement method, the suggested spacing for charges is 33 feet apart for a 100-foot-front contaminated area. The charges should be placed along the upwind edge of the contaminated area (see Figure 5-1, C).

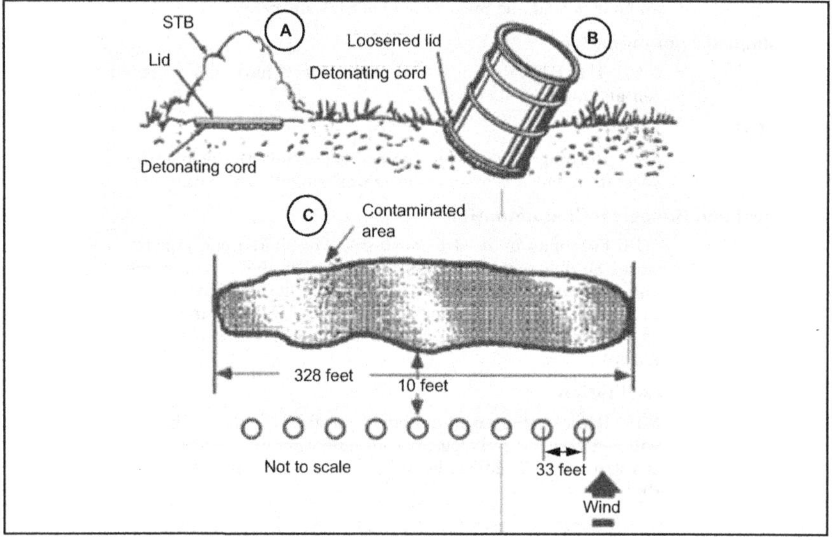


**Figure 5-1. Two Methods of Emplacing Dry STB for Distribution by Detonation**

### Radiological Contamination

5-21. Explosives will not destroy radiological contamination or its hazards.

## FLUSHING

5-22. Flushing with large quantities of water removes contamination. A pressurized stream of hot, soapy water delivered by the PDDA, M17 LDS, or fire trucks can remove significant amounts of an agent. Scrubbing removes even more.

### Chemical and Biological Contamination

5-23. Flushing is not very effective against some kinds of agents, especially thickened agents. However, flushing may speed up the weathering process.

### Radiological Contamination

5-24. Flushing is very effective for removing loose radioactive contamination such as fallout. Rain out, however, can coat surfaces with a film that resists flushing. Such films must be scrubbed. Any contamination removed by flushing and scrubbing will remain radioactive, so control the runoff.

## CLEARING

5-25. To clear passageways, remove the contaminated layers covering the terrain.

### Chemical and Biological Contamination

5-26. Most liquid chemical agents will not penetrate the soil more than 2 inches, and biological agents penetrate even less. Earthmoving equipment or hand tools can be used to scrap aside contamination. By scraping off 2 to 4 inches of earth, a passageway can be created that is free of transfer hazards. However, contamination on either side of the passageway will continue to present a vapor hazard.

### Radiological Contamination

5-27. Fallout does not penetrate the top layer of soil unless it is followed by rain. It can be scraped aside. Move contaminated soil as far away as possible; the piles tend to concentrate radiation. Immediately below a nuclear blast area, the soil may be radioactive to a depth of 5 feet. In most cases, decon of this much earth is not practical.

# FIXED-SITE DECON

5-28. Personnel and equipment need to be decontaminated to reduce the hazard and the spread of contamination, limit casualties, and minimize operational degradation. The goal is to restore the unit to an acceptable level of operational effectiveness. The inability of a fixed site to move away from contamination makes decon more critical. Decon is necessary when protective items cannot be worn in performing the mission or when long-term occupancy is anticipated.

5-29. Decon capabilities may be designed and built into a facility and used if the facility becomes contaminated. Mobile decon equipment may also be available at a fixed site to decon equipment, roads, and buildings. Interior decon may be accomplished with large commercial heaters. Loading docks, entries and exits, and building exteriors can be decontaminated with more conventional methods, such as using an STB dry mix or STB slurry, soap and water, or an M13 DAP. Commanders should identify all systems that are capable of contributing to the decon effort (for example, fire trucks, steam cleaners, and water pumps).

5-30. Decon equipment can be retrofitted to accommodate the process in each area. For instance, rest and relief shelters need rapid, personnel decon systems, and supply operations need a decon system that can handle moderate size pieces of equipment, but at a high volume. Light maintenance needs a system that can decon equipment sensitive to harsh decontaminants or water, and heavy maintenance needs a system that can clean large equipment.

5-31. The commander should designate and train teams that can perform decon for fixed-site operations. The following are fixed-site decon techniques that the teams could apply:

## AIRFIELD

5-32. Decon the area required to sustain the mission (see Figure 5-2). Figure 5-3 illustrates the STB application on the entrance/exit of the airfield hangers. Note that the STB application includes 10 feet on each side of the runway between the entrance and the exit. A transfer hazard will remain a problem until all parts of the airfield have weathered. Buildings' or hangers' entrance/exit and doors must be decontaminated accordingly. Figure 5-3 illustrates a field-expedient method that could be employed at any fixed site to cover the major entry/exit toward the site.

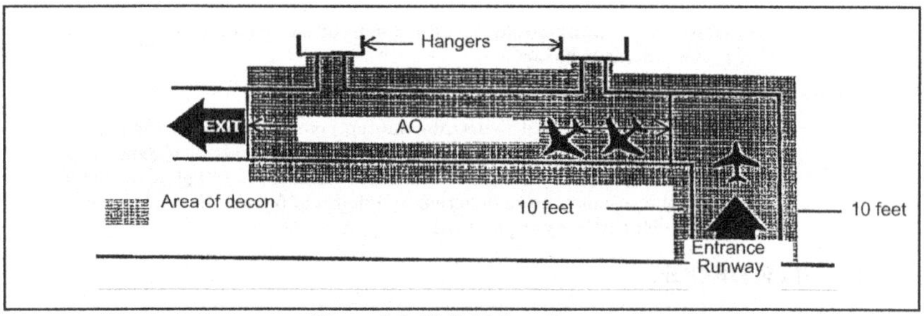


**Figure 5-2. Airfield-Area Decon**

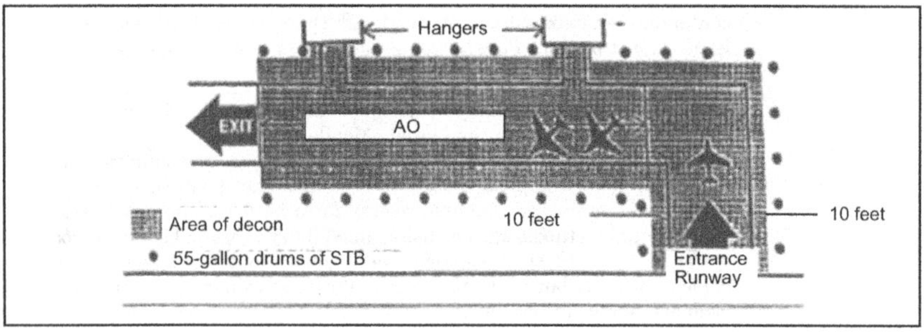


**Figure 5-3. Field-Expedient Airfield Decon**

## HELIPAD

5-33. To decon a helipad site, spray the STB slurry on the entrance/exit approach or the path (see Figure 5-4). Since airborne STB particles can damage sensitive aircraft components, the decontaminated area should be washed with water before landing helicopters.

## POMCUS/MOTOR PARK

5-34. At a prepositioning of material configured to unit sets (POMCUS)/motor-park site, decon entrance/exit approaches to limit contamination. Exposed vehicles in a POMCUS/motor park will become contaminated but most likely there will not be enough personnel to drive the vehicles to a decon station. Decon parked equipment by spraying it with an STB slurry or with hot, soapy water (see Figure 5-5). The M12A1 PDDA crew drives around the parked vehicles and performs vehicle washdown to accelerate the weathering process.

**NOTE: For more detailed guidance on the decon procedures of fixed sites, ports, and airfields, refer to FM 3-4-1.**

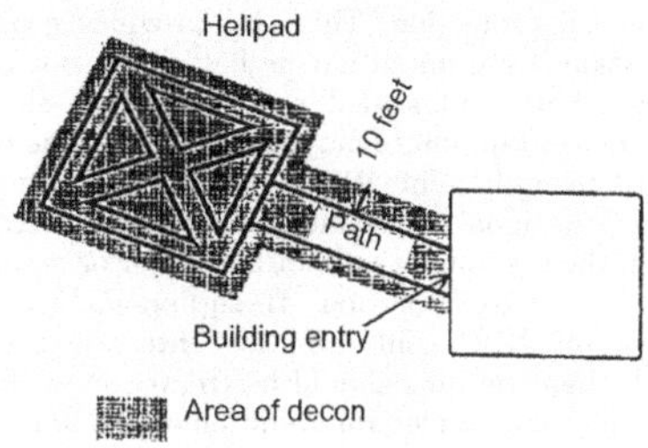


**Figure 5-4. Helipad-Area Decon**

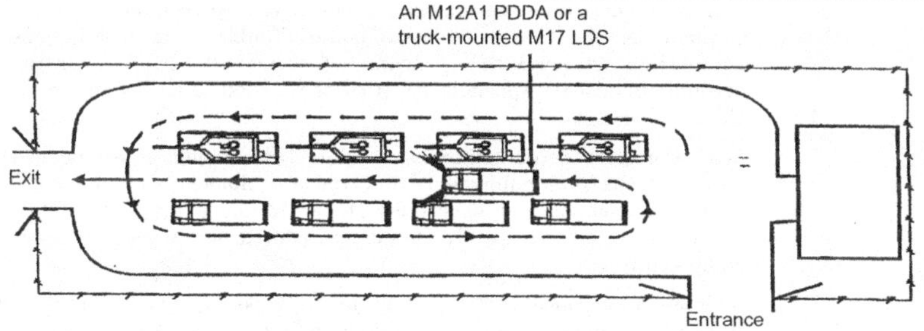


**Figure 5-5. POMCUS/Motor-Park-Site Decon**

# VULNERABLE EQUIPMENT DECON

5-35. Most military equipment has not been critically assessed for its ability to withstand decon operations. As more testing is done, specific decon instructions will be included in the TM for all types of equipment. Some equipment is extremely vulnerable to damage when subjected to decon operations. Electronics and optic equipment are especially vulnerable. Some materials (such as canvas and rubber) tend to absorb chemical agents, and decontaminating absorbent surfaces is extremely difficult, if not impossible. Most decontaminants are highly corrosive and cannot be used on certain materials. They may also render ammunition unserviceable. Do not subject vulnerable equipment to unnecessary decon operations. Radiac meters may measure low levels of radiation that are not dangerous. The commander's operational exposure guidance (OEG) determines the maximum permissible radiation dose rate (usually 0.33 cGy per hour). If this dose rate is exceeded, you must decon.

## ELECTRONICS

5-36. Nonsealed electronic-equipment circuitry can be damaged by corrosive decon materials. Most field electronic equipment is watertight for environmental protection. This also provides good protection against NBC contamination. Contamination probably will not penetrate gasket-equipped protective covers and sealed components on electronic equipment; but if exposed, the contaminants may be present on the outside of cases containing the electronic equipment. Wipe down the outside portions of the equipment case with a designated decontaminant. After decontaminating the outside, wipe down the equipment with either water or an approved solvent to remove traces of the decon solution. If equipment seals appear damaged or if penetration of NBC contamination into the inside of the equipment is suspected, then the unit should be treated as if it was nonsealed. Under no circumstances should electronic equipment be immersed in a decon solution or subjected to a high-pressure application of decon solutions.

5-37. For chemical contamination, wipe all metal electronic cases with DS2. If a chemical agent is identified as HD, allow 30 minutes contact time. Next, wipe the case down thoroughly with a damp cloth, dry it off, and wipe it down with a designated oil (for biological agents, wait 10 minutes). If DS2 is not available or the case is of some material other than metal, wipe the equipment's exterior with a cloth and hot, soapy water or use a nonstandard decontaminant (see Appendix B). If contamination is not extensive, use an IEDK.

5-38. Corrosive decontaminants, such as DS2, should never be used on nonsealed electronic equipment. This type of equipment is often found inside shelter assemblies and helicopters. Refer to the appropriate TMs or to Appendix C for acceptable decon procedures for nonsealed electronic equipment.

5-39. For radiological contamination, brush, wipe, or vacuum contamination from the equipment. Since contamination is not destroyed but just moved from one place to another, control the runoff and treat it is as a hazardous substance.

## OPTICS

5-40. Optic systems are extremely vulnerable to decon materials that might scratch or adversely affect the lenses. Wipe optic systems with a soft, nonabrasive material such as lens-cleaning tissue, cotton wadding, or soft cloth dipped in hot, soapy water. Wipe the optic system with decontaminants. Do not immerse it.

5-41. Hot, soapy water is the preferred decontaminant for chemical and biological contamination. When DS2 is applied to coated lenses, it causes some damage. Likewise, exposing lenses to an STB slurry removes some of the magnesium from the coating. Small optic surfaces can be decontaminated with the IEDK.

5-42. Radiological contamination should be blown off with a stream of air or wiped off with hot, soapy water. Rinse by wiping with a sponge dipped in clean water.

## AMMUNITION

5-43. Decon contaminated ammunition with cool, soapy water. Apply the soapy water with the PDDA or with brushes, mops, rags, or brooms. Cool, soapy water is the preferred decontaminant for all types of contamination.

**WARNING**

**Do not use DS2 or an STB powder on ammunition. It removes critical markings from the ammunition. The STB powder may cause a fire when it comes in contact with a blister agent or DS2. It may also corrode and render the ammunition unserviceable. Do not use nonstandard decontaminants that are corrosive. They also may remove critical markings from the ammunition.**

## CANVAS ITEMS

5-44. These items include load-carrying equipment and web gear. It may be necessary to burn or bury them if they are heavily contaminated. Either an STB dry mix or an STB slurry may be used to decon them. The STB slurry is more effective. In many cases, weathering may be the preferred decon technique because scrubbing canvas frequently imbeds the contamination further and worsens the situation. If the item must be decontaminated, boil it for 1 hour in soapy water. This is the preferred method for chemical and biological decon. Radiological contamination can be removed by brushing it off, then washing. It may also be vacuumed off a surface.

5-45. Salvage of contaminated canvas requires a great amount of logistic and manpower requirements. Other options for disposing of contaminated canvas should be considered against METT-TC. NBC covers protect vulnerable items, but these covers must be buried or destroyed after use.

## FOOD AND WATER

5-46. Most units cannot test the purity of food and water even if they can decon it. Food and water decon usually takes special handling by trained technicians. If your water becomes contaminated, contact medical personnel. Water-treatment specialists test and treat the water in an NBC environment. Direct contaminated-food questions to supporting veterinary personnel.

5-47. Food and water should be kept in sealed containers and protective packaging. Leave food in wrappers. Keep wrapped food in boxes, tins, or other containers. Store these containers in closed refrigerators or vans. See FM 3-3 for further discussion on avoiding contamination.

5-48. Contaminated food should be placed into three groups for handling.

### Group I

5-49. Group I consists of canned and unopened items that have been exposed only to the vapors of an agent. Generally, the items in this group will be safe for issue to personnel after a brief period of outdoor airing to remove clinging vapors.

### Group II

5-50. Group II consists of canned and unopened items that have been contaminated with a liquid chemical agent. Attempts to decon porous packaging materials, such as cardboard or wood, are likely to be unsuccessful and may result in spreading the contamination. To handle such items, strip off the outer contaminated coverings and examine the inner layer to determine if agent penetration has occurred. If it has, continue stripping off layers until an uncontaminated layer is reached. Items packaged for military operations are usually packed in boxes, so this procedure is feasible. When the inner uncontaminated package is reached, place it in group I. If the agent has penetrated to the food, place it in group III.

### Group III

5-51. Group III consists of unpackaged or poorly packaged items that have been exposed to an agent in either a vapor or liquid form. Foodstuffs in this group should be decontaminated only when absolutely necessary.

5-52. If you do not know if the food is contaminated, do not use it. Do not decon items in group III without the help of veterinary personnel.

## CHEMICAL MUNITIONS DISPOSAL AND DECON

5-53. In combat, chemical munitions may be encountered. Captured enemy chemical munitions may have sustained damage or have deteriorated to where they are leaking. Leaking munitions must be decontaminated, evacuated, and disposed of safely by specialized units with properly trained soldiers performing this task. All soldiers must use protective clothing and masks when there is a possibility of exposure to a toxic chemical agent. The MOPP level will be specified by the unit commander. Use detection equipment (CAM, M256A1 detector kit, M8A1 or M22 ACAA, and M8/M9 detector paper) to monitor chemical munitions.

5-54. Unexploded leaking munitions should be treated in the same manner as conventional ordnance. A spot report must be submitted to an explosive ordnance disposal (EOD) unit.

5-55. When disposing of and decontaminating chemical munitions, see Army regulation (AR) 385-10 and TM 10-277 for additional safety guidance.

### HANDLING

5-56. Chemical projectiles should not be handled. A good rule of thumb is to treat them as you would a fused high-explosive (HE) round. Security and protective measures must match those given to conventional munitions. Notify an EOD unit for disposal and decon of chemical munitions. EOD units have the responsibility for sealing, packaging, and disposing of chemical munitions.

### CONTAMINATION CONTROL

5-57. Whenever a chemical projectile is damaged and starts to leak, the round presents the same hazards to personnel in the area as an enemy chemical attack. Only munitions that have not completely detonated or have been severely damaged may exude liquid agent. It is the vapor hazard created when the chemical leakage occurs that poses the greatest threat to personnel. When a leaking chemical munition is discovered, react as if you are under chemical attack.

## DEPLETED URANIUM (DU) DECON

5-58. DU is an extremely dense metal used in munitions to penetrate heavy armor or as a protective shielding (armor packages). DU is also used as equipment components. All components containing radioactive isotopes are listed in TB 43-0116.

5-59. DU exposure and incidents may occur anytime there is damage to the DU armor package, a vehicle is hit with DU munitions, DU munitions are damaged, or equipment components containing DU are damaged. The DU armor package can be damaged during vehicle maneuvers, on-board fires, maintenance activities, or ballistic impacts. DU munition problems may occur during storage, transportation, combat, testing, or manufacturing. DU contamination may be present on the ground in areas where equipment was destroyed or damaged.

5-60. DU contamination may include DU oxides (dust), contaminated shrapnel, munition components, or armor components. DU emits primarily alpha particles; however, beta, gamma, and x-ray ionizing radiation are also emitted. DU contamination can be inhaled, ingested, or injected. DU contamination does not pose an immediate health risk. Consequently, contamination should be removed from personnel or vehicle surfaces when directed by the unit commander based on METT-TC.

5-61. Visual signs that DU contamination is present include heavy, dull-black dust or small round holes. DU contamination can only be verified with a radiac meter. An AN/VDR-2 or AN/PDR-77 with an alpha or beta probe (flat-pancake surface) is used to detect and measure DU contamination. These probes are within the radiological protection officer (RPO) kit.

5-62. When working on or within DU contaminated equipment, soldiers should wear their protective mask. Depending on temperature, protective-clothing availability, DU contamination levels, and tasks to be performed, soldiers can wear the BDO or coveralls or they can roll down their sleeves and blouse their trousers as directed by the unit's chemical or medical personnel.

5-63. General decon procedures follow:

- Use a radiac meter to determine if DU contamination is present.
- Provide protection, including appropriate clothing, for workers as directed by the unit's chemical or medical personnel.
- Identify what is to be decontaminated.
- Obtain necessary equipment and materials (Appendix D).

- Brush, wash, or wipe off contamination with a damp cloth. Use a high-efficiency particulate air filter (HEPA) vacuum cleaner, if available.
- Work from the outside of the contaminated area to the inside.
- Cover fixed contamination with tape, paint, paper, plastic, or other disposable material.
- Use the standard double-bag-and-tag process for hazardous waste. The only contaminated waste generated by DU will be the vacuum-cleaner bags after use on multiple vehicles.
- Continue the decontaminating process until contamination levels are reduced to the appropriate level (see AR 11-9).
- Decon personnel and equipment according to Chapter 4 of this manual.

## CONTAMINATED REMAINS DECON

5-64. Mortuary affairs personnel establish and operate the mortuary affairs decontamination collection point (MADCP). When an NBC event has occurred on the battlefield, there is a very high probability that many deceased personnel were exposed to contaminating agents. If the situation does not lend itself to the determination of a hazard on an individual basis, all remains within the affected area will be treated as if contaminated. If the theater surgeon or his staff determines that biological agents have been employed, all remains will be treated as if contaminated.

5-65. Whenever the threat of NBC warfare exists in a theater of operations, the unified commander will direct the Joint Mortuary Affairs Office (JMAO) to be prepared to handle contaminated remains. The JMAO will serve as the theater central point of coordination for the operation. Joint operations of contaminated remains are described in Joint Publication 4-06.

5-66. Personnel support is required after completing the evacuation mission to the MADCP. A DTD takes about 1 hour. The MADCP site will require a complete detailed decon by a chemical decon unit.

## RADIOLOGICAL DECON

5-67. Radiological contamination may occur in the form of one element. In the following paragraphs, the decon of six specific, commonly found radioactive elements is discussed. The discussion is applicable not only to these elements but also to other elements having similar chemical properties.

### CESIUM

5-68. The common radioisotope of cesium is cesium-137. It emits beta and gamma radiation, decaying to stable barium-137. Cesium-137 is widely used in gamma sources. It occurs in these sources as cesium-chloride pellets. Cesium chloride is a soluble salt. The contamination from a sealed-source leak absorbs water, becomes damp, and creeps. Contamination from a sealed cesium source is best decontaminated by wet procedures unless the contamination is on a porous surface, in which case wet procedures should be preceded by vacuuming. Cesium is known to adsorb from a solution onto glass surfaces. Decontaminating a cesium liquid-contaminated surface is best accomplished by wetting the surface, absorbing the solution with a rag or other absorbent material, and rinsing the area several times with water. If the contamination persists, use a detergent solution and scrub with a brush. A cesium-contaminated solution that has been standing for some time is best decontaminated by absorbing any remaining liquid, treating the surface several times with water (allowing the water to stand on the surface for about 1 minute each time), and then absorbing the liquid from the surface. If the contamination remains, further treatment depends on the surface. Metallic surfaces are treated with strong mineral or oxidizing acids. Waxed surfaces are removed. If contamination still persists, abrasives or other removal techniques are used.

## COBALT

5-69. The common radioisotope of cobalt is cobalt-60, a beta gamma emitter. Metallic cobalt-60 is commonly used in sealed gamma sources. Particles of cobalt dust adhering to small articles are readily removed by ultrasonic cleaners or by dipping the article in a dilute solution of nitric, hydrochloric, or sulfuric acid. Cobalt-dust contamination that exists over a large area is best removed by vacuuming. Sealed cobalt sources may leak as a result of electrolytic action between the cobalt and the container. The result is often a soluble cobalt salt, which creeps and spreads. This is best decontaminated with a detergent or an ethylene diamine tetra acetic acid (EDTA) solution, followed by a treatment with mineral acids. Contamination from solutions containing cobalt may be treated with water.

## PLUTONIUM

5-70. The most common isotope in which plutonium may be present as a contaminant is plutonium-239, an alpha emitter. This isotope is present in the AN/UDM-6 calibration source. Plutonium contamination may be the result of a nuclear weapon's accident, in which case, the plutonium will be scattered as a metal or oxide in a dust form. Both forms of plutonium are insoluble. Aging of plutonium-239 contamination is impractical since it has a 24,000-year half-life. Plutonium contamination that covers a small area is best decontaminated by vacuuming. If contamination remains, the area should be washed with a detergent solution. Any contamination that remains can be sealed in a protective coating of paint, varnish, or plastic. Plutonium oxide or metal dust spread over a large area, such as a field, is best decontaminated by removing the top layer of soil and disposing of it as radioactive waste. Personnel should wear respiratory protection when decontaminating or moving the soil.

## STRONTIUM

5-71. The most common radioisotope of a strontium is strontium-90, a beta emitter. The daughter particle of strontium-90 is yttrium-90, which is also a beta emitter. Strontium-90-yttrium-90 is commonly used in sealed beta sources such as the M6 source. Generally, it is present as chlorine or carbonate. The chlorine is hygroscopic; it absorbs water and creeps out of the container. This contamination is best decontaminated by vacuuming, followed by a treatment with water, a complexing-agent solution, and a mineral acid, in that order. Contamination resulting from a dilution containing strontium is best decontaminated by absorbing the solution and washing the area with a detergent solution. If strontium contamination persists, the top layer of the surface should be removed by abrasives or other removal procedures and a sealing coat should be placed over the surface.

## TRITIUM

5-72. Tritium is the radioisotope of hydrogen and is a weak beta emitter. If it is released to an area as a gas, the best decon method is to flush the area with air. Since inhalation of tritium can present an internal hazard, personnel entering an area containing tritium gas should wear an appropriate self-contained type of breathing apparatus. Objects in an area exposed to tritium for a length of time may absorb the gas and should be disposed of, if possible. They may be degassed, under a vacuum, by flushing with helium or hydrogen. A surface which monitors clean may become contaminated again in a matter of hours by percolation. There is no practical way of removing tritium oxide (T20) from water due to its similarity to natural water.

## URANIUM

5-73. The most probable source of uranium contamination is a nuclear weapon accident in which the fissionable uranium is spread as a metal or oxide dust. The common isotopes of uranium contamination are uranium-235 and uranium-238. This metal or oxide is insoluble and is best removed from a contaminated surface by brushing or vacuuming, followed by a treatment with mineral acids or oxidizing acids and then sealed. Large-area uranium contamination is best decontaminated by removing the top layer of the surface or by sealing it.

# Chapter 6

# Effects of Environmental Conditions

US forces may find themselves anywhere in the world and subjected to the conditions in that region. Weather and terrain conditions will dictate how to fight and what equipment is most effective to support the fighting force under NBC conditions. The conditions in cold- and arctic-weather areas, warmed areas, urban areas, mountains, jungles, and deserts impact how decon operations are performed.

## COLD- AND ARCTIC-WEATHER AREAS

6-1. These regions comprise about 45 percent of the North American continent and about 65 percent of the Eurasian land mass. They are characterized by extreme cold weather and deep snow during winter months. Spring breakup and mud restrict mobility. Whiteout and grayout cause loss of depth perception, making vehicle movement hazardous. Ice fog (clouds of ice crystals) covers troops, vehicles, bivouac areas, and permanent facilities, marking their location. Chemical agents become more persistent in low temperatures (see Table 6-1, page 6-2). As a result, some normally nonpersistent chemicals, such as sarin (GB), become persistent; this nerve agent could remain a transfer hazard for up to 30 days in arctic climates. Normally, persistent agents become more persistent. Although frozen agents do not present a significant problem in solid state, they become hazards when they warm up. Decon is a problem because low temperatures also reduce the effect of decontaminants.

6-2. The freezing points of decontaminants are as follows:

- Water, 32°F.
- STB slurry with antiset, 0°F.
- DS2, -25°F.
- SKD, -50°F.
- IEDK, -50°F.

### DECONTAMINANTS

6-3. When temperatures reach 32°F, water can no longer be used as a decon. In such situations, use DS2. DS2 is effective down to -25°F. For temperatures below -25°F, use STB or HTH as a dry mix (two parts of STB to three parts of earth or snow) for decon. Apply the dry mix by shoveling it on contaminated surfaces or by filling sandbags with the mix and dusting it on the surfaces. Remove the dry mix by brushing, scraping, or using uncontaminated earth or snow to "wash" it off. Other decontaminating methods using nonstandard solvents and fuels may be used, but observe fire safety, protect personnel from corrosives, and take precautions against super-cooling effects. Because of their low freezing points, solvents such as aviation fuel (JP8), diesel fuel, and kerosene may be used to physically remove contamination. These solvents only flush the agent from the surfaces. They generally do not neutralize agents nor do they eliminate agents that soak into surfaces. Nonstandard solvents generally are very flammable and must be handled with care. See Appendix B for specific nonstandard solvents.

**Table 6-1. Freezing Points of Selected Chemical Agents**

| Agents | Symbols | Contact Hazard | Vapor Hazard | Freezing Point |
|---|---|---|---|---|
| | | **Nerve** | | |
| Tabun | GA | Extreme | Low-moderate | +23°F |
| Sarin | GB | Extreme | Extreme | -69°F |
| Soman | GD | Extreme | Probable | -44°F |
| | GF | Extreme | Probable | -22°F |
| | VX | Extreme | Negligible | -60°F |
| | VR-55 | Extreme | Probable | Unknown |
| | TGD | Extreme | Probable | Depends on the percent of thickener |
| | | **Blister** | | |
| Distilled mustard | HD | Extreme | Negligible | +57°F |
| Mustand-lewisite | HL | Extreme | Low | -14°F (Pure)<br>-44°F (Plant purity) |
| Nitrogen mustard | HN-1 | Extreme | Low | -29°F |
| Nitrogen mustard | HN-2 | Extreme | Low | -76°F |
| Lewisite | L | Extreme | Negligible | 0°F |
| Nitrogen mustard | HN-3 | Extreme | Low | +25°F |
| Phosgene oxime | CX | Extreme | Low | +95°F |
| | | **Blood** | | |
| Hydrogen cyanide | AC | Low | Extreme | +8°F |
| Cyanogen chloride | CK | Low | Extreme | +20°F |
| Arsine | SA | Low | Extreme | -177°F |
| | | **Choking** | | |
| Phosgene | CG | Slight | Extreme | -198°F |
| Diphosgene | DP | Slight | Extreme | -71°F |

## EQUIPMENT-DECON DEVICES

6-4. Discussed in the following paragraphs are the different types of equipment decon devices:

### M11 Decon Apparatus

6-5. This apparatus is designed to dispense DS2 at temperatures as low as -15°F. DS2 is effective down to -25°F, but it becomes thicker as the temperature drops. Normally, one nitrogen cylinder is used to charge the M11. In cold weather, a second nitrogen cylinder may be used if inadequate pressure is developed with one. Never use more than two nitrogen cylinders.

### M13 Decon Apparatus

6-6. This apparatus is designed to dispense DS2 at temperatures as low as -15°F. As the temperature drops, the flow of decontaminants through the hose and wand becomes degraded.

### M17 Lightweight Decon System and M12A1 Decon Apparatus

6-7. Both apparatuses have problems in the cold similar to any other water pump. Normal engine "cold-soaking" problems that cause starting difficulties and brittle metal parts have been observed. Also, internal pumps and lines will crack and break when water freezes within these systems. To avoid freeze-up problems, use the shutdown procedures described in TMs 3-4230-228-10 and 3-4230-209-20&P.

## DETECTION AND WARNING DEVICES

6-8. In the following paragraphs, the different types of detection and warning devices are discussed:

### M8/M9 Detector Paper

6-9. M8/M9 detector paper is limited in the cold because only agents in a liquid form can be detected. If the substance is thickened or solidified, collect a sample with a stick or scraper and wipe it onto a sheet of M8/M9 detector paper. Place the sample on a heated surface, such as an operating vehicle or a power generator, to thaw the suspected agent. Because of off gassing, do not do this in a heated vehicle or tent. Put a box or can over the paper to concentrate any vapors. Put an M256A1 detector kit inside the box or can. Heat the suspected agent enough to enable detection and allow agent identification.

### M256A1 Detector Kit

6-10. As temperatures drop toward 32°F, the M256A1 detector kit takes longer to give a positive or negative indication. When temperatures reach freezing, carry the liquid components in the breast pockets of the chemical protective clothing. It may also be necessary to put the test sampler back into the shipping bag after sampling and place it on a heated surface to hasten the reaction. The kit will not work at temperatures below -25°F.

### Chemical-Agent Monitor

6-11. The CAM will have difficulty in detecting agents at low temperatures. Contaminated personnel entering a building or a collective protection shelter will desorb agents as they warm up. Use the CAM in a vestibule or an air lock where the temperature has been raised to that of the building's or shelter's interior. Likewise, check aircraft and vehicles requiring maintenance as they pass through a warmed-up site before entry into a hanger or maintenance bay.

### M8A1 Automatic Chemical-Agent Alarm

6-12. When operating an M8A1 ACAA at temperatures below 20°F, use the M253 winterization kit. Warm-up time with this kit is 50 minutes.

### EQUIPMENT-DECON STATIONS

6-13. Place decon sites in built-up areas and near road junctions and intersections of forest lanes to facilitate access. Divide the decon sites and approach routes into sectors, and assign units to keep them open in spite of mud or heavy snow. With present technology, equipment decon problems in the field are difficult to overcome in an arctic climate. Commanders should consider fighting dirty in arctic regions. Fresh units can be rotated into the field so that dirty units can be moved back to built-up areas to decon.

6-14. Decon apparatuses and water trucks may have to be deployed with empty tanks instead of full ones as in temperate climates. To prevent freezing, it may be necessary to preheat water when loading the water trucks and the tanks of the decon apparatuses and keep it heated until it is used. These vehicles may have to be enclosed and warmed so that the engines will start. Decon apparatuses and water trucks must be drained immediately after use to prevent damage from freezing. Commanders may choose to conduct minimum decon until they can get back to fixed facilities and conduct a more thorough decon.

6-15. Vehicles and personnel covered with contaminated snow must be decontaminated before the snow has a chance to melt and freeze. Such snow forms layers of ice that make contamination difficult to remove. Radioactive fallout that is mixed with snowfall must be removed as soon as possible. Use tree branches (if available) to remove contaminated snow.

6-16. Snow can be used to cover contamination; however, the snow can blow away or the contamination can resurface when tracked vehicles, troop movement, or digging disturbs it. Snow cover provides some protection if left undisturbed, but this protection is too uncertain to rely on.

## WARMED AREAS

6-17. Preventing contamination from entering warmed areas is a challenging problem. For example, if soldiers get frozen agents on their clothing, it will be hard to detect because low temperatures have slowed its effects. However, if the temperature warms or if the soldiers enter a warmed area, the agents become dangerous. Because of this, it may be necessary to set up a warming station for each warmed area to isolate the agents. Otherwise, all occupants may be subjected to hazardous liquids and vapors.

6-18. Additional soldiers and equipment are needed for these warming stations. Therefore, in cold-weather operations, the decon and detection priority must be heated support facilities. FM 3-4 discusses individual and collective protection in detail.

## URBAN AREAS

6-19. The need for decon operations in urban areas may also exist, and it may be easier to support. Water sources are generally available and commercial chemicals may be used as decontaminants. Site security may be simplified because of limited observation and poor fields of fire. Decon operations in urban areas will not differ significantly from similar operations in the field; however, keep the following considerations in mind:

### STREETS AND STRUCTURES

6-20. When buildings are contaminated with persistent chemicals, their value for cover, concealment, and shelter is reduced. Wood and concrete tend to absorb liquid agents, and they may give off toxic vapors for days or weeks. Building decon is very difficult and requires large quantities of decontaminants. The hazard can be reduced by covering the contamination with plastic sheets, STB slurry, sodium silicate, or other substances that cover or absorb the agent. Even though a particular part of a building is not intended for occupation, it may still need to be decontaminated to prevent the contamination from spreading.

6-21. Streets and sidewalks also absorb liquid agents, then give off toxic vapors when heated by the sun. These surfaces may need to be decontaminated several times to reduce hazards. Streets, sidewalks, or other porous surfaces are best decontaminated by weathering if the time and the situation permit.

### CIVILIANS

6-22. Try to locate decon operations away from civilians because they may injure themselves or interfere with the work. Civilians must not be forced to prepare decon sites or to help with decon operations. However, volunteers who have been cleared by the Assistant Chief of Staff, G5 (Civil Affairs) (G5) may be used.

### SUPPLIES AND AMMUNITION

6-23. Store ammunition, as well as other supplies, under cover to protect it from NBC contamination. FM 3-4 explains contamination covers in detail.

### SANITATION SYSTEMS

6-24. Urban areas may have sophisticated sanitation systems. When those systems are destroyed, sanitary conditions become far worse than those in areas where sanitation systems have never existed. Sanitation systems must be maintained to avoid epidemics and to avoid overloading biological-decon capabilities. Contaminated water and residue must be controlled so that they will not create a hazard to the civilian population or interfere with sanitation systems. Engineer support may be necessary to construct controlled runoff areas.

## MOUNTAINS

6-25. Excluding the extremely high, alpine-type mountains, most mountain systems are characterized by heavy woods or jungle, compartments and ridge systems, limited routes, and highly variable weather conditions. All these factors will affect decon operations.

### MOBILITY

6-26. The terrain and the disruption of existing routes may dictate that mountain operations be decentralized. Additional water-carrying vehicles may be needed to support these operations. Forces must be organized to be self-sufficient and should have decon elements attached to them initially. Decon platoons are attached to brigades and decon squads are attached to battalions. When decon elements operate independently, their parent chemical companies may not be able to provide support. Decon elements must be equipped with greater hauling capabilities, especially for water. Administrative and logistical support to the decon element will have to come from the support unit.

### WIND

6-27. Mountains have changing weather and constant winds that promote weathering at a faster rate. In high mountains, calm winds in the ridges and passes are seldom; whereas, in protected valleys, strong winds are rare. Normally, wind velocity increases with altitude. Generally, downwind contamination will be greater because high winds are more frequent. However, variable winds will make predictions unreliable.

### TEMPERATURE

6-28. Cool or cold temperatures have an adverse effect on decon operations. With every 328-foot gain in elevation, the temperature will normally drop $3^{0}$ to $5^{0}$F. The drier the air, the more pronounced this drop will be. At high altitudes, there may be differences of $40^{0}$ to $50^{0}$F between the temperature in the sun and that in the shade. Consequently, the temperature rises swiftly after sunrise and drops quickly after sunset. Low temperatures that normally exist at extremely high altitudes may demand decon procedures and precautions similar to those used in cold-weather operations. Daytime operations can be scheduled to avoid some low-temperature problems.

### SUNLIGHT

6-29. Sunlight destroys most biological agents. Above the timberline, there is little protection from the sun, so the effectiveness of a biological attack is reduced.

## JUNGLE

6-30. Jungle climates vary with the location. Close to the equator, all seasons are nearly alike, with rains throughout the year. Farther from the equator, especially in India and Southeast Asia, jungles have distinct wet (monsoon) and dry seasons. Both zones have high temperatures (averaging $78^{0}$ to $98^{0}$F), heavy rainfall (as much as 400 inches annually), and high humidity (90 percent) throughout the year.

6-31. Transportation is extremely difficult due to natural obstacles; therefore, decon operations will be decentralized as much as possible. Because of high humidity and temperatures, MOPP gear should not be worn for long periods. Since decon crews will be more susceptible to heat stress, staffs should plan for frequent crew rotations and provide enough decon personnel to conduct decon operations.

6-32. Contamination will be retained temporarily in the jungle canopy, reducing the immediate hazard. Later, rains will wash these particles to the ground and concentrate them in low areas. These areas are likely to become contamination "hot spots." Decon sites should be put on high ground during the rainy season.

6-33. Tropical rainstorms will flood decon sites unless the sites are adequately drained. Ground that appears firm may become impassable when it rains. Rain and intense heat will speed the deterioration of many supplies. Do not break the seals on packing material until it is absolutely necessary.

6-34. Rain and high humidity (wetness) degrade the protective abilities of the overgarment. Take care to ensure the maximum length of time of the overgarment's usefulness. Additional quantities of overgarments must be maintained to replace those that have gotten wet. Rain helps the decon process by washing away contamination on exposed surfaces. Rain can also hydrolyze some agents. However, runoff may contaminate the soil.

## TEMPERATURE AND HUMIDITY

6-35. When temperatures rise to the $85^{0}$ to $100^{0}$F range, troops can only continue medium or heavy workloads by reducing their MOPP level. More soldiers will be needed to make up for extra rest periods and more frequent rotation periods during decon operations.

6-36. Frequently oil exposed metal parts and grease wheel bearings to protect them from moisture. Since petroleum, oils, and lubricants (POL) tend to absorb chemical agents, additional decon may be needed. Measures taken to protect electronic communications equipment from moisture will reduce the need for decon since these measures provide good NBC protection. Seal radios with silicone compound and pack them in waterproof containers when they are not in use (remove batteries). Protect handsets and microphones with plastic bags and check the bags frequently for moisture buildup.

6-37. The jungle's high humidity and temperature increase the need for maintenance. Contact teams must operate independently as they move from site to site doing repairs. There may be no decon specialists to help them.

6-38. Biological agents thrive in the heat, humidity, and shade, which are characteristics of the jungle. Therefore, weathering is not a practical means of biological decon.

6-39. Solid decontaminants, such as STB powder, tend to cake and decompose at a faster rate than in temperate climates. Caking is no problem, but the decomposition eventually makes STB powder ineffective.

## TIME

6-40. Many tasks in the jungle take more time than in other environments. More time for movement and security must be allowed, which may leave less time for planning and preparation. In the jungle, more decon personnel will be required to perform decon operations.

### SECURITY

6-41. Decon stations are lucrative targets; therefore, security must be a prime consideration. Relocating a decon site may require a small security force, thus additional personnel will be required.

### RESUPPLY

6-42. Because of the terrain, aerial resupply usually will be common practice. Keep this in mind when determining the amounts of decontaminants to keep on hand since they will compete with ammunition, POL, medical supplies, food, and water for priority of delivery.

### PERSISTENCY

6-43. Sunlight and wind are reduced within thick jungles due to the vegetation. As a result, chemical and biological agents are less likely to disperse and will remain a hazard for an extended time.

## DESERT

6-44. Desert regions are characterized by extreme temperatures with ranges varying between $30^0$ to $130^0$F over a 24-hour period. These regions have long drought periods, which are interrupted by sudden rains that bring flash floods. There are shortages of suitable groundwater. Large areas suitable for tracked-vehicle maneuvers may sometime have impassable ravines; wet, spongy grounds; and sand areas.

### WATER

6-45. The principal problem for decon in the desert is the lack of water. The use of STB slurry and DS2 will burden the logistical system because of the water required for mixing and rinsing. Nonwater-based decontaminants (natural solvents) may be required (see Appendix B of this manual). Contamination avoidance becomes increasingly important in desert operations because of limited water sources. Therefore, maximum use should be made of field-expedient covers, especially at night when an attack by chemical or biological agents is most likely.

### HEAT STRESS

6-46. Heat stress is a critical problem for soldiers working in the desert environment in full MOPP. Operating decon stations in daytime temperatures may require short periods of work followed by long periods of rest. Performing operations at night to avoid heat stress may create light discipline problems.

### WEATHERING

6-47. Weathering is a viable decon option. High daytime temperatures can increase evaporation of liquid contamination. As a result, vapor concentrations will be high but may not last long. Refer to FM 3-4 for MOPP opening/unmasking guidance to estimate the chemical hazards resulting from enemy attacks. If liquid contamination soaks into soft, porous soil (such as loose sand), evaporation is not as quick. Strong winds also increase the evaporation rate. Low temperatures during the night have a reverse effect and tend to increase the persistency of chemical and biological contamination. The sandblasting effect of sandstorms may remove contamination from surfaces facing the storm. Desert sunlight and high temperatures will destroy many chemical and biological agents without additional decon measures.

## BEARINGS AND OTHER CRITICAL MOVING PARTS

6-48. Bearings and other critical moving parts need extra lubrication in the desert. This complicates decon because lubricants tend to absorb chemical agents. After a sandstorm, maintenance must be conducted regardless of the last scheduled maintenance. Therefore, perform decon first and if time is critical, decon only those surfaces that will be touched during maintenance. This will not eliminate vapor hazards.

## Chapter 7

# Aviation

Contamination in the battle space causes a unique problem for aviation. Since an aircraft uses its high mobility to maneuver across the entire battle space, avoiding contamination can be very difficult. Once an aircraft is contaminated, it is very difficult to decon. Therefore, aviation units need to prepare effective standing operating procedures (SOPs) that emphasize contamination avoidance, protection, and methods to decon each type of aircraft using an immediate, an operational, and a thorough decon.

## CONTAMINATION AVOIDANCE

7-1. Contamination avoidance is critical due to the difficulty in decon. Aircrews must be aware of NBC hazards. For instance, an aircraft that hovers or lands in a contaminated area could splash contamination onto itself. Likewise, contaminated passengers or ground crews pose a danger of transferring contaminants into the aircraft where they are difficult to remove. Aircraft could also possibly pick up some contamination by flying through an invisible vapor cloud. Hazards that will only incapacitate ground personnel for a short time can be fatal in the air.

7-2. Some things that can be done to limit contamination are as follows:

- Know what areas are contaminated and avoid these, if possible.
- Pick landing zones that will have a reduced splash effect if an aircraft must land in a contaminated area.
- Limit the spread of contamination into the aircraft from the outside. For example, ground crews at the forward arming and refueling point (FARP) could conduct arming and refueling without requiring the aircrew to exit the aircraft.
- Conduct inspections without touching or shaking items. The contaminated crews should perform this task. Many inspection points can be inspected visually.
- Increase the use of covers when not flying. Use engine covers, flyaway gear, and hatches. If possible, provide overhead cover for parked aircraft.
- Limit the number of aircraft that must operate in a contaminated area or use aircraft already contaminated.
- Line the troop compartment with plastic (a field expedient way to limit the spread of contamination) when carrying contaminated personnel or casualties. A plastic curtain can be fastened between the troop compartment and the flight compartment with tape or Velcro to limit contamination transfer. The aircraft's heater can be used with

the curtain to create an overpressure in the pilot's compartment. This will limit vapors from entering the compartment.

- Apply M9 detector paper to the landing gear of the aircraft. FARP personnel should always check the M9 detector paper before servicing the aircraft. Another piece of M9 detector paper can be placed on the windscreen where the aircrew can see it.

> **CAUTION**
> M9 detector paper will show false positives if exposed to petroleum products.

## DECONTAMINANTS

7-3. Only approved cleaning compounds may be used to decon aircraft. Caustic decontaminants such as DS2, STB, bleaches, and sodium hypochlorite are not considered safe. STB corrodes metal components and the aircraft's skin, and DS2 corrodes rubber, plastic, and acrylic-plastic windows.

7-4. Soap and water, kerosene, JP8, and diesel fuels are approved as decontaminants on selected parts of the aircraft. JP8 is effective in removing some agents from the aircraft's skin and components; however, it does not neutralize the agents. Personnel must use care when handling JP8. When using a cloth soaked in JP8, personnel must avoid wiping internal components near the exhaust. If water is available, use it to rinse off the JP8. Many portions of the aircraft are delicate and cannot stand high-pressure water or extreme hot air.

7-5. Sodium carbonate is a fair decontaminant against G nerve agents, but it is not effective against V nerve and blister agents. The chemical reaction of the V nerve agent and sodium carbonate is extremely slow and produces a product that is very toxic. Most of the field-expedient decontaminants are corrosive and could cause damage to the aircraft's skin or acrylic-plastic windows.

## AIRCRAFT DECON

7-6. When an aircraft becomes contaminated, the aircrew must fly in MOPP gear. However, wearing MOPP gear degrades the aircrew's ability to perform its jobs over time.

7-7. If air and ground crews are careful when operating in a contaminated environment, contamination can be contained on the exterior of the aircraft. The increased airflow over the aircraft's skin increases the rate of evaporation. Still, some agents will remain and continue to be a hazard. Thickened agents evaporate more slowly and may remain a hazard even after prolonged flights. If the interior is contaminated, flying the aircraft with the doors open can help reduce the hazard. The heat from the running engine decontaminates the interior of the engine compartment. However, more active methods of decon may be needed. Use the immediate-, operational-, or thorough-decon method with modifications to perform an aircrew spot decon, an aircraft washdown, and the DAD.

## DECON TYPES

7-8. Table 7-1 shows the types of aircraft decon. The crew of a contaminated aircraft will use the techniques described in previous chapters to decon. Unit personnel conduct the aircrew spot decon and aircraft washdown, but a chemical decon unit usually conducts the DAD.

### Aircrew Spot Decon

7-9. After personal wipe down, aviators may conduct a spot decon on their aircraft (see Table 7-1). The aircrew spot decon provides the same benefits as the operator's spray down. It limits the transfer and spread of contamination by decontaminating the surfaces that must be touched during operations. Spot decon applies to the first six functional activities. The goal is to limit the transfer and spread of contamination while sustaining flight operations.

**Table 7-1. Aircraft Decon Types**

| Decon Types | When, Why, Who Performs Decon | Areas to Decon | Types of Decon Agents |
|---|---|---|---|
| Aircrew spot decon | The crew decontaminates after the immediate decon to allow continued operations | Essential operating surfaces on the aircraft | IEDK, JP8[1] |
| Aircraft washdown | The unit decon PDDE crew or decon unit performs decon within 6-24 hours to reduce contact hazards[2] | The entire exterior surface of the aircraft[3] | Hot, soapy water |
| DAD | The decon units decon when the mission allows to reduce the hazard to negligible risk levels | The entire exterior surface and selected interior surfaces of the aircraft | 10 percent sodium-carbonate solution (10 pounds of sodium carbonate to 12 gallons of water) |

**[1]JP8 should not be used inside the aircraft.**

**[2]An aircraft washdown is most effective if conducted within 1 hour of contamination.**

**[3]An aircrew spot decon is performed to reduce contact hazards inside the aircraft. Do not spray water inside the aircraft.**

7-10. Air or ground crews may perform the aircrew spot decon. Wash exterior surfaces with decontaminants to flush off contamination. Fuel, soap, and water are most commonly used. Exact procedures and areas to decon are identified in the aircraft's maintenance manual. See Table 7-2, pages 7-4 through 7-6, for guidance on the aircrew spot decon.

### Aircraft Washdown

7-11. An aircraft washdown site consists of a single station; with other decon sites, there may be several stations. Before arriving at the site, the aircrew should fly its aircraft at high speeds to help evaporate exterior contamination. If the aircraft's interior has been contaminated, open the doors; however, before arriving at the washdown site, close the doors. At the washdown site, the aircraft will land and reduce its speed to an idle. If the aircrew is contaminated, it should exit the aircraft and perform a MOPP-gear exchange.

### Table 7-2. Aircraft Spot Decon

| Actions and Locations | Areas to Decon | Personnel Responsible | Decontaminants | Procedures | Remarks |
|---|---|---|---|---|---|
| Refueling at FARP | Fuel ports, hatches, and all areas that FARP personnel touch | POL handler | Diesel fuel; JP8; or hot, soapy water | Wipe fuel ports and hatches with a sponge dipped in the decontaminant. Do not allow the decontaminant to enter the fuel system. Control runoff because the agent will not be neutralized. This method simply flushes contamination from the surface. | FARP personnel should conduct an operational decon before servicing aircraft if they are contaminated. |
| Arming at FARP | Armament system | Ammo handler | JP8 or hot, soapy water | Wipe the armament system with a sponge dipped in the decontaminant. Control runoff because the agent will not be neutralized. This method simply flushes contamination from the surface. | **CAUTION** **Take care to prevent certain areas of the armament system from being exposed to the decontaminants.** The TM on the armament system should be checked to know what areas to avoid. |
| Entering and exiting aircraft anywhere | Door handles, steps, ladders, handholds, and all other areas that the aircraft crew is likely to touch | Crew members and FARP personnel | JP8 or hot, soapy water | Apply the decontaminant. Take care to prevent the spread of liquid contamination from the outside of the aircraft to the inside. This includes controlling the contaminated runoff. This method simply flushes contamination from the surface. Precautions might include having the crew chief do the decon, which reduces the chance of contamination spreading into the pilot and copilot compartment. All crew members should diligently practice NBC avoidance. | Procedures should be developed for each type of aircraft. Before entering the aircraft, use an IEDK on boots and gloves. FARP personnel can decon the area for exiting aircrew members. |

**Table 7-2. Aircraft Spot Decon (Continued)**

| Actions and Locations | Areas to Decon | Personnel Responsible | Decontaminants | Procedures | Remarks |
|---|---|---|---|---|---|
| Preflight and postflight checks anywhere. Maintenance inspections at aviation intermediate maintenance (AVIM) facilities | Areas that must be touched as part of the inspection | Aircrew | Diesel fuel; JP8; or hot, soapy water on exterior surfaces. Use hot air on interiors or areas that are not compatible with decontaminants or liquids. | Wipe areas that are required to be touched for preflight and postflight checks with a sponge dipped in the decontaminant. Wash gloves in a decontaminant before touching uncontaminated surfaces. Decon gloves with an IEDK after the inspection is completed. Control runoff because the agent will not be neutralized. This method simply flushes contamination from the surface. | Overheated air should not be used directly on instrumentation. Crews may want to wear wet-weather gear to keep most of the contamination off overgarments. |
| Repair and recovery anywhere | Parts or areas that will need to be decontaminated depending on the situation | Battle-damage repair team | Diesel fuel; JP8; or hot, soapy water on exterior surfaces. Use hot air on interiors or areas that are not compatible with liquid decontaminants. | Decon only those parts that need to be touched during repair. | Contaminated aircraft should not be brought into an uncontaminated area. |
| Cannibalization anywhere on the battle space | Parts or assemblies being cannibalized | Maintenance personnel (someone who knows what assembly or parts are needed) | DS2; diesel fuel; 5 percent chlorine solution; or hot, soapy water on exterior parts or surfaces. Use hot air on interior parts that are not compatible with liquids or corrosive decontaminants. | Decon only those parts or assemblies that need to be touched during cannibalization. | Caustic decontaminants should be used only on those areas that have been removed from the aircraft. Assemblies must be rinsed thoroughly before replacing. |

**Table 7-2. Aircraft Spot Decon (Continued)**

| Actions and Locations | Areas to Decon | Personnel Responsible | Decontaminants | Procedures | Remarks |
|---|---|---|---|---|---|
| Overhaul at AVIM facilities | All areas and equipment required to be worked on during the overhaul | Maintenance personnel and/ or a chemical decon platoon from division or corps | DS2; diesel fuel; 5 percent chlorine solution; or hot, soapy water. Surfaces that are easily destroyed by liquid or corrosion should be decontaminated using hot air. | Wash with diesel fuel and then hot, soapy water. Then rinse. Check for contamination with the M8 detector paper or the CAM. If time permits, allow the equipment to weather to reduce chemical contamination to negligible risk levels. | Caustic decontaminants should be used only on those assemblies that have been removed from the aircraft. Assemblies must be rinsed thoroughly before replacing. This process has the same results as a deliberate decon. |

To remove the BDU, follow the procedures for removing the protective overgarment as described in paragraphs 8-124 and 8-125. Do not remove the patient's identification tags.

### Step 5 - Decon the Patient's Skin

8-130. **Spot decon.** Wash exposed skin surfaces with soap and warm water, to include areas around the neck, wrists, and lower parts of the face. Decon the patient's identification tags and chain, if necessary.

8-131. **Combat medic care.** The combat medic gently cuts away the bandage. He decontaminates the area around the wound and irrigates it with a 0.5 percent chlorine solution. If bleeding begins, he replaces the bandage with a clean one. He replaces the old tourniquet by placing a new one ½ to 1 inch above the old one. He then removes the old tourniquet and decontaminates the patient's skin with a 0.5 percent chlorine solution. He does not remove a splint. He decontaminates the splint by thoroughly rinsing it, to include the padding and cravats, with a 0.5 percent chlorine solution.

**WARNING**

**Do not use an SDK around any wounds. Do not remove splints. The splint will not be removed until the patient has been evacuated to a forward MTF or hospital.**

8-132. **Completeness of decon check.** Check the patient with M8 detector paper or the CAM for completeness of decon.

**NOTE: Other monitoring devices may be used, if available.**

8-133. **Contaminated-waste disposal.** Dispose of contaminated bandages and coverings by placing them in a contaminated-waste bag. Seal the bag and place it in the contaminated-waste dump.

### Step 6 - Transfer the Patient to the Hot Line

8-134. Move the patient to the hot line. Two medics from the clean side of the hot line move the patient into the clean treatment area.

**NOTE: Before decontaminating another patient, each decon team member drinks about half a quart of water. The exact amount of water consumed is increased or decreased according to the work level and temperature (see Table 8-2, page 8-11).**

## AMBULATORY PATIENT

8-135. The decon procedures for an ambulatory patient that is contaminated with radiation are discussed in the following paragraphs. Some procedures can be done with one soldier, while others require more than one.

8-136. All ambulatory patients requiring EMT or AMT in the clean area of the BAS will be decontaminated. Stable patients not requiring treatment at the BAS, but requiring evacuation to the medical company's clearing station or a corps hospital for treatment (for example, a patient with a broken arm), should be evacuated in their protective overgarments and masks by any

where it will pose a vapor hazard and be difficult, if not impossible, to decon later.

**NOTE: Many (but not all) aircraft have been painted with CARC paint. Aircraft will be painted with CARC paint when sent to the appropriate maintenance shop. Contamination stays on top of CARC surfaces where it can be more easily neutralized or removed.**

7-14. An aircraft washdown should be conducted within 6 hours and is most effective if conducted within 1 hour. Some amounts of chemical contamination may remain after the aircraft washdown. The aircrew should continue to wear, as a minimum, its protective masks and rubber gloves for protection until a deliberate decon can be conducted.

7-15. Presently, there is no timely detection method to check the effectiveness of decon for biological contamination. Medical personnel will monitor the aircrew's disposition if biological contamination is suspected.

### Detailed Aircraft Decon

7-16. The goal of a DAD is to reduce contamination hazards to a negligible risk level so that air and ground crews can operate the aircraft without protective measures. The DTD can be conducted at the same time. The decon site should be set up in a similar fashion to the thorough-decon site (see Chapter 4). The decon site consists of the following areas: a predecon staging area, DAD and DTD areas, and a postdecon AA. For safety reasons, it is not possible to decon aircraft and ground vehicles at the same site. A separate DED site can be established in the vicinity to support the contaminated aviation unit's ground vehicles.

7-17. Aviation units must be supported by a chemical unit to set up and operate the DAD site. The chemical unit, assisted by the supported aviation unit, operates the DAD site while the aviation unit sets up and operates the DTD. See Chapter 4 for DTD operations.

7-18. Typically, a DAD is either performed as part of a reconstitution effort or before maintenance operations (after every 25 hours of flight time). When the aircraft is either overhauled or grounded for scheduled maintenance, contamination hazards should be reduced to negligible risk levels by going through a DAD. When components are removed from the aircraft for repair, care must be taken because of residual contamination. Decon these components either before cannibalization or overhaul.

7-19. Chemical units (decon and smoke/decon) are responsible for setting up, operating, and closing the DAD portion of the thorough-decon operation. The site setup is the same for all types of contamination (chemical, biological, and radiological). The DAD area is composed of the following five stations: primary wash, decon-solution application, contact time/interior decon, rinse, and check (see Figure 7-2).

## DECON STATIONS

7-20. The actions that are taken at each of the stations are as follows:

### Station 1 - Primary Wash

7-21. At this station, gross contamination is removed from the aircraft. Spray the aircraft for 2 to 3 minutes with hot, soapy water. Avoid hitting the

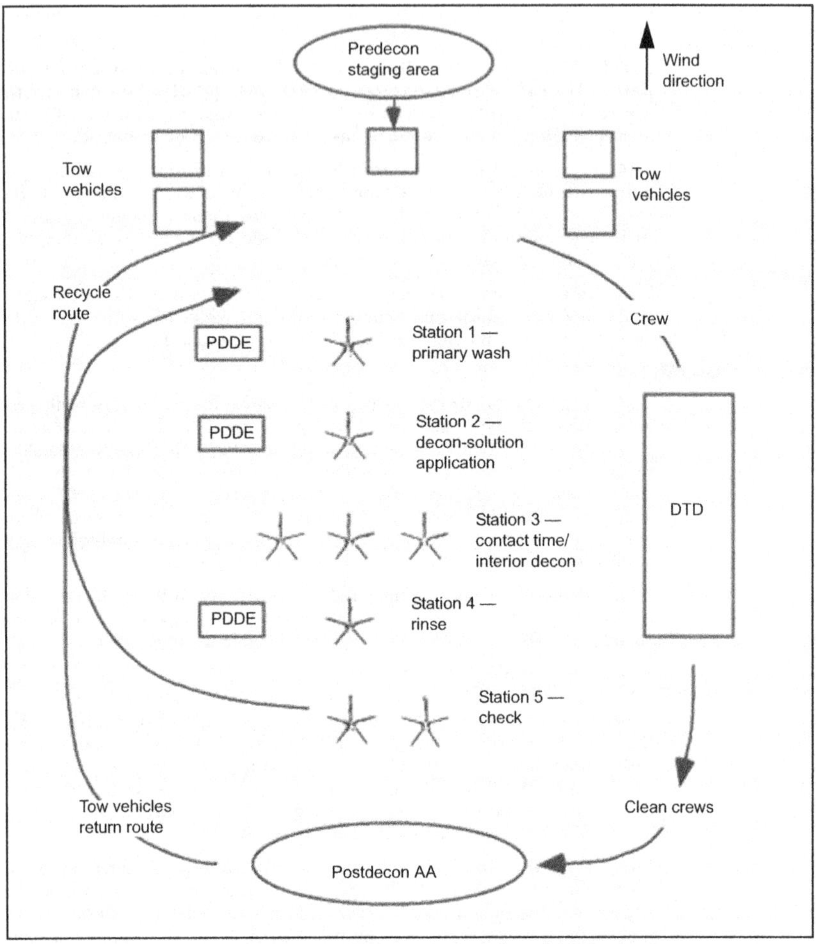


**Figure 7-2. Optimum Layout for a DAD Station**

aircraft's skin at a $90^0$ angle to prevent damage. Also avoid the sensitive areas shown in Figures 7-3 through 7-6, pages 7-10 through 7-13.

### Station 2 - Decon-Solution Application

7-22. At this station, a decon solution is applied using the M12A1 PDDE or the M17 LDS (in the siphinjector mode). After the solution is applied, a member of the scrubbing team scrubs the aircraft's skin. This allows the chemical agent

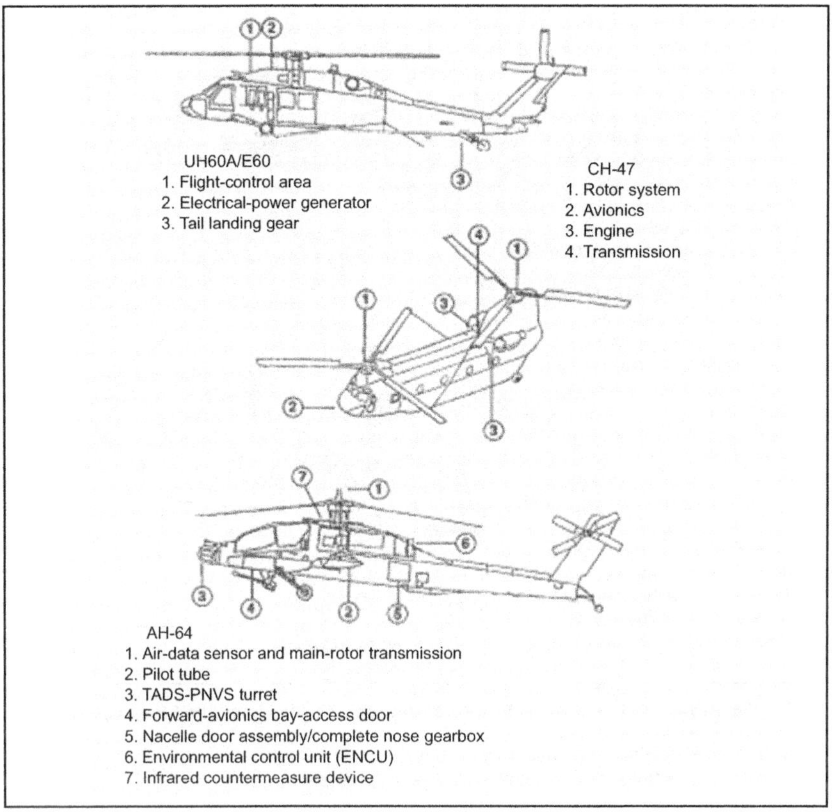


**Figure 7-3. No Direct Water-Pressure Areas for the UH60A/E60, CH-47, and AH-64 Helicopters**

to mix with the decon solution, especially when thickened chemical agents are present. For decontaminants, refer to Appendix B. STB and HTH should not be used because of their corrosive nature.

### Station 3 - Contact Time/Interior Decon

7-23. At this station, the decon solution is allowed to completely neutralize the chemical agent; also, the interior of the aircraft is decontaminated, if necessary. The aircraft will remain in station 3 for no less than 30 minutes. The decon solution reacts with most chemical agents within 5 minutes. However, allowing the decon solution to remain on the contaminated surface for 30 minutes ensures a complete neutralization.

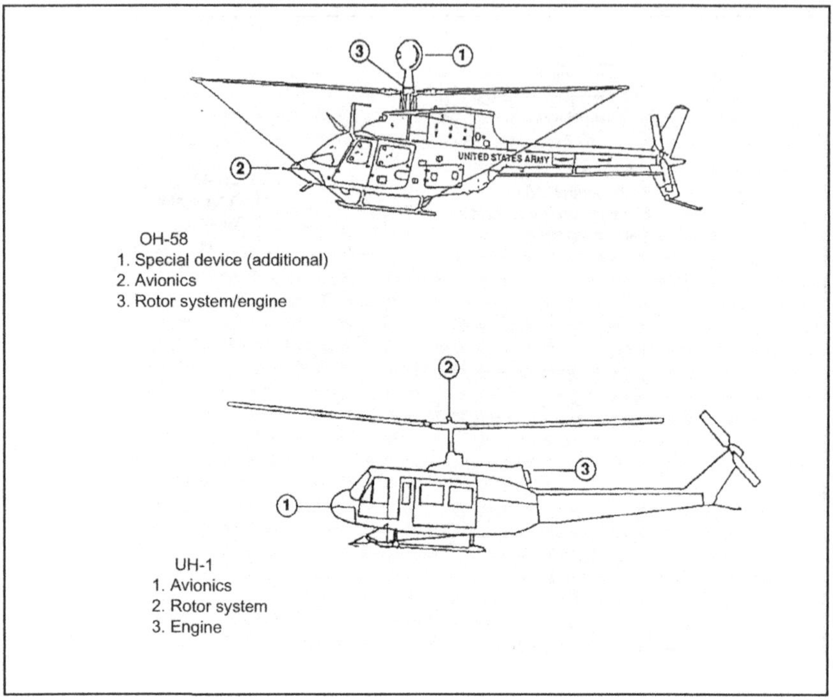


**Figure 7-4. No Direct Water-Pressure Areas for the OH-58 and UH-1 Helicopters**

7-24. While the aircraft is in this station, an attendant inspects the interior of the aircraft for liquid contamination using the CAM and M8 detector paper. If liquid-chemical contamination is identified, he will decon the interior of the aircraft.

| WARNING |
|---|
| **Personnel should not be in the aircraft during this operation.** |

7-25. Decon the aircraft's interior with hot, soapy water as long as maintenance personnel approve the use of it on certain pieces of equipment. Use IEDKs or damp, soapy washcloths to decon optical and electrical equipment and instruments.

**NOTE: The IEDK leaves a charcoal residue on surfaces.**

7-26. Field studies show the Herman-Nelson portable duct heaters to be effective in decontaminating chemically contaminated vehicle and aircraft

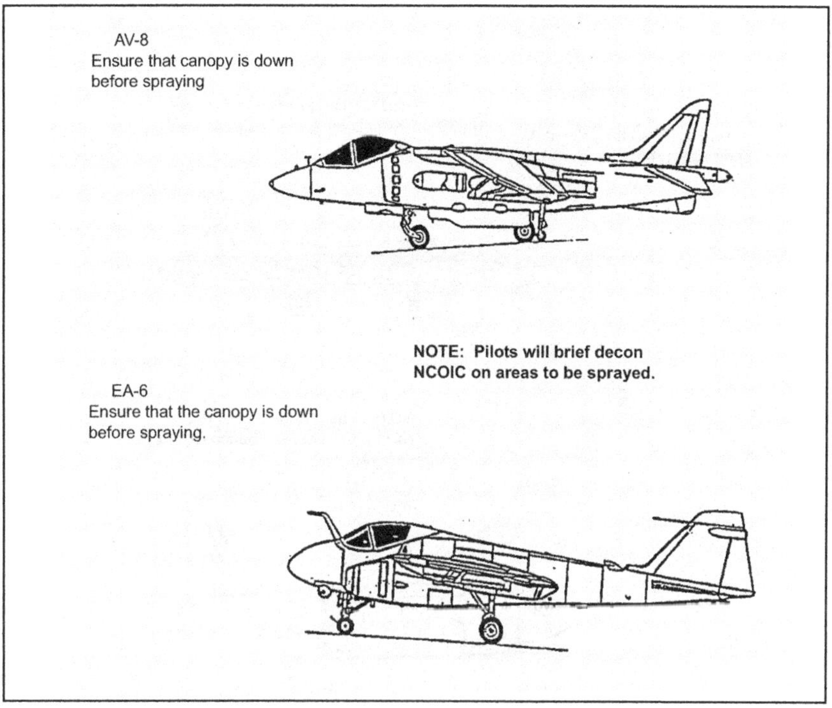


**Figure 7-5. No Direct Water-Pressure Areas for the AV-8 and EA-6**

interiors. However, using a hot-air decon has the following drawbacks: time required, heat damage, and downwind vapor hazards. A hot-air decon for a helicopter or light aircraft requires about 30 to 120 minutes to remove all of the contamination. Times are based on the amount of contamination and interior surface materials. Enough time must be provided to get the contaminated surface hot enough to vaporize the agent. The Herman-Nelson heaters discharge air at a temperature between 150$^{0}$ and 280$^{0}$F. Pressurized containers, such as fire extinguishers, must be removed before starting a hot-air decon. These items will fail catastrophically and could cause severe damage. The heat will also damage sensitive electronic equipment. Additionally, chemical agents irreversibly bind to many materials, particularly gaskets, vinyl coatings, seat belts, and seat cushions. These items should be removed since they cannot be decontaminated. A hot-air decon does not destroy chemical agents. The air exiting the aircraft may contain hazardous levels of the agent. A downwind exclusion area of about 164 yards must be maintained to prevent unwanted casualties.

7-27. While a hot-air decon is an alternate to using hot, soapy water or an IEDK, the technique has many drawbacks. Maintenance personnel must be

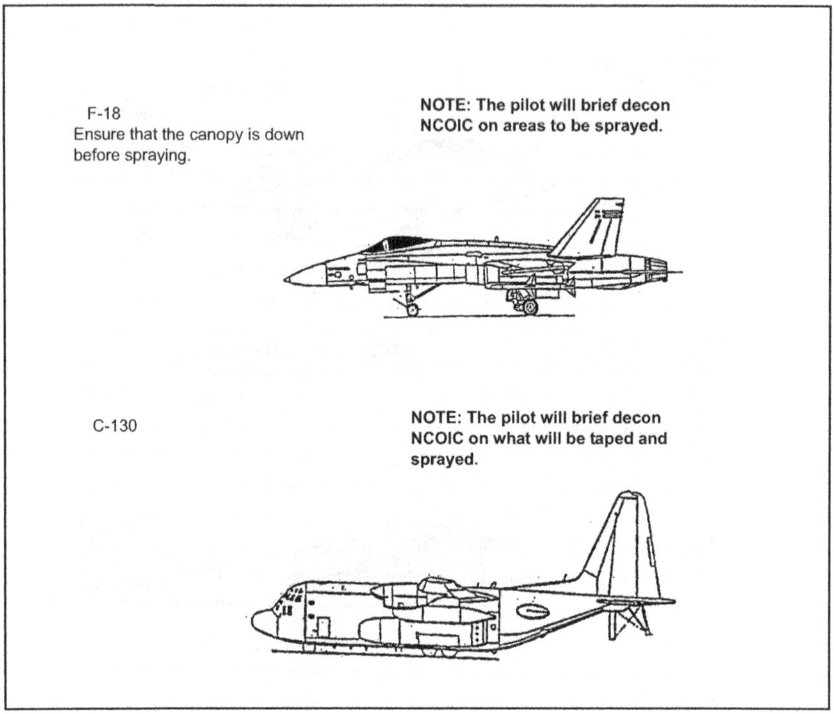


**Figure 7-6. No Direct Water-Pressure Areas for the F-18 and C-130**

consulted before using this technique to ensure that the aircraft is not damaged by the heat.

7-28. For radiological contamination, use a radiac meter to determine the extent and location of contamination inside the aircraft. If there is contamination, determine the intensity of the contamination. If the contamination has an intensity greater than 0.33 cGy, the interior of the aircraft must be decontaminated. Use hot, soapy water to wash the contaminated areas and a sponge to mop up the water and the contamination.

**Station 4 - Rinse**

7-29. At this station, the decon solution is removed from the aircraft. Spray the aircraft with water from top to bottom. Take care not to damage the aircraft's skin. This station will use about 250 gallons of water. Failure to remove all decon solution from the aircraft's skin may cause corrosion.

### Station 5 - Check

7-30. At this station, the completeness of the decon is checked. Determine if the aircraft has a negligible risk or if it still has significant contamination remaining. Detection procedures vary depending on the type of contamination. If significant contamination is found on the aircraft, it will be recycled to station 2 for chemical contamination or station 1 for radiological contamination.

7-31. **Chemical.** Use the CAM to check for the presence of vapors. A one bar or lower reading on the CAM indicates a negligible risk. If the CAM indicates the presence of vapors, use M8 detector paper to check for liquid contamination. If it is suspected that these are producing a false positive, use an M256A1 detector kit to confirm or deny the presence of contamination. If the aircraft has significant contamination remaining, recycle it. Desorption of chemical agents may occur after the decon. On CARC-painted surfaces, the desorption of vapors will stop sooner than on alkyd-painted surfaces.

7-32. **Radiological.** Use the radiac meter to determine if any contamination remains. If so, determine the intensity of the contamination. If the contamination is greater than 0.33 cGy, recycle the aircraft to station 1.

### RECYCLE CRITERIA

7-33. The commander, with the chemical unit leader's help, establishes the recycle criteria before starting a decon operation. If contamination is detected at station 5, the recycle criteria is used to determine which aircraft are returned to station 2 or, if radiologically contaminated, station 1. If the unit has sufficient time and resources available, any aircraft having more contamination than a negligible risk should be recycled. However, time and resources are usually limited and not all aircraft can be recycled.

## AIRCRAFT DECON-SITE SELECTION

7-34. Operational- and thorough-decon sites for aircraft must be selected with care. The decon area must be big enough for the required number of aircraft, have a readily available water source, and allow for adequate drainage. The site should be relatively secure; however, it should be close enough to the AOs and FARP to allow for a reasonably quick turnaround of aircraft. The site must not have less than a 5 percent slope. Decon sites should be integrated into the tactical plan.

## DECON-SITE LAYOUT

7-35. The equipment and personnel requirements for a DAD layout are identified in Table 7-3, page 7-16. Chemical units can establish other layouts based on METT-TC.

7-36. It may not be possible, for a variety of reasons, for a chemical platoon to use the optimum layout. Limited personnel or equipment will affect the layout of the thorough aircraft decon site. The platoon leader will use METT-TC and the concepts outlined in this chapter to establish a DAD station within his capabilities.

**WARNING**

**Chemical unit leaders must consider the impact of the work/rest cycle on their operation to process aircraft. Failure to initiate a work/rest cycle could result in heat casualties and mission failure.**

**Table 7-3. Personnel and Equipment Requirements for an Optimum DAD Layout**

| Stations | Personnel | | Equipment/Supplies |
|---|---|---|---|
| | Decon Platoon | Augmenter | |
| Station 1 - primary wash | 1 squad leader<br>1 PDDE operator<br>2 sprayers | 2 scrubbers | 1 PDDE<br>1 3,000-gallon tank<br>2 65-GPM pumps<br>6 long-handled brushes<br>5 TAP aprons<br>Liquid detergent |
| Station 2 - decon-solution application | 1 squad leader<br>2 appliers<br>2 sprayers<br>1 PDDE operator | 4 appliers | 1 PDDE<br>18 long-handled brushes<br>9 mops with extra heads<br>Liquid detergent |
| Station 3 - contact time/interior decon | 1 NCO (CAM operator) | 2 interior decon assis-tants | 2 AN/VDR-2 or AN/PDR-77<br>3 TAP aprons<br>6 30-gallon containers<br>10 books of M8 detector paper<br>30 sponges<br>8 M256A1 detector kits<br>50 plastic trash bags<br>1 clipboard with a pen<br>1 stopwatch<br>1 CAM |
| Station 4 - rinse | 1 squad leader<br>1 PDDE operator<br>2 sprayers | | 1 PDDE<br>1 3,000-gallon tank<br>3 65-GPM pumps<br>1 TPU<br>2 TAP aprons |
| Station 5 - check | 1 NCO<br>2 CAM operators | | 1 CAM<br>10 M256A1 detector kits<br>20 books of M8 detector paper<br>2 AN/VDR-2 or AN/PDR-77<br>2 M8A1 or M22 ACAA |
| $C^2$ | 1 platoon leader<br>1 platoon sergeant | | 1 HMMWV<br>3 marking kits |
| Aircraft moving team | | 6 drivers<br>18 ground guides | |
| Total personnel | 20 | 32 | |

## Chapter 8

# Patient Evacuation and Decon

Evacuating patients under NBC conditions forces the unit commander to consider to what extent he will commit evacuation assets to enter the contaminated area. Generally, if most or all of a supported force is operating in a contaminated area, most or all of the medical-evacuation assets will operate there also. If it is possible to keep some ambulances free from contamination, every effort should be made to do so.

## EVACUATION CONSIDERATIONS

8-1. On the modern battle space, there are three basic modes of evacuating patients (personnel, ground vehicles, and aircraft). Personnel who physically carry the patients incur a great deal of inherent stress. Cumbersome MOPP gear, climate, increased workloads, and fatigue will greatly reduce the effectiveness of unit personnel.

8-2. If evacuation personnel are to be sent into an area that is contaminated with radiation, an OEG must be established. Radiation-exposure records must be maintained by the supported unit chemical NCO and made available to the commander, staff, and medical leader. Based on the OEG, the commander or medical leader decides which evacuation elements to send into the contaminated environment. Every effort is made to limit the number of evacuation assets that are contaminated; however, a number of these assets will become contaminated in the course of a battle. Therefore, optimize the use of resources, medical and nonmedical, which are already contaminated before employing uncontaminated resources.

8-3. Once a vehicle has entered a contaminated area, it is highly unlikely that it can be spared long enough to undergo a complete decon. This will depend on the contaminant, the tempo of the battle, and the resources available for casualty evacuation. Normally, contaminated vehicles (air and ground) will be confined to dirty environments.

8-4. Ground ambulances should be used instead of air ambulances in contaminated areas. They are more plentiful, are easier to decon, and can be replaced more easily. However, this does not preclude the use of aircraft, if required.

8-5. The patient's medical condition and the relative positions of the contaminated area, the forward line of troops (FLOT), and the threat's air-defense systems determine if and where air ambulances may be used in the evacuation process. One or more air ambulances may be restricted to the contaminated areas. To the greatest extent possible, use ground vehicles to cross the line separating the contaminated and clean areas.

8-6. After the patients are loaded, the medical-evacuation vehicles proceed to a medical-treatment facility (MTF) with a patient-decon station. (For the setup of a patient-decon station, see FM 8-10-7.) The patients are decontaminated and treated at this station. If further evacuation is required, transfer the patients to a clean ground or air ambulance. The routes that the ground vehicles use to cross between contaminated and clean areas should be considered dirty routes and should not be crossed by clean vehicles. The effects of the wind and time on the contaminants must be considered.

8-7. The rotor wash of the helicopters must always be kept in mind when evacuating patients. The intense winds will disturb the contaminants, increasing vapor hazards.

8-8. The helicopter should be allowed to land and reduce to a flat pitch before the patients are brought near. Additionally, the helicopter must not land too close to a decon station because any trace of contaminants in the rotor wash may be spread into the clean treatment area.

8-9. Evacuating patients must continue even under NBC conditions. The medical leader must recognize the constraints that NBC operations place on him, and he must plan and train to overcome these constraints.

## PATIENT DECON

8-10. Patient decon presents special problems for units and combat-health-support (CHS) personnel. Under NBC conditions, contaminated wounded soldiers create increased hazards to rescuers and CHS personnel. In the following paragraph, decon procedures at unit level are discussed:

8-11. On the NBC battle space, two classifications of patients will be encountered—contaminated and uncontaminated. Those contaminated may suffer from the effects of an NBC agent, a conventional wound, or both. For more information on the treatment of NBC patients, see FM 8-9. Some may suffer battle fatigue or heat injuries induced by the stress of NBC conditions and the extended time spent in MOPP4. It is important to follow proper decon procedures to limit the spread of contamination. The most important decon is performed at the contamination site. Later decon may be too late to prevent injury. All agents should be promptly removed from the skin.

8-12. Patient decon must begin at the platoon and company level with the individual soldier. The soldier himself or members of his team must perform immediate decon and administer nerve-agent antidotes and convulsant antidotes, if required. Tag the patient with a DD Form 1380 or a field-expedient tag, noting the time and type of contamination (see Figure 8-1). If available, use the CAM or M8/M9 detector paper to determine the type and concentration of contamination. When the patient's condition and the battle permits, the patient may go through a MOPP-gear exchange (see Chapter 3 of this manual). The MOPP-gear exchange must not cause further injury to the patient.

## PATIENT DECON AT AN MTF

8-13. In the following paragraphs, the types of MTFs are discussed:

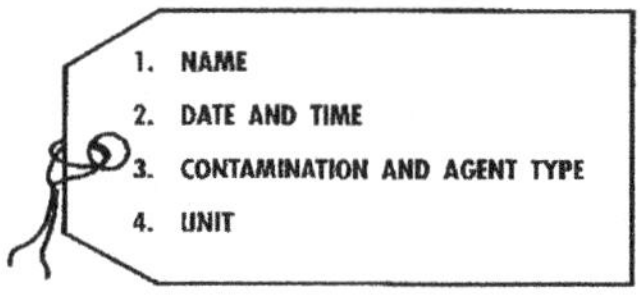


**Figure 8-1. Field-Expedient NBC Patient Tag**

## BATTALION AID STATION (BAS)

8-14. When battle conditions prevent decon procedures forward, the patient may have to be evacuated to the BAS before decon. Contaminated patients arriving at the MTF must be decontaminated before admission into the clean treatment area.

8-15. Patient decon is the systematic removal of clothing and contaminants from a patient who is unable to decon himself. Patient decon is done by an 8-man, patient-decon team consisting of nonmedical personnel from the supported unit at the BAS. The team operates under the supervision of medical personnel to ensure that no further injury is caused to the patient by the decon process.

## MEDICAL COMPANY'S CLEARING STATION

8-16. The medical company's clearing station may receive patients from the BAS or directly from other areas who may be contaminated. The clearing station must also have a patient-decon area. For more information on BAS operations under NBC conditions, see FMs 8-10-4 and 8-10-7. As with the BAS, the clearing station must have at least an 8-man, patient-decon team consisting of nonmedical personnel from the supported units to perform the decon. Procedures for patient decon at the clearing station are the same as for the BAS.

## HOSPITAL

8-17. To the maximum extent possible, hospitals are located away from tactical or logistical targets. Patients evacuated from forward areas should already be decontaminated; however, contaminated patients may arrive from forward MTFs and units located within the geographical area of the hospital and require decon. Patient decon is done by 20 nonmedical personnel from units located in the geographical area/base cluster of the hospital.

8-18. If the hospital does not have CPSs and becomes contaminated with a persistent agent, patients are rerouted to other hospitals. All inpatients are evacuated, if possible, and the hospital is decontaminated. On completion of the decon, the hospital will return to normal operations.

8-19. A hospital with CPSs will decon the areas around the entry to these facilities, then continue receiving and caring for patients. Forward medical facilities and hospitals use the same patient-decon procedures. Several patient-decon stations can be operated at a hospital decon site. All patients arriving at the hospital will be considered contaminated. They must be decontaminated before being admitted into the clean areas. Perform decon as required.

# CHLORINE-SOLUTION PREPARATION FOR PATIENT DECON

8-20. To decon a patient, use an SDK; however, if an SDK is not available, use a chlorine solution. If the chlorine solution is used, two concentrations of it are required. A 5 percent chlorine solution is required to decon gloves, aprons, litters, scissors, patient's hood, and other nonskin contact areas. A 0.5 (½) percent chlorine solution is required to decon the patient's mask, skin, and splints and to irrigate his wounds. To prepare the solutions, use calcium hypochlorite granules or sodium hypochlorite (household bleach) (see Table 8-1).

**Table 8-1. Preparation of Chlorine Solution for Patient Decon**

| HTH Ounces | HTH Spoonfuls* | Household Bleach | Solution Percentage in 5 Gallons of Water |
|---|---|---|---|
| 6 | 5 | 2 quarts | 0.5 |
| 48 | 40 | ** | 5 |

*These measurements are used when bulk HTH is used. To measure this preparation, use the plastic spoon supplied with your meal, ready-to-eat (MRE). Use a heaping spoonful of chlorine (all that the spoon will hold). Do not shake any granules off the spoon before adding to the water.

**Do not dilute in water because household bleach is about a 5 percent solution.

**NOTE: HTH is supplied in a 6-ounce jar in the chemical-agent patient's treatment and decon medical-equipment set.**

**CAUTION**

Only use a 0.5 percent chlorine solution on the patient's skin. Avoid vigorous scrubbing because it can force the agent into the skin.

# CHEMICAL-AGENT PATIENT DECON PROCEDURES

8-21. Decon procedures for chemical-agent patients are discussed in the following paragraphs:

## LITTER PATIENT

8-22. Before patients receive medical treatment in the clean treatment area, the 8-man patient-decon team decontaminates them. Figure 8-2 shows one way to establish the patient-decon station. Place the bandage scissors in a container of 5 percent chlorine solution between each use. The team members decon their gloves and aprons with a 5 percent chlorine solution.

**NOTE: Litter patients requiring emergency medical treatment (EMT) or ambulatory medical treatment (AMT) in the clean area of the MTF will be completely decontaminated. However, a patient requiring immediate evacuation should have only his wound area and MOPP gear spot decontaminated to remove any gross contamination (for example, a stable patient with a partial amputation of a lower extremity). The patient should be evacuated in his MOPP gear.**

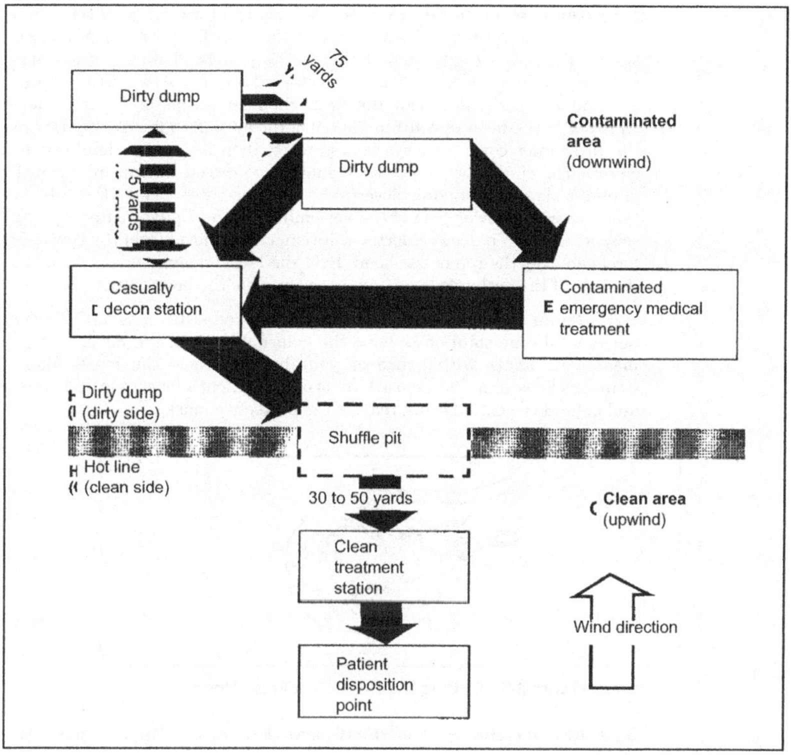


**Figure 8-2. Layout for a Patient-Decon Station and a Clean Treatment Area Without a CPS**

8-23. Decon the patient's skin, bandages, wounds, mask, identification tags with chain, and splints by wiping them with a 0.5 percent chlorine solution. For treatment procedures, refer to FMs 8-9, 8-33, and 8-285.

8-24. Some procedures in the following steps can be done with one soldier, while others require more than one soldier.

### Step 1 - Decon the Patient's Mask and Hood

8-25. **Move the patient to the clothing removal station.** After the patient has been triaged and stabilized (if necessary) by the senior medic in the patient-decon area, move him to the litter stands at the clothing removal station.

8-26. **Decon the hood.** Use either an IEDK or a 5 percent chlorine solution (or household bleach) to wipe down the front, sides, and top of the hood.

8-27. **Remove the hood.** Remove the hood by cutting it with scissors or by loosening it from the mask attachment points for the quick-doff hood or other similar hoods (see Figure 8-3). Before cutting the hood, dip the scissors in a 5 percent chlorine solution. Cut the neck cord and the small string under the voicemitter. Release or cut the hood shoulder straps and unzip the hood zipper. Cut the hood upward to the top of the eye-lens outsert, staying close to the filter-inlet cover and eye-lens outsert, then across the forehead to the outer edge of the other eye-lens outsert. Proceed downward toward the patient's shoulder, staying close to the eye lens outsert and filter-inlet cover, then across the lower part of the voicemitter to the zipper. After dipping the scissors in the 5 percent chlorine solution, cut the hood from the center of the forehead over the top of the head. Fold the left and right sides of the hood to the side of the patient's head, laying the sides of the hood on the litter.

8-28. **Decon the protective mask and exposed skin.** Use an SDK or a 0.5 percent chlorine solution to wipe the external parts of the mask. Cover the mask's air inlets with gauze or your hand to keep the mask filter dry. Continue by wiping the exposed areas of the patient's face, to include the neck and behind the ears. Do not remove the protective mask.

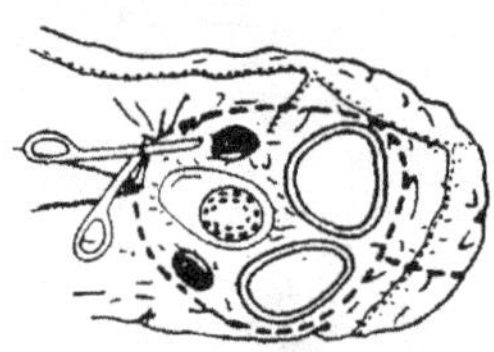

**Figure 8-3. Cutting the Protective Mask Hood**

8-29. **Remove the field medical card (FMC).** Cut the patient's FMC tie wire, allowing the FMC to fall into a plastic bag. Seal the plastic bag and rinse the outside of the bag with a 5 percent chlorine solution. Place the plastic bag under the back of the protective mask's head straps. The FMC will remain with the patient in the contaminated area and a clean copy will be made before the patient is moved to the clean area.

### Step 2 - Remove Gross Contamination From the Patient's Protective Overgarment

8-30. Remove all visible contamination spots from the overgarment with an SDK or a 5 percent chlorine solution.

### Step 3 - Remove the Patient's Personal Effects and Protective Overgarment

8-31. **Remove the patient's personal effects.** Remove the patient's personal effects from his protective overgarment and BDU pockets. Place the articles in a plastic bag, label the bag with the patient's identification, and seal the bag. If the articles are not contaminated, they are returned to the patient. If the articles are contaminated, place them in the contaminated holding area until they can be decontaminated, then return them to the patient.

**CAUTION**

Bandages may have been applied to control severe bleeding; therefore, they are treated like tourniquets. Only medical personnel will remove bandages, tourniquets, and splints.

**NOTE: The overgarment jacket and trousers will be cut simultaneously. One soldier cuts the jacket, while another soldier cuts the trousers.**

8-32. **Cut and remove the overgarment jacket.** Before cutting the overgarment jacket and trousers, dip the scissors in a 5 percent chlorine solution to prevent contamination of the patient's BDU or undergarments. Make two cuts, one up each sleeve from the wrist up to the shoulder and then across the shoulder through the collar (Figure 8-4). Cut around bandages, tourniquets, and splints, leaving them in place. Do not allow your gloves to touch the patient along the cut line. Keep the cuts close to the inside of the arms so that most of the sleeve material can be folded outward. Unzip the jacket and roll the chest sections to the respective sides, with the inner surface outward. Continue by tucking the clothing between the arm and chest. Roll the cut sleeves away from the arms, exposing the black liner.

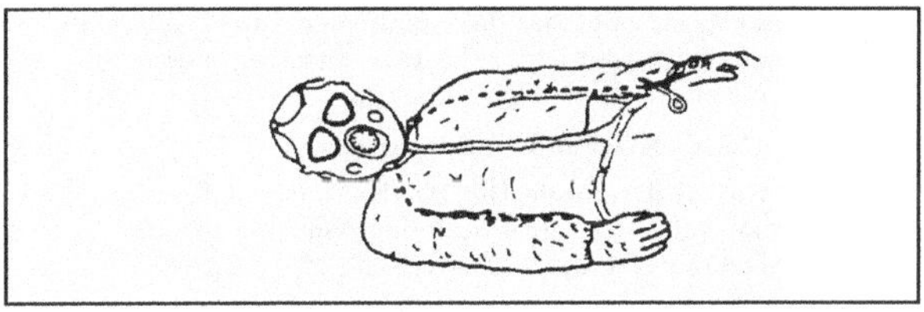

**Figure 8-4. Cutting the Overgarment Jacket**

8-33. **Cut and remove the overgarment trousers.** Cut both trousers legs starting at the ankle as shown in Figure 8-5, page 8-8. Keep the cuts near the inside of the legs, along the inseam, to the crotch. With the left leg, cut all the way to the waist, avoiding the pockets. With the right leg, cut across at the crotch to the left leg cut. Cut around bandages, tourniquets, and splints, leaving them in place. Place the scissors in a 5 percent chlorine solution. Fold the cut trouser halves away from the patient and allow the halves to drop to the litter with the contaminated (green) side down. Roll the inner leg portion under and between the legs.

8-34. **Remove the outer gloves.** Before touching the patient, the patient-decon team decontaminates its gloves with a 5 percent chlorine solution. Lift the patient's arms up and out of the cutaway sleeves unless detrimental to his condition. Grasp the fingers of the gloves, roll the cuffs over the fingers, and turn the gloves inside out. Do not remove the inner cotton gloves at this time. Carefully lower the patient's arms across the chest after the outer gloves have been removed (Figure 8-6, page 8-8). Do not allow the patient's arms to come into contact with the exterior of his overgarment. Drop his gloves into the

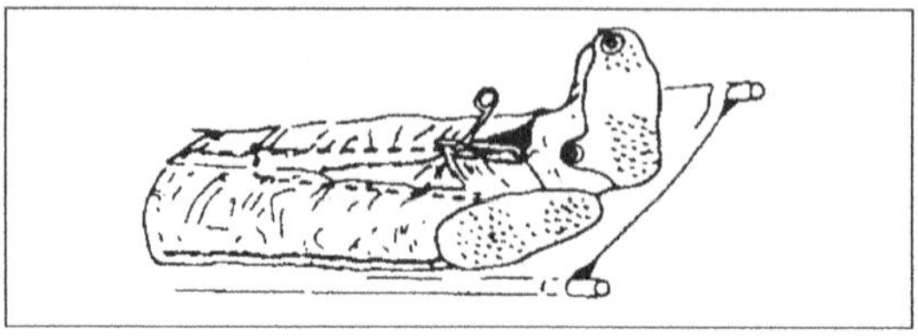

Figure 8-5. Remove the Overgarment Trousers

contaminated-waste bag. The team members decon their gloves with the 5 percent chlorine solution.

8-35. **Remove the overboots.** Cut the overboot laces and fold the lacing eyelets flat outward. While standing at the foot of the litter, hold the patient's heel with one hand. Pull the overboot downward, then toward you to remove it. Remove the two overboots simultaneously. This reduces the likelihood of contaminating one of the combat boots.

8-36. While holding the patient's heels off the litter, have a team member wipe the end of the litter with a 5 percent chlorine solution to neutralize any contamination that was transferred to the litter from the overboots. Lower the patient's heels onto the decontaminated litter. Place the overboots in the contaminated-waste bag. The team members decon their gloves with the 5 percent chlorine solution.

## Step 4 - Remove the Patient's BDU and Undergarments

8-37. **Cut and remove the BDU.** To cut and remove the BDU jacket and trousers, follow the procedures for removing the protective overgarment as described in paragraph 8-32 and 8-33.

8-38. **Remove the combat boots.** Cut the bootlaces along the tongue. Remove the boots by pulling them toward you. Place the boots in the contaminated-waste bag. Do not touch the patient's skin with your contaminated gloves when removing his boots.

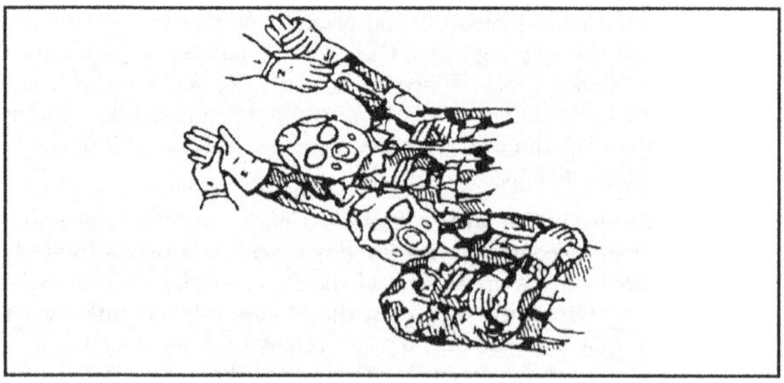

Figure 8-6. Removing the Outer Gloves and Positioning the Arms After the Glove Removal

8-39. **Cut and remove the undergarments.** Follow the procedures for cutting away the protective overgarment and rolling it away from the patient (see paragraph 8-32 and 8-33). If the patient is wearing a brassiere, cut it between the cups. Cut both shoulder straps where they attach to the cups and lay them back off of the shoulders. Remove the socks and cotton gloves. Do not remove the identification tags.

### Step 5 - Transfer the Patient to a Decon Litter

8-40. After the patient's clothing has been cut away, transfer him to a decon litter or a canvas litter with a plastic-sheeting cover. Three decon team members decon their gloves and aprons with a 5 percent chlorine solution. One member places his hands under the patient's legs at the thighs and Achilles tendons, a second member places his arms under the patient's back and buttocks, and a third member places his arms under the patient's shoulders and supports the head and neck. They carefully lift the patient using their knees (not their backs) to minimize back strain. While the patient is elevated, another decon team member removes the litter from the litter stands and replaces it with a decontaminated (clean) litter. The team members carefully lower the patient onto the clean litter. The clothing and overgarments are placed in a contaminated-waste bag and moved to the contaminated-waste dump. The dirty litter is rinsed with the 5 percent chlorine solution and placed in the litter storage area.

### Step 6 - Decon the Patient's Skin

8-41. **Spot decon.** With the patient in a supine position, spot decon the skin by using an SDK or a 0.5 percent chlorine solution. Decon areas of potential contamination, to include areas around the neck, wrists, and lower parts of the face. Decon the patient's identification tags and chain, if necessary.

**NOTE: A complete body wash is not appropriate and may be harmful to the patient. During a complete body wash, the patient would have to be rolled over to reach all areas of the skin. This is not necessary for an adequate decon.**

8-42. **Combat medic care.** The combat medic gently cuts away the bandage. He decontaminates the area around the wound and irrigates it with a 0.5 percent chlorine solution. If bleeding begins, he replaces the bandage with a clean one. He replaces the old tourniquet by placing a new one ½ to 1 inch above the old one. He then removes the old tourniquet and decontaminates the patient's skin with an SDK or a 0.5 percent chlorine solution. He does not remove a splint. He decontaminates the splint by thoroughly rinsing it, to include the padding and cravats, with a 0.5 percent chlorine solution.

**WARNING**

**Do not use an SKD around any wounds. Do not remove splints. Splints will not be removed until the patient has been evacuated to a forward MTF or hospital.**

8-43. **Completeness of decon check.** Check the patient with M8 detector paper or the CAM for completeness of decon.

**NOTE: Other monitoring devices may be used, if available.**

8-44. **Contaminated-waste disposal.** Dispose of contaminated bandages and coverings by placing them in a contaminated-waste bag. Seal the bag and place it in the contaminated-waste dump.

### Step 7 - Transfer the Patient Across the Shuffle Pit

8-45. The patient's clothing has been cut away and his skin, bandages, and splints have been decontaminated. Transfer the patient to the shuffle pit and place the litter on the litter stands. The shuffle pit is wide enough to prevent the decon team members from straddling it while carrying the litter. A third team member will assist with transferring the patient to a clean treatment litter in the shuffle pit. Decon personnel rinse or wipe down their aprons and gloves with a 5 percent chlorine solution.

8-46. The three team members lift the patient off the decon litter (see step 5 for lifting procedures). While the patient is elevated, another team member removes the litter from the stands and returns it to the decon area. A medic from the clean side of the shuffle pit replaces the litter with a clean one. The patient is lowered onto the clean litter. Two medics from the clean side of the shuffle pit move the patient to the clean treatment area. The patient is treated in this area or awaits processing into the CPS. The litter is wiped down with a 5 percent chlorine solution in preparation for reuse. Once the patient is in the air lock of the CPS and the air lock has been purged, his protective mask is removed. Place the mask in a plastic bag and seal it.

**NOTE: Before decontaminating another patient, each decon team member drinks about half a quart of water. The exact amount of water consumed is increased or decreased according to the work level and temperature (see Table 8-2).**

## AMBULATORY PATIENT

8-47. All ambulatory patients requiring EMT or AMT in the clean treatment area of the BAS will be decontaminated. Stable patients not requiring treatment at the BAS, but requiring evacuation to the medical company's clearing station or a corps hospital for treatment (for example, a patient with a broken arm), should be evacuated in their protective overgarments and masks by any available transportation. However, before evacuation, spot remove all thickened agents from their protective clothing.

**NOTE: Place the bandage scissors that are used in this procedure in a container of 5 percent chlorine solution when not in use. Most ambulatory patients will be treated in the contaminated treatment area and returned to duty. Upon removal of an ambulatory patient's clothing, he becomes a litter patient. The BAS and clearing station do not have clothing to replace those that are cut off during the decon process. The patient must be placed in a patient protective wrap (PPW) for protection during evacuation (Figure 8-7, page 8-12).**

8-48. The ambulatory patient is decontaminated and undressed as follows. Some procedures in the following steps can be done with one soldier, while others require more than one.

Table 8-2. Heat-Injury Prevention and Water Consumption

| Criteria | | | Controls | |
|---|---|---|---|---|
| Heat | | Water Intake Quart/Hour | Physical Activity for Soldiers/Trainees | |
| Condition | WBGT Index °F* | | Acclimatized to Work/Rest** | Not Acclimatized to Work/Rest |
| White | 78-81.9 | At least ½ | Continuous | |
| Green | 82-84.9 | At least ½ | 50/10 minutes | Use discretion in planning heavy exercise. |
| Yellow | 85-87.9 | At least 1 | 45/15 minutes | Suspend strenuous exercise during the first 3 weeks of training. Training activities may be continued on a reduced scale after the second week of training. Avoid activity in the direct sun. |
| Red | 88-89.9 | At least 1½ | 30/30 minutes | Curtail strenuous exercise for all personnel with less than 12 weeks of hot-weather training. |
| White | 78-81.9 | At least ½ | Continuous | |
| Green | 82-84.9 | At least ½ | 50/10 minutes | Use discretion in planning heavy exercise. |
| Black | 90 and up | More than 2 | 20/40 minutes | Suspend physical training and strenuous exercise. Essential operational commitments (not for training), where the risk of heat casualties may be warranted, are excluded from suspension. Enforce water intake to minimize expected heat injuries. |

*MOPP gear or body armor adds $10^{0}$F to the wet-bulb gradient-temperature (WBGT) index.
**An acclimatized soldier is one who has worked in the given heat condition for 10 to 14 days.

**NOTE: "Rest" means minimal physical activity. Rest should be accomplished in the shade, if possible. Any activity requiring only minimal physical activity can be performed during rest periods (for examples, training by lecture or demonstration, minor maintenance procedures on vehicles or weapons, and personal-hygiene activities such as skin and foot care).**

### Step 1 - Remove the LCE

8-49. Remove the LCE by unfastening/unbuttoning all connectors or tie straps and then place the equipment in a plastic bag. Place the plastic bag in the designated storage area for later decon.

### Step 2 - Decon the Patient's Mask and Hood

8-50. **Begin the clothing removal process.** After the patient has been triaged and treated (if necessary) by the senior medic in the patient-decon station, the clothing removal process begins.

8-51. **Decon the hood.** Use either an IEDK or a 5 percent chlorine solution (or household bleach) to wipe down the front, sides, and top of the hood.

8-52. **Remove the hood.** Remove the hood by cutting it with scissors or by loosening it from the mask attachment points for the quick-doff hood or other similar hoods (see Figure 8-3, page 8-6). Before cutting the hood, dip the scissors in a 5 percent chlorine solution. Cut the neck cord and the small

To remove the BDU, follow the procedures for removing the protective overgarment as described in paragraphs 8-124 and 8-125. Do not remove the patient's identification tags.

### Step 5 - Decon the Patient's Skin

8-130. **Spot decon.** Wash exposed skin surfaces with soap and warm water, to include areas around the neck, wrists, and lower parts of the face. Decon the patient's identification tags and chain, if necessary.

8-131. **Combat medic care.** The combat medic gently cuts away the bandage. He decontaminates the area around the wound and irrigates it with a 0.5 percent chlorine solution. If bleeding begins, he replaces the bandage with a clean one. He replaces the old tourniquet by placing a new one ½ to 1 inch above the old one. He then removes the old tourniquet and decontaminates the patient's skin with a 0.5 percent chlorine solution. He does not remove a splint. He decontaminates the splint by thoroughly rinsing it, to include the padding and cravats, with a 0.5 percent chlorine solution.

**WARNING**

**Do not use an SDK around any wounds. Do not remove splints. The splint will not be removed until the patient has been evacuated to a forward MTF or hospital.**

8-132. **Completeness of decon check.** Check the patient with M8 detector paper or the CAM for completeness of decon.

**NOTE: Other monitoring devices may be used, if available.**

8-133. **Contaminated-waste disposal.** Dispose of contaminated bandages and coverings by placing them in a contaminated-waste bag. Seal the bag and place it in the contaminated-waste dump.

### Step 6 - Transfer the Patient to the Hot Line

8-134. Move the patient to the hot line. Two medics from the clean side of the hot line move the patient into the clean treatment area.

**NOTE: Before decontaminating another patient, each decon team member drinks about half a quart of water. The exact amount of water consumed is increased or decreased according to the work level and temperature (see Table 8-2, page 8-11).**

## AMBULATORY PATIENT

8-135. The decon procedures for an ambulatory patient that is contaminated with radiation are discussed in the following paragraphs. Some procedures can be done with one soldier, while others require more than one.

8-136. All ambulatory patients requiring EMT or AMT in the clean area of the BAS will be decontaminated. Stable patients not requiring treatment at the BAS, but requiring evacuation to the medical company's clearing station or a corps hospital for treatment (for example, a patient with a broken arm), should be evacuated in their protective overgarments and masks by any

articles in a plastic bag, label the bag with the patient's identification, and seal the bag. If the articles are not contaminated, they are returned to the patient. If the articles are contaminated, place them in the contaminated holding area until they can be decontaminated, then return them to the patient.

8-57. **Remove the overgarment jacket.** Have the patient stand with his feet spread apart at shoulder width. Unsnap the front flap of the jacket and unzip the jacket. If the patient can extend his arms, have him clinch his fist and extend his arms backward at about a 30° angle. Move behind the patient, grasp his jacket collar at the sides of the neck, and peel the jacket off the shoulders at a 30° angle down and away from the patient. Avoid any rapid or sharp jerks, which spread contamination. Gently pull the inside sleeves over the patient's wrists and hands.

8-58. If the patient cannot extend his arms, you must cut the jacket to aid in its removal. Before cutting the overgarment jacket, dip the scissors in a 5 percent chlorine solution to prevent contamination of the patient's BDU or underclothing. As with the litter patient, make two cuts, one up each sleeve from the wrist up to the shoulder and then across the shoulder through the collar. Cut around bandages, tourniquets, and splints, leaving them in place. Do not allow your gloves to touch the patient along the cut line. Peel the jacket back and downward to avoid spreading contamination. Ensure that the outside of the jacket does not touch the patient or his inner clothing.

8-59. **Cut and remove the overgarment trousers.** Unfasten or cut all ties, buttons, or zippers before grasping the trousers at the waist and peeling them down over the patient's combat boots. Again, the trousers are cut to aid in removal. If necessary, cut both trouser legs starting at the ankle. Keep the cuts near the inside of the legs, along the inseam, to the crotch. Cut around all bandages, tourniquets, and splints. Continue to cut up both sides of the zipper to the waist and allow the narrow strip with the zipper to drop between the legs. Peel or allow the trouser halves to drop to the ground. Have the patient step out of the trouser legs one at a time. Place the trousers in the contaminated-waste bag. Place the scissors in a 5 percent chlorine solution.

8-60. **Remove the outer gloves.** Grasp the fingers of the gloves, roll the cuffs over the fingers, and turn the gloves inside out. Do not remove the inner cotton gloves at this time. Drop the gloves into the contaminated-waste bag. Do not allow the patient to touch his clothing or other contaminated objects with his hands.

8-61. **Remove the overboots.** Cut the overboot laces and fold the lacing eyelets flat on the ground. Step on the toe and heel eyelets to hold the overboot on the ground and have the patient step out of it. Repeat this procedure for the other overboot. If the overboots are in good condition, they can be decontaminated and reissued.

8-62. **Remove the patient's cotton glove liners.** Instruct the patient to remove his cotton glove liners to reduce the possibility of spreading contamination. Have the patient grasp the heel of one glove liner with the other gloved hand, peeling it off of his hand. Hold the removed glove by the inside and grasp the heel of the other glove, peeling it off of his hand. Place both gloves in the contaminated-waste bag.

### Step 5 - Check the Patient for Contamination

8-63. After the patient's overgarment has been removed, check his BDU by using M8 detector paper or the CAM. Carefully survey the patient, paying particular attention to discolored areas, damp spots, and tears on the uniform; areas around the neck, wrists, and ears; and bandages, tourniquets, and splints. Remove contaminated spots by using an SDK or a 0.5 percent chlorine solution or, if possible, by cutting away the contaminated area. Always dip the scissors in a 5 percent chlorine solution after each cut. Recheck the area with the detection equipment. If significant contamination is found on the BDU, then remove it and spot decon the skin. To remove the BDU, follow the procedures for removing the overgarment as described in paragraphs 8-57 through 8-59. Do not remove the patient's identification tags.

### Step 6 - Decon the Patient's Skin

8-64. **Spot decon.** Use an SDK or a 0.5 percent chlorine solution to spot decon the skin and areas of potential contamination, to include areas around the neck, wrists, and lower parts of the face. Decon the patient's identification tags and chain, if necessary.

8-65. Have the patient hold his breath, close his eyes, and lift, or assist him with lifting, his mask at the chin. Wipe his face and exposed areas of the skin with an SDK or a 0.5 percent chlorine solution. Starting at the top of the ear and quickly wiping downward, wipe all folds in the skin, ear lobes, upper lip, chin, dimples, and nose. Continue up the other side of the face to the top of the other ear. Wipe the inside of the mask where it touches the face. Have the patient reseal and check his mask.

**CAUTION**

Keep the decon solution out of the patient's eyes.

8-66. **Combat medic care.** The combat medic gently cuts away the bandage. He decontaminates the area around the wound and irrigates it with a 0.5 percent chlorine solution. If bleeding begins, he replaces the bandage with a clean one. He replaces the old tourniquet by placing a new one ½ to 1 inch above the old one. He then removes the old tourniquet and decontaminates the patient's skin with an SDK or a 0.5 percent chlorine solution. He does not remove a splint. He decontaminates the splint by thoroughly rinsing it, to include the padding and cravats, with a 0.5 percent chlorine solution.

8-67. **Contaminated-waste disposal.** Dispose of contaminated bandages and coverings by placing them in a contaminated-waste bag. Seal the bag and place it in the contaminated-waste dump.

### Step 7 - Proceed Through the Shuffle Pit to the Clean Treatment Area

8-68. Have the decontaminated patient proceed through the shuffle pit to the clean treatment area. To ensure that the patient's boots are well decontaminated, have him stir the contents of the shuffle pit with his boots as he crosses it. The patient's combat boots and protective mask will be removed at the entrance of the CPS or clean treatment area.

# BIOLOGICAL-AGENT PATIENT DECON PROCEDURES

8-69. The decon station as established for chemical-agent patients is also used for biologically contaminated patients. The 8-man, patient-decon team is required for biologically contaminated patient-decon procedures.

## LITTER PATIENT

8-70. The decon procedures for a litter patient that is contaminated with a biological agent are described in the following paragraphs. Some procedures can be done with one soldier, while others require more than one.

### Step 1 - Decon the Patient's Mask and Hood

8-71. **Move the patient to the clothing removal station.** After the patient has been triaged and stabilized (if necessary) by the senior medic in the patient-decon area, move him to the litter stands at the clothing removal station.

8-72. **Decon the hood.** Use a 0.5 percent chlorine solution to wipe down the front, sides, and top of the hood.

8-73. **Remove the hood.** Remove the hood by cutting it with scissors or by loosening it from the mask attachment points for the quick-doff hood or other similar hoods (see Figure 8-3, page 8-6). Before cutting the hood, dip the scissors in a 5 percent chlorine solution. Cut the neck cord and the small string under the voicemitter. Release or cut the hood shoulder straps and unzip the hood zipper. Cut the hood upward to the top of the eye-lens outsert, staying close to the filter-inlet cover and eye-lens outsert, then across the forehead to the outer edge of the other eye-lens outsert. Proceed downward toward the patient's shoulder, staying close to the eye-lens outsert and filter-inlet cover, then across the lower part of the voicemitter to the zipper. After dipping the scissors in the 5 percent chlorine solution, cut the hood from the center of the forehead over the top of the head. Fold the left and right sides of the hood to the side of the patient's head, laying the sides of the hood on the litter.

8-74. **Decon the protective mask and exposed skin**. Use a 0.5 percent chlorine solution to wipe the external parts of the mask. Cover the mask's air inlets with gauze or your hand to keep the mask filter dry. Wash the exposed areas of the patient's face, to include the neck and behind the ears, with soap and warm water or wipe them with a 0.5 percent chlorine solution. Do not remove the protective mask.

8-75. **Remove the FMC.** Cut the patient's FMC tie wire, allowing the FMC to fall into a plastic bag. Seal the plastic bag and rinse the outside of the bag with a 0.5 percent chlorine solution. Place the plastic bag under the back of the protective mask's head straps. The FMC will remain with the patient in the contaminated area and a clean copy will be made before the patient is moved to the clean area.

### Step 2 - Remove Gross Contamination From the Patient's Protective Overgarment

8-76. Remove all visible contamination spots from the overgarment by using a 0.5 percent chlorine solution.

### Step 3 - Remove the Patient's Personal Effects and Protective Overgarment

8-77. **Remove the patient's personal effects.** Remove the patient's personal effects from his protective overgarment and BDU pockets. Place the articles in a plastic bag, label the bag with the patient's identification, and seal the bag. If the articles are not contaminated, they are returned to the patient. If the articles are contaminated, place them in the contaminated holding area until they can be decontaminated, then return them to the patient.

| CAUTION |
|---|
| Bandages may have been applied to control severe bleeding; therefore, they are treated like tourniquets. Only medical personnel will remove bandages, tourniquets, and splints. |

**NOTE: The overgarment jacket and trousers will be cut simultaneously. One soldier cuts the jacket, while another soldier cuts the trousers.**

8-78. **Cut and remove the overgarment jacket.** Before cutting the overgarment jacket and trouser, dip the scissors in a 5 percent chlorine solution to prevent contamination of the patient's BDU or undergarments. Make two cuts, one up each sleeve from the wrist up to the shoulder and then across the shoulder through the collar (Figure 8-4, page 8-7). Cut around bandages, tourniquets, and splints, leaving them in place. Do not allow your gloves to touch the patient along the cut line. Keep the cuts close to the inside of the arms so that most of the sleeve material can be folded outward. Unzip the jacket and roll the chest sections to the respective sides, with the inner surface outward. Continue by tucking the clothing between the arm and chest. Roll the cut sleeves away from the arms, exposing the black liner.

8-79. **Cut and remove the overgarment trousers.** Cut both trousers legs starting at the ankle as shown in Figure 8-5, page 8-8. Keep the cuts near the inside of the legs, along the inseam, to the crotch. Cut the left leg all the way to the waist, avoiding the pockets. Cut the right leg across at the crotch to the left leg cut. Cut around bandages, tourniquets, and splints, leaving them in place. Place the scissors in a 5 percent chlorine solution. Fold the cut trouser halves away from the patient and allow the halves to drop to the litter with the contaminated (green) side down. Roll the inner leg portion under and between the legs.

8-80. **Remove the outer gloves.** Before touching the patient, the patient-decon team decontaminates its gloves with a 5 percent chlorine solution. Lift the patient's arms up and out of the cutaway sleeves unless detrimental to his condition. Grasp the fingers of the gloves, roll the cuffs over the fingers, and turn the gloves inside out. Do not remove the inner cotton gloves at this time. Carefully lower the patient's arms across the chest after the outer gloves have been removed (Figure 8-6, page 8-8). Do not allow the patient's arms to come into contact with the exterior of his overgarment. Drop his gloves into the contaminated-waste bag. The team members decon their gloves with the 5 percent chlorine solution.

8-81. **Remove the overboots.** Cut the overboot laces and fold the lacing eyelets flat outward. While standing at the foot of the litter, hold the patient's heel with one hand. Pull the overboot downward, then toward you to remove it. Remove the two overboots simultaneously. This reduces the likelihood of contaminating one of the combat boots.

8-82. While holding the patient's heels off the litter, have a team member wipe the end of the litter with a 0.5 percent chlorine solution to neutralize any contamination that was transferred to the litter from the overboots. Lower the patient's heels onto the decontaminated litter. Place the overboots in the contaminated-waste bag. The team members decontaminate their gloves with the 5 percent chlorine solution.

### Step 4 - Remove the Patient's BDU and Undergarments

8-83. **Cut and remove the BDU.** To cut and remove the BDU jacket and trousers, follow the procedures (described in paragraphs 8-78 and 8-79) for removing the protective overgarment.

8-84. **Remove the combat boots.** Cut the bootlaces along the tongue. Remove the boots by pulling them toward you. Place the boots in the contaminated-waste bag. Do not touch the patient's skin with your contaminated gloves when removing his boots.

8-85. **Cut and remove the undergarments.** Follow the procedures for cutting away the protective overgarment and rolling it away from the patient (see paragraphs 8-78 and 8-79). If the patient is wearing a brassiere, cut it between the cups. Cut both shoulder straps where they attach to the cups and lay them back off of the shoulders. Remove the socks and cotton gloves. Do not remove the identification tags.

### Step 5 - Transfer the Patient to a Decon Litter

8-86. After the patient's clothing has been cut away, transfer him to a decon litter or a canvas litter with a plastic-sheeting cover. Three decon team members decon their gloves and aprons with a 5 percent chlorine solution. One member places his hands under the patient's legs at the thighs and Achilles tendons, a second member places his arms under the patient's back and buttocks, and a third member places his arms under the patient's shoulders and supports the head and neck. They carefully lift the patient using their knees (not their backs) to minimize back strain. While the patient is elevated, another decon team member removes the litter from the litter stands and replaces it with a decontaminated (clean) litter. The patient is carefully lowered onto the clean litter. The clothing and overgarments are placed in a contaminated-waste bag and moved to the contaminated-waste dump. The dirty litter is rinsed with the 0.5 percent chlorine solution and placed in the litter storage area.

### Step 6 - Decon the Patient's Skin

8-87. **Spot decon.** With the patient in a supine position, spot decon the skin by washing it with soap and warm water or by wiping it with a 0.5 percent chlorine solution. Decon areas of potential contamination, to include areas around the neck, wrists, and lower parts of the face. Decon the patient's identification tags and chain, if necessary.

**NOTES:**

**1. Use a 0.5 percent chlorine solution to decon patients suspected of being contaminated with mycotoxins.**

**2. A complete body wash is not appropriate and may be harmful to the patient. During a complete body wash, the patient would have to be rolled over to reach all areas of the skin. This is not necessary for an adequate decon.**

8-88. **Combat medic care.** The combat medic gently cuts away the bandage. He decontaminates the area around the wound and irrigates it with a 0.5 percent chlorine solution. If bleeding begins, he replaces the bandage with a clean one. He replaces the old tourniquet by placing a new one ½ to 1 inch above the old one. He then removes the old tourniquet and decontaminates the patient's skin with a 0.5 percent chlorine solution. He does not remove a splint. He decontaminates the splint by thoroughly rinsing it, to include the padding and cravats, with a 0.5 percent chlorine solution.

**WARNING**

**Do not use an SDK around any wounds. Do not remove splints. The splint will not be removed until the patient has been evacuated to a forward MTF or hospital.**

8-89. **Completeness of decon check.** Check the patient with M8 detector paper or the CAM for completeness of decon.

**NOTE: Other monitoring devices may be used, if available.**

8-90. **Contaminated-waste disposal.** Dispose of contaminated bandages and coverings by placing them in a contaminated-waste bag. Seal the bag and place it in the contaminated-waste dump.

### Step 7 - Transfer the Patient to the Hot Line

8-91. The patient's clothing has been cut away and his skin, bandages, and splints have been decontaminated. Transfer the patient to the hot line and place the litter on the litter stands. A third team member will assist with transferring the patient to a clean treatment litter at the hot line. Decon personnel rinse or wipe down their aprons and gloves with a 5 percent chlorine solution.

8-92. Three team members lift the patient off the decon litter (see step 5 for lifting procedures). While the patient is elevated, another team member removes the litter from the stands and returns it to the decon area. A medic from the clean side of the hot line replaces the litter with a clean one. The patient is lowered onto the clean litter. Two medics from the clean side of the hot line move the patient to the clean treatment area. The patient is treated in this area or awaits processing into the CPS. The litter is wiped down with a 0.5 percent chlorine solution in preparation for reuse. Once the patient is in the air lock of the CPS and the air lock has been purged, his protective mask is removed. Place the mask in a plastic bag and seal it.

**NOTE: Before decontaminating another patient, each decon team member drinks about half a quart of water. The exact amount of water consumed is increased or decreased according to the work level and temperature (see Table 8-2, page 8-11).**

## AMBULATORY PATIENT

8-93. The decon procedures for an ambulatory patient that is contaminated with a biological agent are described in the following paragraphs. Some procedures can be done with one soldier, while others require more than one.

8-94. All ambulatory patients requiring EMT or AMT in the clean treatment area of the BAS will be decontaminated. Stable patients not requiring treatment at the BAS, but requiring evacuation to the medical company's clearing station or a corps hospital for treatment (for example, a patient with a broken arm), should be evacuated in their protective overgarments and masks by any available transportation. However, before evacuation, spot remove all thickened agents from their protective overgarment.

**NOTE: Place the bandage scissors that are used in this procedure in a container of 5 percent chlorine solution when not in use. Most ambulatory patients will be treated in the contaminated treatment area and returned to duty. Upon removal of an ambulatory patient's clothing, he becomes a litter patient. The BAS and clearing station do not have clothing to replace those that are cut off during the decon process. The patient must be placed in a PPW for protection during evacuation (Figure 8-7, page 8-12).**

### Step 1 - Remove the LCE

8-95. Remove the LCE by unfastening/unbuttoning all connectors or tie straps and then place the equipment in a plastic bag. Place the plastic bag in the designated storage area for later decon.

### Step 2 - Decon the Patient's Mask and Hood

8-96. **Begin the clothing removal process.** After the patient has been triaged and treated (if necessary) by the senior medic in the patient-decon station, the clothing removal process begins.

8-97. **Decon the hood.** Use a 0.5 percent chlorine solution to wipe down the front, sides, and top of the hood.

8-98. **Remove the hood.** Remove the hood by cutting it with scissors or by loosening it from the mask attachment points for the quick-doff hood or other similar hoods (Figure 8-3, page 8-6). Before cutting the hood, dip the scissors in a 5 percent chlorine solution. Cut the neck cord and the small string under the voicemitter. Release or cut the hood shoulder straps and unzip the hood zipper. Cut the hood upward to the top of the eye-lens outsert, staying close to the filter-inlet cover and eye-lens outsert, then across the forehead to the outer edge of the other eye-lens outsert. Proceed downward toward the patient's shoulder, staying close to the eye-lens outsert and filter-inlet cover, then across the lower part of the voicemitter to the zipper. After dipping the scissors in the 5 percent chlorine solution, cut the hood from the center of the forehead over the top of the head. Fold the left and right sides of the hood away from the patient's head and remove the hood.

8-99. **Decon the protective mask and exposed skin.** Use a 0.5 percent chlorine solution to decon the external parts of the mask. Cover the mask's air inlets with gauze or your hands to keep the mask filters dry. Wash the exposed areas of the patient's face, to include the neck and behind the ears, with soap and warm water or wipe them with a 0.5 percent chlorine solution. Do not remove the protective mask.

8-100. **Remove the FMC.** Cut the patient's FMC tie wire, allowing the FMC to fall into a plastic bag. Seal the plastic bag and rinse the outside of the bag with a 5 percent chlorine solution. Place the plastic bag under the back of the protective mask's head straps. The FMC will remain with the patient in the contaminated area and a clean copy will be made before the patient is moved to the clean area.

**Step 3 - Remove Gross Contamination From the Patient's Protective Overgarment**

8-101. Remove all visible contamination spots from the overgarment by using a 0.5 percent chlorine solution.

**Step 4 - Remove the Patient's Personal Effects and Protective Overgarment**

8-102. **Remove the patient's personal effects.** Remove the patient's personal effects from his protective overgarment and BDU pockets. Place the articles in a plastic bag, label the bag with the patient's identification, and seal the bag. If the articles are not contaminated, they are returned to the patient. If the articles are contaminated, place them in the contaminated holding area until they can be decontaminated, then return them to the patient.

8-103. **Remove the overgarment jacket.** Have the patient stand with his feet spread apart at shoulder width. Unsnap the front flap of the jacket and unzip the jacket. If the patient can extend his arms, have him clinch his fist and extend his arms backward at about a 30° angle. Move behind the patient, grasp his jacket collar at the sides of the neck, and peel the jacket off the shoulders at a 30° angle down and away from the patient. Avoid any rapid or sharp jerks, which spread contamination. Gently pull the inside sleeves over the patient's wrists and hands.

8-104. If the patient cannot extend his arms, you must cut the jacket to aid in its removal. Before cutting the overgarment jacket, dip the scissors in a 5 percent chlorine solution to prevent contamination of the patient's BDU or undergarment. As with the litter patient, make two cuts, one up each sleeve from the wrist up to the shoulder and then across the shoulder through the collar. Cut around bandages, tourniquets, and splints, leaving them in place. Do not allow your gloves to touch the patient along the cut line. Peel the jacket back and downward to avoid spreading contamination. Ensure that the outside of the jacket does not touch the patient or his inner clothing.

8-105. **Cut and remove the overgarment trousers.** Unfasten or cut all ties, buttons, or zippers before grasping the trousers at the waist and peeling them down over the patient's combat boots. Again, the trousers are cut to aid in removal. If necessary, cut both trouser legs starting at the ankle. Keep the cuts near the inside of the legs, along the inseam, to the crotch. Cut around all bandages, tourniquets, and splints. Continue to cut up both sides of the zipper to the waist and allow the narrow strip with the zipper to drop between the legs.

Peel or allow the trouser halves to drop to the ground. Have the patient step out of the trouser legs one at a time. Place the trousers in the contaminated-waste bag. Place the scissors in a 5 percent chlorine solution.

8-106. **Remove the outer gloves.** Grasp the fingers of the gloves, roll the cuffs over the fingers, and turn the gloves inside out. Do not remove the inner cotton gloves at this time. Drop the gloves into the contaminated-waste bag. Do not allow the patient to touch his clothing or other contaminated objects with his hands.

8-107. **Remove the overboots.** Cut the overboot laces and fold the lacing eyelets flat on the ground. Step on the toe and heel eyelets to hold the overboot on the ground and have the patient step out of it. Repeat this procedure for the other overboot. If the overboots are in good condition, they can be decontaminated and reissued.

8-108. **Remove the patient's cotton glove liners.** Instruct the patient to remove his cotton glove liners to reduce the possibility of spreading contamination. Have the patient grasp the heel of one glove liner with the other gloved hand, peeling it off of his hand. Hold the removed glove by the inside and grasp the heel of the other glove, peeling it off of his hand. Place both gloves in the contaminated-waste bag.

### Step 5 - Check the Patient for Contamination

8-109. After the patient's overgarment has been removed, check his BDU by using M8 detector paper or the CAM. Carefully survey the patient, paying particular attention to discolored areas, damp spots, and tears on the uniform; areas around the neck, wrists, and ears; and bandages, tourniquets, and splints. Remove contaminated spots by washing with soap and water, by wiping with a 0.5 percent chlorine solution or, if possible, by cutting away the contaminated area. Always dip the scissors in a 5 percent chlorine solution after each cut. Recheck the area with the detection equipment. If significant contamination is found on the BDU, then remove it and spot decon the skin. To remove the BDU, follow the procedures for removing the protective overgarment as described in paragraphs 8-103 through 8-105. Do not remove the patient's identification tags.

### Step 6 - Decon the Patient's Skin

8-110. **Spot decon.** Use soap and water or a 0.5 percent chlorine solution to spot decon the skin and areas of potential contamination, to include areas around the neck, wrists, and lower parts of the face. Decon the patient's identification tags and chain, if necessary.

8-111. Have the patient hold his breath, close his eyes, and lift, or assist him with lifting, his mask at the chin. Wash his face and exposed areas of the skin with soap and water or wipe them with a 0.5 percent chlorine solution. Starting at the top of the ear and quickly wiping downward, wipe all folds in the skin, ear lobes, upper lip, chin, dimples, and nose. Continue up the other side of the face to the top of the other ear. Wipe the inside of the mask where it touches the face. Have the patient reseal and check his mask.

| CAUTION |
| --- |
| Keep the decon solution out of the patient's eyes. |

8-112. **Combat medic care.** The combat medic gently cuts away the bandage. He decontaminates the area around the wound and irrigates it with a 0.5 percent chlorine solution. If bleeding begins, he replaces the bandage with a clean one. He replaces the old tourniquet by placing a new one ½ to 1 inch above the old one. He then removes the old tourniquet and decontaminates the patient's skin with an SDK or a 0.5 percent chlorine solution. He does not remove a splint. He decontaminates the splint by thoroughly rinsing it, to include the padding and cravats, with a 0.5 percent chlorine solution.

**NOTE: Use a 0.5 percent chlorine solution to decon ambulatory patients suspected of being contaminated with mycotoxins.**

8-113. **Contaminated-waste disposal.** Dispose of contaminated bandages and coverings by placing them in a contaminated-waste bag. Seal the bag and place it in the contaminated-waste dump.

#### Step 7 - Cross the Hot Line to the Clean Treatment Area

8-114. Have the decontaminated patient cross the hot line to the clean treatment area. The patient's boots must be decontaminated at the hot line before enters the clean area. The patient's combat boots and protective mask will be removed at the entrance of the CPS or clean treatment area.

## RADIOLOGICAL-AGENT PATIENT DECON PROCEDURES

8-115. The decon of patients that are contaminated with radiation is easily accomplished without interfering with the required medical care. The 8-man, patient-decon team is required to perform the decon procedures on patients contaminated with radiation.

**NOTE: Patients must be monitored by using a radiac meter before, during, and after each step of the decon procedure.**

### LITTER PATIENT

8-116. The decon procedures for a litter patient that is contaminated with radiation are discussed in the following paragraphs. Some procedures can be done with one soldier, while others require more than one soldier.

#### Step 1 - Decon the Patient's Mask and Hood

8-117. **Move the patient to the clothing removal station.** After the patient has been triaged and stabilized (if necessary) by the senior medic in the patient-decon area, move him to the litter stands at the clothing removal station.

8-118. **Decon the hood.** Brush contamination off or use soap and water to wipe down the front, sides, and top of the hood.

8-119. **Remove the hood.** Remove the hood by cutting it with scissors or by loosening it from the mask attachment points for the quick-doff hood or other similar hoods (see Figure 8-3, page 8-6). Before cutting the hood, dip the scissors in a 5 percent chlorine solution. Cut the neck cord and the small string under the voicemitter. Release or cut the hood shoulder straps and unzip the hood zipper. Cut the hood upward to the top of the eye-lens outsert, staying close to the filter-inlet cover and eye-lens outsert, then across the forehead to the outer edge of the other eye-lens outsert. Proceed downward toward the patient's shoulder, staying close to the eye-lens outsert and filter-inlet cover, then across the lower part of the voicemitter to the zipper. After dipping the scissors in the 5 percent chlorine solution, cut the hood from the center of the forehead over the top of the head. Fold the left and right sides of the hood to the side of the patient's head, laying the sides of the hood on the litter.

8-120. **Decon the protective mask and exposed skin.** Use soap and water to wipe the external parts of the mask. Cover the mask's air inlets with gauze or your hand to keep the mask filter dry. Wash the exposed areas of the patient's face, to include the neck and behind the ears, with soap and warm water. Do not remove the protective mask.

8-121. **Remove the FMC.** Cut the patient's FMC tie wire, allowing the FMC to fall into a plastic bag. Seal the plastic bag and rinse the outside of the bag with soap and water. Place the plastic bag under the back of the protective mask's head straps. The FMC will remain with the patient in the contaminated area and a clean copy will be made before the patient is moved to the clean area.

### Step 2 - Remove Gross Contamination From the Patient's Protective Overgarment

8-122. Remove all visible contamination spots from the overgarment by brushing off or by washing with soap and water.

### Step 3 - Remove the Patient's Personal Effects and Protective Overgarment

8-123. **Remove the patient's personal effects.** Remove the patient's personal effects from his protective overgarment and BDU pockets. Place the articles in a plastic bag, label the bag with the patient's identification, and seal the bag. If the articles are not contaminated, they are returned to the patient. If the articles are contaminated, place them in the contaminated holding area until they can be decontaminated, then return them to the patient.

**CAUTION**

Bandages may be applied to control severe bleeding; therefore, they are treated like tourniquets. Only medical personnel will remove bandages.

**NOTE: The overgarment jacket and trousers will be cut simultaneously. One soldier cuts the jacket, while another soldier cuts the trousers.**

8-124. **Cut and remove the overgarment jacket.** Before cutting the overgarment jacket and trousers, dip the scissors in a 5 percent chlorine solution to prevent contamination of the patient's BDU or undergarment. Make two cuts, one up each sleeve from the wrist up to the shoulder and then across the shoulder through the collar (Figure 8-4, page 8-7). Cut around bandages, tourniquets, and splints, leaving them in place. Do not allow your gloves to touch the patient along the cut line. Keep the cuts close to the inside of the arms so that most of the sleeve material can be folded outward. Unzip the jacket and roll the chest sections to the respective sides, with the inner surface outward. Continue by tucking the clothing between the arm and chest. Roll the cut sleeves away from the arms, exposing the black liner.

8-125. **Cut and remove the overgarment trousers.** Cut both trousers legs starting at the ankle as shown in Figure 8-5, page 8-8. Keep the cuts near the inside of the legs, along the inseam, to the crotch. Cut the left leg all the way to the waist, avoiding the pockets. Cut the right leg across at the crotch to the left leg cut. Cut around bandages, tourniquets, and splints, leaving them in place. Place the scissors in a 5 percent chlorine solution. Fold the cut trouser halves away from the patient and allow the halves to drop to the litter with the contaminated- (green-) side down. Roll the inner leg portion under and between the legs.

8-126. **Remove the outer gloves.** Before touching the patient, the patient-decon team decontaminates its gloves with a 5 percent chlorine solution. Lift the patient's arms up and out of the cutaway sleeves unless detrimental to his condition. Grasp the fingers of the gloves, roll the cuffs over the fingers, and turn the gloves inside out. Do not remove the inner cotton gloves at this time. Carefully lower the patient's arms across the chest after the outer gloves have been removed (Figure 8-6, page 8-8). Do not allow the patient's arms to come into contact with the exterior of his overgarment. Drop his gloves into the contaminated-waste bag. The team members decon their gloves with the 5 percent chlorine solution.

8-127. **Remove the overboots.** Cut the overboot laces and fold the lacing eyelets flat outward. While standing at the foot of the litter, hold the patient's heel with one hand. Pull the overboot downward, then toward you to remove it. Remove the two overboots simultaneously. This reduces the likelihood of contaminating one of the combat boots.

8-128. While holding the patient's heels off the litter, have a team member wipe the end of the litter with a 0.5 percent chlorine solution to neutralize any contamination that was transferred to the litter from the overboots. Lower the patient's heels onto the decontaminated litter. Place the overboots in the contaminated-waste bag. The team members decon their gloves with the 5 percent chlorine solution.

**NOTE: Patients arriving at the MTF in MOPP gear will only have their MOPP gear removed. They will remain in their BDUs unless contamination is found on them. If contamination is found, follow the procedures for removing the protective overgarment as described in paragraphs 8-124 and 8-125.**

### Step 4 - Check the Patient for Contamination

8-129. After the patient's overgarment has been removed, check his BDU by using M8 detector paper or the CAM. Carefully survey the patient, paying particular attention to discolored areas, damp spots, and tears on the uniform; areas around the neck, wrists, and ears; and bandages, tourniquets, and splints. Remove contaminated spots by washing with soap and water, by wiping with a 0.5 percent chlorine solution or, if possible, by cutting away the contaminated area. Always dip the scissors in a 5 percent chlorine solution after each cut. Recheck the area with the detection equipment. If significant contamination is found on the BDU, then remove it and spot decon the skin.

To remove the BDU, follow the procedures for removing the protective overgarment as described in paragraphs 8-124 and 8-125. Do not remove the patient's identification tags.

### Step 5 - Decon the Patient's Skin

8-130. **Spot decon.** Wash exposed skin surfaces with soap and warm water, to include areas around the neck, wrists, and lower parts of the face. Decon the patient's identification tags and chain, if necessary.

8-131. **Combat medic care.** The combat medic gently cuts away the bandage. He decontaminates the area around the wound and irrigates it with a 0.5 percent chlorine solution. If bleeding begins, he replaces the bandage with a clean one. He replaces the old tourniquet by placing a new one ½ to 1 inch above the old one. He then removes the old tourniquet and decontaminates the patient's skin with a 0.5 percent chlorine solution. He does not remove a splint. He decontaminates the splint by thoroughly rinsing it, to include the padding and cravats, with a 0.5 percent chlorine solution.

**WARNING**

**Do not use an SDK around any wounds. Do not remove splints. The splint will not be removed until the patient has been evacuated to a forward MTF or hospital.**

8-132. **Completeness of decon check.** Check the patient with M8 detector paper or the CAM for completeness of decon.

**NOTE: Other monitoring devices may be used, if available.**

8-133. **Contaminated-waste disposal.** Dispose of contaminated bandages and coverings by placing them in a contaminated-waste bag. Seal the bag and place it in the contaminated-waste dump.

### Step 6 - Transfer the Patient to the Hot Line

8-134. Move the patient to the hot line. Two medics from the clean side of the hot line move the patient into the clean treatment area.

**NOTE: Before decontaminating another patient, each decon team member drinks about half a quart of water. The exact amount of water consumed is increased or decreased according to the work level and temperature (see Table 8-2, page 8-11).**

## AMBULATORY PATIENT

8-135. The decon procedures for an ambulatory patient that is contaminated with radiation are discussed in the following paragraphs. Some procedures can be done with one soldier, while others require more than one.

8-136. All ambulatory patients requiring EMT or AMT in the clean area of the BAS will be decontaminated. Stable patients not requiring treatment at the BAS, but requiring evacuation to the medical company's clearing station or a corps hospital for treatment (for example, a patient with a broken arm), should be evacuated in their protective overgarments and masks by any

available transportation. However, before evacuation, spot remove all thickened agents from their protective overgarment.

**NOTE: Place the bandage scissors that are used in this procedure in a container of 5 percent chlorine solution when not in use. Most ambulatory patients will be treated in the contaminated treatment area and returned to duty. Upon removal of an ambulatory patient's clothing, he becomes a litter patient. The BAS and clearing station do not have clothing to replace those that are cut off during the decon process. The patient must be placed in a PPW for protection during evacuation (Figure 8-7, page 8-12).**

### Step 1 - Remove the LCE

8-137. Remove the LCE by unfastening/unbuttoning all connectors or tie straps and then place the equipment in a plastic bag. Place the plastic bag in the designated storage area for later decon.

### Step 2 - Decon the Patient's Mask and Hood

8-138. **Begin the clothing removal process.** After the patient has been triaged and treated (if necessary) by the senior medic in the patient-decon station, the clothing removal process begins.

8-139. **Decon the hood.** Brush off contamination or use soap and water to wipe down the front, sides, and top of the hood.

8-140. **Remove the hood.** Remove the hood by cutting it with scissors or by loosening it from the mask attachment points for the quick-doff hood or other similar hoods (see Figure 8-3, page 8-6). Before cutting the hood, dip the scissors in a 5 percent chlorine solution. Cut the neck cord and the small string under the voicemitter. Release or cut the hood shoulder straps and unzip the hood zipper. Cut the hood upward to the top of the eye-lens outsert, staying close to the filter-inlet cover and eye-lens outsert, then across the forehead to the outer edge of the other eye-lens outsert. Proceed downward toward the patient's shoulder, staying close to the eye-lens outsert and filter-inlet cover, then across the lower part of the voicemitter to the zipper. After dipping the scissors in the 5 percent chlorine solution, cut the hood from the center of the forehead over the top of the head. Fold the left and right sides of the hood away from the patient's head and remove the hood.

8-141. **Decon the protective mask and exposed skin.** Use soap and water to wipe the external parts of the mask. Cover the mask's air inlets with gauze or your hands to keep the mask filters dry. Wash the exposed areas of the patient's face, to include the neck and behind the ears, with soap and warm water. Do not remove the protective mask.

8-142. **Remove the FMC.** Cut the patient's FMC tie wire, allowing the FMC to fall into a plastic bag. Seal the plastic bag and rinse the outside of the bag with a 5 percent chlorine solution. Place the plastic bag under the back of the protective mask's head straps. The FMC will remain with the patient in the contaminated area and a clean copy will be made before the patient is moved to the clean area.

## Step 3 - Remove Gross Contamination From the Patient's Protective Overgarment

8-143. Remove all visible contamination spots by brushing off or by washing with soap and water.

## Step 4 - Remove the Patient's Personal Effects and Protective Overgarment

8-144. **Remove the patient's personal effects.** Remove the patient's personal effects from his protective overgarment and BDU pockets. Place the articles in a plastic bag, label the bag with the patient's identification, and seal the bag. If the articles are not contaminated, they are returned to the patient. If the articles are contaminated, place them in the contaminated holding area until they can be decontaminated, then return them to the patient.

8-145. **Remove the overgarment jacket.** Have the patient stand with his feet spread apart at shoulder width. Unsnap the front flap of the jacket and unzip the jacket. If the patient can extend his arms, have him clinch his fist and extend his arms backward at about a 30° angle. Move behind the patient, grasp his jacket collar at the sides of the neck, and peel the jacket off the shoulders at a $30^0$ angle down and away from the patient. Avoid any rapid or sharp jerks, which spread contamination. Gently pull the inside sleeves over the patient's wrists and hands.

8-146. If the patient cannot extend his arms, you must cut the jacket to aid in its removal. Before cutting the overgarment jacket, dip the scissors in a 5 percent chlorine solution to prevent contamination of the patient's BDU or undergarment. As with the litter patient, make two cuts, one up each sleeve from the wrist up to the shoulder and then across the shoulder through the collar. Cut around bandages, tourniquets, and splints, leaving them in place. Do not allow your gloves to touch the patient along the cut line. Peel the jacket back and downward to avoid spreading contamination. Ensure that the outside of the jacket does not touch the patient or his inner clothing.

8-147. **Cut and remove the overgarment trousers.** Unfasten or cut all ties, buttons, or zippers before grasping the trousers at the waist and peeling them down over the patient's combat boots. Again, the trousers are cut to aid in removal. If necessary, cut both trouser legs starting at the ankle. Keep the cuts near the inside of the legs, along the inseam, to the crotch. Cut around all bandages, tourniquets, and splints. Continue to cut up both sides of the zipper to the waist and allow the narrow strip with the zipper to drop between the legs. Peel or allow the trouser halves to drop to the ground. Have the patient step out of the trouser legs one at a time. Place the trousers in the contaminated-waste bag. Place the scissors in a 5 percent chlorine solution.

8-148. **Remove the outer gloves.** Grasp the fingers of the gloves, roll the cuffs over the fingers, and turn the gloves inside out. Do not remove the inner cotton gloves at this time. Drop the gloves into the contaminated-waste bag. Do not allow the patient to touch his clothing or other contaminated objects with his hands.

8-149. **Remove the overboots.** Cut the overboot laces and fold the lacing eyelets flat on the ground. Step on the toe and heel eyelets to hold the overboot on the ground and have the patient step out of it. Repeat this procedure for the other overboot. If the overboots are in good condition, they can be decontaminated and reissued.

8-150. **Remove the patient's cotton glove liners.** Instruct the patient to remove his cotton glove liners to reduce the possibility of spreading contamination. Have the patient grasp the heel of one glove liner with the other gloved hand, peeling it off of his hand. Hold the removed glove by the inside and grasp the heel of the other glove, peeling it off of his hand. Place both gloves in the contaminated-waste bag.

### Step 5 - Check the Patient for Contamination

8-151. After the patient's overgarment has been removed, check his BDU by using M8 detector paper or the CAM. Carefully survey the patient, paying particular attention to discolored areas, damp spots, and tears on the uniform; areas around the neck, wrists, and ears; and bandages, tourniquets, and splints. Remove contaminated spots by washing with soap and water. Always dip the scissors in a 5 percent chlorine solution after each cut. Recheck the area with the detection equipment. If significant contamination is found on the BDU, then remove it and spot decon the skin. To remove the BDU, follow the procedures for removing the protective overgarment as described in paragraphs 8-145 through 8-147. Do not remove the patient's identification tags.

### Step 6 - Decon the Patient's Skin

8-152. **Spot decon.** Wash exposed skin surfaces with soap and warm water, to include areas around the neck, wrists, and lower parts of the face. Decon the patient's identification tags and chain, if necessary.

8-153. Have the patient hold his breath, close his eyes, and lift, or assist him with lifting, his mask at the chin. Wipe his face and exposed areas of the skin with soap and water. Starting at the top of the ear and quickly wiping downward, wipe all folds in the skin, ear lobes, upper lip, chin, dimples, and nose. Continue up the other side of the face to the top of the other ear. Wipe the inside of the mask where it touches the face. Have the patient reseal and check his mask.

8-154. **Combat medic care.** The combat medic gently cuts away the bandage. He decontaminates the area around the wound and irrigates it with a 0.5 percent chlorine solution. If bleeding begins, he replaces the bandage with a clean one. He replaces the old tourniquet by placing a new one ½ to 1 inch above the old one. He then removes the old tourniquet and decontaminates the patient's skin with a 0.5 percent chlorine solution. He does not remove a splint. He decontaminates the splint by thoroughly rinsing it, to include the padding and cravats, with a 0.5 percent chlorine solution.

8-155. **Contaminated-waste disposal.** Dispose of contaminated bandages and coverings by placing them in a contaminated-waste bag. Seal the bag and place it in the contaminated-waste dump.

### Step 7 - Transfer the Patient to the Hot Line

8-156. Have the decontaminated patient cross the hot line to the clean treatment area. The patient's boots must be decontaminated at the hot line before he enters the clean area. The patient's combat boots and protective mask will be removed at the entrance of the CPS or clean treatment area.

**NOTE: Before decontaminating another patient, each decon team member drinks about half a quart of water. The exact amount of water consumed is increased or decreased according to the work level and temperature (see Table 8-2, page 8-11).**

# Chapter 9

# Logistics

The materials that are needed to conduct decon operations are identified in this chapter.

## SKIN DECON AND PERSONAL WIPE DOWN

9-1. Each soldier carries his own skin-decon supplies. The SDK is stored in the right cargo pocket of the overgarment trousers or BDU trousers. The basis of issue per soldier is two SDKs (12 packets). Each kit contains six individual decon packets. The unit maintains at least one SDK per assigned person for resupply. Personal wipe down is also done with these kits.

## INDIVIDUAL-GEAR DECON

9-2. The standard decon kit for individual gear is the IEDK. Each soldier carries an IEDK in the right cargo pocket of his overgarment trousers or BDU trousers. The basis of issue per soldier is one IEDK. Each kit contains four individual decon packets, enough to decon two complete sets of individual gear. The unit maintains at least one IEDK per assigned soldier for resupply.

### OPERATOR'S SPRAY DOWN

9-3. The M11 or M13 is the decon apparatus used for the operator's spray down. Fill the M11 with a 1 $^1/_3$-quart can of DS2 before use. The IEDK can be used on equipment that the DS2 may cause damage to by corrosive action.

9-4. Four nitrogen cylinders and two 1 $^1/_3$-quart cans of DS2 should be carried in the vehicle with each M11 (two nitrogen cylinders are authorized per can of DS2). Use the nitrogen cylinders to pressurize the M11 during use. Operating the M11 in below-freezing weather (below 32$^o$F) requires two nitrogen cylinders to expend its contents.

9-5. Units maintain limited spare parts and accessories. They are not required to maintain additional stocks of 1 $^1/_3$-quart cans of DS2. The 5-gallon pails that are authorized to each unit can be used to refill empty M11s until a resupply of 1 $^1/_3$-quart cans of DS2 is received from the battalion.

9-6. The M13 comes with a 3.7 gallon DS2 container.

### MOPP-GEAR EXCHANGE

9-7. Each soldier wears or carries one complete set of MOPP gear. The unit stocks a second complete set, which is sized and identified for each soldier. When a squad or platoon undergoes a MOPP-gear exchange, the unit sends a supply vehicle with replacement MOPP gear and any decontaminants (STB) to rendezvous with the contaminated element at the operational-decon site. (The procedure on how to prepare the STB dry mix is discussed in Appendix B.) Use about 40 pounds of the STB dry mix per squad-sized element. In wartime, 5-pound quantities of the STB dry mix can be put in 1-gallon airtight containers and stored in the company's supply section.

**NOTE: Do not leave a 50-pound drum of STB dry mix open because it loses its effectiveness if left open to the air for extended periods. The unit is resupplied with MOPP gear and decontaminants from its battalion.**

9-8. Units maintain a 5 percent overage of MOPP gear based on their personnel strength or authorization (whichever is greater) to ensure a complete range of sizes and replacement gear. All MOPP gear, serviceable and unserviceable, is handled as organizational clothing and equipment (Class II supplies). The higher HQ is responsible for resupply of these items.

### VEHICLE WASHDOWN

9-9. Like the MOPP-gear exchange, a vehicle washdown is done in the unit's AO. The battalion's PDDE crew, stationed in the battalion trains, conducts the washdown. The crew moves to the operational-decon site, rendezvous with the contaminated element, and conducts the washdown. The crew uses its PDDE to spray about 100 to 150 gallons of hot, soapy water on each vehicle to wash off the gross contamination. For vehicles such as the M1-series armored fighting vehicles, about 200 gallons of water may be required per vehicle. One hundred gallons of water per vehicle provides a 2- to 3-minute wash. To speed up the operation, the crew should heat the water before arriving at the operational-decon site. See Chapters 2 and 3 and Table 9-1 for the planning and coordination of and the requirements for a vehicle washdown.

9-10. The M12A1 PDDA injects detergent into the water as it operates. It uses 2.5 quarts of detergent for every 1,200 gallons of water (about 1 quart of soap per 450 gallons of water). The battalion decon crew should maintain a basic load of liquid all-purpose detergent, which is sufficient to decon 30 percent of the battalion's organic vehicles.

## DETAILED TROOP DECON

9-11. Generally, units conduct DTD in the brigade support area. Battalions conduct it in the division/corps support area. Materials for this technique usually are stocked in the battalion or brigade trains. Reconstitution operations should be closely associated with decon operations. The battalion's assessment and recovery team (ART) ensures that the material and equipment are available for the decon operation as part of the reconstitution effort.

## DETAILED EQUIPMENT DECON

9-12. A chemical platoon must have access to large water sources—rivers, ponds, and public water systems—to conduct the DED. The chemical unit leader estimates the amounts of decontaminants that are needed. The platoon sets up the DED site, supplies the decontaminants, and conducts the DED. The platoon should carry enough decontaminants to service one company/team/battery. A decon platoon that is assigned to support a maneuver company (about 16 vehicles per tank company) should carry a minimum of 4 gallons of liquid detergent, 48 5-gallon cans of DS2, and 10 50-pound drums of STB. The chemical platoon is usually resupplied through its parent unit. Command-assignment relationships can change the resupply channels.

**Table 9-1. Estimated Water Consumption for Decon**

| Required Equipment | Time | Water Consumption |
|---|---|---|
| | | **Operational Decon[1]** |
| M12A1 PDDA or M17 LDS | 1-3 minutes | 100 to 150 gallons per regular vehicle<br>150 to 200 gallons per armored or larger vehicle<br>Example:<br>15 (contaminated vehicles) x 150 (gallons of water) = 2,250 gallons[2] |
| | | **Detailed Equipment Decon[3]** |
| M12A1 PDDA[4] or M17 LDS[4] | See Chapter 4 for the time allowed during a primary wash and rinse. Use the proper decon method. | Vehicles:<br>Station 1 - primary wash, 250 gallons<br>Station 4 - rinse, 200 gallons<br>Armored or larger vehicle:<br>Station 1 - primary wash, 300 gallons<br>Station 4 - rinse, 200 gallons<br>Example:<br>6 vehicles x 450 gallons of water = 2,700 gallons<br>4 tanks x 500 gallons of water = 2,000 gallons<br>Total: 4,700 gallons[2] |
| | | **Detailed Troop Decon** |
| 30-gallon container | | The initial setup requires 258 gallons of water. The water must be exchanged after 10 troops have been decontaminated through the DTD to avoid the transfer of contamination.<br>Station 1 (120 gallons, 4 30-gallon containers). |
| 3-gallon container | | Station 2 (6 gallons, 2 3-gallon containers). |
| 30-gallon container | | Station 4 (180 gallons, 6 30-gallon containers). |
| 3-gallon container | | Station 7 (12 gallons, 4 3-gallon containers).<br>Example:<br>About 150 troops are to be decontaminated through the DTD. You will need 258 gallons of water per every 10 troops.<br>Example:<br>150 (number of troops) ÷ 10 (required water exchange) = 15 (the amount of times the water will need to be exchanged)<br><br>15 x 258 (gallons of water per every 10 troops) = 3,870 gallons of water for 150 troops[2] |

[1]To reduce contamination, conduct the operator's spray down before the operational decon. This process requires less water consumption during a thorough decon.

[2]Always include a 10 percent planning factor to the total estimate of the water consumption for the DED and the DTD. Example: 2,250 (gallons of water for the DTD) x 10 percent = 225 (additional gallons of water). A total of 2,250 + 225 = 2,475 (gallons of water required for the DTD).

[3]The planner should consider vehicle predecon action to estimate water consumption. He should consider vehicles that were processed through operational decon because they will normally increase the weatheriza-tion process and may reduce water usage.

[4]The M17 LDS with two wands uses 14 gallons of water per minute. The M17 LDS with only one wand, which is fitted with an injector, uses 25 gallons per minute. The M12A1 PDDA pumps 25 gallons of water per minute.

9-13. Chemical platoons attached to division engineers may be resupplied through the division engineer battalion instead of their parent company. See Table 9-2 for the decon resources that are available at each organizational level. For the equipment and supplies that are needed for decon operations, see Table 9-3, pages 9-5 through 9-7.

**Table 9-2. Decon Resources Available at Each Organizational Level**

| Organizational Level | Decon Resources |
|---|---|
| Individual soldiers | 2 SDKs<br>1 canteen of water<br>2 IEDKs |
| Operators and crews | 2 SDKs<br>2 IEDKs<br>1 on-board decon apparatus (M11 or M13)<br>Soap and water<br>2 50-pound drums STB |
| Companies | 2 5-gallon pails of DS2<br>2 immersion heaters<br>2 to 3 30-gallon containers<br>6 3-gallon containers<br>6 long-handled brushes<br>6 sponges<br>300 plastic trash bags |
| Battalion PDDE crews | PDDE (M17 LDS)<br>Basic load, liquid detergent |
| Chemical company decon squad | PDDE (M12A1 PDDA/M17 LDS)<br>Basic load, liquid detergent |
| Chemical company decon platoon | PDDE (M12A1 PDDA/M17 LDS)<br>Interior decon equipment<br>Sufficient materials to set up a DTD |

### Table 9-3. Equipment and Supplies Needed for Decon Operations

| Minimum Amounts of Equipment and Supplies Needed for Decon Techniques | | | | | | | Nomenclature | NSN | Class of Supply | Unit of Issue | Basis of Issue (See TOE for Actual Authorization)*** |
|---|---|---|---|---|---|---|---|---|---|---|---|
| SD* | PW* | OS* | MGX** | VW** | DTD** | DED** | | | | | |
| | | 1 | | | | *** | Decon apparatus, DS2, ABC-M11 or M13 DAP | 4230-00-720-1618<br>4230-01-113-4124 | II<br>II | Each<br>Each | 1 per every major equipment |
| | | 2 | | | | | Cylinder, nitrogen filled | 4230-00-775-7541 | II | Box | 2 per can, DS2<br>5 per box |
| 1 | 1 | 1 | 2 | | 35 | 2 | Decon kit,<br>M291 SDK<br>M295 IEDK | 4230-01-0276-1905<br>6850-01-3577-8456 | II<br>II | Each<br>Each | 1 per mask<br>2 per soldier |
| | | 1 | | | | | Fluid-filled container, DS2 | 4230-01-136-8888 | II | Each | 1 per M13 DAP |
| | | | | | 1 | | Filter canister, C2, or filter canister, C2A1 | 4240-01-119-2315<br>4240-01-361-1319 | II | Each | 1 per M40-series mask |
| | | | | | 1 | | Hood, M40 mask | 4240-01-376-3152 | II | Each | 1 per mask |
| | | | 1 | | 2 | | Shears | 5111-00-223-6371 | II | Each | As needed |
| | | | | | | | Knife | 5110-00-240-5943 | II | Each | As needed |
| | | | *** | | | *** | Axe, single bit | 5110-00-293-2336 | II | Each | 1 per most vehicles |
| | | | *** | | *** | *** | Shovel, hand, RD, PT, D handle | 5120-00-293-3336 | VII | Each | 1 each for most vehicles |
| | | | | | | *** | NAAK, MK1 | 6505-01-140-6455 | VII | Each | 3 per individual |
| | | | | | | *** | CANA | 6505-01-274-0951 | VII | Each | 1 per individual<br>5 per combat life saver<br>10 per combat medic |
| | | | | | *** | 3 | Paper, chemical agent, detector, M9 | 6665-00-049-8982 | II | Roll | 1 per squad<br>3 per platoon |

**Table 9-3. Equipment and Supplies Needed for Decon Operations (Continued)**

| Minimum Amounts of Equipment and Supplies Needed for Decon Techniques | | | | | | | Nomenclature | NSN | Class of Supply | Unit of Issue | Basis of Issue (See TOE for Actual Authorization)*** |
|---|---|---|---|---|---|---|---|---|---|---|---|
| SD* | PW* | OS* | MGX** | VW** | DTD** | DED** | | | | | |
| | | | | | *** | *** | Paper, chem cal agent, detector, M8 | 6665-00-050-8529 | II | Book | 6 books per company |
| | | | | | *** | *** | Radiac meter, AN/PDR-2 | 6665-01-222-1425 | VII | Each | Per MTOE |
| | | | | | 2 | *** | Radiac meter, IM-93/UD | 6665-00-752-7759 | VII | Each | Per MTOE |
| | | | | | 1 | 2 | Alarm, chemical agent, M8A1 or M22 | 6665-00-935-6955<br>6665-01-438-6963 | VII<br>VII | Each<br>Each | Per MTOE |
| | | | | | 4 | 2 | CAM | 6665-01-199-4153 | VII | Each | Per MTOE |
| | | | | | 4 | | Detector kit, chemical agent, M256A1 | 6665-01-016-8399 | II | Kit | 1 per squad |
| | | | | | 1 | | Mask sanitizing solution | 6810-00-266-6979 | III | Tube | 4 per 10 masks |
| | | | | | ** | 1 | Decontaminating agent, STB | 6850-00-297-6653 | III | Drum | 2 50-pound drums per company |
| | | | | | | *** | Decontaminating agent, DS2 | 6850-00-753-4870 | II | Can | 5 gallons, 25 cans per decon |
| | | 1 | | | 2 | 24 | Brush, scrub, long-handled | 7920-00-141-5452 | II | Each | As required |
| | | | | | 7 | | Pail, metal, 14-quart | 7240-00-160-0455 | II | Each | As required |
| | | | 1 | | 9 | 4 | Garbage can, galvanized, 30-gallon | 7240-00-160-0440 | II | Each | 2 per company |
| | | | | | | 6 | Mop | 7920-00-224-8756 | | Each | As required |
| | | | | | 5 | | Sponge, cellulose | 7920-00-240-2559 | II | Each | As required |
| | | | 2 | | 4 | 6 | Brush, scrub | 7920-00-240-7171 | II | Each | As required |
| | | | 2 | | 2 | | Towels, paper | 7920-00-823-6931 | II | Box | As required |
| | | | 1 | 1 | 1 | 1 | Detergent, GP, liquid | 7930-00-282-9699 | II | Gallon | As required |

**Table 9-3. Equipment and Supplies Needed for Decon Operations (Continued)**

| Minimum Amounts of Equipment and Supplies Needed for the Decon Techniques | | | | | | | Nomenclature | NSN | Class of Supply | Unit of Issue | Basis of Issue (See TOE for Actual Authorization)*** |
|---|---|---|---|---|---|---|---|---|---|---|---|
| SD* | PW* | OS* | MGX** | VW** | DTD** | DED** | | | | | |
| | | | 1 | | 1 | 1 | Bag, plastic | 8105-00-655-8286 | II | Box | 125 count |
| | | | | ** | | 10 | TAP apron | 8415-00-281-7813 through 8415-00-281-7816 | II | Each | 10 per platoon |
| | | | 1 | | 1 | | Suit, clothing, protective, See CTA 50-909 for clothing tariff | 8415-01-137-1700 through 8415-01-137-1707 | II | Each | 2 per soldier |
| | | | 1 | | 1 | | Glove set, CP, See SB 10-523 for clothing tariff | 8415-01-033-3517 through 8415-01-033-3520 | II | Pair | 1 per soldier |
| | | | | | | | Cover, helmet, chemical protective | 8415-01-111-9028 | II | Each | 1 per soldier |
| | | | 1 | | 1 | | Overboots | 8430-01-048-6305 through 8430-01-049-0887 | II | Pair | 1 per soldier |
| | | | 2 | 2 | 2 | 2 | NBC marking kit | 9905-12-124-5955 | II | Kit | 1 per squad |
| | | | | | 2 | | Immersion heater | | II | Each | 2 per company |
| *Techniques executed by individual soldiers | | | | | | | SD – Skin decon<br>PW – Personal wipe down<br>OS – Operator's spray down | | Amount is consumption rate for 1 soldier | | |
| **Techniques executed by units | | | | | | | MGX – MOPP-gear exchange<br>VW – Vehicle washdown<br><br>DTD – Detailed troop decon<br>DED – Detailed equipment decon | | Amount is consumption rate for 1 platoon (4 vehicles, 40 personnel)<br><br>Amount is consumption rate for 1 company (20 vehicles, 150 personnel) | | |
| ***Amounts vary, depending on situation | | | | | | | | | | | |

# Chapter 10

# Decon Procedures for Individual and Crew-Served Weapons

Decon procedures for light infantry crew-served weapons and some individual major weapons that are critical to the unit's mission are discussed in this chapter. For those units that are not authorized decon devices such as the M11 or M13 DAP, alternative measures are described. Decon procedures are specified for the following weapons: M60 machine gun; M249 squad automatic weapon (SAW); M47 Dragon; tube-launched, optically tracked, wire-guided (TOW) missile (M220 series); 81-millimeter mortar; 60-millimeter mortar; M203 grenade launcher; AT4 light antitank weapon (LAW); LAW (M72 series); and 66-millimeter rocket launcher.

## OPERATIONAL-DECON PROCEDURES USING THE IEDK

10-1. Operational decon is conducted to sustain combat operations, remove gross contamination, and reduce the transfer/spread of contamination. Before decon, soldiers should visually inspect the weapon system for contamination. If liquid contamination is present, follow the procedures prescribed for each weapon system. The protection level required for operational-decon operations is MOPP4.

**NOTE: The United States Marine Corp (USMC) does not possess the IEDK; therefore, it will have to use additional SDKs to decon individual and crew-served weapons.**

10-2. The following procedures reduce the spread/transfer of liquid contamination. These procedures are unique to the IEDK. Using other decon equipment with these procedures could result in a malfunction of the weapons/weapon systems. Powder from the IEDK could affect the mechanical parts of the weapons/weapon systems if not used properly. See the maintenance manual for proper lubrication after decon.

| WARNING |
|---|
| **Do not apply an IEDK to any optic lens. The abrasive effect of the charcoal inside the IEDK will damage the lens.** |

### M60 MACHINE GUN

**NOTE: Ensure that the feed tray of the weapon is closed. This prevents the powder from the IEDK from falling inside the feed tray and jamming the weapon. Do not touch the barrel of the weapon with the chemical protective gloves when it is hot. The heat of the barrel could melt the gloves.**

10-3. To decon the M60 machine gun, follow the steps below:

- Step 1. Decon your gloves using the IEDK.
- Step 2. Use the IEDK to remove all liquid contamination from the bipod, barrel assembly, forearm assembly, and carrying handle. Use a second IEDK if necessary.
- Step 3. Use the IEDK to decon the feed-tray cover, shoulder stock, and trigger assembly.
- Step 4. Repeat step 1.

**NOTE: Do not decon the flash suppressor with the IEDK. Powder debris may fall into the barrel and cause the weapon to malfunction.**

## M249 SQUAD AUTOMATIC WEAPON

**NOTE: Ensure that the feed tray of the weapon is closed. This prevents the powder from the IEDK from falling inside the feed tray and jamming the weapon. Do not touch the barrel of the weapon with chemical protective gloves when it is hot. The heat of the barrel could melt the gloves.**

10-4. To decon the M249 SAW, follow the steps below:

- Step 1. Decon your gloves using the IEDK.
- Step 2. Use the IEDK to remove all liquid contamination from the bipod, barrel assembly, gas regulator, and carrying handle.
- Step 3. Use the IEDK to decon the feed-tray cover, magazine, trigger assembly, and shoulder stock.
- Step 4. Repeat step 1.

**NOTE: Ensure that the weapon is upright when decontaminating the feed-tray cover with the IEDK. This prevents powder from the IEDK from falling into the ammunition feeder. Place the magazine in the well before the decon.**

## M203 GRENADE LAUNCHER

**NOTE: The M203 is not a crew-served weapon; however, this weapon is considered critical for accomplishing the infantry squad's mission. Decon the M16A2 according to STP 21-24-SMCT. Before decontaminating the M203 grenade launcher, ensure that the dust cover is closed and the magazine is stored in the well. Ensure that the barrel assembly is closed.**

10-5. To decon the M203 grenade launcher, follow the steps below:

- Step 1. Decon your gloves using the IEDK.
- Step 2. Use the IEDK to decon the barrel, hand guard, and receiver.
- Step 3. Use the IEDK to decon the lower receiver group and butt stock.
- Step 4. Repeat step 1.

## M47 DRAGON

10-6. Use this procedure only when liquid contamination is present on encased missile or tracker components. If the launcher has been fired and chemical contamination is present, follow the procedures shown in steps 1, 4, and 5 of paragraph 10-7 to decon the tracker. Gather all the contaminated launching tubes in a single area before departing the contaminated area and mark them properly.

**NOTE: Do not use the IEDK on any optic lens (exclude rubber eyecups), battery connector, electrical connectors, or lubricated components. Using the IEDK could cause corrosion if the powder from it is left on the equipment for an extended period of time. However, if these items are stored in a protective container/box, decon the storage container with the IEDK. Rinse or flush the lens with water only if it becomes contaminated. The rubber eyecups can be rinsed with water or decontaminated with the IEDK.**

10-7. To decon the M47 Dragon, follow the steps below:

- Step 1. Decon your gloves using the IEDK.
- Step 2. Use the IEDK to decon the bipod.
- Step 3. Use the IEDK to decon the exterior of the launcher.
- Step 4. Rinse or flush any liquid contamination off the exterior of the day/night sight with water. Use a rag or any available towelettes to dry the sight but do not touch the lens.
- Step 5. Repeat step 1.

## TOW MISSILE (M220 SERIES)

10-8. Gather all the contaminated launching tubes in a single area before departing the contaminated area and mark them properly.

**NOTE: Do not use the IEDK on any optic lens (exclude rubber eyecups), battery connector, electrical connectors, or lubricated components. However, if these items are stored in a protective container/box, decon the storage container with the IEDK or spray it with the M11 DAP. Do not touch or decon the front lens of the night sight. Rinse or flush the lens with water if it becomes contaminated with liquid chemical agent. The rubber eyecups can be rinsed with water.**

10-9. To decon the TOW missile (M220 series), follow the steps below:

- Step 1. Decon your gloves using the IEDK.
- Step 2. Use an M11 DAP filled with bleach or soapy water to spray down the launch tube, traversing unit, and tripod respectively. Protect or cover the day/night sights when spraying.
- Step 3. Use the IEDK to decon the exterior of the day-sight tracker and night-sight tracker. Avoid touching the lens. Once the sights are decontaminated, brush off any powder residue.
- Step 4. Use the IEDK to decon the collimator.
- Step 5. Use the IEDK to decon the exterior of the missile-guidance set and cable surface. Ensure that the electrical connectors are covered before the decon.
- Step 6. Use the IEDK to decon the power conditioner and its cable for the battery component.
- Step 7. Repeat step 1.

## AT4 LIGHT ANTITANK WEAPON

**NOTE: Ensure that the transport safety pin, the cocked-level safety, and the forward safety on the launcher are engaged.**

10-10. To decon the AT4 LAW, follow the steps below:

- Step 1. Decon your gloves using the IEDK.
- Step 2. Use the IEDK to wipe the exterior of the launcher and the front and rear sights.
- Step 3. Repeat step 1.

## 66-MILLIMETER ROCKET LAUNCHER

**NOTE: Ensure that the launcher is in the safe position and is not loaded before decontaminating. Omit step 2 if the launcher has not been fired in a contaminated environment.**

10-11. To decon the 66-millimeter rocket launcher, follow the steps below:

- Step 1. Decon your gloves using the IEDK.
- Step 2. Extend the launcher into the firing position.
- Step 3. Use the IEDK to decon the exterior of the launcher and around the reflecting sight.
- Step 4. Repeat step 3 using the IEDK.
- Step 5. Repeat step 1.

## 81-MILLIMETER MORTAR

**NOTE: Following the decon of the mortar with the IEDK, ensure that the mechanical parts of the mortar are dried. Lubricate the mortar when the mission permits. Do not touch the mortar tube with chemical protective gloves when it is hot. The heat of the tube could melt the gloves. Do not apply the IEDK to any optic lens due to the abrasive effect of the charcoal in the IEDK. Rinse or flush the lens with water, when required.**

10-12. To decon the 81-millimeter mortar, follow the steps below:

- Step 1. Decon your gloves using the IEDK.
- Step 2. Use the IEDK to decon from the blast attenuator device to the lower half of the mortar tube.
- Step 3. Use the IEDK to decon the base plate.
- Step 4. Use the IEDK to decon the bipod.
- Step 5. Use the IEDK to decon the elbow telescope and telescope mount. Do not use the IEDK on the optic lens. Flush the lens with water if contamination is present.
- Step 6. Repeat step 1.

## 60-MILLIMETER MORTAR

**NOTE: Following the decon of the mortar with the IEDK, ensure that the mechanical parts of the mortar are dried. Lubricate the mortar when the mission permits. Do not touch the mortar tube with chemical protective gloves when it is hot. The heat of the tube could**

**melt the gloves. Do not apply the IEDK to any optic lens due to the abrasive effect of the charcoal in the IEDK. Rinse or flush the lens with water, when required.**

10-13. To decon the 60-millimeter mortar, follow the steps below:

- Step 1. Decon your gloves using the IEDK.
- Step 2. Use the IEDK to decon the mortar tube and handgrip. If the mortar is heavily contaminated, you may need to use another IEDK.
- Step 3. Use the IEDK to decon the bipod and base plate.
- Step 4. Use the IEDK to decon the exterior of the telescope. Do not apply the IEDK to the lens. Flush the lens with water if contamination is present.
- Step 5. Repeat step 1.

### JAVELIN

10-14. The decon procedures for the Javelin and its components are listed in Table 10-1, page 10-6.

**WARNING**
**The IEDK only removes a liquid hazard. Decontaminated items may still present a vapor hazard. Do not unmask until it has been determined safe to do so.**

**CAUTION**
Close the lens covers on the Javelin before decontaminating because decon powders on the lens can obscure your vision. Do not use the IEDK on the eyecup because the decon powder entering the eyecup can obscure your vision.

## THOROUGH-DECON PROCEDURES USING THE HTH SOLUTION

10-15. When conducting the DED, the equipment is decontaminated with a 10 percent HTH solution (see Table 10-2, page 10-6) or with pure bleach that is diluted. Use the M256A1 detector kit, CAM, or M8 detector paper to determine the effectiveness of the decon. The crew-served weapons should be decontaminated before the DTD by the individual weapons' crew (the preferred method). Alternatively, the commander could task his own decon team to decon the unit's light weapons. The contaminated unit is responsible for establishing an area about 109 to 164 yards upwind from the DTD and DED sites for decon of the crew-served weapons. Unit leaders supervise this procedure. The unit chemical NCO plans and coordinates the preparation of this area.

10-16. The preferred decontaminant for a thorough decon is a 10 percent HTH solution (see Table 10-2). Use a contact time for the HTH solution of 15 minutes for chemical and biological agents. Prepare the amount of solution required according to the number and size of the weapon systems that need to be decontaminated.

**Table 10-1. Decon Procedures for the Javelin and Its Components**

| Immediate Decon | |
|---|---|
| **Round** | **Command Launch Unit (CLU)** |
| Step 1. Decon your gloves using the IEDK.<br>Step 2. Decon the round by patting it with the IEDK, working from the front to the rear until the entire round is covered.<br>**NOTE: When the shoulder strap is grossly contaminated, discard it.**<br>Step 3. Repeat step 1.<br>Step 4. Discard the IEDK. | Step 1. Decon your gloves using the IEDK.<br>Step 2. Decon the CLU by patting it with the IEDK. Pay special attention to the handgrips, CLU carrying handle, battery cover, round interface-connector protective cover, and switch boots.<br>Step 3. Use a knife to cut the protective-cover lanyard from the protective cover and remove it.<br>Step 4. Repeat step 1.<br>Step 5. Discard the IEDK. |
| **Thorough Decon** | |
| **Predecon Round** | **Predecon CLU** |
| Step 1. Check for contamination.<br>Step 2. Remove the shoulder strap.<br>Step 3. Use a knife to cut the connector boot.<br>Step 4. Use a knife to cut the protective-cover lanyard from the protective cover and remove it.<br>Step 5. Use a knife to scrape about 1/4 inch of the shoulder pad away from the launcher-tube assembly (LTA).<br>Step 6. Ensure that the protective cover is installed on the round/CLU interface connector. | **NOTE: Ensure that the on/off switch is in the off position. If the connector was not covered before contamination, place the protective cover on the connector before decon. After the decon, remove the protective cover and allow the connector to weather. The foam around the connector may be cut away as required.**<br>Step 1. Check for contamination.<br>Step 2. Remove the rubber from the handles and remove the lanyards from the CLU.<br>Step 3. Ensure that the protective covers are installed on the connectors. |
| **Decon LTA** | **Decon CLU** |
| Step 1. Submerge one half of the LTA in the decon solution.<br>Step 2. Submerge the other half of the LTA in the decon solution.<br>Step 3. Rinse one half of the LTA by submerging it in the rinse solution.<br>Step 4. Rinse the other half of the LTA by submerging it in the rinse solution.<br>Step 5. Use the CAM to check the LTA, ensuring that all contaminants have been removed.<br>Step 6. Repeat steps 1 through 5 above if the LTA is still contaminated. | Step 1. Open the lens covers.<br>Step 2. Submerge the CLU in the decon solution.<br>Step 3. Rinse the CLU by submerging it in the rinse solution.<br>Step 4. Use the CAM to check the CLU, ensuring that all contaminants have been removed.<br>Step 5. Repeat steps 1 through 4 above if the CLU is still contaminated. If only the switch boots remain contaminated, remove them by using a knife. |

**Table 10-2. HTH Mixture**

| HTH Solution | Amount of HTH per 1 Gallon of Water | Amount of HTH per 5 Gallons of Water |
|---|---|---|
| 5 percent | 9.6 ounces | 48 ounces |
| 10 percent | 12 ounces | 60 ounces |

**NOTE: Refer to Appendix B for instruction on how to use sodium hypochlorite (household bleach) and HTH, and see Table 10-2 for instructions on how to prepare the 5 and 10 percent HTH solution.**

10-17. Follow the procedures below during the DED and DTD. Always wear chemical gloves when performing these procedures.

- Use the HTH solution to decon up to 10 weapon systems. Use the required number of rags on a weapon system and then dispose of them properly to avoid transferring the contamination.
- Do not apply the HTH solution or bleach on electrical connectors and lens.
- Use bleach or a 10 percent HTH solution on the exterior of the optic systems that are not sensitive to bleach and then rinse them with water.

10-18. Once the HTH solution has been used on the weapon systems, thoroughly dry and lubricant them. Avoid spilling the HTH solution or bleach on the chemical overgarment.

## MATERIALS

10-19. Listed in the following paragraphs are the different types of materials that will be needed to clean the weapon systems:

### Cleaning Rags

10-20. The recommended number of clean rags to use with the HTH solution on each weapon system follows:

- M60 machine gun/M249 SAW (three rags)
  - 1: Bipod and barrel.
  - 1: Forearm, carrying handle, and rear sight.
  - 1. Feed-tray cover, shoulder stock, pistol grip, and trigger assembly.
- M203 grenade launcher (two rags)
  - 1: M16A1/2 barrel, hand guard, receiver group, and butt stock.
  - 1: M203 receiver group, sight, and exterior of the barrel.
- M47 Dragon (three rags)
  - 1: Exterior of the sight.
  - 1: Bipod (if weapon is not fired).
  - 1: Round (canister) (if weapon is not fired).

**NOTE: Keep the electrical connector covered before the decon. The Dragon's rubber eyepiece can be removed and submerged in the decontaminant.**

- TOW missile (M220 series) (seven rags)
  - 1: Exterior of the night sight.
  - 1: Exterior of the day sight.
  - 1: Launch tube.
  - 1: Traversing unit.
  - 1: Tripod assembly.
  - 1: Battery-power conditioner and cable. Ensure that the battery is in a waterproof container or case.
  - 1: Missile-guidance set and its cable.

**NOTE: Lock the bridge clamp after removing the round. Do not get the HTH solution on the electrical connector under the bridge clamp or on the post-amplifier cable.**

- 66-millimeter rocket launcher (one rag)
- AT4 LAW (one rag)
- 81-millimeter mortar (four rags)
  - 1: Blast denature device and tube.
  - 1: Bipod.
  - 1: Base plate.
  - 1: Exterior of the telescope.
- 60-millimeter mortar (three rags)
  - 1: Tube (barrel) and handgrip.
  - 1: Base plate and bipod.
  - 1: Exterior of the telescope.

### Water

3-21. Use five gallons of water (as a minimum) for the HTH mix. Change the solution after 10 weapons are decontaminated.

### Buckets

3-22. Use a minimum of two buckets for the procedures below.

### Trash Bags

3-23. Dispose of the contaminated rags in plastic trash bags.

### NBC Marking Kit

3-24. Mark the contaminated area once the decon procedures are completed.

### Lubricants

3-25. Use the required lubricants for the weapon systems. Follow the procedures specified in the applicable TMs.

## PROCEDURES

3-26. The procedures on how to decon weapons and weapon systems with the HTH solution are described below:

- Step 1. Submerge the rag or cloth in the HTH solution and ensure that the material is completely dampened.
- Step 2. Wipe down the weapons and weapon systems with the rag until the contamination is no longer suspected.
- Step 3. Submerge the rag or cloth into a container of clean water (use hot water when available) and wipe down the weapons and weapon systems once more to remove the HTH or bleach solution.
- Step 4. Completely dry the weapons and weapon systems once step 3 is completed.
- Step 5. Lubricate the weapons and weapon systems.

## Appendix A

# Conversions and Measurements

This appendix contains an English/Metric conversion chart (see Table A-1). It also contains the measurements and weights of decontaminant containers that are replacements for used containers (see Table A-2, page A-2).

**Table A-1. English/Metric Conversion Chart**

| To Convert | To | Multiply By | To Convert | To | Multiply By |
|---|---|---|---|---|---|
| Feet | Centimeters<br>Miles<br>Kilometers<br>Meters | 30.48<br>0.00019<br>0.00003<br>0.3048 | Gallons | Milliliters<br>Cubic meters | 3,785<br>0.00379 |
| Cubic yards | Cubic meters | 0.76456 | Quarts (fluid) | Millimeters<br>Liters | 946.358<br>0.946333 |
| Cubic feet | Cubic meters<br>Liters | 0.02831702<br>28.32 | Ounces | Grains<br>Drams<br>Pounds<br>Grams<br>Kilograms | 437.6<br>16<br>0.0625<br>28.34952<br>0.02835 |
| Yards | Meters | 0.9144 | Pounds | Grams<br>Kilograms<br>Metric tons | 453.6<br>0.453592<br>0.0004536 |
| Inches | Centimeter<br>Kilometers<br>Millimeters | 2.54000508<br>$2.54000508 \times 10^{-5}$<br>25.4000508 | Ounces (fluid) | Milliliters<br>Liters | 29.57353<br>0.02957 |
| | | | Square yards | Square meters | 0.836 |

**NOTE: For temperatures, Celsius = 5/9 (F° - 32°) and Fahrenheit = 9/5 (C° + 32°)**

**Table A-2. Measurements and Weights of Decontaminant Containers**

| Containers | Measurements | Weights |
|---|---|---|
| Drum, 55 gallons, 16 gauge<br>NSN 8110-00-597-2353 | Volume—12 cubic feet<br>Length—35 inches<br>Width—27.5 inches | Empty—70 pounds<br>Filled with water—526 pounds<br>Filled with STB (slurry)—620 pounds<br>Filled with DS2—507 pounds |
| Drum, 55 gallons, 18 gauge<br>NSN 8110-00-292-9783 | Volume—12 cubic feet<br>Length—35 inches<br>Width—27.5 inches | Empty—50 pounds<br>Filled with water—506 pounds<br>Filled with STB (slurry)—600 pounds<br>Filled with DS2—487 pounds |
| Ash and garbage can, 32 gallons (GI can)<br>NSN 7240-00-160-0440 | Volume—7 cubic feet<br>Length—26.5 inches<br>Width—20 inches | Empty—33 pounds<br>Filled with water—300 pounds<br>Filled with STB (slurry)—353 pounds<br>Filled with DS2—288 pounds |
| Gasoline can, 5 gallons<br>NSN 7240-00-178-8286 | Volume—1 cubic foot<br>Length—18.50 inches<br>Width—6.75 inches | Empty—10.5 pounds<br>Filled with water—52 pounds<br>Filled with STB (slurry)—66 pounds<br>Filled with DS2—50 pounds |
| Pail, 5 gallons, DS2<br>NSN 7240-00-575-2243 | Volume—1cubic foot<br>Length—13.75 inches<br>Width—11.12 inches | Empty—6 pounds<br>Filled with water—48 pounds<br>Filled with STB (slurry)—61 pounds<br>Filled with DS2—46 pounds |
| Drum, 8 gallons, STB, 16 gauge<br>NSN 6850-00-297-6693 | Volume—1.4 cubic feet<br>Length—14 inches<br>Width—14 inches | Empty—11 pounds<br>Filled with water—78 pounds<br>Filled with STB (slurry)—91 pounds<br>Filled with DS2—71 pounds |

## Appendix B

# Decontaminants

The decontaminants listed in Tables B-1 through B-3, pages B-1 through B-14, are grouped according to their classification: standard, nonstandard, or natural. Each is identified by name and national stock number (NSN) (where applicable). Also listed are brief directions on preparing the decontaminants and any appropriate cautions and information on application, coverage, substitute solutions, and related effectiveness of the decontaminants.

**Table B-1. Standard Decontaminants Available in the Supply System**

| Decontaminants | Use | Remarks | Cautions | Preparation |
|---|---|---|---|---|
| Decontaminating solution number 2; DS2 in a 1$^{1}/_{3}$-quart can (NSN 6850-00-753-4827); DS2 in a 14-liter container for the M13 DAP; (NSN 4230-01-136-8888); DS2 in a 5-gallon pail (NSN 6850-00-753-4870) | Bio<br>Chem | Is effective against all known toxic chemical and biological agents (except bacterial spores) if sufficient contact time is allowed.<br>Must remain in contact with contaminated surface for about 30 minutes.<br>Must be rinsed off with water. Recheck the item for contamination.<br>Can be used at temperatures above -15°F.<br>Is used with the M11 or M13 DAP or can be applied with brooms and swabs.<br>Is most effective when the application is accompanied by scrubbing action. | Is extremely irritating to the eyes and skin. Must wear protective mask and rubber gloves. If DS2 contacts skin, wash area with water. Do not inhale vapors. Will cause a green or black color change upon contact with M8 detector paper and cause a false/positive reading with M9 detector paper.<br>Will ignite spontaneously on contact with STB or HTH.<br>Should not be spilled on the chemical protective overgarment.<br>Is a combustible liquid with a flash point of 160°F. Do not confuse it with a fire extinguisher. Spraying DS2 on surfaces above 168°F will ignite it.<br>Must not be used on individual mask (damages Mylar diaphragm in voicemitter assembly). | Mixing is not required.<br>Issued in ready-to-use solutions. |

**Table B-1. Standard Decontaminants Available in the Supply System (Continued)**

| Decontaminants | Use | Remarks | Cautions | Preparation |
|---|---|---|---|---|
| Decontaminating solution number 2 (continued) | | | Will corrode aluminum, cadmium, tin, and zinc; will soften leather. May soften, remove, or discolor paint. Rinse well after use and oil metal surfaces.<br>Is ineffective against bacterial spores. | |
| Supertropical bleach (NSN 6850-00-297-6653) | Bio<br>Chem | Is effective against lewisite, V and G agents, and biological agents.<br>Must remain in contact with contaminated surface for at least 30 minutes; then, wash off with clear water.<br>Does not effectively decon mustard if it has solidified at low temperatures.<br>Should be applied several times to porous surfaces. | Will ignite spontaneously on contact with liquid blister agent or DS2.<br>Will give off toxic vapors on contact with G agent.<br>Is not recommended for ship use. Store on top deck only.<br>Is corrosive to most metals and damaging to most fabrics (rinse thoroughly and oil metal surfaces).<br>Should not be inhaled or allowed to touch the skin. When preparing a slurry, wear a protective mask or respiratory protective device.<br>Should be stored in an unheated warehouse away from combustibles and metals subject to corrosion. | Slurry paste: mix one 50-pound drum of STB with 6 gallons of water. Slurry paste consists of about equal parts (by weight) of STB and water.<br>Dry mix: mix two shovels of STB to three shovels of earth or inert material (ashes).<br>Slurry mix, chemical: mix will consist of 40 parts of STB to 60 parts of water (by weight). To mix in the M12A1 PDDA, use 1,300 pounds of STB, 225 gallons of water, 12½ pounds of antiset, 24 ounces of antifoam.<br>Slurry mix, biological: mix will consist of 7 parts of STB to 93 parts of water (by weight). To mix in the M12A1 PDDA, use 150 pounds of STB, 225 gallons of water, 12½ pounds of antiset, and 24 ounces of antifoam.<br>Camouflage: lampblack or dye mixes may be added for camouflage. |

**Table B-1. Standard Decontaminants Available in the Supply System (Continued)**

| Decontaminants | Use | Remarks | Cautions | Preparation |
|---|---|---|---|---|
| Mask sanitizing solution | Bio<br>Chem | Use on a previously cleaned mask with filter elements/canisters removed.<br>Place the mask face up; attach the canteen to the mask at the drinking tube. Drain one canteen full of sanitizing solution through the mask. Rinse the mask with two canteens of clear water.<br>Immerse the mask and outserts in the sanitizing solution. Agitate the mask for 5 minutes. Rinse it twice in clear water, agitating 2 to 3 minutes each time.<br>Dry all parts of the mask and reassemble.<br>Use one gallon of solution for every 10 masks. | | Fill a standard plastic canteen to the shoulder with water. Add a 0.5-gram tube of calcium hypochlorite from the water-purification kit (NSN 6810-00-266-6976). Cover the canteen and shake vigorously for 30 seconds.<br>Mix bulk quantities as follows: add 2.0 grams of calcium hypochlorite from a 6-ounce jar (NSN 6810-00-255-0471) to 1 gallon of water.<br>Use a ratio of about 1 pound of soap per gallon of water for smaller amounts of solution. Mix 2 pints of detergent to 450 gallons of water in the M12A1 PDDA. |
| Soaps and detergents: detergent, GP, liquid (NSN 7930-00-282-9699) | Rad<br>Bio<br>Chem | Scrub or wipe the contaminated surface with a hot, soapy water solution or immerse the item in the solution. | Are effective in physically removing contamination. However, casualty-producing levels of contamination may remain in the runoff water and must be considered contaminated. | Mix 75 pounds of powdered soap in 350 gallons of water. If powdered soap is not available, use bar laundry soap (75 pounds of soap cut into 1-inch pieces and dissolved in 350 gallons of hot water).<br>Use a ratio of about 1 pound of soap per gallon of water for smaller amounts of soap solution. Mix 2 pints of detergent to 450 gallons of water in an M12A1 PDDA. |

**Table B-1. Standard Decontaminants Available in the Supply System (Continued)**

| Decontaminants | Use | Remarks | Cautions | Preparation |
|---|---|---|---|---|
| Calcium hypochlorite (high-test hypochlorite [HTH] or high-test bleach [HTB]) (NSN 6810-01-225-2682) (25 pounds), (NSN 6810-00-225-0472) (100 pounds) | Bio<br>Chem | Is effective against mustards, lewisite, V agents, and all biological agents, including bacterial spores.<br>Will react rapidly (within 5 minutes) with mustards and lewisite.<br>Should be allowed a 15-minute contact time for biological agents.<br>Will act faster than STB. Can be used as a dry mix or slurry. | Precautions are the same as for STB.<br>Pure calcium hypochlorite will burn on contact with VX, HD, or DS2.<br>Agent is more corrosive than STB. Will destroy clothing, has a toxic vapor, and will burn the skin.<br>Protective mask and rubber gloves are the minimum protective equipment needed when handling. | Chemical: mix 5 pounds of decontaminant to 6 gallons of water (10 percent solution).<br>Biological: mix 1 pound of decontaminant to 6 gallons of water (2 percent solution).<br>PDDE: mix a slurry of 1 part decontaminant to 2 parts water (any heavier slurry will clog the decon apparatus).<br>HTH should only be used if STB is not available. |

**Table B-2. Nonstandard Decontaminants Available in the Supply System**

| Decontaminants | Use | Remarks | Cautions | Preparation |
|---|---|---|---|---|
| Oxidizing agents (nitric acid, aqua regia, sodium dichromate, and potassium permanganate) | Rad | Are effective in dissolving surfaces containing absorbed radioactive contamination.<br>Are applied to a surface or the item is dipped. Exposure must be limited due to the corrosive nature of the solution. Rinse thoroughly with water and detergent and then with clear water. | Are extremely corrosive. Use only under the supervision of a trained individual.<br>Will require the use of a neoprene or rubber protective apron, gloves, boots, and safety glasses when handling. (Rubber offers only limited protection.) | Aqua regia is prepared by mixing 3 parts of concentrated hydrochloric acid and 1 part of concentrated nitric acid. Other oxidizing agents do not require mixing. |
| Iodine water purification tablets | Bio | Are used when it is impractical to boil drinking water. Two iodine tablets per canteen are effective against most biological agents. | None | None |

**Table B-2. Nonstandard Decontaminants Available in the Supply System (Continued)**

| Decontaminants | Use | Remarks | Cautions | Preparation |
|---|---|---|---|---|
| Complexing (chelating) agents (versenc, citric acid, sequesterene, sodium citrate, tartanic acid, sodium oxalate, sodium tartrategoxalic acid, orthophosphoric acid, and similar agents) | Rad | Will aid in the physical removal of absorbed contamination on surfaces.<br>Should be applied as a film over the surface using the PDDE, firefighting apparatus, or tree or garden sprayer.<br>Should be allowed a 30-minute contact time and then flushed with water. | Do not neutralize contamination. Runoff will be contaminated. | Mix 3 to 5 percent of the agent (by weight) in water. |
| Disinfectant chlorine (NSN 6840-00-270-8172) | Bio | Use to decon utensils, mess gear, exteriors of sealed containers, and food products that can withstand soaking. Dispose of any food or vegetables that are damaged and any outer leaves that are bruised or torn. Do not cut or peel fruits and vegetables before disinfecting them. Leave items in the solution for 30 minutes and stir occasionally. Rinse thoroughly in potable water. | None | Dissolve one package of disinfectant in 20 gallons of warm potable water (100°F). |

**Table B-2. Nonstandard Decontaminants Available in the Supply System (Continued)**

| Decontaminants | Use | Remarks | Cautions | Preparation |
|---|---|---|---|---|
| Disinfectant chlorine (continued) | Bio | Use the solution only once.<br>Prepare an emergency solution by mixing one level MRE spoonful of calcium hypochlorite (water disinfecting powder) to each 10 gallons of water. If liquid chlorine bleach is available, use about $^{1}/_{3}$-canteen cup of 5 percent chlorine bleach to each 10 gallons of water.<br>Make fresh solutions for rinsing and disinfecting utensils for each 100 persons. | None | |
| Ethylene oxide | Bio | Is effective against all microorganisms, to include bacterial spores.<br>Should be applied in the strength of 30 pounds for every 1,000 cubic feet.<br>Should be allowed a 6-hour contact time (contact time must be doubled for each 20°F drop in temperature below 75°F).<br>Should be used in an airtight enclosure. | Is flammable. Not recommended for interior use. | None |
| Formalin (formaldehyde) | Bio | Agent is effective against all microorganisms, to include bacterial spores.<br>Agent is used for interior decon of relatively close areas.<br>Vapors should remain 16 hours in a closed structure; then, aerate until the odor is no longer objectionable. | Vapors are very toxic.<br>Agent will curl and discolor paper. Leaves a white residue.<br>A self-containing breathing apparatus is required in closed spaces.<br>Personnel handling or spraying the agent should wear impermeable protective clothing. | No mixing is required. However, less residue remains and less aeration is required if the mixture of 5 parts formalin and 3 parts methanol is used. (Use this mixture at a rate of 4/5 quart per 1,000 cubic feet of space). |

**Table B-2. Nonstandard Decontaminants Available in the Supply System (Continued)**

| Decontaminants | Use | Remarks | Cautions | Preparation |
|---|---|---|---|---|
| Formalin (formaldehyde) (continued) | Bio | Optimum conditions for spraying formalin are 70° to 80°F with an 85 percent relative humidity. The minimum effective relative humidity is 70 percent. At 60°F, the exposure should be increased to 24 hours.<br>Agent is applied as a vapor from standard insecticide sprayers or is vaporized by heat or a bubbling steam from a pan. | Personnel entering an area containing formalin vapors should—<br>• Wear a protective mask.<br>• Wear washable outer clothing, fastened to prevent vapors from entering at wrists, ankles, or neck.<br>• Remove outer clothing after emerging from vapors.<br>• Shower and put on clean clothing as soon as possible.<br>Vapors are not flammable; open flame should not be used for vaporizing when methanol has been added to the agent. When steam is used, the source of the steam should be outside the area being decontaminated. | |
| Detrochlorite | Bio | Agent is a thickened bleach that is very useful on vertical surfaces.<br>Coverage is 1 gallon per 8 square yards. Apply by means of the PDDA. Allow 30 minutes contact time, then rinse with water. | Agent is very corrosive.<br>See *Preparation* column. | Mix by weight 19.3 percent diatomaceous earth, 0.5 percent anionic wetting agent, 2.9 percent calcium hypochlorite (70 percent available chlorine), and 77.3 percent water.<br>Mix wetting agent and diatomaceous earth with water before adding the calcium hypochlorite. Mixing the wetting agent and calcium hypochlorite in a dry and undiluted state may cause an explosion. |

**Table B-2. Nonstandard Decontaminants Available in the Supply System (Continued)**

| Decontaminants | Use | Remarks | Cautions | Preparation |
|---|---|---|---|---|
| Peracetic acid (PAA) | Bio | Agent is effective against all microorganisms, to include bacterial spores.<br>A contact time of 10 minutes must be allowed.<br>Item should be wiped with a rag or swab. Remove excess acid and aerate for 10 to 15 minutes or until no objectionable odor remains.<br>Small items should be immersed for 10 minutes. Remove excess acid and aerate until no objectionable odor remains. | Fumes are highly irritating.<br>Prolonged exposure will damage most material.<br>A 40 percent solution has a low flash point (105°F); a 3 percent solution is nonflammable.<br>Burns and blistered on the skin will occur.<br>Agent must be stored in original containers under refrigeration to prevent decomposition.<br>Protective mask and clothing are required.<br>A violent explosion may result if heavy metal ions come in contact with the agent.<br>Prolonged exposure will corrode iron and deteriorate rubber, plastic, and leather. | Available as a 40 percent solution. Mix 1 quart of PAA to 3½ gallons of water (add PAA to the water). |
| Hyamine (benzethonium chloride) | Bio | Agent is effective against all bacteria.<br>A contact time of 5 to 30 minutes should be allowed. | Agent is very toxic.<br>Estimated fatal dose to man is 1 to 3 grams.<br>Care should be taken when mixing to avoid inhalation of powder.<br>Agent is not to be used on aircraft or ships. | Use a 0.1 to 1 percent solution (1 pound of hyamine for every 12 gallons of water yields a 1 percent solution). |
| Carbon dioxide and ethylene oxide mixture (with more than 87 percent ethylene oxide) | Bio | Should be applied in the strength of 30 pounds for every 1,000 cubic feet.<br>Should be allowed a 12-hour contact time (doubled for each 20°F drop in temperature below 75°F).<br>Is nonflammable and is recommended for interior use.<br>Should be used in an airtight enclosure. | Will blister the skin.<br>Aerate items next to the skin 18 to 24 hours. | None |

**Table B-2. Nonstandard Decontaminants Available in the Supply System (Continued)**

| Decontaminants | Use | Remarks | Cautions | Preparation |
|---|---|---|---|---|
| Sodium hypochlorite solution (household bleach) | Bio<br>Chem | Is effective against blister and V agents and all biological agents.<br>Will react rapidly (within 5 minutes) with blister and V agents.<br>Should be allowed a 10- to 15-minute contact time for biological agents.<br>Should be applied undiluted with brooms, brushes, or swabs.<br>Is the preferred decon for ship use. A 5:1 concentration is recommended.<br>Has a limited storage problem. | Is harmful to the skin and clothing if undiluted. Remove from the skin and clothing by flushing with water.<br>Is corrosive to metals unless rinsed, dried, and lubricated after decon.<br>Should be stored in a cool place. | For chemical decon, no mixing is required.<br>For biological decon, dilute by adding 2 parts bleach to 10 parts water.<br>For decon of cotton clothing and utensils, dilute 2 cups of bleach to 1 gallon of water.<br>For application, dilute half and half with water and spray from the PDDE. |
| 2-Propanone (acetone) | Chem | Freezing point is -203°F; boiling point is 133°F (evaporates rapidly).<br>Good decontaminant for use in arctic regions.<br>Commonly obtained as fingernail polish remover or paint thinners.<br>Scrubbing increases its effectiveness. | Is extremely flammable.<br>Does not neutralize agents.<br>Is effective for dissolving and flushing agent by physically removing it | None |
| Dietyl ether | Chem | Freezing point is -241°F; boiling point is 93°F.<br>Good decontaminant for use in arctic regions.<br>Available through medical supply facilities.<br>Scrubbing increases its effectiveness. | Is the same as for 2-propanone. | None |

**Table B-2. Nonstandard Decontaminants Available in the Supply System (Continued)**

| Decontaminants | Use | Remarks | Cautions | Preparation |
|---|---|---|---|---|
| Ethylene glycol | Chem | Scrub on contaminated surfaces and rinse thoroughly. | Removes contamination, but does not neutralize it. Therefore, runoff residue must be considered contaminated. | Mix equal amounts of solution and water. |
| Solvents (gasoline, JP-4, diesel fuel, kerosene, and similar solvents) | Chem | Scrub on contaminated surfaces and rinse thoroughly. | Is the same as for ethylene glycol.<br>May damage materials such as rubber and plastic. | None |
| Sodium hydroxide (caustic soda or lye) (NSN 6810-00-174-6581) (100 pounds) | Bio<br>Chem | Is effective against G agents, lewisite, and all biological agents, including bacterial spores.<br>Will neutralize G agents on contact.<br>Should be allowed to remain in contact with the chemically contaminated surface for about 15 minutes. | Agent will damage the skin, eyes, and clothes and can cause upper respiratory or lung damage if inhaled.<br>Full rubber protective clothing, gloves, boots, and mask are required when using.<br>Affected area must be washed immediately with large amounts of water and flushed with diluted acetic acid or vinegar. Remove affected clothing. If eyes are involved, flush them at once with large amounts of warm water and seek medical attention.<br>Runoff from decon operations is highly corrosive and toxic. Drain runoff into a sump and bury.<br>Equipment must be flushed with large amounts of clear water. | Small amount—10 pounds of lye to 12 gallons of water (10 percent solution). Mix in an iron or steel container (never aluminum, zinc, or tin). Add lye to the water to prevent boiling and splattering due to heat being emitted. Do not handle mixing container with bare hands.<br>Large amount (PDDE use)—Prepare a solution of 227 grams (½ pound) of lye for each gallon of water. Pump 350 gallons of water into the tank unit. Connect the tank unit, pump unit, and heater together. Heat the water to 122°F. Disconnect the heater unit and add 175 pounds of lye to the heated water. Circulate the solution with the pump unit until all the lye is dissolved. The temperature will increase noticeably. Use while hot. |

**Table B-2. Nonstandard Decontaminants Available in the Supply System (Continued)**

| Decontaminants | Use | Remarks | Cautions | Preparation |
|---|---|---|---|---|
| Sodium hydroxide (caustic soda or lye) (NSN 6810-00-174-6581) (100 pounds) (continued) | | | Sodium hydroxide—<br>• Is not recommended for ship use. Store on top deck only.<br>• Is corrosive to most metals.<br>• Is not recommended if less toxic caustic decontaminants are available. Will cause a red color change upon contact with M8 detector paper.<br>Agent's effectiveness is directly proportional to the strength of the solution. | Simultaneous mixing and applying—Sprinkle dry lye on the contaminated area and then dissolve it with a spray of steam or hot water. Do not wash the lye off the surface while applying the steam or hot water.<br>Paint removal—1 pound of lye per 2½ gallons of water is capable of removing an average coat of paint from about 11 square yards of surface. This solution is effective in removing paint on which chemical contamination has absorbed. (Can substitute calcium hydroxide, potassium hydroxide, or trisodium phosphate for sodium hydroxide.) |
| Sodium carbonate (washing soda, soda ash, sal soda, or laundry soda) | Chem | Is effective against G agents and CN.<br>Will react rapidly with G agents, normally within 5 minutes.<br>Is the preferred decontaminant for ship use. The recommended concentration is 5 percent by weight. There is no storage problem.<br>Should be used with a hot solution to decon CN effectively. | Should not be used for VX. It cannot detoxify VX and creates extremely toxic by-products.<br>Does not dissolve mustard agents or detoxify them. | Mix 10 pounds of washing soda to 12 gallons of water (10 percent solution). |
| Potassium hydroxide (caustic potash) | Bio<br>Chem | Is the same as for sodium hydroxide. | Is the same as for sodium hydroxide. | Is the same as for sodium hydroxide. |
| Hexachloramelamine | Chem | Is effective against mustard agents. | May require the use of a protective mask and rubber gloves when used.<br>Is corrosive to metal. | Is not soluble in water but is soluble in organic solvents such as gasoline, kerosene, and paint thinner. |

**Table B-2. Nonstandard Decontaminants Available in the Supply System (Continued)**

| Decontaminants | Use | Remarks | Cautions | Preparation |
|---|---|---|---|---|
| Ammonia or ammonium hydroxide (household ammonia) | Chem | Is effective against G agents.<br>Is slower acting than sodium hydroxide or potassium hydroxide. | May require the use of a self-contained breathing apparatus or special purpose mask. | Ammonium hydroxide is a water solution of ammonia. No further mixing is required. |
| Dichloramine-B and Dichloramine-T | Chem | Is effective against mustard agents. | May require the use of a protective mask and rubber gloves when used.<br>Is corrosive to metal. | Is not soluble in water, but is soluble in certain organic solvents.<br>Is normally mixed as a 10 percent solution in dichloroethane. |
| Perchloroethylene (tetrachloroethylene) | Chem | Freezing point is -8°F; boiling point is 250°F.<br>Agent—<br>• Is good for use in arctic climates.<br>• Is a nonflammable, synthetic solvent widely used in dry cleaning plants.<br>• Dissolves H and V agents but not G.<br>• Has a low toxicity.<br>Scrubbing increases its effectiveness. | Physically dissolves and removes contamination, but does not neutralize it. | Requires no mixing (practically insoluble in water). |
| Acids (sulfuric acid, hydrochloric acid, oxalic acid, and similar acids) | Rad | Are effective solvents for rust and mineral deposits holding radioactive material on metal surfaces.<br>Should be allowed a 1-hour contact time.<br>Must be flushed with water, scrubbed with a water-detergent solution, and flushed again with water. | Are difficult to handle and are harmful to the body, especially the eyes.<br>Can produce boiling and splattering of the solution when mixed.<br>May require the use of rubber boots, gloves, aprons, and goggles when used.<br>Will require the use of respiratory protection when used in closed areas.<br>Can cause serious injury when they come in contact with the body. Flush the area immediately with water. Use a 5 percent solution of water and baking soda (sodium bicarbonate). | None |

### Table B-3. Natural Decontaminants

| Decontaminants | Use | Remarks | Cautions | Preparation |
|---|---|---|---|---|
| Water | Rad<br>Bio<br>Chem | Can be used to flush contamination from surfaces. | Is effective in physically removing contamination, but does not neutralize it. Do not use water on lewisite. | Hot water with soap makes water more effective in removing agents. Boiling for 15 minutes (30 minutes at high altitude) destroys biological agents. |
| Steam | Rad<br>Bio<br>Chem | Is more effective when accompanied by scrubbing. | Is effective in physically removing contamination, but does not neutralize it. | None |
| Absorbents (earth, sawdust, ashes, rags, and similar materials) | Chem | Are used to remove gross contamination, physically, from surfaces. | The contamination is transferred from the surface to the absorbent. The absorbent becomes contaminated and must be disposed of accordingly.<br>Sufficient contamination to produce casualties may well remain on surfaces. | |
| Sealants (concrete, asphalt, earth, paint, and similar materials) | Rad<br>Bio<br>Chem | Sealants are used to physically seal in or shield contamination. They are effective as follows:<br>• 12 inches of earth provides a good protection from fallout (3 inches will reduce the dose rate about one half).<br>• 1 inch of asphalt or concrete completely absorbs alpha and beta radiation. | A break in the surface of the sealant will expose the contamination.<br>Contaminated areas covered with sealants must be marked with appropriate NBC warning signs. | None |

**Figure B-3. Natural Decontaminants (Continued)**

| Decontaminants | Use | Remarks | Cautions | Preparation |
|---|---|---|---|---|
| Sealants (concrete, asphalt, earth, paint, and similar materials) (continued) | Rad<br>Bio<br>Chem | • 1/4 inch of grout shields alpha and bata radiation.<br>• 4 inches of earth provides good protection from chemical contamination.<br>Burying items contaminated with biological agents is an effective means of sealing off contamination. | A break in the surface of the sealant will expose the contamination.<br>Contaminated areas covered with sealants must be marked with appropriate NBC warning signs. | None |
| Weather/time | Rad<br>Bio<br>Chem | UV light kills most bio-organism agents, organisms, and radiation decay overtime. | Should be used when time and the mission permits. | None |
| Burning | Bio<br>Chem | Must be approved by the first general officer in the chain of command. | Creates downwind hazards.<br>Requires that sentries be posted to keep people out of the danger area. | None |

## Appendix C

# Decon of Specific Surfaces and Materials

Table C-1, pages C-1 through C-7, lists specific surfaces or materials and explains briefly how to best decon each for chemical, biological, or radiological contamination.

**Table C-1. Decon Procedures for Specific Surfaces and Materials**

| Surface or Material | Types of Contamination and How to Decon | | |
|---|---|---|---|
| | **Chemical** | **Biological** | **Radiological** |
| Asphalt roads (applicable to small vital areas only) | Flush with water.<br>Spray with an STB slurry from the PDDE.<br>Cover with STB (pure form). When liquid contamination is visible and personnel are nearby, use the STB dry mix.<br>Weather.<br>Cover small areas or paths across roads with 4 inches of earth. | Weather (remain masked).<br>Wet with water (will help prevent secondary aerosols but does not decon).<br>Apply 2 percent household-bleach solution.<br>Spray with an STB slurry from the PDDE.<br>Pour, spray, or spread oil on the surface (will help pro vent secondary aerosols, but does not decon). | Brush or sweep.<br>Flush with water (this may drive some of the contamination into the surface; waste must be controlled).<br>Clean with a vacuum. |
| Roofs | Follow the same directions as for asphalt roads. | Follow the same directions as for asphalt roads.<br>Apply detrochlorite; leave on at least 30 minutes, then flush with water. | Follow the same directions as for asphalt roads. |
| Brick and stone roads (applicable to small vital areas only) | Weather.<br>Spray with an STB slurry from the PDDE or apply with brushes and brooms. Leave on for 24 hours, then flush with water.<br>Wash with soapy water, preferably hot.<br>Cover small areas or paths across roads with 4 inches of earth. | Follow the same directions as for asphalt roads. | Follow the same directions as for asphalt roads.<br>Use abrasion (sand blasting). This provides direct and complete removal of contaminated dust; however, sand and equipment being used becomes contaminated. |

### Table C-1. Decon Procedures for Specific Surfaces and Materials (Continued)

| Surface or Material | Types of Contamination and How to Decon | | |
|---|---|---|---|
| | Chemical | Biological | Radiological |
| Brick and stone buildings, bunkers, gun emplacements, and tank obstacles | Spray with an STB slurry from the PDDE or apply with brushes and brooms. Leave on for 24 hours, then flush with water.<br>Use STB (pure form) or an STB dry mix around buildings where waste water runs.<br>Wash with soapy water, preferably hot.<br>Weather. | Follow the same directions as for asphalt roads.<br>Apply an STB slurry to vertical surfaces by manual means or with the PDDE. The slurry may be left on exteriors. | Follow the same directions as for brick and stone roads. |
| Concrete roads (applicable to small vital areas only) | Spray with an STB slurry from the PDDE.<br>Cover with an STB slurry or STB dry mix.<br>Weather.<br>Cover small areas or paths across roads with 4 inches of earth.<br>Scrape the layer of contaminated earth to the side of the road. | Follow the same directions as for asphalt roads. | Follow the same directions as for brick and stone roads. |
| Earth (applicable to small vital areas only): roads, gun emplacements, bivouac areas, pathways, and bomb craters | Spray with an STB slurry from the PDDE.<br>Cover with STB (pure form). When liquid contamination is visible and personnel are nearby, use an STB dry mix.<br>Weather.<br>Cover small areas or paths across roads with 4 inches of earth.<br>Scrape the layer of contaminated earth to the side of the road. | Follow the same directions as for asphalt roads.<br>Burn. | Earth moving (removal): Try to control contaminated dust, as equipment may become contaminated. Consider waste disposal.<br>Sealing (with earth): Does not create a waste-disposal problem; however, equipment may be contaminated. |
| Grass and low vegetation: fields and open terrain | Burn.<br>Spray with an STB slurry from the PDDE.<br>Cover with STB (pure form) or an STB dry mix.<br>Explode drums of STB.<br>Clear paths through area using detonating cord or other detonating devices. | Burn.<br>Follow the same directions as for asphalt roads. | Follow the same directions as for earth. |

**Table C-1. Decon Procedures for Specific Surfaces and Materials (Continued)**

| Surface or Material | Types of Contamination and How to Decon | | |
|---|---|---|---|
| | **Chemical** | **Biological** | **Radiological** |
| Undergrowth, tall grass (applicable to small vital areas only): meadows, jungles, and forests | Burn (may cause a downwind vapor hazard).<br>Spray slurry with PDDE.<br>Explode drums of STB.<br>Clear a path with detonating cord, bangalore torpedoes, or demolition snakes. | Burn.<br>Follow the same directions as for sand. | Follow the same directions as for earth. |
| Sand (applicable to small vital areas only): beaches and deserts | Flush with water.<br>Use DS2.<br>Spread STB (pure form) or spray an STB slurry over the surface.<br>Weather.<br>Cover paths with roofing paper.<br>Scrape off 2 to 4 inches of contaminated top layer. | Burn. | Follow the same directions as for earth. |
| Fabrics: canvas, covers, tarpaulins, tentage, mask carriers, web gear, and clothing | Cotton: **NOTE: Not for MOPP gear.**<br>Immerse in boiling, soapy water for 1 hour (1 pound of soap to 10 gallons of water) and stir.<br>Use a 5 percent sodium-carbonate solution for G agents.<br>Immerse in boiling water for 1 hour.<br>Launder by standard methods.<br>Use an STB slurry.<br>Weather except for V agents.<br><br>Woolen:<br>Do not use DS2, as it is not recommended.<br>Immerse in warm, soapy water (100°F) for 1 hour or longer with light agitation.<br>Dry items slowly because the fabric may shrink. | Cotton:<br>Boil in water for 15 minutes.<br>Autoclave for 45 minutes at 253°F.<br>Immerse in a 2 percent household-bleach solution for 30 minutes and rinse immediately.<br>Launder (destroys or inactivates all but highly resistant spores).<br><br>Woolen:<br>Do not use DS2, as it is not recommended.<br>Launder (fabric may shrink). | Cotton and woolen:<br>Using DS2 is not recommended.<br>Brushing removes contamination dust, but it presents a dust hazard to personnel.<br>Laundering is the most practical procedure; however, the fabric may shrink. Try to control waste. |

**Table C-1. Decon Procedures for Specific Surfaces and Materials (Continued)**

| Surface or Material | Types of Contamination and How to Decon | | |
|---|---|---|---|
| | **Chemical** | **Biological** | **Radiological** |
| Leather: boots, gloves, and other items | Scrub with hot, soapy water and rinse.<br>Immerse in warm, soapy water at 120°F for 4 hours and rinse.<br>Use a 5 percent sodium-carbonate solution for G agents.<br>Air. | Immerse in a 2 percent household-bleach solution and rinse.<br>Immerse in a 2 percent PAA for 10 minutes, rinse, and air 10 to 15 minutes.<br>Wipe with a 2 percent PAA, remove excess agent, and air 10 to 15 minutes. | Brush.<br>Flush with water or soapy water. |
| Glass, windows | Use an IEDK or DS2.<br>Wash with hot, soapy water.<br>Rinse with clear water or an organic solvent.<br>Blot off surface.<br>Air.<br>Weather. | Use an IEDK.<br>Wash with soap and water.<br>Wipe with a disinfectant solution or a 2 percent PAA (see similar procedures on page C-7 for mess gear). | Use an IEDK.<br>Wash with a detergent.<br>Flush with water.<br>Wipe with solvents. |
| Glass, lens | Wash with hot, soapy water.<br>Rinse with clear water or an organic solvent.<br>Blot off surface.<br>Air.<br>Weather. | Wash with hot, soapy water.<br>Wipe with alcohol or household bleach. | Brush or wipe (be careful so as not to scratch the lens).<br>Use compressed air to blow contamination from surface. |
| Metal (unpainted), ammunition | Wipe with soapy water.<br>Wipe with organic solvent and dry.<br>Air. | Wipe with soapy water.<br>Wipe with a 2 percent household-bleach solution.<br>Air. | Brush or wipe. |
| Metal (unpainted), machinery | Use DS2.<br>Follow the same directions as for ammunition. | Use DS2.<br>Wipe with a 2 percent PAA, rinse, and air 10 to 15 minutes. | Brush or wipe.<br>Wash with a detergent.<br>Flush with water. |
| Metals (painted): vehicles, weapons, and equipment | Use DS2 (may soften paint).<br>Wash with hot, soapy water and rinse.<br>Spray with an STB slurry from the PDDE then remove it in 1 hour and oil the surface.<br>Weather.<br>Air.<br>Use an IEDK to decon individual gear. | Wash with a detergent and a high-pressure water system.<br>Apply detrochlorite. Leave it on for 30 minutes then remove by washing with a stream of water.<br>Steam clean using a detergent.<br>Use a household-bleach solution.<br>Use a 2 percent PAA. | Brush or wipe.<br>Wash.<br>Use organic solvents, caustic agents (not on aluminum or magnesium surfaces), complexing agents (of small value on weathered surfaces), or abrasives. |

**Table C-1. Decon Procedures for Specific Surfaces and Materials (Continued)**

| Surface or Material | Types of Contamination and How to Decon | | |
|---|---|---|---|
| | Chemical | Biological | Radiological |
| Wood (unpainted surfaces): buildings, vehicle bodies, boxes, crates, and similar items | Apply an STB slurry with the PDDE, brooms, or swabs. Let the slurry remain 12 to 24 hours then flush with water. Repeat the application and flush again.<br>Scrub with hot, soapy water and rinse.<br>Weather. | Apply detrochlorite. Leave it on for at least 30 minutes then flush with water.<br>Apply an STB slurry to vertical surfaces. The slurry may be left on interiors.<br>Weather (sun and rain eliminate most microorganisms within 1 day).<br>Burn. | Wash the interior with large amounts of water (some contamination may soak into surfaces). |
| Wood (painted surfaces): buildings and boxes | Apply an STB slurry with the PDDE, brooms, or swabs. Let the slurry remain 12 to 24 hours then flush with water.<br>Scrub with hot, soapy water and rinse.<br>Use DS2 and rinse. DS2 may soften paint.<br>Weather. | Follow the same directions as for wood (unpainted surfaces). | Wash the interior with large amounts of water.<br>Wipe the contamination from the surface. |
| Plastics (opaque): insulation, telephones, and panel boards | Use DS2 (may soften or damage some plastics).<br>Wash with hot, soapy water and rinse.<br>Weather.<br>Air. | Follow the same directions as for glass lens. | Wash with a detergent.<br>Flush with water.<br>Wipc or bruch. |
| Plastics (transparent): eyepieces and airplane canopies | Use DS2 (may soften or damage some plastics).<br>Wash with hot, soapy water and rinse.<br>Weather.<br>Air. | Follow the same directions as for glass lens. | Wash with a detergent.<br>Flush with water.<br>Wipe or brush. |
| Nonsealed electronic equipment | Use a mild, evaporative solvent such as alcohol.<br>Use DS2.<br>Use a mild detergent and water.<br>Rinse with distilled water.<br>Keep the solution out of the adjustable switches, connectors, and relays.<br>Use hot-air blowers. | Use a mild, evaporative solvent such as alcohol.<br>Use DS2.<br>Use a mild detergent and water.<br>Rinse with distilled water.<br>Keep the solution out of the adjustable switches, connectors, and relays.<br>Use hot-air blowers. | Use hot-air blowers.<br>Wipe or brush. |

**Table C-1. Decon Procedures for Specific Surfaces and Materials (Continued)**

| Surface or Material | Types of Contamination and How to Decon | | |
|---|---|---|---|
| | **Chemical** | **Biological** | **Radiological** |
| Rubber (impermeable): aprons, suits, and other items | Spray with DS2 and rinse after 30 minutes.<br>Immerse in soapy water (just below the boiling point) for 1 hour. Do not agitate. Rinse with clear water and hang to dry.<br>Use a 10 percent sodium-carbonate solution for G agents, rinse, and air.<br>Apply hot, soapy water with brushes and rinse.<br>Spray with an STB slurry from the PDDE.<br>Wash off, after a few minutes, with clear water. | Follow the same directions as for leather. | Brush.<br>Scrub or flush with water or soapy water. |
| Rubber (natural and synthetic): gloves and boots | Spray with a 10 percent mixture of HTH and rinse.<br>Immerse in STB slurry for 4 hours, rinse, and dry.<br>Use an IEDK in emergencies.<br>Air. | Follow the same directions as for leather. | Brush.<br>Scrub or flush with water or soapy water. |
| Rubber: mask facepieces and other rubber articles coming in direct contact with the skin | Use IEDK in emergencies.<br>Wash with warm, soapy water. | Follow the same directions as for leather. | Brush.<br>Scrub or flush with water or soapy water. |
| Rubber: tires, hoses, mats, and insulation | Spray with a 10 percent mixture of HTH and rinse.<br>Apply an STB slurry. Allow the slurry to remain at least 30 minutes then flush with clear water (may be left on tires).<br>Apply hot, soapy water.<br>Air.<br>Weather. | Follow the same directions as for leather. | Brush.<br>Scrub or flush with water or soapy water. |
| Water | Use trained water-purification personnel to decon. | Boil small amounts 15 minutes.<br>Chlorinate using a chlorination kit.<br>Add iodine water-purification tablets to small amounts. | Flocculation (requires special chemicals to remove suspended matter).<br>Ion exchange (removes radians from solution). |

**Table C-1. Decon Procedures for Specific Surfaces and Materials (Continued)**

| Surface or Material | Types of Contamination and How to Decon | | |
|---|---|---|---|
| | Chemical | Biological | Radiological |
| Mess gear and canned rations | Immerse in boiling, soapy water for 30 minutes and rinse.<br>Immerse in boiling water for 30 minutes.<br>Spray with DS2.<br>Wash in hot, soapy water; rinse; and air. | Wash with soap and water, then immerse in disinfectant solution (disinfectant, chlorine or $^1/_3$ cup of household bleach per 10 gallons of water).<br>Boil in water for 15 minutes (not effective on toxins and bacterial spores.) Immerse in a 5 percent sodium-carbonate solution (4 pounds of washing soda to 10 gallons of water) and rinse with potable water.<br>Immerse in a household-bleach solution (2 gallons of bleach to 25 gallons of water) for 30 minutes, rinse, and air for 10 to 15 minutes.<br>Immerse in an HTH solution (½ pound to 25 gallons of water) for 30 minutes and rinse.<br>Immerse in a 2 percent PAA for 10 minutes, rinse, and air 10 to 15 minutes. | Wash with soap and water and rinse.<br>Brush and then wipe contamination from surfaces and containers. |
| Food: not canned or protected by impermeable container | Do not consume food that is suspected to be contaminated with chemical agents until veterinary personnel approve it. | Boil water 15 minutes. Cook food thoroughly.<br>Immerse in or spray with a 2 percent household-bleach solution. (Packaged food or food that is peeled or pared can be immersed or sprayed with the solution.) | Wash or trim contamination from unpackaged food. |
| Food: canned, bottled, or protected by impermeable container | Follow the same directions as for mess gear and canned rations. | Follow the same directions as for mess gear and canned rations. | Follow the same directions as for mess gear and canned rations. |
| Paper currency | Destroy by burning. Do not decon. | Destroy by burning. Do not decon. | Destroy by burning. Do not decon. |
| Coins | Wipe with soapy water.<br>Wipe with an organic solvent. | Use UV rays.<br>Wash with soapy water.<br>Wipe with a 2 percent household-bleach solution.<br>Air. | Brush or wipe.<br>Wash with a detergent.<br>Flush with water. |

## Appendix D

# Decon Kits, Apparatuses, and Equipment

Various materials and equipment are used in decon operations. Some are simple to use and are readily available to individual soldiers. Others are very complex to use and are available only to specially trained teams. Table D-1, pages D-1 and D-2, lists the decon equipment and materials for the different use levels; Table D-2, pages D-3 through D-5, lists the detection equipment and materials for the different use levels.

**Table D-1. Decon Equipment and Materials**

| Item and Description | Use | Limitations | Reference |
|---|---|---|---|
| | **Individual** | | |
| Decontaminating kit, skin, M291 SDK, (NSN 4230-01-276-1905) (20 kits per box) | To decon your skin, completely, through physical removal, absorption, and neutralization of toxic agent with no long-term effects<br>**NOTE: Use this kit for both actual combat and training purposes.** | Is for external use only.<br>**WARNING**<br>**Keep decon powder out of eyes, cuts, or wounds. It could slightly irritate the skin or eyes.** | TM 3-4230-229-10 |
| Decon kit, individual equipment, M295, IEDK, (NSN 6850-01-357-8456) | To decon your chemical protective gloves, mask, hood, overboots, LCE, and weapon. | **WARNING**<br>**Do not use for skin decon. Keep off the skin and out of wounds, eyes, and mouth.** | TM 3-4230-235-10 |
| Decontaminating apparatus, portable, M13 DAP, (3.7 gal), (NSN 4230-01-133-4124). It has a 3.7-gallon disposable DS2 container. The apparatus can be mounted to a standard 5-gallon fuel can mount on vehicles and equipment. | To spray DS2 on surfaces of vehicles and equipment. | Cannot refill DS2 container. See Table A-2, page A-2, for a list of replacement containers. | TM 3-4230-214-12&P<br>TM 43-0001-26-1 |

**Table D-1. Decon Equipment and Materials (Continued)**

| Item and Description | Use | Limitations | Reference |
|---|---|---|---|
| **Battalion Decon Crew and Chemical Company** | | | |
| Decontaminating apparatus, power driven, skid mounted, multipurpose, integral, 500 gallon, M12A1 PDDA, (NSN 4230-00-926-9488), LIN F81880. Apparatus includes pump unit, tank unit, and M2 water heater (all mounted on skids). | To spray decon agent, STB slurries, and solutions as well as hot, soapy water rinses during field decon operations.<br>To pump water or foam to fight fires, de-ice items, wash vehicles, and pump various fluids. | Do not use with defoliants, herbicides, or insecticides. | LO 3-4230-209-12<br>LO 5-2805-259-12<br>TM 3-4230-209-20&P<br>TM 43-0001-26-1 |
| Decontaminating system, lightweight, M17 LDS, (NSN 4230-01-251-8702). The M17 is a portable pump and water-heating unit for producing hot water and steam. The system incorporates a 1,580- to 3,000-gallon collapsible water tank, two wand assemblies, and connecting hoses. | To perform operational and thorough decon of vehicles and equipment. Can also be used for troop showers, as necessary. | | TM 3-4230-228-10 |

**Table D-2. Detection Equipment and Materials**

| Item and Description | Use | Limitations | Reference |
|---|---|---|---|
| | | **Individual** | |
| Paper, chemical agent, detector, VGH, M8, (NSN 6665-00-050-8529). Paper is issued in a book of 25 sheets, perforated for easy removal. A color-comparison bar chart is printed on the inside front cover. | To detect the presence of liquid V, G, and H chemical agents | Cannot be used to detect vapors or chemical agents in water, DS2, or petroleum products. May cause false readings. | TM 3-6665-254-12<br>TM 3-6665-268-10<br>TM 3-6665-307-10<br>TM 43-0001-26-1 |
| Paper, chemical agent, detector, M9, (NSN 6665-01-226-5589). Paper is issued in a 7-ounce dispenser box that contains one 30-foot roll of 2-inch-wide detector paper and plastic storage bags. The paper has an adhesive back for attaching to equipment and clothing. | To detect the presence of liquid V, G, and H chemical agents | Cannot be used to detect vapors or chemical agents in water. Will not stick to dirty, oily, or greasy surfaces. Contamination indications cannot be read under red light or by color-blind soldier. The following can cause false readings:<br>• Temperatures above 125°F.<br>• Brake fluid.<br>• Aircraft cleaning compound.<br>• DS2.<br>• Petroleum products.<br>• Insect repellent. | TM 3-6665-311-10 |
| | | **Company** | |
| Chemical-agent alarm, automatic, M22 ACAA, (NSN 6665-01-438-6983) | To detect chemical nerve agents in the air. | **WARNING**<br>**Radiation hazard. Contains beta emitters.** | TM 3-6665-321-12&P |
| Chemical-agent alarm, automatic, M8A1 ACAA, (NSN 6665-01-105-5623). It can be vehicle-mounted, back-packed, or ground-emplaced. | To detect chemical nerve agents in the air. | **WARNING**<br>**Radiation hazard. Contains Americium (AM241).** | TM 3-6665-312-12&P |

Table D-2. Detection Equipment and Materials (Continued)

| Item and Description | Use | Limitations | Reference |
|---|---|---|---|
| Chemical agent monitor system (CAM) (NSN 6665-01-199-4153) | To search out areas; to search and locate contamination on personnel, equipment, ship's structure, aircraft, land vehicles, buildings, and terrain; and to monitor for effectiveness of decon. Can also be used for monitoring collective protection. The CAM responds to nerve- and blister-agent vapors down to the lowest concentration that could affect personnel over a short period. | Cannot be used as a detector. The CAM is a monitor and not a detector. Since it is a monitor, it can become contaminated and overloaded (saturated). The CAM can only report conditions at the front of the inlet probe. It is, therefore, a point monitor only and cannot give a realistic assessment of the vapor hazard over an area from one position.<br>**WARNING**<br>**Beta radiation hazard.** | TM 3-6665-331-10 |
| Detector kit, chemical agent, M256A1 (NSN 6665-01-438-6983). It contains M8 detector paper for liquid agents and samplers/detectors for vapors. | To detect liquid G, V, and H chemical agents using M8 detector paper and to detect and determine the type of vapor (G, V, H, and AC) using samplers/detectors. | | TM 3-6665-307-10 |

**Table D-2. Detection Equipment and Materials (Continued)**

| Item and Description | Use | Limitations | Reference |
|---|---|---|---|
| Radiac set, AN/VDR-2 (NSN 6665-01-222-1425) | To measure gamma radiation dose rates from 0.01 uGy/hr to 100 Gy/hr; to detect and display the level of beta particle dose rates from 0.1 uGy/hr to 5 cGy/hr; and to measure, store, and display accumulated dose rates from 0.01 uGy to 9.99 Gy. The instrument consists of a radiac meter with an internal sensor for obtaining dose rates during both mounted and dismounted operations. It has a second sensor housed in a probe and attached to the radiac meter with a cable and input connector. It is used for monitoring personnel, supplies, and equipment. The radiac set uses a presettable, an audible, and a visual warning device integral to the radiac meter. The system is air-transportable and organic to all units. | | TM 11-6665-251-10 |
| Computer indicator, radiac, CP696/PDR-75; detector, radiac/ DT236/PDR-75; radiac set, AN/PDR-75, (NSN 6665-01-211-4217) | To measure the accumulated neutron and gamma radiation dose recorded by the DT236. A person who may be exposed to radiation from tactical nuclear weapons wears the DT236 on his wrist. | | |

## Appendix E

# DTD Station Charts for Chemical Unit Leaders and NBC NCOs

This appendix provides station charts for the DTD. They can be laminated and posted at each station. They provide the station number and on the reverse side, instructions for the station supervisor/attendant. They are not meant to be all-inclusive and do not include setup procedures or equipment requirements. They are an easy guide for soldiers going through a decon and for station supervisors/attendants.

# STATION 1

# INDIVIDUAL-
# GEAR
# DECON

# TASK: Remove contamination to a negligible risk from individual gear (LCE, mask carrier, helmet, and weapon)

- The soldier—
  - Decontaminates his own gloves with hot, soapy water or an STB slurry.
  - Decontaminates his hose and canister (if wearing the M42 mask) with either hot, soapy water and a sponge or an STB slurry mix.
  - Decontaminates his own gear by washing and scrubbing it for 4 minutes in a decontaminant container with hot, soapy water or an STB slurry mix.

- The attendant—
  - Takes the gear to the equipment checkpoint and places it on the "dirty" side of the contamination control line (engineer tape).
  - Waits as the monitor holds the detection device 1 inch from the gear and checks it for residual contamination.
  - Recycles any gear that is above one bar.
  - Places the gear on the clean side of the contamination control line if it passes the check.
  - Carries the gear to the reissue point.

NOTE: Change the slurry mix after 20 soldiers have decontaminated their gear and the rinse water after 10 soldiers or when it appears dirty (place the waste in the sump).

# STATION 2

# OVERBOOT AND HOOD DECON

# TASK: Neutralize gross contamination on overboots, lower trouser legs, protective mask, and hood

- The soldier walks into the shuffle pit and spreads his legs apart (double-shoulder width), bends at the waist, and uses his hands to thoroughly rub the STB dry mix or the STB slurry on his overboots and lower trouser legs.
- Buddy #1 cuts the shoulder straps and drawcord on Buddy #2's hood. He pulls Buddy #2's hood inside out over the front of the mask, being careful not to touch the exposed neck or head. Buddy #1 gathers Buddy #2's hood in one hand and using a cutting tool, cuts away the hood as close as possible to the eye-lens outsert, voicemitter, and inlet-valve cover. Make sure nothing is left dangling below the bottom of the mask. If a replacement hood is not available, decon Buddy #2's hood in the same manner as for a MOPP-gear exchange.
- The attendant—
  - Instructs Buddy #1 on how to roll Buddy #2's hood to ensure that it is accomplished to standard.
  - Will monitor the shuffle pit and add more STB after 10 soldiers have processed through it.

# STATION 3

# OVERGARMENT REMOVAL

# TASK: Remove contaminated overgarment before the agent penetrates material and touches the undergarments or the skin

The attendant—

- Assists the soldier in removing his overgarment.
- Cuts or removes the M9 detector paper from around the soldier's wrist.
- Unfastens the Velcro closure over the jacket zipper.
- Unties the waist cord and unfastens the Velcro straps on the soldier's jacket.
- Unfastens the back snaps.
- Cuts or removes the M9 detector paper from the soldier's trousers, if binding.
- Unfastens the Velcro straps and zipper on the cuffs of the soldier's trousers.
- Unfastens the front waist snaps and unzips the front zipper. He has the soldier lift one leg and point that foot down and bend slightly at the knees for stability.
- Grasps the cuff of the elevated boot with a hand on each side and pulls the cuff in an alternating, jerking motion until the soldier can step out of the trouser leg. The process is repeated on the other leg.

# STATION 4

# OVERBOOT AND GLOVE REMOVAL

# TASK: Remove contaminated overboots and gloves to limit the spread of contamination (the overboots and gloves may also be decontaminated for reissue, if servicable)

- Engineer tape is used to mark the liquid-contamination control line.
- The attendant unfastens or cuts the elastic closures on the soldier's overboots.
- The soldier stands next to and faces the liquid-contamination control line.
- The soldier steps back about 12 inches from the liquid-contamination control line.
- The attendant steps on the back of the soldier's overboot and instructs him to lift his heel and work his foot out of the overboot and step across the liquid-contamination control line. Repeat the process on the other foot.
- The attendant discards the overboots.
- The soldier holds the fingertips of his gloves and partially slides his hands out.
- The attendant removes the soldier's gloves.
- The attendants—
    - Put the gloves and overboots into the containers on the "dirty" side of the control line.
    - Submerge the gloves and overboots in the container of hot, soapy water, ensuring that no water remains inside when removed from the container.
    - Submerge the gloves and overboots in the STB/HTH solution and thoroughly scrub them.
    - Rinse the scrubbed items and check for holes, tears, or punctures. If found, discard the items.
    - Place the usable items on a poncho or a plastic tarp to air dry.
    - Place the air-dried usable items into plastic trash bags along with an M256A1 detector kit. If the kit shows contamination remaining, the attendants can recycle the items or discard them. However, if the kit shows no contaminated, they can be reused.

# STATION 5

# MONITOR

# TASK: Identify contamination on personnel (spot-decon capability and medical aid are provided, as required)

- The monitor checks the soldiers for contamination.
- The aidman checks the soldiers for any chemical-agent symptoms and treats as necessary.
- The monitor asks the soldiers about any damage to their MOPP gear that was identified at stations 2, 3, or 4.

NOTE: If all the liquid contamination is absorbed into the clothing, the M8 detector paper will be negative even though there is a hazard.

# STATION 6

# MASK REMOVAL

## TASK: Remove the mask without contaminating the soldier

- The attendant pulls the hood over the front of the mask, grabs the mask by the voicemitter cover, and pulls the mask off the soldier while he holds his breath.
- The attendant holds the mask open so that the soldier can remove the inserts without touching the outside of the mask if the mask has optical inserts.
- The soldier walks upwind 16 feet, crosses the vapor-contamination control line, and then resumes breathing.
- The attendant brings the mask to station 7.

NOTE: No vapor hazard is expected beyond the vapor-hazard contamination control line.

- The chemical unit positions the M8A1 or M22 ACAA upwind of the site.
- The soldier moves straight ahead while his mask, which may still be emitting vapors, is held on the "dirty" side of the vapor-contamination control line and taken to the next station.

# STATION 7

# MASK
# DECON
# POINT

# TASK: Remove all contamination from the protective masks

The attendants—

- Discard the winterization kits, if used.
- Remove the eye-lens outserts and the hoods if the hoods were not cut off at station 2.
- Remove or discard the filters or canisters and dispose of them in the properly marked containers.
- Wash the masks, hoods, eye-lens outserts, and hoses on the M42 and M43 masks in hot, soapy water.
- Rinse these items in clean water.
- Dip them in the sanitizing solution and agitate for 5 minutes.
- Rinse them in clean water again.
- Add one tube of mask sanitizing solution (calcium hypochlorite) to each quart of water.
- Wipe the masks with rags until they are almost dry.
- Discard each gallon of mask sanitizing solution into a sump after every 10 masks.
- Check for contamination with a CAM.
- Recycle the masks if they are still contaminated and then decon their gloves.
- Take the unassembled masks to the reissue point if they are not contaminated.

# STATION #8

# REISSUE POINT

# TASK: Reissue the masks with their components to the soldiers

- The unit NBC NCO reissues the masks with their components to the soldiers.
- The soldiers will reassemble the masks in the postdecon AA.
- The unit NBC NCO provides the soldiers with replacement parts for all types of protective masks.
- The soldiers pick up individual gear and move to the postdecon AA.

# Glossary

| | |
|---|---|
| **AA** | assembly area |
| **AC** | hydrogen cyanide |
| **ACAA** | automatic chemical-agent alarm |
| **AD** | air defense |
| **AFM** | Air-Force Manual |
| **AFR** | Army-Force Regulation |
| **AMT** | ambulatory medical treatment |
| **AO** | area of operation |
| **APC** | armored personnel carrier |
| **AR** | Army regulation |
| **ART** | assessment and recovery team |
| **attn** | attention |
| **AVIM** | aviation intermediate maintenance facility |
| **BAS** | battalion aid station |
| **BDO** | battle-dress overgarment |
| **BDU** | battle-dress uniform |
| **BFV** | Bradley fighting vehicle |
| **bio** | biological |
| **$C^2$** | command and control |
| **CAM** | chemical-agent monitor |
| **CANA** | convulsant antidote for nerve agent |
| **CARC** | chemical-agent-resistant coating |
| **CG** | phosgene |
| **cGy/hr** | centigray per hour |
| **chem** | chemical |
| **CHS** | combat health support |
| **CK** | cyanogen chloride |
| **Class II Supplies** | clothing, individual equipment, tentage, organizational tool sets and kits, hand tools, maps, and administrative and housekeeping supplies and equipment |

| | |
|---|---|
| **Classes of Supply** | the grouping of supplies by type into 10 categories to facilitate supply management and planning |
| **CLU** | command launch unit |
| **CMO** | chemical officer |
| **CN** | chloroacetophenone |
| **CP** | chemical protective |
| **CPS** | collective protective shelter |
| **CPU** | chemical protective undergarment |
| **CRDEC** | Chemical Research Development and Engineering Center |
| **CSS** | combat service support |
| **CTA** | common table of allowance |
| **CVCUS** | combat vehicle crewman uniform system |
| **CX** | phosgene oxime |
| **DA** | Department of the Army |
| **DAD** | detailed aircraft decon |
| **DAP** | decontaminating apparatus, portable |
| **DD** | Department of Defense |
| **DED** | detailed equipment decon |
| **DISCOM** | Division-support command |
| **DP** | diphosgene |
| **DS2** | decontaminating solution number 2 |
| **DTD** | detailed troop decon |
| **DU** | depleted uranium |
| **EDTA** | ethylene diamine tetra acetic acid |
| **EIC** | end-item code |
| **EMT** | emergency medical treatment |
| **ENCU** | environmental control unit |
| **EOD** | explosive ordnance disposal |
| **F** | Fahrenheit |
| **FARP** | forward arming and refueling point |
| **FLOT** | forward line of troops |
| **FM** | field manual |
| **FMC** | field medical card |
| **FMFM** | Fleet Marine Force Manual |

| | |
|---|---|
| **FRAGO** | fragmentary order |
| **FSB** | forward-support battalion |
| **G4** | Assistant Chief of Staff, G4 (Logistics) |
| **G5** | Assistant Chief of Staff, G5 (Civil Affairs) |
| **GA** | tabun |
| **GB** | sarin |
| **GD** | soman |
| **GF** | soman |
| **GI** | Government issue |
| **GP** | general purpose |
| **GPM** | gallons per minute |
| **Gy** | gray |
| **HD** | distilled mustard |
| **HE** | high explosive |
| **HEMTT** | heavy expanded-mobility tactical truck |
| **HEPA** | high-efficiency particulate air filter |
| **HL** | mustard-lewisite |
| **HMMWV** | high-mobility multipurpose wheeled vehicle |
| **HN-1** | nitrogen mustard |
| **HN-2** | nitrogen mustard |
| **HN-3** | nitrogen mustard |
| **HP** | horsepower |
| **HQ** | headquarters |
| **HTB** | high-test bleach |
| **HTH** | high-test hypochlorite |
| **IEDK** | individual equipment decon kit |
| **JMAO** | Joint Mortuary Affairs Office |
| **JP8** | aviation fuel |
| **L** | lewisite |
| **LAW** | light antitank weapon |
| **LCE** | load-carrying equipment |
| **LDS** | Lightweight Decontaminating System |
| **LIN** | line-item number |
| **LO** | lubrication order |

**LTA** launcher-tube assembly

**MADCP** mortuary affairs decontamination collection point

**METT-TC** mission, enemy, terrain, troops, time available, and civilian consideration

**MGX** MOPP-gear exchange

**MOPP** mission-oriented protective posture

**MP** military police

**MRE** meals, ready-to-eat

**MTF** medical-treatment facility

**MTOE** modification table of organization and equipment

**NATO** North Atlantic Treaty Organization

**NAVMC** Navy/Marine Corp

**NBC** nuclear, biological, and chemical

**NBC5 report** once the NBC reports are posted on the situation map, an NBC5 report showing the contaminated area is prepared by the division

**NCO** noncommissioned officer

**NCOIC** noncommissioned officer in charge

**NSN** national stock number

**OEG** operational exposure guidance

**OPORD** operation order

**OS** operator's spray down

**PAA** peracetic acid

**PALADIN** artillery howitzer

**PDDA** power-driven decon apparatus

**PDDE** power-driven decon equipment

**PL** platoon leader

**PLL** prescribed load list

**PNVS** Pilot Night-Vision System

**POL** petroleum, oils, and lubricants

**POMCUS** prepositioning of material configured to unit sets

**PPW** patient protective wrap

**PSG** platoon sergeant

**psi** pounds per square inch

**PT** pointed

**PW** personal wipe down

| | |
|---|---|
| **rad** | radiological |
| **RD** | rounded |
| **RPO** | radiological protection officer |
| **S4** | Supply Officer (US Army) |
| **SA** | arsine |
| **SAW** | squad automatic weapon |
| **SB** | supply bulletin |
| **SD** | skin decon |
| **SDK** | skin decontaminating kit |
| **SMCT** | soldiers' manual of common tasks |
| **SOI** | signal operating instructions |
| **SOP** | standing operating procedure |
| **STANAG** | Standardized North Atlantic Treaty Organization Agreement |
| **STB** | supertropical bleach |
| **STP** | soldier training publication |
| **T20** | tritium oxide |
| **TADS** | Tactical Air-Defense System |
| **TAP** | toxicological apron, protective |
| **TB** | technical bulletin |
| **TC** | training circular |
| **TGD** | thickened soman |
| **THD** | thickened distilled mustard |
| **TM** | technical manual |
| **TO** | technical order |
| **TOE** | table of organization and equipment |
| **TOW** | tube launched, optically tracked, wire guided |
| **TPU** | tank and pump unit |
| **TRADOC** | United States Army Training and Doctrine Command |
| **uGy** | microgray |
| **US** | United States |
| **USMC** | United States Marine Corp |
| **UV** | ultraviolet |
| **VDR-2** | radiac meter |
| **VR-55** | soman |

| | |
|---|---|
| **VW** | vehicle washdown |
| **VX** | soman |
| **WBGT** | wet-bulb gradient temperature |
| **WO** | warning order |

# Bibliography


AR 11-9. *The Army Radiation Safety Program.* 28 May 1999.

AR 385-10. *The Army Safety Program.* 23 May 1988.

CRDEC Technical Report. CRDEC-CR-87036. *Project VALID Final Comprehensive Report. Herman Nelson Heater Utilization Study.* December 1986. AD-B108506.

CRDEC Technical Report CRDEC-TR-175. *Decontamination Effectiveness of Water Streams.* March 1991. AD-B154567.

CTA 50-909. *Field and Garrison Furnishings and Equipment.* 1 August 1993.

DA Form 2028. *Recommended Changes to Publications and Blank Forms.* 1 February 1974.

DA Form 2404. *Equipment Inspection and Maintenance Worksheet.* 1 April 1979.

DD Form 1380. *US Field Medical Card.* December 1991.

DO49 Technical Report DPB/JOD-91/025. *MOPP Exchange Evaluation, Operational.* September 1991. AB-B158128.

DO49 Technical Report DPG/TA-86-015. *Standard and Nonstandard Decontaminants, and Decontamination Efficiency (U).* May 1987. AD-C041660.

DO49 Technical Report DPG/TA-88/018. *Effects on High Water Pressure on Composite Materials.* June 1988. AD-B124033.

DO49 Technical Report DPG/TA-88/023. *Commercial and Field Expedient Chemical Warfare Agent Decontamination.* March 1988. AD-B121231.

DO49 Technical Report DPG/TA-88/030. *Decontamination on Selected Military Equipment: US Army Ammunition Stocks.* September 1988. AD-B128325.

DO49 Technical Report DPG/TA-89024. *Decontaminative Effects of Soil.* November 1989. AD-B140334.

DO49 Technical Report DPG/TA-90/007. *Water Quality Requirements for Decontamination.* December 1989. AD-B140336.

DO49 Technical Report DPG/TA-90/014. *Mission Oriented Protective Posture (MOPP) Exchange Procedures Evaluation.* January 1990. (No DDIC Number Assigned).

DO49 Technical Report DPG/TA-90/025. *Human Factors Evaluation of the Mission Orientated Protective Posture (MOPP) Exchange Procedures.* July 1990. AD-B148421.

FM 3-3. *Chemical and Biological Contamination Avoidance (FMFM 11-7).* 16 November 1992.

FM 3-3-1. *Nuclear Contamination Avoidance (FMFM 11-18).* 9 September 1994.

FM 3-4. *NBC Protection (FMFM 11-9)*. 29 May 1992.

FM 3-4-1. *Fixed Site Protection*. 16 August 1989.

FM 3-6. *Field Behavior of NBC Agents (Including Smoke and Incendiaries) (AFM 105-7; FMFM 7-11-H)*. 3 November 1986.

FM 3-7. *NBC Field Handbook*. 29 September 1994.

FM 3-9. *Potential Military Chemical / Biological Agents and Compounds (AFR 355-7)*. 12 December 1990.

FM 3-100. *Chemical Operations Principles and Fundamentals*. 8 May 1996.

FM 3-101. *Chemical Staffs and Units*. 19 November 1993.

FM 8-9. *NATO Handbook on the Medical Aspects of NBC Defensive Operations*. 1 February 1996.

FM 8-10-4. *Medical Platoon Leader's Handbook*—Tactics, Techniques, and Procedures. 16 November 1990.

FM 8-10-7. *Health Service Support in a Nuclear, Biological, and Chemical Environment*. 22 April 1993.

FM 8-33. *Control of Communicable Diseases Manual*. 9 April 1996.

FM 8-285. *Treatment of Chemical Agent Casualties and Conventional Military Chemical Injuries (FMFM 11-11)*. 22 December 1995.

FM 21-10. *Field Hygiene and Sanitation*. 22 November 1988.

FM 21-11. *First Aid for Soldiers*. 27 October 1988.

FM 100-5. *Operations*. 14 June 1993.

FM 100-9. *Reconstitution*. 13 January 1992.

FM 100-10. *Combat Service Support*. 3 October 1995.

Joint Publication 3-11. *Joint Doctrine for Nuclear, Biological, and Chemical (NBC) Defense*. 10 July 1995.

Joint Publication 3-12. *Doctrine for Joint Nuclear Operations*. 18 December 1995.

Joint Publication 4-06. *Joint Tactics, Techniques, and Procedures for Mortuary Affairs in Joint Operations*. 28 August 1996.

LO 3-4230-209-12. *Decontaminating Apparatus, Power-Driven, Skid-Mounted, Multipurpose, Nonintegral, 500 Gallon, ABC-M12A1*. 21 June 1974.

LO 5-2805-259-12. *Engine, Gasoline, 20 HP, Military Standard Models 4A084-2, 4A084-3, 4A084-4, (LI-03524B-10-1A)*. 30 January 1989.

SB 10-523. *Size Tariff for Clothing, equipage, and Footwear*. 1 August 1998.

STANAG 2426 NBC. *NBC Contamination Control Policy for NATO Forces (Edition 1)*. 8 February 1995.

STP 21-1-SMCT. *Soldier's Manual of Common Tasks Skill Level 1*. 1 October 1994.

STP 21-24-SMCT. *Soldier's Manual of Common Tasks Skill Levels 2/3/4*. 1 October 1992.

TB 43-0116. *Identification of Radioactive Items in the Army*. 1 April 1998.

TC 3-4-1. *Chemical Agent Monitor Employment*. 17 December 1991.

TM 3-4230-209-20&P. *Unit Maintenance Manual Including Repair Parts and Special Tools List Decontaminating Apparatus: Power-Driven, Skid-Mounted, 500 Gallon, M12A1 (NSN 4230-00-926-9488)*. 28 June 1986.

TM 3-4230-214-12&P. *Operator's and Unit Maintenance Manual Including Repair Parts and Special Tools List for Decontaminating Apparatus: Decontaminating Apparatus: Portable, 14 Liter, M13 (Practice) (NSN 4230-01-345-5172)*. 15 August 1992.

TM 3-4230-228-10. *Operator's Manual for Decontaminating Apparatus: Power Driven, Lightweight, M17 (NSN 4230-01-251-8702)*. 15 May 1989.

TM 3-4230-229-10. *Operator's Manual for Decontaminating Kit, Skin: M291, (NSN 4320-01-251-8702) (SS010-AA-MMO-1-0; TO 11D1-1-131; TM 4230-10/1)*. 2 October 1989.

TM 3-4230-235-10. *Operator's Manual for Decontamination Kit, Individual Equipment: M295 (NSN 6850-01-357-8456) (EIC: Y67)*. 20 February 1994.

TM 3-6665-254-12. *Operator's and Organizational Maintenance Manual: Detector Kit, Chemical Agent, ABC-M18A2 (NSN 6665-00-903-4767)*. 16 February 1976.

TM 3-6665-268-10. *Operator's Manual for Sampling Kit, CBR Agent 34*. 6 September 1968.

TM 3-6665-307-10. *Operator's Manual for Chemical Agent Detector Kit, M256 (NSN 6665-01-016-8399) and M256A1 (NSN 6665-01-133-4964)*. 1 September 1985.

TM 3-6665-311-10. *Operator's Manual for Paper, Chemical Agent Detector: M9 (NSN 6665-01-226-5589 (TO 11H2-2-21)*. 31 August 1998.

TM 3-6665-312-12&P. *Operator's and Organizational Maintenance Manual Including Repair Parts and Special Tools List for M8A1 Automatic Chemical Agent Alarm (NSN 6665-01-105-5623) and Auxiliary Equipment M10 Power Supply (NSN 6665-00-859-2225); M10A1 Power Supply (NSN 6665-00-093-2739); M228 High Profile Mounting Kit (NSN 6665-00-859-2212) and M182 Low Profile Mounting Kit (NSN 6665-00-110-9492) (TO 11H2-17-1)*. 17 March 1985.

TM 3-6665-321-12&P. *Operator's and Unit Maintenance Manual (including Repair Parts and Special Tools List ) for Alarm, Chemical Agent, Automatic: M22 (NSN 6665-01-438-6963) (EIC: Y14) and Auxiliary Equipment Power Supply, Chemical Agent Automatic Alarm: M28 (6130-01-438-6960) (EIC: Y40) Mounting Kit, Chemical Agent Automatic Alarm: M281 (NSN 6665-01-438-6959) (EIC: Y38) Alarm Unit, Chemical Agent Automatic Alarm: ABCA-M42 (NSN 6665-00-859-2215) (EIC: 399) (TO 11H2-23-1; MC 10434A-12&P; EE168-DB-OMP-010)*. March 98.

TM 3-6665-331-10. *Operator's Manual for Chemical Agent Monitor (CAM) (TO 11H2-20-1)*. 12 June 1992.

TM 10-277. *Chemical, Toxicological, and Missile Fuel Handlers Protective Clothing.* 1 November 1980.

TM 11-6665-251-10. *Operator's Manual for Radiac Set AN/VDR-2 (NSN 6665-01-222-1425).* 1 March 1988.

TM 43-0001-26-1. *Army Equipment Data Sheets: Chemical Defense Equipment.* 12 May 1982.

NAVMC 10245. *Equipment Maintenance Request.* February 1994.

# Index

## RANGER CREED

**R**ecognizing that I volunteered as a Ranger, fully knowing the hazards of my chosen profession, I will always endeavor to uphold the prestige, honor, and high esprit de corps of the Rangers.

**A**cknowledging the fact that a Ranger is a more elite Soldier who arrives at the cutting edge of battle by land, sea, or air, I accept the fact that as a Ranger my country expects me to move further, faster, and fight harder than any other soldier.

**N**ever shall I fail my comrades I will always keep myself mentally alert, physically strong, and morally straight and I will shoulder more than my share of the task whatever it may be, one hundred percent and then some.

**G**allantly will I show the world that I am a specially selected and well trained Soldier. My courtesy to superior officers, neatness of dress, and care of equipment shall set the example for others to follow.

**E**nergetically will I meet the enemies of my country. I shall defeat them on the field of battle for I am better trained and will fight with all my might. Surrender is not a Ranger word. I will never leave a fallen comrade to fall into the hands of the enemy and under no circumstances will I ever embarrass my country.

**R**eadily will I display the intestinal fortitude required to fight on to the Ranger objective and complete the mission, though I be the lone survivor.

## STANDING ORDERS, ROGERS' RANGERS

1. Don't forget nothing.
2. Have your musket clean as a whistle, hatchet scoured, sixty rounds powder and ball, and be ready to march at a minute's warning.
3. When you're on the march, act the way you would if you was sneaking up on a deer. See the enemy first.
4. Tell the truth about what you see and what you do. There is an army depending on us for correct information. You can lie all you please when you tell other folks about the Rangers, but don't never lie to a Ranger or officer.
5. Don't never take a chance you don't have to.
6. When we're on the march we march single file, far enough apart so one shot can't go through two men.
7. If we strike swamps, or soft ground, we spread out abreast, so it's hard to track us.
8. When we march, we keep moving till dark, so as to give the enemy the least possible chance at us.
9. When we camp, half the party stays awake while the other half sleeps.
10. If we take prisoners, we keep' em separate till we have had time to examine them, so they can't cook up a story between' em.
11. Don't ever march home the same way. Take a different route so you won't be ambushed.
12. No matter whether we travel in big parties or little ones, each party has to keep a scout 20 yards ahead, 20 yards on each flank, and 20 yards in the rear so the main body can't be surprised and wiped out.
13. Every night you'll be told where to meet if surrounded by a superior force.
14. Don't sit down to eat without posting sentries.
15. Don't sleep beyond dawn. Dawn's when the French and Indians attack.
16. Don't cross a river by a regular ford.
17. If somebody's trailing you, make a circle, come back onto your own tracks, and ambush the folks that aim to ambush you.
18. Don't stand up when the enemy's coming against you. Kneel down, lie down, hide behind a tree.
19. Let the enemy come till he's almost close enough to touch, then let him have it and jump out and finish him up with your hatchet.

*--MAJOR ROBERT ROGERS, 1759*

## RANGER HISTORY

The history of the American Ranger is a long and colorful saga of courage, daring and outstanding leadership. It is a story of men whose skills in the art of fighting have seldom been surpassed. Only the highlights of their numerous exploits are told here.

Rangers primarily performed defensive missions until Benjamin Church's Company of Independent Rangers from Plymouth Colony proved successful in raiding hostile Indians during King Phillip's War in 1675. In 1756, Major Robert Rogers, a native of New Hampshire, recruited nine companies of American colonists to fight for the British during the French and Indian War. Ranger techniques and methods of operation were an inherent characteristic of the American frontiersmen; however, Major Rogers was the first to capitalize on them and incorporate them into the fighting doctrine of a permanently organized fighting force.

The method of fighting used by the first Rangers was further developed during the Revolutionary War by Colonel Daniel Morgan, who organized a unit known as "Morgan's Riflemen". According to General Burgoyne, Morgan's men were "....the most famous corps of the Continental Army, all of them crack shots."

Francis Marion, the "Swamp Fox," organized another famous Revolutionary War Ranger element known as "Marion's Partisans." Marion's Partisans, numbering anywhere from a handful to several hundred, operated both with and independent of other elements of General Washington's Army. Operating out of the Carolina swamps, they disrupted British communications and prevented the organization of loyalists to support the British cause, substantially contributing to the American victory.

The American Civil War was again the occasion for the creation of special units such as Rangers. John S. Mosby, a master of the prompt and skillful use of cavalry, was one of the most outstanding Confederate Rangers. He believed that by resorting to aggressive action he could compel his enemies to guard a hundred points. He would then attack one of the weakest points and be assured numerical superiority.

With America's entry into the Second World War, Rangers came forth to add to the pages of history. Major William O. Darby organized and activated the 1st Ranger Battalion on June19, 1942 at Carrickfergus, North Ireland. The members were all hand-picked volunteers; 50 participated in the gallant Dieppe Raid on the northern coast of France with British and Canadian commandos. The 1st, 3rd, and 4th Ranger Battalions participated with distinction in the North African, Sicilian and Italian campaigns. Darby's Ranger Battalions spearheaded the Seventh Army landing at Gela and Licata during the Sicilian invasion and played a key role in the subsequent campaign which culminated in the capture of Messina. They infiltrated German lines and mounted an attack against Cisterna, where they virtually annihilated an entire German parachute regiment during close in, night, bayonet, and hand-to-hand fighting.

The 2nd and 5th Ranger Battalions participated in the D-Day landings at Omaha Beach, Normandy. It was during the bitter fighting along the beach that the Rangers gained their official motto. As the situation became critical on Omaha Beach, the division commander of the 29th Infantry Division stated that the entire force must clear the beach and advance inland. He then turned to Lieutenant Colonel Max Schneider, Commander of the 5th Ranger Battalion, and said, "Rangers, lead the way." The 5th Ranger Battalion spearheaded the breakthrough and thus enabled the allies to drive inland away from the invasion beaches.

The 6th Ranger Battalion, operating in the Pacific, conducted Ranger type missions behind enemy lines which involved reconnaissance and hard-hitting, long-range raids. They were the first American contingent to return to the Philippines, destroying key coastal installations prior to the invasion. A reinforced company from the 6th Ranger Battalion formed the rescue force which liberated American and allied prisoners of war from the Japanese prison camp at Cabanatuan.

Another Ranger-type unit was the 5307th Composite Unit (Provisional), organized and trained as a long-range penetration unit for employment behind enemy lines in Japanese occupied Burma. The unit commander was Brigadier General (later Major General) Frank D. Merrill. Its 2,997 officers and men became popularly known as "Merrill's Marauders."

The men composing Merrill's Marauders were volunteers from the 5th, 154th, and 33rd Infantry Regiments and from other Infantry regiments engaged in combat in the southwest and South Pacific. These men responded to a call from then Chief of Staff, General George C. Marshall, for volunteers for a hazardous mission. These volunteers were to have a high state of physical ruggedness and stamina and were to come from jungle-trained and jungle-tested units.

Prior to their entry into the Northern Burma Campaign, Merrill's Marauders trained in India under the overall supervision of Major General Orde C. Wingate, British Army. There, they were trained from February to June 1943 in long-range penetration tactics and techniques of the type developed and first employed by General Wingate. The operations of the Marauders were closely coordinated with those of the Chinese 22nd and 38th Divisions in a drive to recover northern Burma and clear the way for the construction of Ledo Road, which was to link the Indian railhead at Ledo with the old Burma Road to China. The Marauders marched and fought through jungle and over mountains from Hukwang Valley in northwest Burma to Myitkyina and the Irrawaddy River. In 5 major and 30 minor engagements, they met and defeated the veteran soldiers of the Japanese 18th Division. Operating in the rear of the main force of the Japanese, they prepared the way for the southward advances of the Chinese by disorganizing supply lines and communications. The climax of the Marauder's operations was the capture of Myitkyina Airfield, the only all-weather strip in northern Burma. This was the final victory of "Merrill's Marauders," which disbanded in August 1944. Remaining personnel were consolidated into the 475th Infantry Regiment, which fought its last battle on February 3, and 4,1945, at Loi-Kang Ridge, China. This Infantry Regiment would serve as the forefather of today's 75th Ranger Regiment.

Shortly after the outbreak of the Korean War in June 1950, the 8th Army Ranger Company was formed of volunteers from American units in Japan. The Company was trained in Korea and distinguished itself in combat during the drive to the Yalu River, performing task force and spearhead operations. In November 1950 during the massive Chinese intervention, this small unit, though vastly outnumbered, withstood five enemy assaults on its position.

In September 1950, a Department of the Army message called for volunteers to be trained as Airborne Rangers. In the 82nd Airborne Division, five thousand regular Army paratroopers volunteered, and from that number nine hundred men were selected to form the initial eight Airborne Ranger Companies. An additional nine companies were formed from volunteers of regular Army and National Guard Infantry Divisions. These seventeen Airborne Ranger companies were activated and trained at Fort Benning, Georgia, with most receiving additional training in the mountains of Colorado.

In 1950 and 1951, some 700 men of the 1st, 2nd, 3rd, 4th, 5th, and 8th Airborne Ranger companies fought to the front of every American Infantry Division in Korea. Attacking by land, water, and air, these six Ranger companies conducted raids, deep penetrations, and ambush operations against North Korean and Chinese forces. They were the first Rangers in history to make a combat jump. After the Chinese intervention, these Rangers were the first Americans to re-cross the 38th parallel. The 2nd Airborne Ranger Company was the only African American Ranger unit in the history of the American Army. The men of the six Ranger companies who fought in Korea paid the bloody price of freedom. One in nine of this gallant brotherhood died on the battlefields of Korea.

Other Airborne Ranger companies led the way while serving with infantry divisions in the United States, Germany, and Japan. Men of these companies volunteered and fought as members of line infantry units in Korea. One Ranger, Donn Porter, would be posthumously awarded the Medal of Honor. Fourteen Korean War Rangers became general officers and dozens became colonels, senior noncommissioned officers, and leaders in civilian life. They volunteered for the Army, the Airborne, the Rangers, and for combat. The first men to earn and wear the coveted Ranger Tab, these men are the original Airborne Rangers.

In October 1951, the Army Chief of Staff, General J. Lawton Collins directed, "Ranger training be extended to all combat units in the Army." The Commandant of the Infantry School was directed to establish a Ranger Department. This new department would develop and conduct a Ranger course of instruction. The objective was to raise the standard of training in all combat units. This program was built upon what had been learned from the Ranger Battalions of World War II and the Airborne Ranger companies of the Korean conflict.

During the Vietnam Conflict, fourteen Ranger companies consisting of highly motivated volunteers served with distinction from the Mekong Delta to the DMZ. Assigned to separate brigade, division, and field force units, they conducted long-range reconnaissance and exploitation operations into enemy-held areas providing valuable combat intelligence. Initially designated at LRRP, then LRP companies, these units were later designated as C, D,E,F,G,H,I,K,L,M,N,O and P (Ranger) 75th Infantry.

Following Vietnam, recognizing the need for a highly trained and highly mobile reaction force, the Army Chief of Staff, General Abrams directed the activation of the first battalion-sized Ranger units since World War II, the 1st and 2nd Battalions (Ranger), 75th Infantry. The 1st Battalion was trained at Fort Benning, Georgia and was activated February 8, 1974 at Fort Stewart, Georgia with the 2nd Battalion being activated on October 3, 1974. The 1st Battalion is now located at Hunter Army Airfield, Georgia and the 2nd Battalion at Fort Lewis, Washington.

The farsightedness of General Abrams' decision, as well as the combat effectiveness of the Ranger battalions, was proven during the United States' invasion of the island of Grenada in October 1983 to protect American citizens there, and to restore democracy. As expected, Rangers led the way! During this operation, code named "Urgent Fury," the Ranger battalions conducted a daring, low level airborne assault (from 500 feet) to seize the airfield at Point Salines, and then continued operations for several days to eliminate pockets of resistance, and rescue American medical students.

As a result of the demonstrated effectiveness of the Ranger battalions, the Department of the Army announced in 1984 that it was increasing the strength of Ranger units to its highest level in 40 years. To do this, it activated another Ranger battalion as well as a Ranger Regimental Headquarters. These new units, the 3rd Battalion (Ranger), 75th Infantry, and Headquarters Company (Ranger), 75th Infantry, have increased the Ranger strength of the Army to over 2,000 soldiers actually assigned to Ranger units. On February 3, 1986, the 75th Infantry was renamed the 75th Ranger Regiment.

On December 20, 1989, the 75th Ranger Regiment was once again called upon to demonstrate its effectiveness in combat. For the first time since its reorganization in 1984, the Regimental Headquarters and all three Ranger battalions were deployed on Operation "Just Cause" in Panama. During this operation, the 75th Ranger Regiment spearheaded the assault into Panama by conducting airborne assaults onto Torrijos/Tocumen Airport and Rio Hato Airfield to facilitate the restoration of democracy in Panama, and protect the lives of American citizens. Between December 20, 1989 and January 7, 1990, numerous follow-on missions were performed in Panama by the Regiment.

Early in 1991, elements of the 75th Ranger Regiment deployed to Saudi Arabia in support of Operation Desert Storm.

In August 1993, elements of the 75th Ranger Regiment deployed to Somalia in support of Operation Restore Hope, and returned November 1993.

In 1994, elements of the 75th Ranger Regiment deployed to Haiti in support of Operation Uphold Democracy.

In 2000 – 2001, elements of the 75th Ranger Regiment deployed to Kosovo in support of Operation Joint Guardian.

Since September 11, 2001, the 75th Ranger Regiment has led the way in the Global War on Terrorism. In October 2001, elements of the 75th Ranger Regiment deployed to Afghanistan in support of Operation Enduring Freedom. In March 2003, elements of the Regiment deployed in support of Operation Iraqi Freedom.

The performance of these Rangers significantly contributed to the overall success of these operations and upheld the Ranger tradition of the past. As in the past, the Regiment stands ready to execute its mission to conduct special operations in support of the United States' policies and objectives.

## RANGER MEDAL OF HONOR WINNERS

| | | | |
|---|---|---|---|
| Millett, Lewis L. Sr. | Captain | Feb 7 1951 | Co. E 2/27th Infantry |
| Porter, Donn F.* | Sergeant | Sept 7 1952 | Co. G 2/14th Infantry |
| Mize, Ola L. | Sergeant | June 10-11 1953 | Co. K 3/15th Infantry |
| Dolby, David C. | Staff Sergeant | May 21 1966 | Co. B 1/8th (ABN) Calvary |
| Foley, Robert F. | Captain | Nov 5 1966 | Co. A 2/27th Infantry |
| Zabitosky, Fred M. | Staff Sergeant | Feb 19 1968 | 5th Special Forces |
| Bucha, Paul W. | Captain | May 16-19 1968 | Co. D 3/187 Infantry |
| Rabel, Laszlo* | Staff Sergeant | Nov 13 1968 | 74th Infantry (LRRP) |
| Howard, Robert L. | Sergeant First Class | Dec 30 1968 | 5th Special Forces |
| Law, Robert D. * | Specialist 4 | Feb 22 1969 | Co. I 75th Infantry (Ranger) |
| Kerrey, J. Robert | Lieutenant | Mar 14 1969 | Seal Team 1 |
| Doane, Stephen H.* | 1st Lieutenant | Mar 25 1969 | Co. B 1/5th Infantry |
| Pruden, Robert J.* | Staff Sergeant | Nov 22 1969 | Co. G 75th Infantry (Ranger) |
| Littrell, Gary L. | Sergeant First Class | April 4-8 1970 | Advisory Team 21 (Ranger) |
| Lucas, Andre C.* | Lt Colonel | Jul 1-23 1970 | HHC 2/506 Infantry |
| Gordon, Gary I. * | Master Sergeant | Oct 3 1993 | Task Force Ranger |
| Shughart, Randall D. * | Sergeant First Class | Oct 3 1993 | Task Force Ranger |

*Awarded posthumously

## SUMMARY OF CHANGES

| | |
|---|---|
| **THROUGHOUT** | UPDATED doctrine, technology, terms; clarified verbiage |
| **CHAPTER 1** | UPDATED duties and responsibilities<br>ADDED FO |
| **CHAPTER 2** | UPDATED WARNO<br>ADDED sample WARNO<br>UPDATED OPORD<br>ADDED terrain model |
| **CHAPTER 3** | ADDED PLOT-CR<br>ADDED risk estimate distances |
| **CHAPTER 4** | IDENTIFIED TL positions during movement |
| **CHAPTER 5** | REORGANIZED combat operations |
| **CHAPTER 7** | ADDED section on antennas |
| **CHAPTER 8** | UPDATED aircraft |
| **CHAPTER 9** | UPDATED rope bridge equipment<br>UPDATED sequence and responsibilities |
| **CHAPTER 10** | UPDATED equipment and descriptions<br>UPDATED climbing commands |
| **CHAPTER 11** | NA |
| **CHAPTER 12** | ADDED first aid techniques |
| **CHAPTER 13** | ADDED figure 'Lighting a time fuse with a match'<br>ADDED expedient explosives<br>ADDED table of US demolitions<br>ADDED breaching and timber cutting charges |
| **CHAPTER 15** | NEW chapter covering vehicle convoy operations |
| **APPENDIX A** | NEW - Resources |
| **APPENDIX B** | NEW - Quick Reference Cards |

# TABLE OF CONTENTS

# Chapter 1
# LEADERSHIP

Leadership, the most essential element of combat power, gives purpose, direction, and motivation in combat. The leader balances and maximizes maneuver, firepower, and protection against the enemy. This chapter discusses how he does this by exploring the principles of leadership (BE, KNOW, DO); the duties, responsibilities, and actions of an effective leader; and the leader's assumption of command.

1-1. **PRINCIPLES**. The principles of leadership are BE, KNOW, and DO.

*BE*

- Technically and tactically proficient
- Able to accomplish to standard all tasks required for the wartime mission.
- Courageous, committed, and candid.
- A leader with integrity.

*KNOW*

- The four major factors of leadership and how they affect each other are--
  ---Led
  ---Leader
  ---Situation
  ---Communications
- Yourself, and the strengths and weaknesses in your character, knowledge, and skills. Seek continual self-improvement, that is, develop your strengths and work to overcome your weaknesses.
- Your Rangers, and look out for their well-being by training them for the rigors of combat, taking care of their physical and safety needs, and disciplining and rewarding them.

*DO--*

- Seek responsibility and take responsibility for your actions; exercise initiative; demonstrate resourcefulness; and take advantage of opportunities on the battlefield that will lead to you to victory; accept fair criticism, and take corrective actions for your mistakes.
- Assess situations rapidly, make sound and timely decisions, gather essential information, announce decisions in time for Rangers to react, and consider the short- and long-term effects of your decision.
- Set the example by serving as a role model for your Rangers. Set high but attainable standards; be willing do what you require of your Rangers; and share dangers and hardships with them.
- Keep your subordinates informed to help them make decisions and execute plans within your intent, encourage initiative, improve teamwork, and enhance morale.
- Develop a sense of responsibility in subordinates by teaching, challenging, and developing them. Delegate to show you trust them. This makes them want more responsibility.
- Ensure the Rangers understand the task; supervise them, and ensure they accomplish it. Rangers need to know what you expect: when and what you want them to do, and to what standard.
- Build the team by training and cross-training your Rangers until they are confident in their technical and tactical abilities. Develop a team spirit that motivates them to go willingly and confidently into combat.
- Know your unit's capabilities and limitations, and employ them accordingly.

1-2. **DUTIES, RESPONSIBILITIES, AND ACTIONS**. To complete all assigned tasks, every Ranger in the patrol must do his job. Each must accomplish his specific duties and responsibilities and be a part of the team.

*PLATOON LEADER (PL)*

- Is responsible for what the patrol does or fails to do. This includes tactical employment, training, administration, personnel management, and logistics. He does this by planning, making timely decisions, issuing orders, assigning tasks, and supervising patrol activities. He must know his Rangers and how to employ the patrol's weapons. He is responsible for positioning and employing all assigned or attached crew-served weapons and employment of supporting weapons.
- Establishes time schedule using backwards planning. Considers time for execution, movement to the objective, and the planning and preparation phase of the operation.
- Takes the initiative to accomplish the mission in the absence of orders. Keeps higher informed by using periodic situation reports (SITREP).
- Plans with the help of the platoon sergeant (PSG), squad leaders, and other key personnel (team leaders, FO, attachment leaders).
- Stays abreast of the situation through coordination with adjacent patrols and higher HQ; supervises, issues FRAGOs, and accomplishes the mission.
- If needed to perform the mission, requests more support for his patrol from higher headquarters.
- Directs and assists the platoon sergeant in planning and coordinating the patrol's sustainment effort and casualty evacuation (CASEVAC) plan.
- During planning, receives on-hand status reports from the platoon sergeant and squad leaders.
- Reviews patrol requirements based on the tactical plan.
- Ensures that all-round security is maintained at all times.
- Supervises and spot-checks all assigned tasks, and corrects unsatisfactory actions.
- During execution, positions himself where he can influence the most critical task for mission accomplishment; usually with the main effort, to ensure that his platoon achieves its decisive point
- Is responsible for positioning and employing all assigned and attached crew-served weapons.
- Commands through his squad leaders IAW the intent of the two levels higher commanders.
- Conducts rehearsals.

*PLATOON SERGEANT (PSG)*

The PSG is the senior NCO in the patrol and second in succession of command. He helps and advises the patrol leader, and leads the patrol in the leader's absence. He supervises the patrol's administration, logistics, and maintenance, and he prepares and issues paragraph 4 of the patrol OPORD.

*DUTIES*

- Organizes and controls the patrol CP IAW the unit SOP, patrol leader's guidance, and METT-TC factors.
- Receives squad leader's requests for rations, water, and ammunition. Work with the company first sergeant or XO to request resupply. Directs the routing of supplies and mail.
- Supervises and directs the patrol medic and patrol aid-litter teams in moving casualties to the rear.
- Maintains patrol status of personnel, weapons, and equipment; consolidates and forwards the patrol's casualty reports (DA Forms 1155 and 1156); and receives and orients replacements.
- Monitors the morale, discipline, and health of patrol members.
- Supervises task-organized elements of patrol:
  - — Quartering parties.
  - — Security forces during withdrawals.
  - — Support elements during raids or attacks.
  - — Security patrols during night attacks.
- Coordinates and supervises patrol resupply operations.
- Ensures that supplies are distributed IAW the patrol leader's guidance and direction.
- Ensures that ammunition, supplies, and loads are properly and evenly distributed (a critical task during consolidation and reorganization).
- Ensures the casualty evacuation plan is complete and executed properly.

- Ensures that the patrol adheres to the platoon leader's time schedule.
- Assists the platoon leader in supervising and spot-checking all assigned tasks, and corrects unsatisfactory actions

*ACTIONS DURING MOVEMENT AND HALTS*

- Takes actions necessary to facilitate movement.
- Supervises rear security during movement.
- Establishes, supervises, and maintains security during halts.
- Knows unit location.
- Performs additional tasks as required by the patrol leader and assists in every way possible. Focuses on security and control of patrol.

*ACTIONS AT DANGER AREAS*

- Directs positioning of near-side security (usually conducted by the trail squad or team).
- Maintains accountability of personnel.

*ACTIONS ON THE OBJECTIVE AREA*

- Assists with ORP occupation.
- Supervises, establishes, and maintains security at the ORP.
- Supervises the final preparation of men, weapons, and equipment in the ORP IAW the patrol leader's guidance.
- Assists the patrol leader in control and security.
- Supervises the consolidation and reorganization of ammunition and equipment.
- Establishes, marks, and supervises the planned CCP, and ensures that the personnel status (to include WIA/KIA) is accurately reported to higher.
- Performs additional tasks assigned by the patrol leader and reports status to platoon leader.

*ACTIONS IN THE PATROL BASE*

- Assists in patrol base occupation.
- Assist in establishing and adjusting perimeter.
- Enforces security in the patrol base.
- Keeps movement and noise to a minimum.
- Supervises and enforces camouflage.
- Assigns sectors of fire.
- Ensures designated personnel remain alert and equipment is maintained to a high state of readiness.
- Requisitions supplies, water, and ammunition, and supervises their distribution.
- Supervises the priority of work and ensures its accomplishment.
  - Security plan.
    - Ensures crew-served weapons have interlocking sectors of fire.
    - Ensures Claymores are emplaced to cover dead space.
    - Ensures range cards and sector sketch are complete.
  - Alert plan.
  - Evacuation plan.
  - Withdrawal plan.
  - Alternate patrol base.
  - Maintenance plan.
  - Hygiene plan.
  - Messing plan.
  - Water plan.
  - Rest plan.
- Performs additional tasks assigned by the patrol leader and assists him in every way possible.

*SQUAD LEADER (SL)*

Is responsible for what the squad does or fails to do. He is a tactical leader who leads by example.

*DUTIES*

- Completes casualty feeder reports and reviews the casualty reports completed by squad members.
- Directs the maintenance of the squad's weapons and equipment.

- Inspects the condition of Rangers' weapons, clothing, and equipment.
- Keeps the PL and PSG informed on status of squad.
- Submits ACE report to PSG.

*ACTIONS THROUGHOUT THE MISSION*

- Obtains status report from team leaders and submits reports to the PL and PSG.
- Makes a recommendation to the PL/PSG when problems are observed.
- Delegates priority task to team leaders, and supervises their accomplishment IAW squad leader's guidance.
- Uses initiative in the absence of orders.
- Follows the PL's plan and makes recommendations.

*ACTIONS DURING MOVEMENT AND HALTS*

- Ensures heavy equipment is rotated among members and difficult duties are shared.
- Notifies PL of the status of the squad.
- Maintains proper movement techniques while monitoring route, pace, and azimuth.
- Ensures the squad maintains security throughout the movement and at halts.
- Prevents breaks in contact.
- Ensures subordinate leaders are disseminating information, assigning sectors of fire, and checking personnel.

*ACTION IN THE OBJECTIVE AREA*

- Ensures special equipment has been prepared for actions at the objective.
- Maintains positive control of squad during the execution of the mission.
- Positions key weapons systems during and after assault on the objective.
- Obtains status reports from team leaders and ensures ammunition is redistributed and reports status to PL.

*ACTIONS IN THE PATROL BASE*

- Ensures patrol base is occupied according to the plan.
- Ensures that his sector of the patrol base is covered by interlocking fires; makes final adjustments, if necessary.
- Sends out LP or OPs in front of assigned sector. (METT-TC dependent).
- Ensures priorities of work are being accomplished, and reports accomplished priorities to the PL and PSG.
- Adheres to time schedule.
- Ensures personnel know the alert and evacuation plans, and the locations of key leaders, OPs, and the alternate patrol base.

*WEAPONS SQUAD LEADER*

Is responsible for all that the weapons squad does or fails to do. His duties are the same as those of the squad leader. Also, he controls the machine guns in support of the patrol's mission. He advises the PL on employment of his squad.

*DUTIES*

- Supervises machine gun teams to ensure they follow priorities of work.
- Inspects machine gun teams for correct range cards, fighting positions, and understanding of fire plan.
- Supervises maintenance of machine guns, that is, ensures that maintenance is performed correctly, that deficiencies are corrected and reported, and that the performance of maintenance does not violate the security plan.
- Assists PL in planning.
- Positions at halts and danger areas and according to the patrol SOP any machine guns not attached to squads.
- Rotates loads. Machine gunners normally get tired first.
- Submits ACE report to PSG.
- Designates sectors of fire, principal direction of fire (PDF), and secondary sectors of fire for all guns.
- Gives fire commands to achieve maximum effectiveness of firepower:
  - — Shifts fires.
  - — Corrects windage or elevation to increase accuracy.
  - — Alternates firing guns.
  - — Controls rates of fire and fire distribution.
- Knows locations of assault and security elements, and prevents fratricide.
- Reports to the PL.

*TEAM LEADER (TL)*

Controls the movement of his fire team and the rate and placement of fire. To do this, leads from the front and uses the proper commands and signals. Maintains accountability of his Rangers, weapons, and equipment. Ensures his Rangers maintain unit standards in all areas, and are knowledgeable of their tasks and the operation. The following checklist outlines specific duties and responsibilities of team leaders during mission planning and execution. The TL leads by example:

*ACTIONS DURING PLANNING AND PREPARATION*

- Warning Order.
  - Assists in control of the squad.
  - Monitors squad during issuance of the order.
- OPORD Preparation.
  - Posts changes to schedule.
  - Posts and updates team duties on warning order board.
  - Submits ammunition and supply requests.
  - Picks up ammunition and supplies.
  - Distributes ammunition and special equipment.
  - Performs all tasks given in the SL's special instructions paragraph.
- OPORD Issuance and Rehearsal.
  - Monitors squad during issuance of the order.
  - Assists SL during rehearsals.
- Takes actions necessary to facilitate movement.
- Enforces rear security.
- Establishes, supervises, and maintains security at all times.
- Performs additional tasks as required by the SL, and assists him in every way possible, particularly in control and security.

*ACTION IN THE ORP*

- Assists in the occupation of the ORP.
- Helps supervise, establish, and maintain security.
- Supervises the final preparation of Rangers, weapons, and equipment in the ORP IAW the SL's guidance.
- Assists in control of personnel departing and entering the ORP.
- Reorganizes perimeter after the reconnaissance party departs.
- Maintains communication with higher headquarters.
- Upon return of reconnaissance party, helps reorganize personnel and redistribute ammunition and equipment; ensures accountability of all personnel and equipment is maintained.
- Disseminates PIR to his team.
- Performs additional tasks assigned by the SL.

*ACTIONS IN THE PATROL BASE*

- Inspects the perimeter to ensure team has interlocking sectors of fire; prepares team sector sketch.
- Enforces the priority of work and ensures it is properly accomplished.
- Performs additional tasks assigned by the SL and assists him in every way possible.

*MEDIC*

Assists the PSG in directing aid and litter teams; monitors the health and hygiene of the platoon.

*DUTIES*

- Treats casualties, conducts triage, and assists in CASEVAC under the control of the PSG.
- Aids the PL or PSG in field hygiene matters. Personally checks the health and physical condition of platoon members.
- Requests Class VIII (medical) supplies through the PSG.
- Provides technical expertise to and supervision of combat lifesavers.
- Ensures casualty feeder reports are correct and attached to each evacuated casualty.
- Carries out other tasks assigned by the PL or PSG.

| |
|---|
| *RADIO OPERATOR*<br>Is responsible for establishing and maintaining communications with higher headquarters and within the patrol.<br>*DUTIES DURING PLANNING*<br>• Enters the net at the specified time.<br>• Ensures that all frequencies, COMSEC fills, and net IDs, are preset in squad/platoon radios.<br>• Informs SL and PL of changes to call signs, frequencies, challenge and password, and number combination based on the appropriate time in the ANCD.<br>• Ensures the proper function of all radios and troubleshoots and reports deficiencies to higher.<br>• Weatherproofs all communications equipment.<br>*DUTIES DURING EXECUTION*<br>• Serves as en route recorder during all phases of the mission.<br>• Records all enemy contact and reports it to higher in a SALUTE format.<br>• Reports all OPSKEDs to higher.<br>• Consolidates and records all PIR. |
| *FORWARD OBSERVER (FO)*<br>Works for the PL. Serves as the eyes of the FA and mortars. Is mainly responsible for locating targets, and for calling for and adjusting indirect fire support. Knows the terrain where the platoon is operating; knows the tactical situation. Knows the mission, the concept, and the unit's scheme of maneuver and priority of fires.<br>*DUTIES DURING PLANNING*<br>• Selects targets to support the platoon's mission based on the company OPORD, platoon leader's guidance, and analysis of METT-TC factors.<br>• Prepares and uses situation maps, overlays, and terrain sketches.<br>*DUTIES DURING EXECUTION*<br>• Informs the FIST headquarters of platoon activities and of the fire support situation<br>• Selects new targets to support the platoon's mission based on the company OPORD, the platoon leader's guidance, and an analysis of METT-TC factors.<br>• Calls for and adjusts fire support.<br>• Operates as a team with the radio operator.<br>• Selects OPs.<br>• Maintains communications as prescribed by the FSO.<br>• Maintains the current 8-digit coordinate of his location at all times. |

1-3. **ASSUMPTION OF COMMAND**. Any platoon/squad member might have to take command of his element in an emergency, so every Ranger must be prepared to do so. During an assumption of command, situation permitting, the Ranger assuming command accomplishes the following tasks (not necessarily in this order) based on METT-TC:

- *INFORMS* the unit's subordinate leaders of the command and notifies higher.
- *CHECKS* security.
- *CHECKS* crew-served weapons.
- *PINPOINTS* location.
- *COORDINATES* and *CHECKS* equipment.
- *CHECKS* personnel status.
- *ISSUES* FRAGO (if required).
- *REORGANIZES* as needed, maintaining unit integrity when possible.
- *MAINTAINS* noise and light discipline.
- *CONTINUES* patrol base activities, especially security, if assuming command in a patrol base.
- *RECONNOITERS* or, at the least, conducts a map reconnaissance.
- *FINALIZES* plan.
- *EXECUTES* the mission.

NOTES

# Chapter 2
# OPERATIONS

This chapter provides techniques and procedures used by Infantry platoons and squads throughout the planning and execution phases of tactical operations. Specifically, it discusses the troop-leading procedures, combat intelligence, combat orders, and planning techniques and tools needed to prepare a platoon to fight. These topics are time sensitive and apply to all combat operations. When they have time, leaders can plan and prepare in depth. If they have less time, they must rely on previously rehearsed actions, battle drills, and standing operating procedures (SOPs).

## 2-1. TROOP-LEADING PROCEDURES

Troop leading procedures comprise the following steps. They are what a leader does to prepare his unit to accomplish a tactical mission. The TLP starts when the leader is alerted for a mission or receives a change or new mission. He can perform Steps 3 through 8 in any order, or at the same time. (He can also use the tools of the tactician shown in Figure 2-1):

| | |
|---|---|
| 1. Receive the mission. | 5. Reconnoiter. |
| 2. Issue a warning order. | 6. Complete the plan. |
| 3. Make a tentative plan. | 7. Issue the complete order. |
| 4. Initiate movement. | 8. Supervise. |

**Figure 2-1. TOOLS OF THE TACTICIAN RELATIONSHIP**

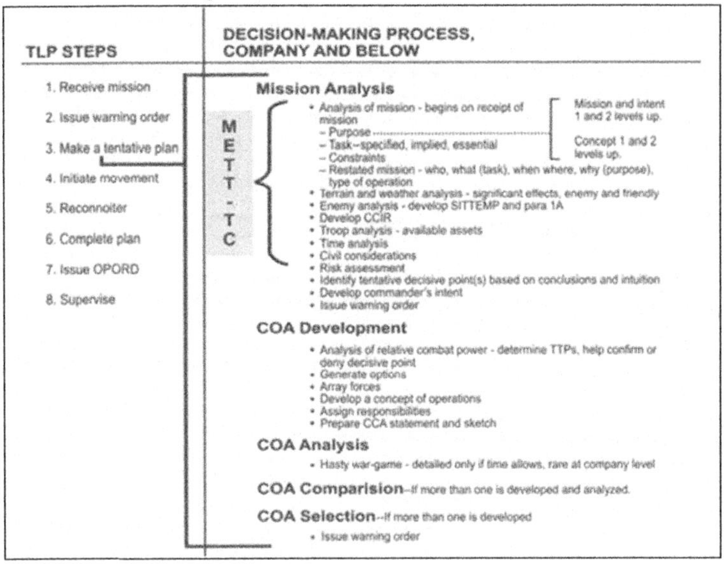

a. **STEP 1--RECEIVE THE MISSION.** The leader may receive the mission in a warning order, an operation order (OPORD), or a fragmentary order (FRAGO). He should use no more than one third of the available time for his own planning and for issuing his OPORD. The remaining two thirds is for subordinates to plan and prepare for the operation. Leaders should also consider other factors such as available daylight and travel time to and from orders and rehearsals.

b. **STEP 2--ISSUE A WARNING ORDER.** The leader provides initial instructions in a warning order. The warning order contains enough information to begin preparation as soon as possible. The warning order mirrors the five-paragraph OPORD format. A warning order may include--

- The mission or nature of the operation (mission statement).
- Time and place for issuance of the operation (coordinating instructions).
- Who is participating in the operation (coordinating instructions).
- Time of the operation (timeline).

c. **STEP 3--MAKE A TENTATIVE PLAN.** The leader uses METT-TC to develop an estimate of the situation, which he will use as the basis for his tentative plan. This set of actions is referred to as the leader's mission analysis:

(1) **Conduct a detailed mission analysis.**

(a) *Concept and Intent.* Higher commanders' concept and intent two levels up. This information is found in paragraph 1b for two levels up and in paragraphs 2 and 3 for one higher.

(b) *Unit Tasks.* Tasks that are clearly stated in the order (Specified Tasks) or tasks that become apparent as the OPORD is analyzed (Implied Tasks).

**EXAMPLE SPECIFIED TASKS**

- Retain hill 545 to prevent envelopment of B Co.
- Provide one squad to the 81-mm platoon to carry ammo.
- Establish an OP vic GA124325 NLT 301500 NOV 89.

**EXAMPLE IMPLIED TASKS**

- Provide security during movement.
- Conduct resupply operations.
- Coordinate with adjacent units.

(c) *Unit Limitations.* The leader next determines all control measures or instructions in the OPORD that restrict his freedom of action; these are called limitations. In every operation, there are some limitations on the unit.

**COMMON LIMITATIONS**

Graphic control measures.
Cross the LD at 100030 OCT 94 (time).
MOPP4 in effect (uniform and environmental considerations).
ADA weapons status, tight; warning status, yellow (rules of engagement).

(d) *Mission-Essential Task(s).* After reviewing all the above factors, the leader identifies his mission-essential task(s). Failure to accomplish a mission-essential task results in the unit's failure to accomplish its primary purpose for that operation. The mission essential task should be found in the maneuver paragraph.

(e) *Restated Mission.* The restated mission statement becomes the focus for the remainder of the estimate process. This is a clear, concise statement of the mission essential task(s) to be accomplished by the unit and the purpose to be achieved. The mission statement will state WHO, WHAT (the task), WHEN (the critical time), WHERE (usually a grid coordinate), and WHY (the purpose the unit must achieve). Some examples of restated missions follow:

| EXAMPLE RESTATED MISSIONS |
|---|
| (WHO) 1st Platoon attacks (WHAT) to seize (WHERE) HILL 482 vic NB 457371 (OBJ BLUE) (WHEN) NLT 09050OZ Dec 92 L 482 (WHY) to enable the company's main effort to destroy enemy command bunker. |
| (WHO) 1st Platoon, C Company defends (WHAT) to destroy from (WHERE) AB163456 to AB163486 to AB123486 to AB123456 (WHEN) NLT 28153OZ Oct 97 (WHY) to prevent enemy forces from enveloping B Company, 1-66 Infantry (L) from the south. |

(2) Analyze the situation and develop a course of action. Each COA must be--suitable, acceptable, feasible, distinguishable, and complete.

(3) Upon developing a COA, the unit leader will assign C2 headquarters, complete generic task organization assigning all organic and attached elements, and prepare COA statement and sketch.

(4) With the restated mission from Step 1 to provide focus, the leader continues the estimate process using the remaining factors of METT-TC.

(a) What is known about the **ENEMY**?

| | |
|---|---|
| ***Composition*** | This is an analysis of the forces and weapons that the enemy can bring to bear. Determine what weapons systems they have available, and what additional weapons and units are supporting him. |
| ***Disposition*** | The enemy's disposition is how he is arrayed on the terrain, such as in defensive positions, in an assembly area, or moving in march formation. |
| ***Strength*** | Percentage strength. |
| ***Recent Activities*** | Identify recent and significant enemy activities that may indicate future intentions. |
| ***Reinforcement Capabilities*** | Determine positions for reserves and estimated time to counterattack or reinforce. |
| ***Possible Courses of Action*** | Determine the enemy's possible COAs. This will help the leader determine how to best array his own forces against the enemy and fight the battle. |

(b) How will **TERRAIN** and **WEATHER** affect the operation? Analyze terrain using OAKOC.

- **Observation and Fields of Fire**. Determine locations that provide the best observation and fields of fire along friendly and enemy avenues of approach, near the objective, and on key terrain. The analysis of fields of fire is mainly concerned with the ability to cover the terrain with direct fire.
- **Avenues of Approach**. Analyze and identify friendly and enemy avenues of approach. Identify avenues of approach en route to the objective, on and around the objective, and for possible enemy counterattack. Also, consider aerial and subterranean avenues.

| | |
|---|---|
| ***Offensive Considerations (Friendly)*** | • How can these avenues support my movement?<br>• What are the advantages/disadvantages of each? (Consider enemy, speed, cover, and concealment.)<br>• What are the likely enemy counterattack routes? |

| | |
|---|---|
| ***Offensive Considerations (Enemy)*** | • How can the enemy use these approaches?<br>• Which avenue is most dangerous? Least? (Prioritize each approach.)<br>• Which avenues would support a counterattack? |

- **Cover and Concealment**. The analysis of cover and concealment is often inseparable from the fields of fires and observation. Weapon positions must have both to be effective and to be survivable. Infantry units are capable of improving poor cover and concealment by digging in and camouflaging their positions. When moving, the terrain is used to provide cover and concealment.
- **Obstacles**. Identify the existing and reinforcing obstacles and hindering terrain that will affect mobility.
- **Key Terrain**. Key terrain is any location or area that the seizure, retention, or control of affords a marked advantage to either combatant. Using the map, aerial photos, and information already gathered, look for key terrain that dominates avenues of approach or the objective area. Next, look for decisive terrain that if held or controlled will have an extraordinary impact on the mission.

(5) Analyze courses of action (war-game). This analysis is conducted by war-gaming friendly courses of action against the enemy's most probable courses of action. The leader can--and should--war-game with his subordinates.

(6) Compare courses of action. The leader compares the COAs and selects the one that is most likely to accomplish the assigned mission. He considers the advantages and disadvantages for each COA. He also considers how the critical events impact on COAs.

(7) Make a decision. The leader selects the COA that he believes has the best chance of accomplishing the mission.

d. **STEP 4--START NECESSARY MOVEMENT**. The unit may need to begin movement while the leader is still planning or forward reconnoitering. This step may occur anytime during the TLP.

e. **STEP 5--RECONNOITER**. If time allows, the leader makes a personal reconnaissance. When time does not allow, the leader must make a map or aerial photo reconnaissance. Sometimes the leader must rely on others, such as scouts, to conduct the reconnaissance.

f. ***STEP 6--COMPLETE THE PLAN***. The leader completes his plan based on the reconnaissance and any changes in the situation.

g. **STEP 7--ISSUE THE COMPLETE ORDER**. Platoon and squad leaders normally issue oral operations orders to aid subordinates in understanding the concept of the mission. Leaders may require subordinates to repeat part of the order, demonstrate it on a terrain model, or sketch their understanding of the operation. Leaders should also quiz their Rangers to ensure that all Rangers understand the mission.

h. **STEP 8--SUPERVISE AND REFINE**. The leader supervises the unit's preparation for combat by conducting rehearsals and inspections.

(1) ***Rehearsals***. Rehearsals include the practice of having squad leaders brief their planned actions in execution sequence to the platoon leader. The leader should conduct rehearsals on terrain that resembles the actual ground, and in similar light conditions. At a minimum, the leader should conduct a backbrief (least preferred), full force rehearsal (most preferred), or reduced force rehearsal (key leaders), and a rock drill.

(a) *Purpose*. The leader uses rehearsals to--

- Practice essential tasks (improve performance).
- Reveal weaknesses or problems in the plan.
- Coordinate the actions of subordinate elements.
- Improve Ranger understanding of the concept of the operation (foster confidence in Rangers).

(b) *Times and Tasks*. The platoon may begin rehearsals of battle drills and other SOP items before the receipt of the operation order. Once the order has been issued, it can rehearse mission specific tasks. Some important tasks to rehearse include--

- Actions on the objective.
- Assaulting a trench, bunker, or building.
- Actions at the assault position.
- Breaching obstacles (mine and wire).
- Using special weapons or demolitions.
- Actions on unexpected enemy contact.

(c) *Types.*

- Backbrief.
  - -- Key leaders sequentially brief the actions required during operation.
  - -- Patrol leader controls.
  - -- Conducted twice: right after FRAGO (confirmation brief), and again after subordinates develop own plan.
- Reduced force.
  - -- Conducted when time is key constraint.
  - -- Conducted when security must be maintained.
  - -- Key leaders normally attend.
  - -- Mock-ups, sand tables, and small scale replicas used.
- Full force.
  - -- Most effective type.
  - -- First executed in daylight and open terrain.
  - -- Secondly conduct in same conditions as operation.
  - -- All Rangers participate.
  - -- May use force on force.
- Techniques.
  - -- Force on force.
  - -- Map (limited value and limited number of attendees).
  - -- Radio (cannot mass leaders; confirms communications).
  - -- Sand table or terrain model (key leaders; includes all control measures).
  - -- Rock drill (similar to sand table/terrain model; subordinates actually move themselves). Very effective for coordinating timing and sequence of events--especially when air assets are used.

(d) *Inspections.* Squad leaders should conduct initial inspections shortly after receipt of the warning order. The platoon sergeant spot-checks throughout the unit's preparation for combat. The platoon leader and platoon sergeant make a final inspection of--

- Weapons and ammunition.
- Uniforms and equipment.
- Mission-essential equipment.
- Soldier's understanding of the mission and their specific responsibilities.
- Communications.
- Rations and water.
- Camouflage.
- Deficiencies noted during earlier inspections.

2-2. **COMBAT INTELLIGENCE**. Gathering information is one of the most important aspects of conducting patrolling operations. The following details what information to collect and how to report it:

a. **Reports**. All information must be quickly, completely, and accurately reported. Use the SALUTE report format for reporting and recording information.

| | |
|---|---|
| **S**IZE | – Seven enemy personnel |
| **A**CTIVITY | – Traveling SW |
| **L**OCATION | – GA123456 |
| **U**NIT/UNIFORM | – OD uniforms with red six-point star on left shoulder |
| **T**IME | – 210200JAN99 |
| **E**QUIPMENT | – Carry one machine gun and one rocket launcher |

b. **Field Sketches**. Try to include a sketch with each report. Include only any aspects of military importance such as targets, objectives, obstacles, sector limits, or troop dispositions and locations (use symbols from FM 1-02). Use notes to explain the drawing, but they should not clutter the sketch.

c. **Captured Documents**. The leader collects documents and turns them in with his reports. He marks each document with the time and place of capture.

d. **Prisoners**. If prisoners are captured during a patrolling operation, they should be treated IAW the Geneva Convention and handled by the 5-S rule:

(1) Search
(2) Silence
(3) Segregate
(4) Safeguard
(5) Speed to rear

e. **Debriefs**. Immediately upon return from a mission, the unit is debriefed. The intelligence officer will generally have a unit-specific format for debriefing a patrol.

2-3. **WARNING ORDER**. Warning orders give subordinates advance notice of upcoming operations. This gives them time to prepare. A warning order must be brief, but complete. *A warning order only authorizes execution when it clearly says so.* (Figure 2-2 shows an example format; Figure 2-3 shows an example warning order.)

**Figure 2-2. WARNING ORDER FORMAT**

## WARNING ORDER ________

*Roll call, pencil/pen/paper, RHB, map, protractor, leader's monitor, hold all questions till the end
References: Refer to higher headquarters' OPORD, and identify map sheet for operation.
Time Zone Used throughout the Order: (Optional)
Task Organization: (Optional; see paragraph 1c.)

1. **SITUATION** (Higher's OPORD para 1a[1]--[3])

a. **Enemy Forces**. Include significant changes in enemy composition, dispositions, and courses of action. Information not available for inclusion in the initial WARNO can be included in subsequent warning orders. (who, what, where)

b. **Friendly Forces**. (Optional) only address if essential to the WARNO.

(1) Higher commander's mission (who, what, where, why).
(2) Higher commander's Intent. (Higher's [go to mapboard] OPORD para 1b[2], give task and purpose)

c. **Attachments and Detachments**. Initial task organization, only address major unit changes.

(1) Orient relative to each point on the compass.
(2) Box in the entire AO with grid lines.
(3) Trace each Zone using boundaries.
(4) Familiarize by identifying 3 natural (terrain) and 3 man-made features in each zone.
(5) Point out the enemy and friendly locations on the map board.

2. **MISSION**. State mission twice (who, what, when, where, and why) and concisely state task and purpose.

3. **EXECUTION**.

a. **Concept of Operation**. Provide as much information as available. The concept should describe the form of maneuver or defensive technique, critical events, decisive point of the operation and why it is decisive, task and purpose for the main and supporting efforts, purposes of the warfighting systems, such as engineers, fire support, intelligence etc), acceptable risk, and the desired endstate. After the desired endstate is stated, leaders should talk through the concept of the operation like a story. Use a simple sketch to assist in relaying the information.

b. **Tasks to Maneuver Units**. Provide specified tasks to subordinate units. These are tactical instructions on how to execute the mission for each element in task organization. Planning guidance consists of tasks assigned to subordinate units, special teams, and key individuals.

c. **Tasks to Combat Support Units**. See paragraph 3b.

d. **Coordinating Instructions**. Include any information available at that time. If you know it, then at least cover--

- Uniform and equipment common to all. Consider the factors of METT-TC and tailor the load for each Ranger.
- Time line. (State 4 W's and specified and implied times. Reverse plan. Use 1/3-2/3 rule.)
- CCIR.
- Risk guidance.
- Deception guidance.
- Specific priorities, in order of completion.
- Guidance on orders and rehearsals.
- Orders group meeting (attendees, location, and time).
- Earliest movement time and degree of notice.

4. **SERVICE SUPPORT**. Include any known logistics preparation for the operation.

a. Special equipment. Identify requirements and any coordination measures the unit needs to take to transfer the special equipment. (State the equipment you will use, need, or want for the mission.)

b. Transportation. Identify requirements and any coordination needed to pre-position assets. (State unit's means of infil/exfil.)

5. **COMMAND AND SIGNAL**.

a. **Command**. State the chain of command if different from unit SOP.

b. **Signal**. Identify current SOI edition, and pre-position signal assets to support operation.

- Give subordinates guidance on tasks to complete for preparation of the OPORD and the mission.
- Give time, place, and uniform of the OPORD.
- Give a time hack and ask for questions.

NOTES

Figure 2-3. EXAMPLE WARNING ORDER

## WARNING ORDER (SQUAD)

ROLL CALL, CAMP DARBY SPECIAL (CDS) 1:50,000 MAP, PEN, PAPER, PENCIL, PROTRACTOR, RANGER HANDBOOK (RHB), HOLD ALL QUESTIONS, AND TLs MONITOR AND TASK-ORGANIZE

1. **SITUATION**.

a. **Enemy**.

WHO: The Aragon Liberation Front (ALF).
WHAT: Has been conducting aggressive offensive operations.
WHERE: In Zones A, B, C.

b. **Friendly**:

(1) **Higher's Mission**.

WHO: 2nd PLT, C CO.
WHAT: Conducts area ambushes to destroy enemy personnel and equipment.
WHERE: Zone D.
WHY: To prevent enemy logistical resupply.

(2) **Higher Commander's Intent**.

- Rapid infiltration into Zone D.
- Find, fix, and finish enemy forces in Zone D.
- Enemy personnel and equipment located and targeted.
- Squads ready to conduct future combat operations.

c. **Attachments/Detachments**. MG Tm #1/2 PLT 160630-190630.

2. **MISSION**.

WHO: 1st SQD, 2nd PLT
WHAT: Conducts a point ambush to destroy enemy personnel and equipment.
WHERE: GA 1805 7878.
WHEN: NLT 25 1830 DEC 2007.
WHY: To prevent enemy logistical resupply and gather PIR.

3. **EXECUTION**.

a. **Concept of Operation**. (Orient Rangers to sketch or terrain model). We are currently located at Camp Darby, GA 1962 7902, which is on a ridge. Our ground tactical plan will begin when we will move generally southwest to our ORP at Grid GA 187778. Terrain feature there is a draw for 1,000 meters, by foot and it will take 1 hour. Here will we finalize the preparing of M, W, and E. We will then move generally southwest to our objective at Grid GA 184774. Terrain feature there is a ridge for 200 meters, by foot, and it will take us 30 minutes to an hour. The reason for this is stealth on the objective while we occupy our positions.

b. **Tasks to Maneuver Units**.

(1) HQs: HQ is responsible for command and control and communications. Commo/HQ is 2nd in the OOM. M240 will be placed in a position to provide supporting fires in the kill zone. RTO will be the recorder en route and during actions on the objective. SL will be located where he can best control his squad and observe the kill zone. SL is responsible for dissemination of information.

(2) ATM: ATM will be 1st OOM/responsible for navigation, and for clearing and securing the far side at all linear danger areas. ATM is flank security for AOO, will provide early warning and seal off objective/1-2 man EPW, and search TM/1-2 man aid and litter TM/1-2 man demo TM/1 man (P), 1 man (A) ORP clearing TM/ 2-2 man flank security. TM for AOO/ 1 compassman, 1 pace man, or ATL will be the sec. TM LDR along the most likely avenue of approach.

(3) BTM: BTM 3RD OOM/responsible for securing the near side of all linear danger areas/BTM. Is assault for AOO. Must provide suppressive fire and assault thru kill zone/1-2 man EPW and search team/1-2 man aid and litter team/1-2 man demolitions team; 1 man (P); 1 man (A) ORP clearing team; 1-2 man S/O team; 1-2 man security team for reentry; 1 compassman; 1 pace man; and BTL. You will be the assault element leader for AOO.

c. **Tasks to Combat Support Units**. None.

d. **Coordinating Instructions**.

(1) Uniform and equipment common to all. At a minimum, each Ranger wears or carries--

(a) A ballistic helmet w/camouflaged band and luminous tape; clean and serviceable ACUs, camelback, black belt w/buckle, blackened combat boots, serviceable OD or black socks, serviceable OD/brown T-shirt, ID tags w/breakaway lanyard around neck (silenced). Right cargo pocket; (RHB, CDS [map], protractor, pencil, pen, and paper) waterproofed, left cargo pocket; patrol cap w/ luminous tape, ID card in right breast pocket, SL will carry sterile fire support overlay in left breast pocket.

(b) LCE, consisting of pistol belt, suspenders, 2 ammo pouches, seven 30-round magazines, one first aid dressing with pouch, one compass and pouch, two 1-quart canteens w/covers and cups, one whistle, and four M67 fragmentation grenades. All are tied down IAW 4th Bn SOP, except that the pace cord is tied to suspenders.

(c) Rucksack, complete; one set serviceable ACUs, two 2-quart canteens w straps; E-tool with cover; two pair socks; two brown T-shirts; weapons cleaning kit; one waterproof bag; a sewing kit; a patrol cap; MREs; poncho and liner; wet weather top; personal hygiene kit; one brown towel; and camouflage stick. RTO also packs and ties off the ANCD.

(d) Optional bug juice, 550 cord, foot powder, black tape, knife, all other items will be approved by SL 15 minutes prior to completion time of OPORD.

(2) Time schedule.

| WHEN | WHAT | WHERE | WHO | REF |
|---|---|---|---|---|
| *0630 | WARNING ORDER | BAY AREA | All | 3D1 |
| 0700 | INITIAL INSPECTION | BAY AREA | All | |
| *0730 | Required Ammo/Supplies | CO TOC | BTL | 4B1 |
| *0745 | P/U-Ammo/Supplies | CO TOC | BTL/Detail | 4B1 |
| 0800 | Test Fire | T/F AREA | All | |
| *0830 | S-2/S-3 Coord | CO BAY AO | SL/RTO | 3D1 |
| *0830 | Fire Spt Coord | BN TOC | SL/ RTO | 3A2 |
| 0850 | Adj Unit Coord | BN TOC | ATL/compass man | 3D1 |
| *0900 | Enter Net | BAY AREA | RTO | 5B1 |
| 0930 | Complete Plan | BAY AREA | All | |
| *1000 | Issue OPORD | BAY AREA | All | 3D1 |
| 1300 | Rehearsal | BAY AREA | All | |
| 1400 | Final Inspection | BAY AREA | All | |
| 1430 | Conduct Move | BAY AREA | All | |
| 1800 | In ORP | GA | All | |
| 2000 | In Position | GA 184774 | All | |
| *2300 | Mission Complete | GA 186874 | All | 3D1 |
| *0200 | Reenter FFU | GA | All | |
| 0500 | S-2 Debrief | BN TOC | All | |
| * Specified time | | | | |

4. **SERVICE AND SUPPORT**.

a. **Special Equipment**.

(1) HQs: One hand-held radio and extra batteries; one AN/PRC-119 radio and extra batteries, one ANCD, one AN/PSN-11 and extra batteries, one M-22 binos, one AN/PVS-14 and extra batteries; one sheets overlay material, three sheets carbon paper; two HE smoke, two red smoke, one M240-B MG complete w/tripod, flex mount, T&E, and Pintle, one AN/PVS-4 and extra batteries, one AN/PVS-7D, and extra batteries.

(2) ATM: One hand-held radio and extra batteries, one M-24 binos, two AN/PVS-14 and extra batteries, two HE smoke, two yellow smoke, one thermite grenade, two gags, two flex cuffs, two sandbags, one CLS bag, one pole-less litter, two M18A1 claymore mines, one AT-4, one LAW, one M-203 vest, two M-249 SAW pouches, one small metal flashlight, and extra batteries secured to rifle.

(3) BTM: One hand-held radio and extra batteries, one M-24 binos, two AN/PVS 14 and extra batteries, two HE smoke, two yellow smoke, one thermite grenade; two each gags, flex cuffs, and sandbags; one CLS bag, one pole-less Litter, two M18A1 claymore mines, one AT-4, one LAW, one M-203 vest, two M249 SAW pouches, one small metal flashlight, and extra batteries secured to rifle.

b. **Ammo**. (per weapon system).

| M16/M4 | M249 | M203 | M240 | SL AND TLs |
|---|---|---|---|---|
| 210 Rounds<br>5.56-MM Ball | 625 Rounds<br>5.56-MM (4:1 Mix) | 16 Rounds 40-MM HEDP<br>5 Rounds 40-MM Shot<br>4 Rounds 40-MM Illumination<br>2 Rounds 40-MM TP-T | 825 Rounds<br>7.62-MM (4:1 Mix) | 42 Rounds<br>5.56-MM<br>Tracer |

c. **Transportation**. NA

5. **COMMAND AND SIGNAL**.

a. **Command**.

| PERSONNEL | POS | SOC |
|---|---|---|
| **HQ/SUPPORT** | | |
| BARNETT | SL | 1 |
| MARENGO | RTO | 4 |
| CASELY | MG | 11 |
| WADE | AG | 12 |
| **ATM/SECURITY** | | |
| TRUITT | ATL | 3 |
| SILCOX | GRN | 5 |
| WISE | R | 6 |
| EDMUNDS | R | 7 |
| **BTM/ASSAULT** | | |
| SPENCER | BTL | 2 |
| CHRISTENSEN | GRN | 8 |
| RUFUS | R | 9 |
| ROBINSON | R | 10 |

| SUBORDINATE GUIDANCE: | |
|---|---|
| ATL: | Write paragraph 1 of squad OPORD, and for drawing all sketches, terrain models, routes, and nonsterile fire support overlays. |
| BTL: | You are second in the chain of command and in charge at all times during my absence.<br>Write paragraph 4 of squad OPORD Write the supply, DX, and ammo lists, draw and issue all items, ensure that everyone conducts a test fire, and all equipment is tied down IAW 4th RTBn SOP, and update the squad status card and hand receipt. |
| RTO: | Write paragraph 5 of squad OPORD, ensure all radios are operational with proper freqs loaded, and ensure we enter the net on time. |
| TL: | TLs: You will update the Warning Order board with all the correct information. As a task is accomplished, you will line it out. Also, post your succession of command (SOC), duty positions, and job descriptions (specialty teams and key individuals). Come and see me for further guidance at the conclusion of this WARNO. |

b. **Signal**.

| | |
|---|---|
| TACTICAL FREQUENCY | 62,000 |
| MEDEVAC FREQUENCY | 35,000 |
| CHALLENGE | BAYONET |
| PASSWORD | MILLETT |
| RUNNING PASSWORD | RANGER |
| NUMBER COMBO | 7 |
| TIME, PLACE, AND UNIFORM FOR OPORD. TIME HACK. ANY QUESTIONS? | |

NOTES

2-4. **OPERATIONS ORDER**. An operations order (OPORD) is a directive issued by a leader to his subordinates in order to effect the coordinated execution of a specific operation. A five-paragraph format (shown below) is used to organize the briefing, to ensure completeness, and to help subordinate leaders understand and follow the order. Use a terrain model or sketch along with a map to explain the order. When possible, such as in the defense, give the order while observing the objective. The platoon/squad leader briefs his OPORD orally off notes that follow the five-paragraph format. Before the issuance of the OPORD, the leader ensures that the following resources are in place: pencil, pen, paper; RHB; map; protractor; leader's monitor. Then he calls roll and says "*Please hold all questions till the end.*" Figure 2-4 shows an example OPORD format. Figure 2-5 shows an OPORD shell.

**Figure 2-4. EXAMPLE OPORD FORMAT**

## OPERATIONS ORDER

[Plans and orders normally contain a code name and are numbered consecutively within a calendar year.]

*References:* [The heading of the plan or order includes a list of maps, charts, datum, or other related documents the unit will need to understand the plan or order. The user need not reference the SOP, but may refer to it in the body of the plan or order. He refers to a map by map series number (and country or geographic area, if required), sheet number and name, edition, and scale, if required. "*Datum*" refers to the mathematical model of the earth that applies to the coordinates on a particular map. It is used to determine coordinates. Different nations use different datum for printing coordinates on their maps. The datum is usually referenced in the marginal information of each map.]

*Time zone used throughout the order:* The time zone used throughout the order (including annexes and appendixes) is the time zone applicable to the operation. Operations across several time zones use Zulu time.

*Task organization:* Describe the allocation of forces to support the commander's concept. Task organization may be shown in one of two places: preceding paragraph one, or in an annex, if the task organization is long and complicated.

**Weather and light data, and general forecast (Only discuss what will affect your patrol. Note effects on friendly and enemy personnel and equipment.):*

High
Moonrise
Sunrise
Low
Moonset
Sunset
Wind Speed
Moon Phase
BMNT
Wind Direction
Percent Illumination
EENT

**This is the information the leader received from higher's OPORD.*

*Terrain*: Make a statement concerning the effects terrain will have on both friendly and enemy forces in the area of operation using the OAKOC format.

1. **SITUATION**.

a. **Enemy Forces**. The enemy situation in higher headquarters' OPORD (paragraph 1a) is the basis for this, but the leader refines this to provide the detail required by his subordinates.

(1) Enemy's composition, disposition, strength.

(2) Recent activities.

(3) Known/suspected locations and capabilities.

(4) Describe the enemy's most likely and most dangerous course of action on a map.

b. **Friendly Forces**. This information is in paragraphs 1b, 2, and 3 of higher headquarters' OPORD.

(1) State the mission of the Higher Unit (2 levels up).

(2) State the mission of the Higher Unit (1 level up).

(3) State intent 2 levels up.

(4) State locations of units to the left, right, front, and rear. State those units' tasks and purposes; and say how those units will influence your unit's patrol.

(a) Show other units locations on map board.

(b) Include statements about the influence each of the above patrols will have on your mission, if any.

- Obtain this information from higher's OPORD. It gives each leader an idea of what other units are doing and where they are going. This information is in paragraph 3a (1) (execution, concept of the operation, maneuver).
- Also include any information obtained when the leader conducts adjacent unit coordination.

c. **Attachments and Detachments**. Do not repeat information already listed under Task Organization. Try to put all information in the Task Organization. However, when not in the Task Organization, list units that are attached or detached to the headquarters that issues the order. State when attachment or detachment is to be effective if different from when the OPORD is effective, such as on order, on commitment of the reserve). Use the term "*remains attached*" when units will be or have been attached for some time.

2. **MISSION**. State the mission derived during the planning process. There are no subparagraphs in a mission statement. Include the 5 W's: Who, What (task), Where, When, and Why (purpose).

- Make it a clear and concise statement and read it twice.
- Go to map and point out the exact location of the OBJ and point out the unit's present location.

3. **EXECUTION**. State the commander's intent.

a. **Concept of Operation**. The concept statement should be concise and clear. It describes how the unit will accomplish its mission from start to finish. The number of subparagraphs, if any, is based on what the leader considers appropriate for the level of leadership and complexity of the operation. Generally, the concept should describe the form of maneuver or defensive technique and critical events; identify what the decisive point of the operation is and why; describe the task and purpose for the main effort and supporting efforts; state the purpose of the warfighting functions, such as engineers, fire support, or intelligence; define acceptable risk; and identify the desired endstate. After stating the desired endstate, leaders should talk through the concept of the operation like a story, using a simple sketch to help them convey the information. Figure 1 shows the six warfighting functions (WFF):

**Figure 1. Warfighting functions.**

- Fire support
- Movement and Maneuver
- Protection
- Command and Control
- Intelligence
- Sustainment (formerly called "CSS")

(1) ***Maneuver***. The maneuver paragraph details the mechanics of the operations. This subparagraph specifically addresses all subordinate units and attachments by name, giving each its mission in the form of a task and purpose. The main effort's mission must be supported by the subordinate unit's missions. Actions on the objective will comprise the majority of this paragraph and should include a detailed plan for engagement/disengagement criteria, an alternate plan in the event of compromise or unplanned movement of enemy forces, and a withdrawal plan.

(2) ***Fires***. State scheme of fires to support the overall concept. This paragraph should state which maneuver unit has priority of fires. Fires should be planned using the PLOT-CR format (purpose, location, observer, trigger, communication method, resources). A target list worksheet and overlay are referenced here, if applicable. Specific targets are discussed and pointed out on the terrain model (see Chapter 3, Fire Support).

(3) ***Casualty Evacuation***.

(a) *Plan*. State a detailed CASEVAC plan during each phase of the operation. Include CCP locations, tentative extraction points, and methods of extraction.

b. **Tasks to Maneuver Units**. Clearly state the missions or tasks for each maneuver unit that reports directly to the headquarters issuing the order. List the units in the task organization, including reserves. Use a separate subparagraph for each maneuver unit. Only state the tasks that are necessary for comprehension, clarity, and emphasis. Place tactical tasks that affect two or more units in Coordinating Instructions (subparagraph 3d). Platoon leaders task their subordinate squads. Those squads may be tasked to provide any of the following special teams: reconnaissance and security, assault, support, aid and litter, EPW and search, clearing, and demolitions. Detailed instructions may also be given to the platoon sergeant, RTO, compassman, and paceman.

(1) ***Tasks to Combat Support Units***. Add subparagraphs here only as necessary. List CS units in subparagraphs from the task organization. Use CS subparagraphs to list only those specific tasks that CS units must accomplish and that are not specified or implied elsewhere. Include organization for combat if not clear from task organization.

c. **Coordinating Instructions**. This is always the last subparagraph under paragraph 3. List only the instructions that apply to two or more units, and which are seldom covered in unit SOPs. Refer the user to an annex for more complex instructions. The following information is required:

*1. Time schedule (state time, place, uniform, and priority of rehearsals, backbriefs, inspections, and movement).*

*2. Give the commander's critical information requirements (CCIR), which includes priority intelligence requirements (PIR), essential elements of friendly information (EEFI), and friendly force information requirements (FFIR):*

*a. PIR is intelligence that the commander must have for planning and decision making.*

*b. EEFI are critical aspects of friendly operations that, if known by the enemy, would compromise, lead to failure, or limit success of the operation.*

*c. FFIR include what the commander needs to know about friendly forces available for the operation. It can include personnel status, ammunition status, and leadership capabilities.*

*3. Risk reduction control measures. These are measures unique to the operation. They supplement the unit SOP and can include mission-oriented protective posture, operational exposure guidance, vehicle recognition signals, and fratricide prevention measures.*

*4. Rules of engagement (ROE).*

*5. Environmental considerations.*

*6. Force protection*

*7. Movement plan. Use terrain model, sketch, or both. State azimuths, directions, and grid coordinates. Use names of real Rangers on the sketches.*

*a. Sketch out the order of movement, the formation, and the movement technique.*

*b. Sketch actions at halts (long and short).*

*c. Routes (use terrain model):*

*1) Orient the terrain model (North, South, East, and West).*

*2) Identify the grid lines, both North-South and East-West.*

*3) Brief the legend.*

*4) Box in the AO with distinguishable natural or man-made terrain features such as roads, ridgelines, rivers, and streams.*

*5) Brief all the routes depicted on the terrain model from start to finish using specific azimuths and distance.*

*a) The primary route from the start point to the ORP, from the ORP to the objective, and from the objective to the patrol base while describing the terrain over which the squad/platoon will travel. (Include near/far side rally points for danger areas.)*

*b) The alternate route from the start point to the ORP, from the ORP to the objective, and from the objective to the patrol base while describing the terrain over which the squad/platoon will travel. (Include near/far side rally points for danger areas.)*

*6) Brief the fire support plan (if not given in fires paragraph). When covering fires the leader should cover the following (PLOT-CR):*

- *Purpose.*
- *Location and terrain feature to the targets.*
- *Observer.*
- *Trigger.*
- *Communication method.*
- *Resources.*

*7) State where the sterile fire support overlay is being carried.*

*a) Rally points and actions at all rally points (include grid location and terrain reference, and use sketch).*

*b) Actions at danger areas. This is a general plan for unknown linear, small open areas and large open areas; specific plan for all known danger areas that unit will encounter along the route (use sketch).*

*c) Actions on enemy contact; battle drills (use sketch).*

4. **SERVICE SUPPORT**. Address service support in the areas shown below as needed to clarify the service support concept. Subparagraphs can include:

a. **General**. Refer to the SOPs that govern the sustainment operations of the unit. Provide current and proposed company trains locations, casualty, and damaged equipment collection points and routes.

b. **Materiel and Services**.

(1) ***Supply***.

Class I--Rations plan.
Class III--Petroleum.
Class V--Ammunition.
Class VII--Major end items.
Class VIII--Medical.
Class IX--Repair parts.
Distribution Methods.

(2) ***Transportation***.

(3) ***Services***. (Laundry and showers.)

(4) ***Maintenance***. (Weapons and equipment.)

(a) *Medical Evacuation and Hospitalization*. Method of evacuating dead and wounded, friendly and enemy personnel. Include priorities and location of CCP.

(b) *Personnel Support*. Method of handling EPWs and designation of the EPW collection point.

5. **COMMAND AND SIGNAL**.

This paragraph states where command and control facilities and key leaders are located during the operation. Use the OPORD shell to save time and organize your thoughts while preparing an OPORD.

a. **Command**.

- Location of the higher unit commander and CP.
- Location of key personnel and CP during each phase of the operation.

- Succession of command.
- Adjustments to the patrol SOP.

b. **Signal**.
- SOI index in effect.
- Methods of communication in priority.
- Pyrotechnics and signals, to include arm and hand signals (demonstrate).
- Code words such as OPSKEDs.
- Challenge and password (use behind friendly lines).
- Number combination (use forward of friendly lines).
- Running password.
- Recognition signals (near/far and day/night).

c. **Special Instructions to RTOs**.

d. **Actions after Issuance of OPORD--**
(1) Issue annexes.
(2) Give time hack.
(3) ASK for questions.

**Figure 2-5. OPORD SHELL**

## OPORD

DTG:

REFERENCE:

TASK ORGANIZATION:

| 1ST PLT | 2ND PLT | TM CONTROL: |
|---|---|---|

1. **SITUATION**.

BMNT: SR: SS: EENT: MR: MS: HI: LO: ILLUM:

a. **Enemy**.

Terrain | End state

(1) Higher mission:

Cdr's intent:

(2) Units to N, S, E, W.

Terrain | End state

2. **MISSION**:

3. **EXECUTION**:

a. **Concept of Operations**.

**Cdr's intent**:

| | | | | | |
|---|---|---|---|---|---|
| (ME):<br>T:<br>P: | (SE1):<br>T:<br>P: | (SE2):<br>T:<br>P: | (SE3):<br>T:<br>P: | (SE4):<br>T:<br>P: | (SE5):<br>T:<br>P: |

b. **Maneuver**

PHASE 1:
ME:
SE1:
SE2:
SE3:
SE4:
SE5:
FIRES:
MTRS:
C2:

PHASE 2:
ME:
SE1:
SE2:
SE3:
SE4:
SE5:
FIRES:
MTRS:
C2:

PHASE 3:
ME:
SE1:
SE2:
SE3:
SE4:
SE5:
FIRES:
MTRS:
C2:

PHASE 4:
ME:
SE1:
SE2:
SE3:
SE4:
SE5:
FIRES:
MTRS:
C2:

c. **Tasks to Maneuver Units**.

(1) Risk-reduction control measures.
Rules of engagement.
(2) Environmental considerations.
(3) Non-Airborne personnel.

(ME): (SE1): (SE2):

(SE3): (SE4): (SE5):

d. **Tasks to Subordinate Units**

Mortars

Coordinating instructions:

PIR:

EEFI:

FFIR:

Timeline.

Coordinating Instructions.

| WHEN | WHO | WHAT | WHERE |
|---|---|---|---|
| | | | |
| | | | |
| | | | |
| | | | |
| | | | |
| | | | |
| | | | |
| | | | |
| | | | |
| | | | |
| | | | |
| | | | |
| | | | |
| | | | |
| | | | |
| | | | |

4. **SERVICE SUPPORT**.

a. **General**.

b. **Material and Services**.

(1) Supply:

Class I.

Class III:

Class V:

Class VIII:

Class IX:

(2) LOGSTAT.

(3) The chain of command is responsible to ensure that ammo lasts between resupplies.

c. **Medical Evacuation**.

(1) Air MEDEVAC is on call through [this] frequency. JP capable; use standard 9-line, 24-hr coverage.

(2) Establish friendly CCPs in AAs. Adjust CCPs as mission progresses.

(3) Troop medical clinic located vic [this location].

(4) Precedence for evacuation of WIA is US military, host nation military, host nation civilian, and enemy military.

(5) Precedence for evacuation of KIA is same. Evacuation of KIA will be as per mission constraints.

d. **Personnel**.

(1) Method of handling EPWs:

**ANNEXES**:

5. **COMMAND AND SIGNAL**.

a. **Command**.

(1) TF CP:

(2) Succession of Command:

b. **Signal**.

(1) Challenge and Password:

(2) Number Combination:

(3) Running Password:

(4) Recognition:

(a) Far:

(b) Near:

(7) Code word:

(8) OPSKED:

(9) Reports. SALUTE to higher on enemy contact.

2-5. **FRAGMENTARY ORDER**. A FRAGO is an abbreviated form of an operations order, usually issued on a day-to-day basis that eliminates the need for restating information contained in a basic operations order. It is issued after an OPORD to change or modify that order or to execute a branch or sequel to that order. Figure 2-6 shows an annotated FRAGO format:

**Figure 2-6. ANNOTATED FRAGO FORMAT**

**FRAGMENTARY ORDER**________

*Time Zone referenced throughout order:*

*Task Organization:*

*Weather and Light Data:*

| | | |
|---|---|---|
| High: | BMNT: | Moonrise: |
| Low: | Sunrise: | Moonset: |
| Wind Speed: | Sunset: | % Illum: |
| Wind Direction: | EENT: | |
| Forecast: | | |

*Terrain* (changes that will effect operation in new area of operations):

OAKOC: Observation and fields of fire, avenues of approach, key terrain, observation, and cover and concealment.

1. **SITUATION**. (Brief changes from base OPORD specific to this day's operation.)

a. **Enemy Situation**.

(1) Composition, disposition and strength.

(2) Capabilities.

(3) Recent activities.
(4) Most likely COA.

b. **Friendly Situation**.
(1) Higher mission.
(2) Adjacent patrols task/purpose.
(3) Adjacent patrol objective/route (if known).

2. **MISSION** (Who, what [task], when, where, why [purpose]—from higher HQ's maneuver paragraph).

3. **EXECUTION**

a. **Concept of Operation**. (Explain how unit will accomplish mission in general terms. Identify mission essential task, designate the main effort and how the supporting efforts support the main effort.)

(1) ***Maneuver***. (Assign task/purpose and discuss actions on the objective in detail. Use a sketch or terrain model to brief.)

(2) ***Fires***. (In support of today's mission, portions that do not change are not briefed.) PLOT-CR:
(a) Purpose.
(b) Location.
(c) Observer.
(d) Trigger.
(e) Communication method.
(f) Resource.

(3) ***CASEVAC***.

b. **Tasks to Maneuver Units**. (List unique tasks that apply to each squad.)

c. **Coordinating Instructions**.

(1) ***Timeline***.
(a) Hit time:
(b) ORP time:
(c) Movement time from PB:
(d) Final inspection:
(e) Rehearsals:
(f) FRAGO complete:

(2) ***Movement Plan***. (Use a sketch to brief.)
(a) Routes (primary and alternate):
(b) OOM:
(c) Formations:
(d) Movement technique:

(3) ***PIR***. (Specific to this mission):

(4) ***Rehearsal Plan***.

(5) ***Patrol Base Plan***. (If not IAW SOP):
(a) Teams:
(b) Occupation plan:
(c) Operations plan (security plan, alert plan, Black and Gold):
(d) Priorities of work:

(6) ***Air Assault Plan***. (Provided with higher's FRAGO, if applicable):
(a) Number, type, and ACL aircraft:
(b) PZ grid, DOL, and PZ posture time:
(c) Load time/lift time/flight time:
(d) Number of lifts and composition:
(e) Air checkpoints en route:
(f) LZ grid/DOL/LZ time:
(g) Actions after getting off aircraft:
(h) Actions on contact on LZ:

(7) ***Linkup Plan***. (If applicable):

(a) Time of linkup:
(b) Location of linkup site:
(c) Stationary element:
(d) Moving element:
(e) Rally points:
(f) Actions at linkup point:
(g) Near/far recognition signals (day and night):

4. **SUPPORT**. (Only cover changes from base order that apply for today's mission).

a. **Material and Services**.

(1) Any changes in classes of supply:
(2) Resupply plan:
(3) Water resupply plan:
(4) Aerial resupply plan (if applicable) :
(5) Truck plan:
(6) Maintenance issues specific to plan:

b. **Medical Evacuation Plan**. (Specific to mission):

(1) CCP point and markings:
(2) Aid and litter duties (if not SOP):

5. **COMMAND AND SIGNAL**.

a. **Command**.

(1) Location of Company CP:
(2) Location of PL:
(3) Location of key leaders:
(4) Succession of command:

b. **Signal**.

(1) Location of radios:

(a) During movement:
(b) During actions on the objective:

(2) SOI in effect:

| | |
|---|---|
| (a) Bn command freq: | Co CP call sign: |
| (b) Bn MEDEVAC freq: | PL call sign: |
| (c) Retrans freq: | PSG call sign: |
| (d) Company freq: | 1 SL call sign: |
| (e) Platoon freq: | 2 SL call sign: |
| | 3 SL call sign: |
| | WSL call sign: |

(3) Pyro signals in use:
(4) Running password/number combination:
(5) OPSKEDs in effect:

(a) The field FRAGO should take no more than 40 minutes to issue, with 30 minutes for the target. The proposed planning guide is as follows:

| | |
|---|---|
| Paragraphs 1 and 2: | 5 minutes |
| Paragraph 3: | 20 to 30 minutes |
| Paragraphs 4 and 5: | 5 minutes |

(b) The FRAGO should focus on actions on the objective. The PL may use subordinates to prepare para 1, 4, 5 and routes and fires for the FRAGO. It is acceptable for subordinates to brief the portions of the FRAGO they prepare.

(c) Use of sketches and a terrain model are critical to allow rapid understanding of the operation/FRAGO.

(d) Rehearsals are critical as elements of the constrained planning model. The FRAGO used in conjunction with effective rehearsals reduces preparation time and allows the PL more time for movement and recon.

(e) Planning in a field environment will necessarily reduce the amount of time leaders have for in-depth mission planning. The TLP gives leaders a framework to plan missions and produce orders when time is short.

2-6. **ANNEXES**. Operation order annexes are necessary to complete the plan and to provide greater clarity and understanding during complex or critical aspects of the operation. Information issued in annex form include the: aerial re-supply, truck movement, air assault, patrol base, small boat, linkup, and stream crossing annexes. Annexes are prepared only if the subject is not addressed thoroughly enough in the OPORD; brevity remains the standard. Annexes are always issued after the operation order.

## AIR MOVEMENT ANNEX

1. **SITUATION**.
   a. **Enemy Situation**.
      (1) Enemy air capability.
      (2) Enemy ADA capability.
      (3) Include in weather: % Illum, illum angle, NVG window, ceiling, and visibility.
   b. **Friendly Situation**.
      (1) Unit(s) supporting operation.
      (2) Friendly ADA status.
2. **MISSION**.
3. **EXECUTION**.
   a. **Concept of Operation**.
   b. **Subunit Missions**.
   c. **Coordinating instructions**.
      (1) ***Pickup Zone***.
         (a) Name/Number/
         (b) Coordinates/
         (c) Load Time/
         (d) Takeoff Time/
         (e) Markings/
         (f) Control/
         (g) Landing Formation/
         (h) Approach/Departure Direction.
         (i) Alternate PZ Name/Number.
         (j) Penetration Points.
         (k) Extraction Points.
      (2) ***Landing Zone***.
         (a) Name/Number.
         (b) Coordinates.
         (c) H-Hour.
         (d) Markings.
         (e) Control.
         (f) Landing Formation/Direction.
         (g) Alt LZ Name/Number.
         (h) Deception Plan.
         (i) Extraction LZ.
      (3) ***Laager Site***.
         (a) Communications.
         (b) Security Force.
      (4) ***Flight Routes and Alternates***.
      (5) ***Abort Criteria***.
      (6) ***Down Aircraft/Crew*** (designated area of recovery [DAR]).
      (7) ***Special Instructions***.
      (8) ***Cross-FLOT Considerations***.
      (9) ***Aircraft Speed***.

(10) ***Aircraft Altitude***.
(11) ***Aircraft Crank Time***.
(12) ***Rehearsal Schedule/Plan***.
(13) ***Actions on Enemy Contact*** (en route and on the ground).

4. **SERVICE SUPPORT**.
   a. **Forward Area Refuel/Rearm Points**.
   b. **Classes I, III, and V** (specific).

5. **COMMAND AND SIGNAL**.
   a. **Command**.
   b. **Signal**.
      (1) Air/ground call signs and frequencies.
      (2) Air/ground emergency code.
      (3) Passwords/number combinations.
      (4) Fire net/Quickfire net.
      (5) Time zone.
      (6) Time hack.

## AERIAL RESUPPLY ANNEX

1. **SITUATION**.
   a. **Enemy Forces** (include weather).
   b. **Friendly Forces**.
   c. **Attachments and Detachments**.

2. **MISSION**.

3. **EXECUTION**.
   a. **Concept of Operation**.
      (1) ***Maneuver***.
      (2) ***Fires***.
   b. **Tasks to Combat Units**.
      (1) ***Command and Control***.
      (2) ***Security***.
      (3) ***Marking***.
      (4) ***Recovery/Transport***.
   c. **Tasks to Combat Support Units**.
   d. **Coordinating Instructions**.
      (1) Flight Route.
         (a) General.
         (b) Checkpoints.
         (c) Communication checkpoint (CCP).
            • Marking of CCP.
            • Report time.
         (d) Heading from CCP.
      (2) Landing/Drop Zone.
         (a) Location.
            • Primary.
            • Alternate.
         (b) Marking.
            • Near.
            • Far.

(3) Drop Information.
  (a) Date/time of resupply (and alternates).
  (b) Code letter at DZ/LZ.
  (c) Length of DZ in seconds or dimensions of LZ.
  (d) Procedures for turning off DZ/LZ.
  (e) Formation, altitude, and air speed.
    - En route.
    - At DZ/LZ.
(4) Actions on enemy contact during resupply.
(5) Abort Criteria: En route and at DZ/LZ.
(6) Actions at DZ/LZ.
  - Rehearsals.

4. **SERVICE SUPPORT**.

5. **COMMAND AND SIGNAL**.
  a. **Command**.
    (1) Location of patrol leader.
    (2) Location of assistant patrol leader.
    (3) Location of members not involved in resupply.
  b. **Signal**.
    (1) Air to ground call-signs and frequencies (primary and alternate).
    (2) Long range visual signals.
    (3) Short range visual signals.
    (4) Emergency procedures and signals.
    (5) Air drop communication procedures.
    (6) Code words.

## PATROL BASE ANNEX

1. **SITUATION**
  a. Enemy Forces
  b. Friendly Forces
  c. Attachments and Detachments

2. **MISSION**.

3. **EXECUTION**.
  a. **Concept of Operation**.
    (1) ***Maneuver***.
    (2) ***Fires***.
  b. **Tasks to Combat Units**.
    (1) ***Teams***.
      - Security
      - Recon
      - Surveillance
      - LP/OPs
    (2) ***Individuals***.
  c. **Tasks to Combat Support Units**.
  d. **Coordinating instructions**.
    (1) Occupation plan.
    (2) Operations plan.
      - Security Plan.
      - Alert Plan.

- Priority of work.
- Evacuation plan.
- Alternate patrol base (used when primary is unsuitable or compromised).

4. **SERVICE SUPPORT**.
   a. Water plan.
   b. Maintenance plan.
   c. Hygiene plan.
   d. Messing plan.
   e. Rest plan.

5. **Command and Signal**.
   a. **Command**.
      (1) Location of patrol leader.
      (2) Location of assistant patrol leader.
      (3) Location of patrol CP.
   b. **Signal**.
      (1) Call signs and frequencies.
      (2) Code words.
      (3) Emergency signals.

## SMALL BOAT ANNEX

1. **SITUATION**.
   a. **Enemy Forces**.
      (1) **Weather**.
         (a) Tide.
         (b) Surf.
         (c) Wind.
      (2) **Terrain**.
         (a) River width.
         (b) River depth and water temperature.
         (c) Current.
         (d) Vegetation.
      (3) **Identification, Location, Activity, and Strength**.
   b. **Friendly Forces** (unit furnishing support).
   c. **Attachments and Detachments**.
   d. **Organization for Movement**.

2. **MISSION**.

3. **EXECUTION**.
   a. **Concept of Operation**.
      (1) ***Maneuver***.
      (2) ***Fires***.
   b. **Tasks to Combat Units**.
      (1) Security.
      (2) Tie-down teams.
         (a) Load equipment.
         (b) Secure equipment.
      (3) Designation of coxswains and boat commanders.
      (4) Selection of navigator(s) and observer(s).

c. **Coordinating Instructions**.
   (1) Formations and order of movement.
   (2) Route and alternate route.
   (3) Method of navigation.
   (4) Actions on enemy contact.
   (5) Rally points.
   (6) Embarkation plan.
   (7) Debarkation plan.
   (8) Rehearsals.
   (9) Time schedule.

4. **SERVICE SUPPORT**.
   a. **Ration Plan**.
   b. **Weapons and Ammunition**.
   c. **Uniform and Equipment**.
      (1) Method of distribution of paddles and life jackets.
      (2) Disposition of boat, paddles and life jackets upon debarkation.

5. **COMMAND AND SIGNAL**.
   a. **Command**.
      (1) Location of patrol leader.
      (2) Location of assistant patrol leader.
   b. **Signal**.
      (1) Signals used between boats and in boats.
      (2) Code words.

## STREAM CROSSING ANNEX

1. **SITUATION**.
   a. **Enemy Forces**.
      (1) ***Weather***.
      (2) ***Terrain***.
         (a) River width.
         (b) River depth and water temperature.
         (c) Current.
         (d) Vegetation.
         (e) Obstacles.
      (3) ***Enemy*** (location, identification, activity).
   b. **Friendly Forces**.
   c. **Attachments and Detachments**.

2. **MISSION**.

3. **EXECUTION**.
   a. **Concept of Operation**.
      (1) ***Maneuver***.
      (2) ***Fires***.
   b. **Tasks to Combat Units**.
      (1) ***Elements***.
      (2) ***Teams***.
      (3) ***Individuals***.

c. **Tasks to Combat Support Units**.
d. **Coordinating Instructions**.
(1) Crossing procedure/techniques.
(2) Security.
(3) Order of crossing.
(4) Actions on enemy contact.
(5) Alternate plan.
(6) Rallying points.
(7) Rehearsal plan.
(8) Time schedule.

4. **SERVICE SUPPORT**.

5. **COMMAND AND SIGNAL**.
a. **Command**.
(1) Location of patrol leader.
(2) Location of assistant patrol leader.
(3) Location of CP.
b. **Signal**.
(1) Emergency signals.
(2) Signals.

## TRUCK ANNEX

1. **SITUATION**.
a. **Enemy Forces**.
b. **Friendly Forces**.
c. **Attachments and Detachments**.

2. **MISSION**.

3. **EXECUTION**.
a. **Concept of Operation**.
(1) ***Maneuver***.
(2) ***Fires***.
b. **Tasks to Combat Units**.
c. **Tasks to Combat Support Units**.
d. **Coordinating Instructions**.
(1) Time of departure and return.
(2) Loading plan and order of movement.
(3) Route (primary and alternate).
(4) Air Guards.
(5) Actions on enemy contact (vehicle ambush) during movement, loading, and downloading.
(6) Actions at the detrucking point.
(7) Rehearsals.
(8) Vehicle speed, separation, and recovery plan.
(9) Broken vehicle instructions.

4. **SERVICE SUPPORT**.

5. **COMMAND AND SIGNAL**.
   a. **Command**. Location of PL and PSG
   b. **Signal**.
      (1) Radio call signs and frequencies
      (2) Code words

2-7. **COORDINATION CHECKLISTS**. The following checklists are items that a platoon/squad leader must check when planning for a combat operation. In some cases, he will coordinate directly with the appropriate staff section, in most cases this information will be provided by the company commander or platoon leader. The platoon/squad leader, to keep him from overlooking anything that may be vital to his mission, may carry copies of these checklists:

Figure 2-7. Intelligence coordination checklist
Figure 2-8. Operations coordination checklist
Figure 2-9. Fire support coordination checklist
Figure 2-10. Coordination with forward unit checklist
Figure 2-11. Adjacent unit coordination checklist
Figure 2-12. Rehearsal area coordination checklist
Figure 2-13. Army aviation coordination checklist
Figure 2-14. Vehicular movement coordination checklist

**Figure 2-7. INTELLIGENCE COORDINATION CHECKLIST**

## INTELLIGENCE COORDINATION CHECKLIST

In this coordination, the leader is informed of any changes in the situation as given in the operation order of mission briefing. He must keep himself constantly updated to ensure the plan is sound:

1. Identification of enemy unit.
2. Weather and light data.
3. Terrain update.
   a. Aerial photos.
   b. Trails and obstacles not on map.
4. Known or suspected enemy locations.
5. Weapons.
6. Probable course of action.
7. Recent enemy activities.
8. Reaction time of reaction forces.
9. Civilians on the battlefield.
10. Update to CCIR.

**Figure 2-8. OPERATIONS COORDINATION CHECKLIST**

## OPERATIONS COORDINATION CHECKLIST

This coordination occurs with the platoon leader/company commander so that the platoon/squad leader can confirm his mission and operational plan, receive any last-minute changes to his mission or plan, and to update his subordinates or issue a FRAGO, if required:

1. Mission backbrief.
2. Identification of friendly units.
3. Changes in the friendly situation.
4. Route selection, LZ/PZ/DZ selection.
5. Linkup procedures.
6. Transportation/movement plan.

7. Resupply (in conjunction with S4).
8. Signal plan.
9. Departure and re-entry of forward units.
10. Special equipment requirements.
11. Adjacent units operating in the area of operations.
12. Rehearsal areas.
13. Method of insertion/extraction.

**Figure 2-9. FIRE SUPPORT COORDINATION CHECKLIST**

## FIRE SUPPORT COORDINATION CHECKLIST

The platoon/squad leader will normally coordinate the following with the platoon forward observer (FO):

1. Mission backbrief.
2. Identification of supporting unit.
3. Mission and objective.
4. Route to and from the objective (include alternate routes).
5. Time of departure and expected time of return.
6. Unit target list (from fire plan).
7. Type of available support (artillery, mortar, naval gunfire, and aerial support to include Army, Navy, and Air Force) and their locations.
8. Ammunition available (to include different fuses).
9. Priority of fires.
10. Control measures.
    a. Checkpoints.
    b. Boundaries.
    c. Phase lines.
    d. Fire support coordination measures.
    e. Priority targets (target list).
    f. RFA (restrictive fire area).
    g. RFL (restrictive fire line).
    h. NFA (no-fire areas).
    i. Precoordinated authentication.
11. Communication (include primary and alternate means, emergency signals and code words).

**Figure 2-10. COORDINATION WITH FORWARD UNIT CHECKLIST**

## COORDINATION WITH FORWARD UNIT CHECKLIST

A platoon/squad that requires foot movement through a friendly forward unit must coordinate with that unit's commander for a safe and orderly passage. If no time and place has been designated for coordination with the forward unit, the platoon/squad leader should set a time and place when he coordinates with the S-3. He must talk with someone at the forward unit who has the authority to commit that unit in assisting the platoon/squad during departure. Coordination entails a two-way exchange of information:

1. Identification (yourself and your unit).
2. Size of platoon/squad.
3. Time(s) and place(s) of departure and return, location(s) of departure point(s), ERRP, and detrucking points.
4. General area of operations.
5. Information on terrain and vegetation.
6. Know or suspected enemy positions or obstacles.
7. Possible enemy ambush sites.
8. Latest enemy activity.
9. Detailed information on friendly positions such as crew-served weapons, FPF.

10. Fire and barrier plan.
   a. Support the unit can furnish. How long and what can they do?
      (1) Fire support.
      (2) Litter teams.
      (3) Navigational signals and aids.
      (4) Guides.
      (5) Communications.
      (6) Reaction units.
      (7) Other.
   b. Call signs and frequencies.
   c. Pyrotechnic plan.
   d. Challenge and password, running password, number combination.
   e. Emergency signals and code words.
   f. If the unit is relieved, pass the information to the relieving unit.
   g. Recognition signals.

**Figure 2-11. ADJACENT UNIT COORDINATION CHECKLIST**

## ADJACENT UNIT COORDINATION CHECKLIST

Immediately after the operation order of mission briefing, the platoon/squad leader should check with other platoon/squad leaders who will be operating in the same areas. If the leader is not aware of any other units operating is his area, he should check with the S3 during the operations coordination. The S3 can help arrange this coordination if necessary. The platoon/squad leaders should exchange the following information with other units operating in the same area:

1. Identification of the unit.
2. Mission and size of unit.
3. Planned times and points of departure and reentry.
4. Route(s).
5. Fire support and control measures.
6. Frequencies and call signs.
7. Challenge and password, running password, number combination.
8. Pyrotechnic plan.
9. Any information that the unit may have about the enemy.
10. Recognition signals.

**Figure 2-12. REHEARSAL AREA COORDINATION CHECKLIST**

## REHEARSAL AREA COORDINATION CHECKLIST

This coordination is conducted with the platoon leader/company commander to facilitate the unit's safe, efficient and effective use of rehearsal area prior to its mission:

1. Identification of your unit.
2. Mission.
3. Terrain similar to objective site.
4. Security of the area.
5. Availability of aggressors.
6. Use of blanks, pyrotechnics, and ammunition.
7. Mock-ups available.
8. Time the area is available (preferably when light conditions approximate light conditions of patrol).
9. Transportation.
10. Coordination with other units using area.

Figure 2-13. ARMY AVIATION COORDINATION CHECKLIST

## ARMY AVIATION COORDINATION CHECKLIST

This coordination is conducted with the platoon leader/company commander and/or S3 Air to facilitate the time, detailed and effective use of aviation assets as they apply to your tactical mission:

1. **SITUATION**.
   a. **Enemy Situation**.
      (1) Enemy air capability.
      (2) Enemy ADA capability.
      (3) Include in Weather: Percent illum, illum angle, NVG window, ceiling, and visibility.
   b. **Friendly Situation**.
      (1) Unit(s) supporting operation, Axis of movement/corridor/routes.
      (2) Friendly ADA status.

2. **MISSION**.

3. **EXECUTION**.
   a. **Concept of the Operation**. Overview of what requesting unit wants to accomplish with the air assault/air movement.
   b. **Tasks to Combat Units**.
      (1) Infantry.
      (2) Attack aviation.
   c. **Tasks to Combat Support Units**.
      (1) Artillery.
      (2) Aviation (lift).
   d. **Coordinating Instructions**.
      (1) ***Pickup Zone***.
      - Direction of landing.
      - Time of landing/flight direction.
      - Location of PZ/alternate PZ.
      - Loading procedures.
      - Marking of PZ (panel, smoke, SM, lights).
      - Flight route planned (SP, ACP, RP).
      - Formations: PZ, en route, LZ.
      - Code words:
        - PZ secure (prior to landing), PZ clear (lead bird and last bird).
        - Alternate PZ (at PZ, en route, LZ), names of PZ/alt PZ.
      - TAC air/artillery.
      - Number of pax per bird and for entire lift.
      - Equipment carried by individuals.
      - Marking of key leaders.
      - Abort criteria (PZ, en route, LZ).

      (2) ***Landing Zone***.
      - Direction of landing.
      - False insertion plans.
      - Time of landing (LZ time).
      - Location of LZ and Alternate LZ.
      - Marking of LZ (panel, smoke, SM, lights).
      - Formation of landing.
      - Code words, LZ name, alternate LZ name.
      - TAC air/artillery preparation, fire support coordination.
      - Secure LZ or not?

4. **SERVICE AND SUPPORT**.
   a. Number of aircraft per lift and number of lifts.
   b. Refuel/rearm during mission or not?
   c. Special equipment/aircraft configuration for weapons carried by unit personnel.
   d. Bump plan.
5. **COMMAND AND SIGNAL**.
   a. Frequencies, call signs and code words.
   b. Locations of air missions commander ground tactical commander, and air assault task force commander.

**Figure 2-14. VEHICULAR MOVEMENT COORDINATION CHECKLIST**

## VEHICULAR MOVEMENT COORDINATION CHECKLIST

This is coordinated with the supporting unit through the platoon sergeant/first sergeant to facilitate the effective, detailed, and efficient use of vehicular support and/or assets:

1. Identification of the unit.
2. Supporting unit identification.
3. Number and type of vehicles and tactical preparation.
4. Entrucking point.
5. Departure/loading time.
6. Preparation of vehicles for movement.
   a. Driver responsibilities.
   b. Platoon/squad responsibilities.
   c. Special supplies/equipment required.
7. Availability of vehicles for preparation/rehearsals/inspection (time and location).
8. Routes.
   a. Primary.
   b. Alternate.
   c. Checkpoints.
9. Detrucking points.
   a. Primary.
   b. Alternate.
10. March internal/speed.
11. Communications (frequencies, call signs, codes).
12. Emergency procedures and signals.

## 2-8. DOCTRINAL TERMS

**TACTICAL TASK**--A clearly defined, measurable activity accomplished by individuals and organizations. Tasks are specific activities that contribute to the accomplishment of encompassing missions or other requirements. A task should be definable, measurable, and decisive (achieve the purpose).

| Actions by Friendly Force | | Effects on Enemy Force |
|---|---|---|
| Assault | Follow and Assume | Block |
| Attack-by-Fire | Follow and Support | Canalize |
| Breach | Linkup | Contain |
| Bypass | Occupy | Defeat |
| Clear | Reconstitution | Destroy |
| Combat Search and Rescue | Reduce | Disrupt |
| Consolidation and Reorganization | Retain | Fix |
| Control | Secure | Interdict |
| Counterreconnaissance | Seize | Isolate |
| Disengagement | Support-by-Fire | Neutralize |
| Exfiltrate | Suppress | Penetrate |
| | | Turn |

**PURPOSE** (in order to)--The desired or intended result of the tactical operation stated in terms related to the enemy or the desired situation. The why of the mission statement. The most important component of the mission statement.

| | | |
|---|---|---|
| Allow | Divert | Prevent |
| Cause | Enable | Protect |
| Create | Envelop | Support |
| Deceive | Influence | Surprise |
| Deny | Open | |

**OPERATION**--A military action or the carrying out of a military action to gain the objectives of any battle or campaign.

| Types of operations include: | Forms of Offensive Maneuver: | Types of Defensive Techniques: | Other Operations: |
|---|---|---|---|
| Movement to Contact | Envelopment | Area Defense | Reconnaissance Ops |
| Search and Attack | Frontal Attack | Mobile Defense | Security Ops |
| Attack | Infiltration | Retrograde Ops | Information Ops |
| Ambush | Penetration | Delay | Combined Arms Breach |
| Demonstration | Turning Movement | Withdrawal | Passage of Lines |
| Feint | | Retirement | Relief in Place |
| Raid | | | River Crossing Ops |
| Spoiling Attack | | | Troop Movement |
| Exploitation | | | Admin |
| Pursuit | | | Approach March |
| | | | Road March |

2-9. **TERRAIN MODEL**. The terrain model (Figure 2-15) is an outstanding visual means used during the planning process to communicate the patrol routes and also detailed actions on the objective. At a minimum, the model is required to during planning to display routes to the objective and to highlight prominent terrain features the patrol will encounter during movement. A second terrain model is usually constructed of the objective area, enlarged to a sufficient size and detail to brief the patrol's actions on the objective.

a. **Checklist**.

(1) North-seeking arrow.
(2) Scale.
(3) Grid lines.
(4) Objective location.
(5) Exaggerated terrain relief, water obstacles.
(6) Friendly patrol locations.
(7) Targets (indirect fires, including grid and type of round).
(8) Routes, primary and alternate.
(9) Planned RPs (ORP, L/URP, RP).
(10) Danger areas (roads, trails, open areas).
(11) Legend.
(12) Blowup of objective area.

b. **Construction**. The following are some field-expedient techniques to aid in terrain model construction:

(1) Use a 3 x 5 card, MRE box, or piece of paper to label the objective or key sites.

(2) Grid lines can be made using string from the guts of 550 cord or colored tape. (Grid lines are identified by writing numbers on small pieces of paper.)

(3) Trees and vegetation are replicated by using moss, green or brown spray paint, pine needles, crushed leaves, or cut up grass.

(4) Water is designated by blue chalk, blue spray paint, blue yarn, tin foil, or MRE creamer.

(5) North seeking arrows are made from sharpened twigs, pencils, or colored yarn.

(6) Enemy positions are designated using red yarn, M-16 rounds, toy Rangers, or poker chips.

(7) Friendly positions such as security elements, support by fire, and assault elements are made using M-16 rounds, toy Rangers, poker chips, small MRE packets of sugar and coffee, or pre-printed acetate cards.

(8) Small pieces of cardboard or paper can identify target reference points and indirect fire targets. Ensure grids are shown for each point.

(9) Breach, support by fire, and assault positions are made using colored yarn or string so that these positions can be easily identified.

(10) Bunkers and buildings are constructed using MRE boxes or tongue depressors/sticks.

(11) Perimeter wire is constructed from a spiral notebook.

(12) Key phase lines are constructed with colored string or yarn.

(13) Trench lines are replicated by colored tape or yarn, by digging a furrow and coloring it with colored chalk or spray paint.

***Note:*** All symbols used on the terrain model must be clearly identified in the legend.

Figure 2-15. TERRAIN MODEL

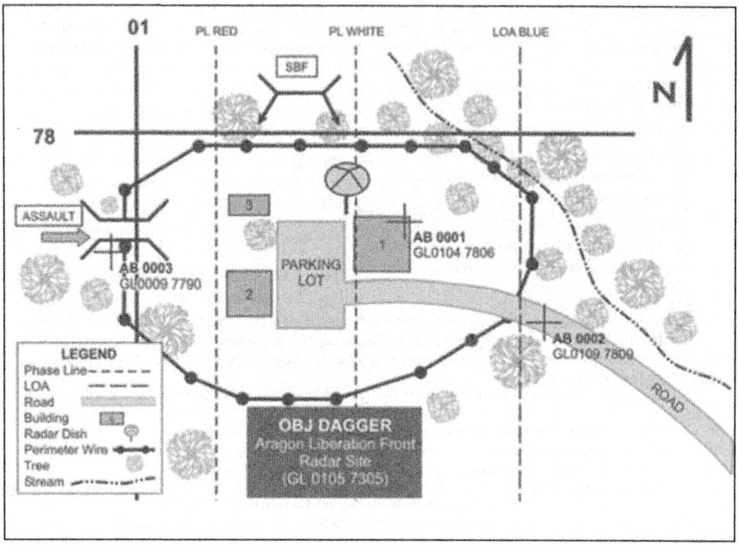


NOTES

# Chapter 3
# FIRE SUPPORT

Indirect fire support can greatly increase the combat effectiveness and survivability of any Infantry unit. The ability to plan for and effectively use this asset is a task that every Ranger and small unit leader should master. Fire support assets can assist a unit suppress, fix, destroy, or neutralize the enemy. Employment of indirect fire support should be considered throughout every offensive and defensive operation. This chapter discusses planning, tasks, capabilities, risk estimate distances, target overlays, close air support, elements and sequence of call for fire, and example call-for-fire transmissions.

3-1. **PLANNING**. Planning is the continual process of selecting targets on which fires are prearranged to support a phase of the commander's plan. The principles of planning follow:

- Consider what the commander wants to do.
- Plan early and continuously.
- Exploit all available targeting assets.
- Use all available lethal and nonlethal fire support means.
- Use the lowest echelon able to furnish effective support.
- Observe all fires.
- Use the most effective fire support asset available.
- Provide adequate fire support.
- Avoid unnecessary duplication.
- Provide for safety of friendly forces and installations.
- Provide for flexibility.
- Furnish the type of fire support requested.
- Consider the airspace.
- Provide rapid and effective coordination.
- Keep all fire support informed.

3-2. **TASKS**.

a. **All Operations**.
   (1) Locate targets.
   (2) Integrate all available assets.
   (3) Destroy, neutralize, or suppress all enemy direct and indirect fire systems.
   (4) Provide illumination and smoke.
   (5) Provide fires in support of JAAT and SEAD missions.
   (6) Deliver scatterable mines.
   (7) Prepare for future operations.
   (8) Provide positive clearance of fires.

b. **Offensive Operations**.
   (1) Support the movement to contact, chance contact.
   (2) Soften enemy defenses before the attack by arranging short, violent preparations, where required.
   (3) Provide support during the attack by attacking high payoff targets.
   (4) Plan for deep and flanking fires.
   (5) Plan fires during consolidation.
   (6) Provide counterfires.

c. **Defensive Operations**.
   (1) Execute all plans as the commander intends.
   (2) Disorganize, delay, and weaken the enemy before the attack.
   (3) Provide counterfire.
   (4) Provide fires in support of planned engagement areas (EAs).
   (5) Attack high payoff targets.
   (6) Plan fires in support of barriers and obstacle plans.
   (7) Plan for deep, flanking, and rear fires.

(8) Provide fires in support of counterattacks.
(9) Plan final protective fire (FPFs).

3-3. **CAPABILITIES**. Tables 3-1 and 3-2 show capabilities of field artillery and mortars.

**Table 3-1. CAPABILITIES OF FIELD ARTILLERY**

| WEAPON | MAX RANGE (meters) | MIN RANGE (meters) | MAX RATE Rds per Min | Burst Radius | SUSTAINED RATE Rds per Min |
|---|---|---|---|---|---|
| 105-mm Howitzer M119, Towed | 14,000 m | 0 | 6 for 2 min | 35 m | 3 Rds for 30 min, then 1 rd per min |
| 155-mm Howitzer M198, Towed | 18 m,100 m 30,000 m (RAP) | 0 | 4 for 3 min 2 for 30 min | 50 m | 1 rd per min Temp Dependent |
| 155-mm Howitzer M109A6 SP | 18 m, 100 m 30,000 m (RAP) | 0 | 4 for 3 min | 50 m | 1 for 60 min 0.5 |

**Table 3-2. CAPABILITIES OF MORTARS**

| WPN | MUNITIONS AVAILABLE | MAX HE RANGE | MIN RANGE (meters) | MAX RATE (Rds per Min) | BURST RADIUS | SUSTAINED RATE (Rds per Min) |
|---|---|---|---|---|---|---|
| 60mm | HE, WP, Illum | 3,500 m | 70 m (HE) | 30 for 4 min | 30 m | 20 |
| 81mm | HE, WP, Illum | 5,600 m | 70 m (HE) | 25 for 2 min | 38 m | 8 |
| 120mm | HE, Smoke, Illum | 7,200 m | 180 m (HE) | 15 for 1 min | 60 m | 5 |

**DANGER**

**DANGER CLOSE**

1. **When the target is within 600 meters of any friendly troops (for mortars and field artillery), announce DANGER CLOSE in the Method of Engagement portion of the call for fire.**
2. **When adjusting 5-inch or smaller naval guns on targets within 750 meters, announce DANGER CLOSE. For larger naval guns, announce DANGER CLOSE for targets within 1,000 meters. Failure to adhere to this guidance can result in fratricide.**
3. **Avoid making corrections using the bracketing method of adjustment, because doing so can cause serious injury or death. Use only the creeping method of adjustment during DANGER CLOSE missions. Make corrections of no more than 100 meters by creeping the rounds to the target.**

3-4. **RISK ESTIMATE DISTANCES**. RED applies to combat only. Minimum safe distances (Table 3-3) apply to training IAW AR 350-1. RED takes into account the bursting radius of particular munitions and the characteristics of the delivery system and associates this combination with a percentage representing the likelihood of becoming a casualty, that is, the percentage of risk. RED is defined as the minimum distance friendly troops can approach the effects of friendly fires without suffering appreciable casualties of 0.1% PI or higher.

> **WARNING**
>
> **[Commanders] use risk-estimate distance formulas to determine acceptable risk levels in combat only. Specifically, use them to identify the risk to your Rangers at various distances from their targets. Risk estimate distances apply only in combat. In training, use minimum safe distances (MSD).**

a. **Casualty Criterion**. The casualty criterion is the 5-minute assault criterion for a prone soldier in winter clothing and helmet. Physical incapacitation means that a soldier is physically unable to function in an assault within a 5-minute period after an attack. A PI value of less than 0.1 percent can be interpreted as being less than or equal to one chance in one thousand.

**Table 3-3. RISK ESTIMATE DISTANCES FOR MORTARS AND CANNON ARTILLERY**

| | Risk Estimate Distances (Meters) | | | | | |
|---|---|---|---|---|---|---|
| | 10 % PI | | | 0.1 % PI | | |
| **System** | 1/3 range | 2/3 range | Max range | 1/3 range | 2/3 range | Max range |
| 60-mm mortar | 60 | 65 | 65 | 100 | 150 | 175 |
| 81-mm mortar | 75 | 80 | 80 | 165 | 185 | 230 |
| 120-mm mortar | 100 | 100 | 100 | 150 | 300 | 400 |
| 105-mm howitzer | 85 | 85 | 90 | 175 | 200 | 275 |
| 155-mm howitzer | 100 | 100 | 125 | 200 | 280 | 450 |
| 155-mm DPICM | 150 | 180 | 200 | 280 | 300 | 475 |

b. **Risk**. Using echelonment of fires within the specified RED for a delivery system requires the unit to assume some risks. The maneuver commander determines by delivery system how close he will allow fires to fall in proximity to his forces. The maneuver commander makes the decision for this risk level, but he relies heavily on the FSO's expertise.

3-5. **TARGET OVERLAYS**.

a. Figure 3-1 shows contents of fire support overlay.

**Figure 3-1. CONTENTS OF FIRE SUPPORT OVERLAY**

- Unit and official capacity of person making overlay.
- Date the overlay was prepared.
- Map Sheet Number.
- Effective Period of Overlay (DTG).
- Priority target.
- ORP Location.
- Call signs and frequencies. (PRI/ALT)
- Routes--Primary/Alternate.
- Phase Lines/Checkpoints used by the patrol.
- Spares.
- Index marks to position overlay on map.
- Objective.
- Target Symbols.
- Description, location and remarks column, complete.

b. Using a PLOT-CR checklist helps ensure the leader's fire support plan is complete:
   (1) Purpose--Purpose of the planned indirect fires.
   (2) Location--Planned targets with an 8-digit grid (minimum).
   (3) Observer--Planned observer to see the impact of the rounds and adjust.
   (4) Trigger--Method to initiate fires.
   (5) Communication--Method to communicate between the observer and the supporting unit.
   (6) Resources--Planned allocated resource for each target.
c. Sterile overlay includes---
   (1) Index marks to position overlay on map.
   (2) Target symbols.
d. Target overlay symbols (Figure 3-2) include--
   (1) Point target:
   (2) Linear target:
   (3) Circular target:

**Figure 3-2. TARGET OVERLAY SYMBOLS--POINT (TOP), LINEAR (CENTER), AND CIRCULAR (BOTTOM)**

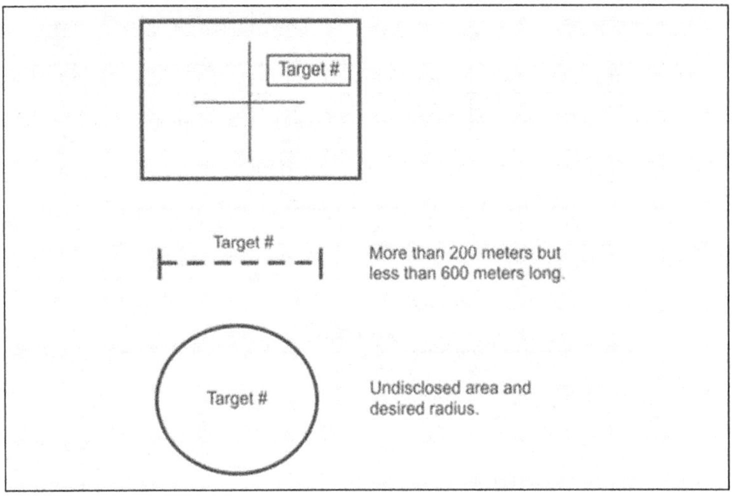


3-6. **CLOSE AIR SUPPORT**. The two types of close air support requests are planned and immediate. Planned requests are processed by the Army chain to corps for approval. Immediate requests are initiated at any level and processed by the battalion S-3 FSO, and air liaison officer.

a. Format for requesting immediate CAS:
   (1) Observer identification.
   (2) Warning order (request close air).
   (3) Target location (grid).

(4) Target description. (Description must include, as a minimum: type and number of targets, activity or movement; point or area targets, include desired results on target and time on target.

(5) Location of friendly forces.

(6) NAV details (elevation).

(7) Threats--ADA, small arms, and so on.

(8) Hazards--friendly aircraft in area.

(9) Wind direction.

b. Close air support capabilities are shown in Table 3-4.

**Table 3-4. CLOSE AIR SUPPORT**

| AIRCRAFT | SERVICE | CHARACTERISTICS |
|---|---|---|
| A-10 * | AF | Specialized CAS aircraft; subsonic; typical load 6,000 lbs, max load 16,000 lbs; 30-mm gun. |
| F-16 * | AF | A multirole aircraft; complements the F-4 and F-15 in an air-to-air role. Most accurate ground delivery system in the inventory; supersonic; typical load 6,000 lbs, max load 10,600 lbs. |
| F-18 * | N/MC | A multirole fighter scheduled to replace the F-4. A wide variety of air-to-surface weapons. Typical load 7,000 lbs, max load 17,000 lbs, 20-mm gun mounted in the nose; air-to-air missiles. |
| AC-130 | AF/R | A specialized CAS/RACO aircraft, propeller driven. Two models: The "A" model has two 40-mm guns, two 20-mm guns and two 7.62-mm miniguns. The similar "H" model has no 7.62-mm miniguns, and has a 105-mm howitzer in place of one of the 40-mm guns. Both models have advanced sensors and a target-acquisition system, including FLIR and low light TV. Very accurate. Vulnerable to enemy air defense systems, so must operate in a low-threat environment. |
| *FM capability. | | |

3-7. **CALL FOR FIRE**.

a. **Observer's Identification - Call Signs**.

b. **Warning Order**.

(1) ***Type of mission***.

(a) Adjust fire.

(b) Fire for effect.

(c) Suppress.

(d) Immediate suppression/immediate smoke.

(2) ***Size of element to Fire for Effect***. When the observer does not specify what size element to fire, the battalion FDC will decide.

(3) ***Method of Target Location***.

(a) Polar plot.

(b) Shift from a known point.

(c) Grid.

(4) **Location of Target**.

(a) *Grid Coordinate - 6 Digit*. 8 digit if greater accuracy is required.

(b) *Shift from a Known Point*. Send observer-target (OT) direction:

- Mils (nearest 10).
- Degrees.
- Cardinal direction.
- Send lateral shift (right/left, nearest 10 mils).
- Send range shift (add/drop, nearest 100 mils).
- Send vertical shift (up/down). Use only if it exceeds 35 mils (nearest 5 mils).

(c) *Polar Plot*. Send direction. (Nearest 10 mils)

- Send distance (nearest 100 mils).
- Send vertical shift (nearest 5 mils).

(5) ***Description of Target***.

(a) Type.

(b) Activity.

(c) Number.

(d) Degree of protection.

(e) Size and shape (length/width or radius).

(6) ***Method of Engagement, Type of Adjustment***. When the observer does not request a specific type of fire control adjustment, area fire is issued.

(a) Area fire--moving target.

(b) Precision fire--point target.

(7) ***Danger Close***. Use when friendly troops are within--

(a) 600 mils for mortars.

(b) 600 mils for artillery.

(c) 750 mils for naval guns 5 inches or smaller.

(8) ***Mark***. Use to orient observer or to indicate targets.

(9) ***Trajectory***.

(a) Low angle (standard).

(b) High angle (mortars or if requested).

(10) ***Ammunition***. Use HE quick used unless observer specifies otherwise.

(a) Projectile (HE, illum, ICM, smoke, and so on).

(b) Fuse (quick, time, and so on).

(c) Volume of fire (observer may request the number of rounds to be fired).

(11) ***Distribution***.

(a) 100 mil sheaf (standard).

(b) Converged sheaf (used for small hard targets).

(c) Special sheaf (any length, width, and attitude).

(d) Open sheaf (separate bursts).

(e) Parallel sheaf (linear target).

(12) ***Method of Fire and Control***.

(a) *Method of Fire*. Specific guns and a specific interval between rounds. Normally adjust fire, one gun is used with a 5-second interval between rounds.

(b) *Method of Control*.

- AT MY COMMAND - FIRE--Remains in effect until observer announces, "*Cancel at my command*."
- CANNOT OBSERVE--Observer can't see the target.
- TIME ON TARGET--Observer tells FDC when he wants the rounds to impact.
- CONTINUOUS ILLUMINATION. Calculated by the FDC, otherwise observer indicates interval between rounds in seconds.
- COORDINATED ILLUMINATION. Observer may order the interval between illum and HE shells.
- CEASE LOADING--Stop loading rounds.
- CHECK FIRING--Stop firing at once.
- CONTINUOUS FIRE--Load and fire as fast as possible.
- REPEAT--Fire another round(s) with or without adjustments.

(13) ***Authentication***. Challenge and reply.

(14) ***Message to Observer***.

(a) Battery(ies) to fire for effect.

(b) Adjusting battery.

(c) Changes to the initial call for fire.

(d) Number of rounds (per tube) to be fired for effect.

(e) Target numbers.

(f) Additional information:

- Time of flight. Moving target mission.
- Probable error in range. 38 meters or greater (normal mission).
- Angle T. 500 mils or greater.

c. **Correction of Errors**. When FDC has made an error when reading back he fire support data, the observer announces "*CORRECTION*" and transmits the correct data in its entirety.

d. **Call-for-Fire Transmissions**. Example call-for-fire transmissions are shown in Table 3-5.

**Table 3-5. EXAMPLE CALL-FOR-FIRE TRANSMISSIONS.**

| *GRID MISSION* | |
|---|---|
| **OBSERVER** | **FIRING UNIT** |
| F24, this is J42, ADJUST FIRE, OVER. | J42, this is F24, AJUST FIRE, OUT. |
| GRID WM180513, DIRECTION 0530, OVER. | GRID WM180513, DIRECTION 0530, OUT. |
| Infantry platoon dug in, OVER. | Infantry platoon dug in, OUT. |
| | SHOT OVER. |
| SHOT OUT. | |
| | SPLASH, OVER. |
| SPLASH OUT. | |
| End of mission, 15 casualties, platoon dispersed, OVER. | End of mission, 15 casualties, platoon dispersed, OUT. |
| *SHIFT FROM KNOWN POINT* | |
| **OBSERVER** | **FIRING UNIT** |
| J42, THIS IS F24, ADJUST FIRE, SHIFT AB1001, OVER. | F24, THIS IS J42, ADJUST FIRE, SHIFT AB1001, OUT. |
| DIRECTION 2420, RIGHT 400, ADD 400, OVER. | DIRECTION 2420, RIGHT 400, ADD 400, OUT. |
| FIVE T-72 TANKS AT POL SITE, OVER. | FIVE T-72 TANKS AT POL SITE, AUTHENTICATE JULIET NOVEMBER, OVER. |
| I AUTHENTICATE TANGO, OVER. | |
| | SHOT, OVER. |
| SHOT OUT. | SPLASH, OVER. |
| SPLASH OUT. | |
| END OF MISSION, 2 TANKS DESTROYED, 3 IN WOODLINE, OVER | END OF MISSION, 2 TANKS DESTROYED, 3 IN WOODLINE, OUT |

| *POLAR* | |
|---|---|
| **OBSERVER** | **FIRING UNIT** |
| *J42, this is F24, adjust fire, polar OVER.* | F24, this is J42, adjust fire, polar, OUT |
| *Direction 2300, distance 4,000, OVER.* | Direction 2300, distance 4,000, OUT. |
| *Infantry platoon dug in, OVER* | Infantry platoon dug in, OUT |
| | Shot OVER. |
| *Shot OUT.* | Splash, OVER. |
| *Splash OUT.* | |
| *End of mission, 15 casualties, platoon dispersed, OVER.* | End of mission, 15 casualties, platoon dispersed, OUT. |

NOTES

## Chapter 4
# MOVEMENT

To survive on the battlefield, stealth, dispersion, and security must be enforced in all tactical movements. The leader must be skilled in all movement techniques.

4-1. **FORMATIONS**. Movement formations are comprised of elements and Soldiers in relation to each other. Fire teams, squads and platoons use several movement formations. Formations provide the leader control based on a METT-TC analysis. Leaders position themselves where they can best command and control formations. The formations below allow the fire team leader to lead by example, *"Follow me, and do as I do."* All Soldiers in the team must be able to see their leader. The formations in Figure 4-1 reflect fire team formations. Squad formations are very similar with more Soldiers. Squads can operate in lines and files similar to fire teams. When squads operate in wedges or in echelon, the fire teams use those formations and simply arrange themselves in column or one team behind the other. Squads may also use the vee, with one team forming the lines of the vee, and the SL at the apex for command and control. Platoons work on the same basic formations as the Squads. When operating as a platoon, the platoon leader must carefully select the location for his machine guns in the movement formation.

**Figure 4-1. FORMATIONS**

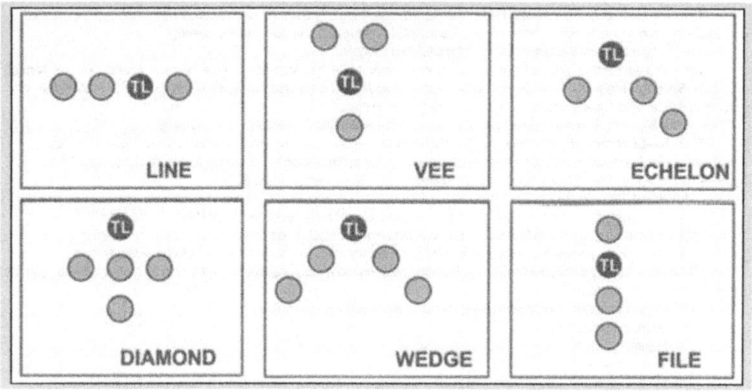


a. **Techniques**. A movement technique is the method a unit uses to traverse terrain. There are three movement techniques: traveling, traveling overwatch, and bounding overwatch. The selection of a movement technique is based on the likelihood of enemy contact and the need for speed. Factors to consider for each technique are control, dispersion, speed, and security. Movement techniques are not fixed formations. They refer to the distances between Soldiers, teams, and squads that vary based on METT-TC. Soldiers must be able to see their fire team leaders. The platoon leader should be able to see his lead squad leader. Leaders control movement with hand-and-arm signals and use radios only when needed.

b. **Standards**.

(1) Unit moves on designated route or arrives at specified location IAW OPORD maintaining accountability of all assigned/attached personnel.

(2) Unit uses movement formation and technique ordered by the leader based on METT-TC.

(3) Leaders remain oriented (within 200 meters) and follow planned route unless METT-TC dictates otherwise.

(4) Unit maintains 360 degree security and remains 100% alert during movement.

(5) Unit maintains 360 degree security and a minimum of 75% security during halts.

(6) If contact with the enemy is made, it is made with the smallest element possible.

(7) Control measures are used during movement such as head counts, rally points, or phase lines.

c. **Fundamentals**.

(1) ***Land Navigation***. Mission accomplishment depends on successful land navigation. The patrol should use stealth and vigilance to avoid chance contact. Designate a primary and alternate compass and pace man per patrol.

***Note:*** The point man will not be tasked to perform compass or pace duties. The point man's sole responsibility is forward security for the element.

(2) ***Avoidance of Detection***. Patrols must use stealth, and use the cover and concealment of the terrain to its maximum advantage. Whenever possible, move during limited visibility in order to maximize technological advantages gained by night vision devices and to hinder the enemy's ability to detect the patrol. Exploit the enemy's weaknesses, and attempt to time movements to coincide with other operations that distract the enemy. The enemy threat and terrain determines which of the three movement techniques will be used:

(a) Fire teams maintain visual contact, but the distance between them is such that the entire patrol does not become engaged if contact is made. Fire teams can spread their formations as necessary to gain better observation to the flanks. Although widely spaced, men retain their relative position in their wedge and follow their team leader. Only in extreme situations should the file be used.

(b) The lead squad must secure the front along with assuming responsibility for navigation. For a long movement, the PL may rotate the lead squad responsibilities. The fire team/squad in the rear is charged with rear security.

(c) Vary movement techniques to meet the changing situation.

(d) With the exception of fire team leaders, leaders move inside their formations where they can maintain the best control.

(3) ***Security***. The patrol must use both active and passive security measures constantly. Assign subunits responsibility for security at danger areas, patrol bases, and most importantly in the objective area.

(4) ***Fire Support***. Plan fire support (mortars, artillery, tactical air, attack helicopter, naval gunfire).

(5) ***Three-Dimensional Battlefield***. 360 Degree Security is achieved through high and low security. Within a fire team, squad, and so on, the leader must assign appropriate sectors of fire to their subordinate in order to ensure all aspects of the battlefield are covered. This includes trees, multiple storied structures, tunnels, sewers, ditches.

d. **Movement Techniques**.

(1) The traveling technique is used when enemy contact is not likely but speed is necessary.

(2) The traveling overwatch technique is used when enemy contact is possible.

(3) The bounding overwatch technique is used when enemy contact is likely, or when crossing a danger area.

e. **Traveling**. In the traveling technique, the distance between individuals is about 10 meters with 20 meters between squads. It has the following characteristics:

(1) More control than traveling overwatch but less than bounding overwatch.

(2) Minimum dispersion.

(3) Maximum speed.

(4) Minimum security.

f. **Traveling Overwatch**. The traveling overwatch technique is the basic movement technique. The distance between individuals is about 20 meters, between teams about 50 meters.

(1) In platoon traveling overwatch, the lead squad must be far enough ahead of the rest of the platoon to detect or engage any enemy before the enemy observes or fires on the main body. However, it must be close enough to be supported by the platoon's small arm's fires. This is normally between 50 to 100 meters, depending on terrain, vegetation, and light and weather conditions.

(2) In a column formation, only the lead squad should use the traveling overwatch; however, if greater dispersion is desired, all squads may use it.

(3) In other formations, all squads use traveling overwatch unless the platoon leader specifies otherwise.

(4) Traveling overwatch has the following characteristics:

- Good control.
- Good dispersion.
- Good speed.
- Good security forward.

g. **Bounding Overwatch**. In the bounding overwatch technique (Figure 4-2), the distance between men remains about 20 meters. The distance between teams and squads varies.

(1) The squad or platoon has a bounding element and an overwatch element. The bounding element moves while the overwatch element occupies an overwatch position that can cover the route of the bounding element by fire. Each bound is within supporting range of the overwatch element.

(2) There are two types of bounding, successive and alternating (Figure 4-2). Successive is nothing more than one squad moving to a position, then the overwatching squad moving to a position generally online with the first squad. Alternating bounding is when one squad moves into position, then the overwatching squad moves to a position in front of the first squad.

**Figure 4-2. SQUAD BOUNDING OVERWATCH**

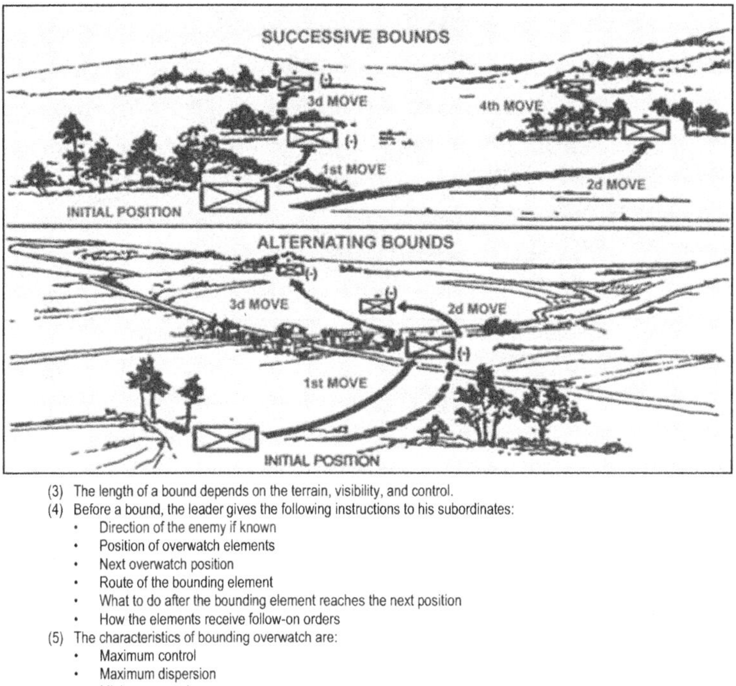


(3) The length of a bound depends on the terrain, visibility, and control.

(4) Before a bound, the leader gives the following instructions to his subordinates:

- Direction of the enemy if known
- Position of overwatch elements
- Next overwatch position
- Route of the bounding element
- What to do after the bounding element reaches the next position
- How the elements receive follow-on orders

(5) The characteristics of bounding overwatch are:

- Maximum control
- Maximum dispersion
- Minimum speed
- Maximum security

h. **Platoon Bounding Overwatch** (Figure 4-3).

(1) ***Method One***. When platoons use bounding overwatch, one squad bounds and one squad overwatches; the third squad awaits orders. Forward observers stay with the overwatching squad to call for fire. Platoon leaders normally stay with the overwatching squad who use machine guns and attached weapons to support the bounding squad.

(2) ***Method Two***. Another way is to have one squad use bounding overwatch and have the other two squads use traveling or traveling overwatch technique.

**Figure 4-3. PLATOON BOUNDING OVERWATCH**

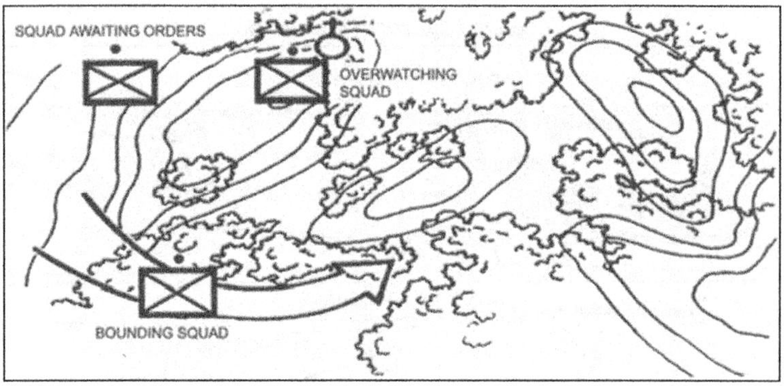


(3) ***Movement Considerations***. When deciding where to move the bounding element, consider--

- Where the enemy is likely to be.
- The mission.
- The routes to the next overwatch position.
- The weapons ranges of the overwatching unit.
- The responsiveness of the rest of the unit.
- The fields of fire at the next overwatch position.

4-2. **TACTICAL MARCHES**. Platoons conduct two types of marches with the company: foot marches and motor (road) marches.

a. **Purpose/General**. A successful foot march is when troops arrive at their destination at the prescribed time, physically able to execute their tactical mission.

b. **Standard**.

(1) The unit crosses the start point and release point at the time specified in the order.

(2) The unit follows the prescribed route, rate of march, and interval without deviation unless required otherwise by enemy action or higher headquarters action.

c. **Fundamentals**.

(1) Effective control.

(2) Detailed planning.

(3) Rehearsals.

d. **Considerations**.

(1) ***METT-TC***.

- Mission ....................................Task and purpose.
- Enemy .....................................Intentions, capabilities, and course of action.
- Terrain and weather.................Road condition/trafficability, and visibility.

- Troops/equipment....................Condition of Soldiers and their loads, numbers and types of weapons and radios.
- Time..........................................Start time, release time, rate of march, time available.
- Civilians ...................................Movement through populated areas, refugees, OPSEC.

(2) ***Task Organization***:
- Security...................................Advance and trail teams.
- Main body ................................Two remaining line squads and weapons squad.
- Headquarters ...........................Command and control.
- Control measures.

(3) ***Start Point and Release Point*** (given by higher):
- (a) Checkpoints.............................At checkpoints report to higher and use checkpoint to remain oriented.
- (b) Rally or rendezvous points ......Used when elements become separated.
- (c) Location of leaders ..................Where they can best control their elements.
- (d) Commo plan ............................Locations of radios, frequencies, call signs, and OPSKEDs.
- (e) Dispersion between Soldiers--
  - 3 to 5 meters / day.
  - 1 to 3 meters / night.

(4) ***March Order***. May be issued as an OPORD, FRAGO, or annex to either (must use operational overlay or strip map)"
- (a) Formations and order of movement
- (b) Route of march........................Assembly area, start point, release point, rally points, checkpoints, break/halt points.
- (c) Start point time, release point time, and rate of march.
- (d) March interval for squads, teams, and individuals.
- (e) Actions on enemy contact........Air and ground.
- (f) Actions at halts.
- (g) Fires ........................................Detailed plan of fire support for the march.
- (h) Water supply plan.
- (i) MEDEVAC plan.

o. **Duties and Responsibilities**.

(1) ***Platoon Leader***:
- Before .........................................Issues warning order and FRAGO; inspects and supervises.
- During .........................................Ensures unit makes movement time; ensures interval is maintained and that unit remains oriented; maintains security; checks condition of men; and enforces water discipline and field sanitation.
- After ............................................Ensures men are prepared to accomplish their mission, supervises SLs, and ensures medical coverage is provided to men as needed.

(2) ***Platoon Sergeant***:
- Before ......................................Assists PL, makes recommendations, and enforces uniform and packing lists.
- During ......................................Controls stragglers, helps platoon leader maintain proper interval and security.
- At Halts ....................................Enforces security, ensures welfare of men, enforces field sanitation and litter discipline, and enforces use of preventive medicine.
- After.........................................Coordinates for water, rations, and medical supplies. Recovers casualties.

(3) ***Squad Leaders***:
- Before ......................................Provides detailed instruction to TLs, inspects boots and socks for serviceability and proper fit, inspects adjustment of equipment, inspects to ensure canteens are full, and ensures equal distribution of loads.
- During ......................................Controls squad, maintains proper interval between men and equipment, enforces security, remains oriented.

- At Halts....................................Ensures security is maintained, provides men for water resupply as detailed. Physically checks the men in his squad, ensures they drink water, and that they change socks as necessary. Rotates heavy equipment. (Units should plan this in detail to avoid confusion before, during, and after halts.)
- After.........................................Occupies squad sector of assembly area, conducts foot inspection, reports condition of men to PL, and prepares men to accomplish the mission.

(4) ***Security Squad***:
- Lead Team.
  -- Serves as point element for platoon.
  -- Recons route to SP.
  -- Calls in checkpoints.
  -- Provides early warning.
  -- Maintains rate of march.
  -- Moves 10 to 20 meters in front of main body.
- Trail Team.
  -- Provides rear security.
  -- Moves 10 to 20 meters behind main body.

(5) ***Medic***:
- Assesses and treats march casualties.
- Advise chain of command on evacuation and transportation requirements of casualties.

(6) ***Individual***:
- Maintains interval.
- Follows TL's examples.
- Relays hand and arm signals; remains alert during movement and at halts.
- Remains alert during movement and at halts.

4-3. **MOVEMENT IN LIMITED VISIBILITY CONDITIONS**. During hours of limited visibility, a platoon will use surveillance, target acquisition, and night observation (STANO) devices to enhance their effectiveness. Leaders must be able to control, navigate, maintain security, and move during limited visibility.

a. **Control**. When visibility is poor, the following methods aid in control.
(1) Use of night vision devices.
(2) Leaders move closer to the front.
(3) Platoon reduces speed.
(4) Use of luminescent tape on equipment.
(5) Reduce intervals between men and elements.
(6) Conducts headcount regularly.

b. **Navigation**. While navigating during limited visibility the same techniques are used as during the day; however, leaders exercise more care to keep the patrol oriented.

c. **Security**.
(1) Enforce strict noise and light discipline.
(2) Use radio-listening silence.
(3) Use camouflage.
(4) Use terrain to avoid detection by enemy surveillance or night vision devices.
(5) Make frequent listening halts; Conduct SLLS (Stop, Look, Listen, Smell).
(6) Mask the sounds of movement when possible. (Rain, wind, and flowing water will mask the sounds of movement).

d. **Rally Points**. Plan actions to be taken at rally points in detail. All elements must maintain communications at all time. There are two techniques for actions at rally points:

(1) ***Minimum Force***: Patrol members assemble at the rally point, and the senior leader assumes command. When the minimum force (designated in the OPORD) is assembled and organized the patrol will continue the mission.

(2) ***Time Available***: The senior leader determines if the patrol has enough time remaining to accomplish the mission.

e. **Actions at Halts**. During halts, post security and cover all approaches into the sector with key weapons.

(1) ***Short Halt***. Typically 1-2 minutes long. Soldiers seek immediate cover and concealment and take a knee. Leaders assign sectors of fire.

(2) ***Long Halt***. More than 2 minutes. Soldiers assume the prone position behind cover and concealment. (Ensure soldiers have clear fields of fire.) Leaders assign sectors of fire.

4-4. **DANGER AREAS**. A danger area is any place on a unit's route where the leader determines his unit may be exposed to enemy observation or fire. Some examples of danger areas are open areas, roads and trails, built-up areas, enemy positions, and natural and man-made obstacles. Bypass danger areas whenever possible.

a. **Standards**.

(1) The unit prevents the enemy from surprising the main body.
(2) The unit moves all personnel and equipment across the danger area.
(3) The unit prevents decisive engagement by the enemy.

b. **Fundamentals**.

(1) Designate near and far side rally points.
(2) Secure near side, left and right flank, and rear security.
(3) Recon and secure the far side.
(4) Cross the danger area.
(5) Plan for fires on all known danger areas.

c. **Technique for Crossing Danger Areas**.

(1) ***Linear Danger Area*** (LDA, Figure 4-4) ***Actions for a Squad***:

**STEP 1:** The alpha team leader (ATL) observes the linear danger area and sends the hand and arm signal to the SL who determines to bound across.

**STEP 2:** SL directs the ATL to move his team across the LDA far enough to fit the remainder of the squad on the far side of the LDA. Bravo team moves to the LDA to the right or left to provide an overwatch position prior to A team crossing.

**STEP 3:** SL receives the hand and arm signal that it is safe to move the rest of the squad across (B team is still providing overwatch).

**STEP 4:** SL moves himself, radio operator, and B team across the LDA. (A team provides overwatch for squad missions.)

**STEP 5:** A team assumes original azimuth at SL's command or hand and arm signal.

Figure 4-4. LINEAR DANGER AREA

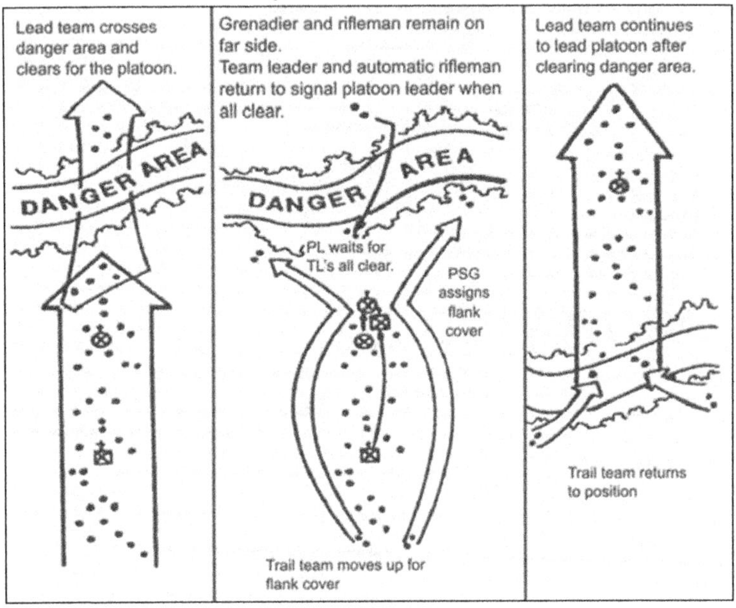


(2) ***LDA Crossing for a Platoon***:

(a) The lead squad halts the platoon, and signals danger area.

(b) The platoon leader moves forward to the lead squad to confirm the danger area and decides if current location is a suitable crossing site.

(c) The platoon leader confirms danger area/crossing site and establishes near and far side rally points.

(d) On the platoon leader's signal, the A team of the lead squad establishes an overwatch position to the left of the crossing site. Prior to crossing, the compassman with the lead two squads confirm azimuth and pace data.

(e) B team of the lead squad establishes an overwatch position to the right of the crossing site.

(f) Once overwatch positions are established, the platoon leader gives the second squad in movement the signal to bound across by fire team.

(g) Once across, the squad is now lead in movement and continues on azimuth

(h) Once Stop, Look, Listen, and Smell (SLLS) is conducted, squad leader signals platoon leader all clear.

- Day time—hand and arm signal such as thumbs up.
- Night time—Clandestine signal such as IR, red lens.

(i) Platoon leader receives all clear and crosses with radio operator, FO, WSL, and 2 gun teams.

(j) Once across, PL signals the 3rd squad in movement to cross.

(k) PSG with medic and one gun team crosses after 2nd squad is across (sterilizing central crossing site).

(l) PSG signals security squad to cross at their location.

***Notes:*** Platoon leader will plan for fires at all known LDA crossing sites.
Nearside security in overwatch will sterilize signs of the patrol .

(3) ***Danger Area (Small/Open)***

(a) The lead squad halts the platoon and signals danger area.

(b) The PL moves forward to the lead squad to confirm the danger area.

(c) The platoon leader confirms danger area and establishes near and far side rally points.

(d) The PL designates lead squad to bypass danger area using the detour-bypass method.

(e) Paceman suspends current pace count and initiates an interim pace count. Alternate pace/compass man moves forward and offsets compass 90 degrees left or right as designated and moves in that direction until clear of danger area.

(f) After moving set distance (x-meters as instructed by PL). Lead squad assumes original azimuth, and primary pace man resumes original pace.

(g) After the open area, the alternate pace/compass man offsets his compass 90 degrees left or right and leads the platoon/squad the same distance (x-meters) back to the original azimuth.

(4) ***Danger Areas (Series)***: A series of danger areas is two or more danger areas within an area that can be either observed or covered by fire.

- Double linear danger area (use linear danger area technique and cross as one LDA).
- Linear/small open danger area (use by-pass/contour technique. Figure 4-5).
- Linear/large open danger area (use platoon wedge when crossing).

***Note:*** A series of danger areas is crossed using the technique which provides the most security.

(5) ***Danger Area (Large)***:

(a) Lead squad halts the platoon, and signals danger area.

(b) The platoon leader moves forward with RTO and FO and confirms danger area.

(c) The platoon leader confirms danger area and establishes near and far side rally points.

(d) PL designates direction of movement.

(e) PL designates change of formation as necessary to ensure security.

***Notes:*** Before point man steps into danger area, PL and FO adjust targets to cover movement.
If far side of danger area is within 250 meters, PL establishes overwatch, and designates lead squad to clear wood line on far side.

Figure 4-5. SMALL OPEN AREA

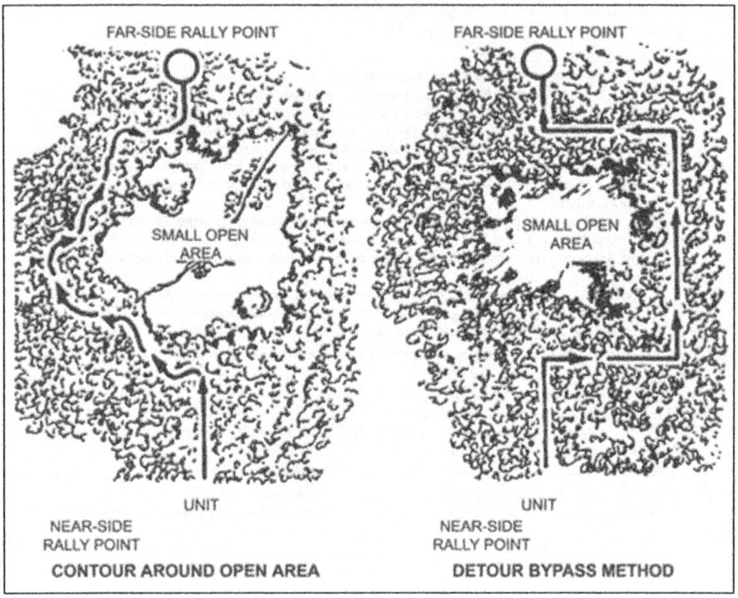


NOTES

## Chapter 5

# PATROLS

Infantry platoons and squads primarily conduct two types of patrols: reconnaissance and combat. This chapter describes the principles of patrolling, planning considerations, types of patrols, supporting tasks, patrol base, and movement to contact (FM 7-8, FM 3-0, and FM 101-5-1). In this chapter, the terms "*element*" and "*team*" refer to the squads, fire teams, or buddy teams that perform the tasks as described.

5-1. **PRINCIPLES**. All patrols are governed by five principles:

a. **Planning**. Quickly make a simple plan and effectively communicate it to the lowest level. A great plan that takes forever to complete and is poorly disseminated isn't a great plan. Plan and prepare to a realistic standard and rehearse everything.

b. **Reconnaissance**. Your responsibility as a Ranger leader is to confirm what you think you know, and to learn that which you do not already know.

c. **Security**. Preserve your force as a whole. Every Ranger and every rifle counts; anyone could be the difference between victory and defeat.

d. **Control**. Clear understanding of the concept of the operation and commander's intent, coupled with disciplined communications, to bring every man and weapon available to overwhelm the enemy at the decisive point.

e. **Common Sense**. Use all available information and good judgment to make sound, timely decisions.

5-2. **PLANNING**. This paragraph provides the planning considerations common to most patrols. It discusses task organization, initial planning and coordination, completion of the plan, and contingency planning.

a. **Task Organization.** A patrol is a mission, not an organization. To accomplish the patrolling mission, a platoon or squad must perform specific tasks, for example, secure itself, cross danger areas, recon the patrol objective, breach, support, or assault. As with other missions, the leader tasks elements of his unit in accordance with his estimate of the situation, identifying those tasks his unit must perform and designating which elements of his unit will perform which tasks. Where possible, in assigning tasks, the leader should maintain squad and fire team integrity. The chain of command continues to lead its elements during a patrol. Squads and fire teams may perform more than one task in an assigned sequence; others may perform only one task. The leader must plan carefully to ensure that he has identified and assigned all required tasks in the most efficient way. Elements and teams for platoons conducting patrols include--

(1) Elements common to all patrols:

(a) *Headquarters Element*. The headquarters consists of the platoon leader (PL), RTO, platoon sergeant (PSG), FO, RTO, and medic. It may include any attachments that the PL decides that he or the PSG must control directly.

(b) *Aid and Litter Team.* Aid and litter teams are responsible for buddy aid and evacuation of casualties.

(c) *Enemy Prisoner of War Team.* EPW teams control enemy prisoners using the five S's and the leader's guidance.

(d) *Surveillance Team.* The surveillance team keeps watch on the objective from the time that the leader's reconnaissance ends until the unit deploys for actions on the objective. They then rejoin their parent element.

(e) *En Route Recorder.* Part of the HQ element, maintains communications with higher and acts as the recorder for all CCIR collected during the mission.

(f) *Compass Man.* The compass man assists in navigation by ensuring the patrol remains on course at all times. Instructions to the compass man must include initial and subsequent azimuths. As a technique, the compass man should preset his compass on the initial azimuth before the unit moves out, especially if the move will be during limited visibility conditions. The platoon or squad leader should also designate an alternate compass man.

(g) *Point/Pace Man.* As required, the PL designates a primary and alternate point man and a pace man for the patrol. The pace man aids in navigation by keeping an accurate count of distance traveled. The point man selects the actual route through the terrain, guided by the compass man or team leader. In addition, the point man also provides frontal security.

(2) Elements common to all combat patrols:

(a) *Assault Element.* The assault element seizes and secures the objective and protects special teams as they complete their assigned actions on the objective.

(b) *Security Element.* The security element provides security at danger areas, secures the ORP, isolates the objective, and supports the withdrawal of the rest of the patrol once actions on the objective are complete. The security element may have separate security teams, each with an assigned task or sequence of tasks.

(c) *Support Element*. The support element provides direct and indirect fire support for the unit. Direct fires include machine guns, medium and light antiarmor weapons, small recoilless rifles. Indirect fires available may include mortars, artillery, CAS, and organic M203 weapon systems.

(d) *Demolition Team*. Demolition teams are responsible for preparing and detonating the charges to destroy designated equipment, vehicles, or facilities on the objective.

(e) *EPW and Search Teams*. The assault element may provide two-man (buddy teams) or four-man (fire team) search teams to search bunkers, buildings, or tunnels on the objective. These teams will search the objective or kill zone for any PIR that may give the PL an idea of the enemy concept for future operations. Primary and alternate teams may be assigned to ensure enough prepared personnel are available on the objective.

(f) *Breach Element*. The breach team conducts initial penetration of enemy obstacles to seize a foothold and allow the patrol to enter an objective. This is typically done IAW METT-TC and the steps outlined in the "*Conduct an Initial Breach of a Mined Wire Obstacle*" battle drill in Chapter 6 of this Handbook.

(3) Elements common to all reconnaissance patrols:

(a) *Reconnaissance Team*. Reconnaissance teams reconnoiter the objective area once the security teams are in position. Normally these are two-man teams (buddy teams) to reduce the possibility of detection.

(b) *Reconnaissance and Security Teams*. R&S teams are normally used in a zone reconnaissance, but may be useful in any situation when it is impractical to separate the responsibilities for reconnaissance and security.

(c) *Security Element*. When the responsibilities of reconnaissance and security are separate, the security element provides security at danger areas, secures the ORP, isolates the objective, and supports the withdrawal of the rest of the platoon once the recon is complete. The security element may have separate security teams, each with an assigned task or sequence of tasks.

b. **Initial Planning and Coordination**. Leaders plan and prepare for patrols using the troop-leading procedures and the estimate of the situation, as described in Chapter 2. Through an estimate of the situation, leaders identify required actions on the objective (mission analysis) and plan backward to departure from friendly lines and forward to reentry of friendly lines. Because patrolling units act independently, move beyond the direct-fire support of the parent unit, and operate forward of friendly units, coordination must be thorough and detailed. Coordination is continuous throughout planning and preparation. PLs use checklists to preclude omitting any items vital to the accomplishment of the mission.

(1) ***Coordination with Higher Headquarters***. This coordination includes Intelligence, Operations, and Fire Support IAW Chapter 2, *Coordination Checklists* (**page 2-34**). This initial coordination is an integral part of Step 3 of Troop-Leading Procedures, *Make a Tentative Plan*.

(2) ***Coordination with Adjacent Units***. The leader also coordinates his unit's patrol activities with the leaders of other units that will be patrolling in adjacent areas at the same time, IAW Chapter 2-7, *Coordination Checklists* (**page 2-34**).

c. **Completion of Plan**. As the PL completes his plan, he considers--

(1) ***Specified and Implied Tasks***. The PL ensures that he has assigned all specified tasks to be performed on the objective, at rally points, at danger areas, at security or surveillance locations, along the route(s), and at passage lanes. These make up the maneuver and tasks to maneuver units subparagraphs of the Execution paragraph.

(2) ***Key travel and Execution Times***. The leader estimates time requirements for movement to the objective, leader's reconnaissance of the objective, establishment of security and surveillance, completion of all assigned tasks on the objective, and passage through friendly lines. Some planning factors are--

- Movement: Average of 1 Kmph during daylight hours in Woodland Terrain; Average limited visibility ½ Kmph. Add additional time for restrictive, or severely restrictive terrain such as mountains, swamps, or thick vegetation.
- Leader's recon: NLT 1.5 hr.
- Establishment of security and surveillance: 0.5 hr.

(3) ***Primary and Alternate Routes***. The leader selects primary and alternate routes to and from the objective. The return routes should differ from the routes to the objective. The PL may delegate route selection to a subordinate, but is ultimately responsible for the routes selected.

(4) ***Signals***. The leader should consider the use of special signals. These include hand-and-arm signals, flares, voice, whistles, radios, and infrared equipment. Primary and alternate signals must be identified and rehearsed so that all Soldiers know their meaning.

(5) ***Challenge and Password Forward of Friendly Lines***. The challenge and password from the unit's ANCD must not be used beyond the FLOT.

(a) *Odd-Number System.* The leader specifies an odd number. The challenge can be any number less than the specified number. The password will be the number that must be added to it to equal the specified number, for example, the number is 7, the challenge is 3, and the password is 4.

(b) *Running Password.* ANCDs may also designate a running password. This code word alerts a unit that friendly Soldiers are approaching in a less than organized manner and possibly under pressure. The number of Soldiers approaching follows the running password. For example, if the running password is "*Ranger,*" and five friendly Soldiers are approaching, they would say "*Ranger five.*"

(6) ***Location of Leaders***. The PL considers where he and the PSG and other key leaders are located during each phase of the mission. The PL positions himself where he can best control the actions of the patrol. The PSG is normally located with the assault element during a raid or attack to help the PL control the use of additional assaulting squads, and will assist with securing the OBJ. The PSG will locate himself at the CCP to facilitate casualty treatment and evacuation. During a reconnaissance mission, the PSG will stay behind in the ORP to facilitate the transfer of Intel to the higher headquarters, and control the recon elements movement into and out of the ORP.

(7) ***Actions on Enemy Contact***. Unless required by the mission, the unit avoids enemy contact. The leader's plan must address actions on chance contact at each phase of the patrol mission. The unit's ability to continue will depend on how early contact is made, whether the platoon is able to break contact successfully (so that its subsequent direction of movement is undetected), and whether the unit receives any casualties because of the contact. The plan must address the handling of seriously wounded Soldiers and KIAs. The plan must also address the handling of prisoners who are captured because of chance contact and are not part of the planned mission.

(8) ***Contingency Plans***. The leader leaves his unit for many reasons throughout the planning, coordination, preparation, and execution of his patrol mission. Each time the leader departs the patrol main body, he must issue a five-point contingency plan to the leader left in charge of the unit. The contingency plan is described by the acronym GOTWA, as follows. The patrol leader will additionally issue specific guidance stating what tasks are to be accomplished in the ORP in his absence:

- G: Where the leader is GOING.
- O: OTHERS he is taking with him.
- T: TIME he plans to be gone.
- W: WHAT to do if the leader does not return in time.
- A: The unit's and the leader's ACTIONS on chance contact while the leader is gone.

(9) ***Rally Points***. The leader considers the use and location of rally points. A rally point is a place designated by the leader where the unit moves to reassemble and reorganize if it becomes dispersed. Soldiers must know which rally point to move to at each phase of the patrol mission should they become separated from the unit. They must also know what actions are required there and how long they are to wait at each rally point before moving to another.

(a) *Criteria.* Rally points must be--

- Easily identifiable in daylight and limited visibility.
- Show no signs of recent enemy activity.
- Covered and concealed.
- Away from natural lines of drift and high-speed avenues of approach.
- Defendable for short periods of time.

(b) *Types.* The most common types of rally points are initial, en route, objective, and near-and-far-side rally points.

(10) ***Objective Rally Point***. The ORP is typically 200 to 400m from the objective, or at a minimum, one major terrain feature away. Actions at the ORP include--

- Conduct SLLS and pinpoint location.
- Leaders Recon of the Objective.
- Issuing a FRAGO, if needed.
- Making final preparations before continuing operations, for example, recamouflaging, preparing demolitions, lining up rucksacks for quick recovery. Preparing EPW bindings, first aid kits, litters, and inspecting weapons.
- Accounting for Soldiers and equipment after actions at the objective are complete.
- Reestablishing the chain of command after actions at the objective are complete.
- Disseminating information from reconnaissance, if contact was not made.

(11) ***Leader's reconnaissance of the objective***. The plan must include a leader's reconnaissance of the objective once the platoon or squad establishes the ORP. Before departing, the leader must issue a 5-point contingency plan. During his

reconnaissance, the leader pinpoints the objective, selects reconnaissance, security, support, and assault positions for his elements, and adjusts his plan based on his observation of the objective. Each type of patrol requires different tasks during the leader's reconnaissance. The platoon leader will bring different elements with him. (These are discussed separately under each type of patrol). The leader must plan time to return to the ORP, complete his plan, disseminate information, issue orders and instructions, and allow his squads to make any additional preparations. During the Leader's Reconnaissance for a Raid or Ambush, the PL will leave surveillance on the OBJ.

(12) ***Actions on the objective***. Each type of patrol requires different actions on the objective. Actions on the objective are discussed under each type of patrol.

5-3. **RECONNAISSANCE PATROLS**. Recon patrols are one of the two types of patrols. provide timely and accurate information on the enemy and terrain. They confirm the leader's plan before it is executed. Units on reconnaissance operations collect specific information (*priority intelligence requirements* [PIR]) or general information (*information requirements* [IR]) based on the instructions from their higher commander. The two types of recon patrols discussed here are area and zone. This section discusses the fundamentals of reconnaissance, task standards for the two most common types of recon, and actions on the objective for those types of recon.

a. **Fundamentals of Reconnaissance**. In order to have a successful area reconnaissance, the platoon leader must apply the fundamentals of the reconnaissance to his plan during the conduct of the operation.

(1) ***Gain all required information***. The parent unit tells the patrol leader (PL) what information is required. This is in the form of the IR and PIR. The platoon's mission is then tailored to what information is required. During the entire patrol, members must continuously gain and exchange all information gathered, but cannot consider the mission accomplished unless all PIR has been gathered.

(2) ***Avoid detection by the enemy***. A patrol must not let the enemy know that it is in the objective area. If the enemy knows he is being observed, he may move, change his plans, or increase his security measures. Methods of avoiding detection are--

(a) Minimize movement in the objective area (area reconnaissance).

(b) Move no closer to the enemy than necessary.

(c) If possible, use long-range surveillance or night vision devices.

(d) Camouflage, stealth, noise, and light discipline.

(e) Minimize radio traffic.

(3) ***Employ security measures***. A patrol must be able to break contact and return to the friendly unit with what information is gathered. If necessary, break contact and continue the mission. Security elements are emplaced so that they can overwatch the reconnaissance elements and suppress the enemy so the reconnaissance element can break contact.

(4) ***Task organization***. When the platoon leader receives the order, he analyzes his mission to ensure he understands what must be done. Then he task organizes his platoon to best accomplish the mission IAW METT-TC. Recons are typically squad-sized missions.

b. **Task Standards**.

(1) ***Area Reconnaissance***. The area recon patrol collects all available information on PIR and other intelligence not specified in the order for the area. The patrol completes the recon and reports all information by the time specified in the order. The patrol is not compromised.

(2) ***Zone Reconnaissance***. The zone recon patrol determines all PIR and other intelligence not specified in the order for its assigned zone. The patrol reconnoiters without detection by the enemy. The patrol completes the recon and reports all information by the time specified in the order.

c. **Actions on the Objective, Area Reconnaissance** (Figure 5-1).

(1) The element occupies the ORP as discussed in the section on occupation of the ORP. The RTO calls in spare for occupation of ORP. The leader confirms his location on map while subordinate leaders make necessary perimeter adjustments.

(2) The PL organizes the platoon in one of two ways: separate recon and security elements, or combined recon and security elements.

(3) The PL takes subordinates leaders and key personnel on a leader's recon to confirm the objective and plan.

(a) Issues a 5-point contingency plan before departure.

(b) Establishes a suitable release point that is beyond sight and sound of the objective if possible, but that is definitely out of sight. The RP should also have good rally point characteristics.

(c) Allow all personnel to become familiar with the release point and surrounding area.

(d) Identifies the objective and emplaces surveillance. Designates a surveillance team to keep the objective under surveillance. Issues a contingency plan to the senior man remaining with the surveillance team. The surveillance team is positioned with one man facing the objective, and one facing back in the direction of the release point.

(e) Takes subordinate leaders forward to pinpoint the objective, emplace surveillance, establish a limit of advance, and choose vantage points.

(f) Maintains commo with the platoon throughout the leader's recon.

(4) The PSG maintains security and supervises priorities of work in the ORP.

(a) Reestablishes security at the ORP.

(b) Disseminates the PLs contingency plan.

(c) Oversees preparation of recon personnel (personnel recamouflaged, NVDs and binos prepared, weapons on safe with a round in the chamber).

(5) The PL and his recon party return to the ORP.

(a) Confirms the plan or issues a FRAGO.

(b) Allows subordinate leaders time to disseminate the plan.

(6) The patrol conducts the recon by long-range observation and surveillance if possible.

(a) R&S elements move to observation points that offer cover and concealment and that are outside of small-arms range.

(b) Establishes a series of observation posts (OP) if information cannot be gathered from one location.

(c) Gathers all PIR using the SALUTE format.

(7) If necessary, the patrol conducts its recon by short-range observation and surveillance.

(a) Moves to an OP near the objective.

(b) Passes close enough to the objective to gain information.

(c) Gathers all PIR using the SALUTE format.

(8) R&S teams move using a technique such as the cloverleaf method to move to successive OP's. In this method, R&S teams avoid paralleling the objective site, maintain extreme stealth, do not cross the limit of advance, and maximize the use of available cover and concealment.

(9) During the conduct of the recon, each R&S team will return to the release point when any of the following occurs:

- They have gathered all their PIR.
- They have reached the limit of advance.
- The allocated time to conduct the recon has elapsed.
- Contact has been made.

(10) At the release point, the leader will analyze what information has been gathered and determine if he has met the PIR requirements.

(11) If the leader determines that he has not gathered sufficient information to meet the PIR requirements, or if the information he and the subordinate leader gathered differs drastically, he may have to send R&S teams back to the objective site. In this case, R&S teams will alternate areas of responsibilities. For example, if one team reconnoitered from the 6 – 3 – 12, then that team will now recon from the 6 – 9 – 12.

(12) The R&S element returns undetected to the ORP by the specified time.

(a) Disseminates information to all patrol members through key leaders at the ORP, or moves to a position at least one terrain feature or one kilometer away to disseminate. To disseminate, the leader has the RTO prepare three sketches of the objective site based on the leader's sketch and provides the copies to the subordinate leaders to assist in dissemination.

(b) Reports any information requirements and/or any information requiring immediate attention to higher headquarters, and departs for the designated area.

(13) If contact is made, move to the release point. The recon element tries to break contact and return to the ORP, secure rucksacks, and quickly move out of the area. Once they have moved a safe distance away, the leader will inform higher HQ of the situation and take further instructions from them.

(a) While emplacing surveillance, the recon element withdraws through the release point to the ORP, and follows the same procedures as above.

(b) While conducting the reconnaissance, the compromised element returns a sufficient volume of fire to allow them to break contact. Surveillance can fire an AT-4 at the largest weapon on the objective. All elements will pull off the objective and move to the release point. The senior man will quickly account for all personnel and return to the ORP. Once in the ORP, follow the procedures previously described.

Figure 5-1. ACTIONS ON THE OBJECTIVE, AREA RECONNAISSANCE

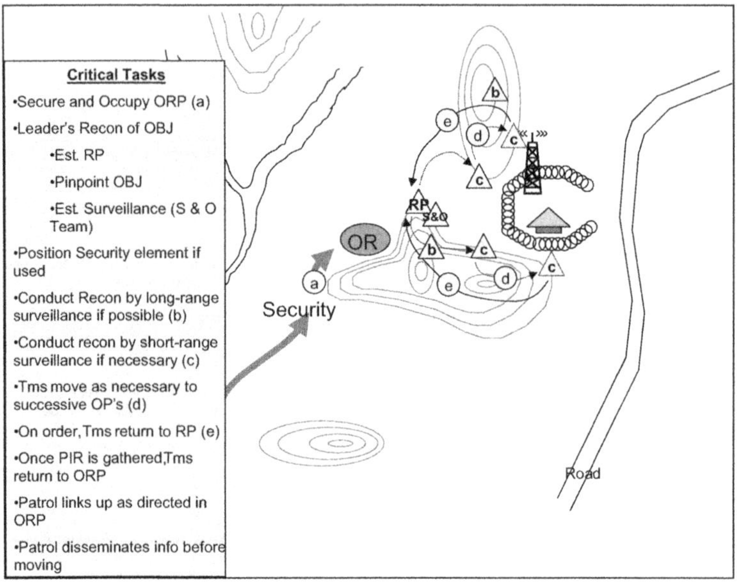


d. **Actions on the Objective, Zone Reconnaissance**.

(1) The element occupies the initial ORP as discussed in the section occupation of the ORP. The radio operator calls in spare for occupation of ORP. The leader confirms his location on map while subordinate leaders make necessary perimeter adjustments.

(2) The recon team leaders organize their recon elements.

(a) Designate security and recon elements.

(b) Assign responsibilities (point man, pace man, en route recorder, and rear security), if not already assigned.

(c) Designates easily recognizable rally points.

(d) Ensure local security at all halts.

(3) The patrol recons the zone.

(a) Moves tactically to the ORPs.

(b) Occupies designated ORPs.

(c) Follows the method designated by the PL (fan, converging routes, or box method, Figure 5-2).

**Figure 5-2. COMPARISON OF ZONE RECONNAISSANCE METHODS**

| *FAN METHOD* | *CONVERGING ROUTES METHOD* | *BOX METHOD* |
|---|---|---|
| • Uses a series (fan) of ORPs.<br>• Patrol establishes security at first ORP.<br>• Each recon element moves from ORP along a different fan-shaped route. Route overlaps with that of other recon elements. This ensures recon of entire area.<br>• Leader maintains reserve at ORP.<br>• When all recon elements return to ORP, PL collects and disseminates all info before moving to next ORP. | • PL selects routes from ORP thru zone to a linkup point at the far side of the zone from the ORP.<br>Each recon element moves and recons along a specified route. They converge (link up) at one time and place. | • PL sends recon elements from the first ORP along routes that form a box.<br>He sends other elements along routes throughout the box. All teams link up at the far side of the box from the ORP. |

(d) The recon teams reconnoiter.

- During movement, the squad will gather all PIR specified by the order.
- Recon team leaders will ensure sketches are drawn or digital photos are taken of all enemy hard sites, roads, and trails.
- Return to the ORP, or link up at the rendezvous point on time.
- When the squad arrives at new rendezvous point or ORP, the recon team leaders report to the PL with all information gathered.

(e) The PL continues to control the recon elements.

- PL moves with the recon element that establishes the linkup point.
- PL changes recon methods as required.
- PL designates times for the elements to return to the ORP or to linkup.
- PL collects all information and disseminates it to the entire patrol. PL will brief all key subordinate leaders on information gathered by other squads, establishing one consolidated sketch if possible, and allow team leaders time to brief their teams.
- PL and PSG account for all personnel.

(f) The patrol continues the recon until all designated areas have been reconned, and returns undetected to friendly lines.

5-4. **COMBAT PATROLS**. Combat patrols are the second type of patrol. Combat patrols are further divided into raids and ambushes Units conduct combat patrols to destroy or capture enemy Soldiers or equipment; destroy installations, facilities, or key points; or harass enemy forces. Combat patrols also provide security for larger units. This section describes overall combat patrol planning considerations, task considerations for each type of combat patrol, and finally actions on the objective for each type.

a. **Planning Considerations (General)**. In planning a combat patrol, the PL considers the following:

(1) ***Tasks to Maneuver Units***. Normally the platoon headquarters element controls the patrol on a combat patrol mission. The PL makes every try to maintain squad and fire team integrity as he assigns tasks to subordinates units.

(a) The PL must consider the requirements for assaulting the objective, supporting the assault by fire, and security of the entire unit throughout the mission.

- For the assault on the objective, the PL considers the required actions on the objective, the size of the objective, and the known or presumed strength and disposition of the enemy on and near the objective.
- The PL considers the weapons available, and the type and volume of fires required to provide fire support for the assault on the objective.
- The PL considers the requirement to secure the platoon at points along the route, at danger areas, at the ORP, along enemy avenues of approach into the objective, and elsewhere during the mission.
- The PL will also designate engagement/disengagement criteria.

(b) The PL assigns additional tasks to his squads for demolition, search of EPWs, guarding of EPWs, treatment and evacuation (litter teams) of friendly casualties, and other tasks required for successful completion of patrol mission (if not already in the SOP).

(c) The PL determines who will control any attachments of skilled personnel or special equipment.

(2) ***Leader's Reconnaissance of the Objective***. In a combat patrol, the PL has additional considerations for the conduct of his reconnaissance of the objective from the ORP.

(a) *Composition of the leader's reconnaissance party*. The platoon leader will normally bring the following personnel.

- Squad leaders to include the weapons squad leader.
- Surveillance team.
- Forward observer.
- Security element (dependent on time available).

(b) *Conduct of the leader's reconnaissance*. In a combat patrol, the PL considers the following additional actions in the conduct of the leader's reconnaissance of the objective.

- The PL designates a release point approximately half way between the ORP and this objective. The PL posts the surveillance team. Squads and fire teams separate at the release point, and then they move to their assigned positions.
- The PL confirms the location of the objective or kill zone. He notes the terrain and identifies where he can emplace claymores to cover dead space. Any change to his plan is issued to the squad leaders (while overlooking the objective if possible).
- If the objective is the kill zone for an ambush, the leader's reconnaissance party should not cross the objective; to do so will leave tracks that may compromise the mission.
- The PL confirms the suitability of the assault and support positions and routes from them back to the ORP.
- The PL issues a five-point contingency plan before returning to the ORP.

b. **Ambush**.

(1) ***Planning Considerations***. An ambush is a surprise attack from a concealed position on a moving or temporarily halted target. Ambushes are classified by category--hasty or deliberate; type--point or area; and formation--linear or L-shaped. The leader uses a combination of category, type, and formation in developing his ambush plan. The key planning considerations include--

(a) Coverage of entire kill zone by fire.

(b) METT-TC.

(c) Use of existing or reinforcing obstacles, including Claymores, to keep the enemy in the kill zone.

(d) Security teams (typically equipped with hand-held antitank weapons such as AT-4 or LAW; Claymores; and various means of communication.

(e) Protect the assault and support elements with claymores or explosives.

(f) Use security elements or teams to isolate the kill zone.

(g) Assault through the kill zone to the limit of advance (LOA). (The assault element must be able to move quickly through its own protective obstacles.)

(h) Time the actions of all elements of the platoon to preclude loss of surprise. In the event any member of the ambush is compromised, he may immediately initiate the ambush.

(i) When the ambush must be manned for a long time, use only one squad to conduct the entire ambush and determining movement time of rotating squads from the ORP to the ambush site.

(2) ***Categories***.

(a) *Hasty Ambush*. A unit conducts a hasty ambush when it makes visual contact with an enemy force and has time to establish an ambush without being detected. The actions for a hasty ambush must be well rehearsed so that Soldiers know what to do on the leader's signal. They must also know what action to take if the unit is detected before it is ready to initiate the ambush.

(b) *Deliberate Ambush*. A deliberate ambush is conducted at a predetermined location against any enemy element that meets the commander's engagement criteria. The leader requires the following detailed information in planning a deliberate ambush: size and composition of the targeted enemy, and weapons and equipment available to the enemy.

(3) ***Types***.

(a) Point ambush. In a point ambush, Soldiers deploy to attack an enemy in a single kill zone.

(b) Area ambush. In an area, Soldiers deploy in two or more related point ambushes.

(4) ***Formations***. (Figure 5-3).

(a) *Linear Ambush*. In an ambush using a linear formation, the assault and support elements deploy parallel to the enemy's route. This positions both elements on the long axis of the kill zone and subjects the enemy to flanking fire. This formation can be used in close terrain that restricts the enemy's ability to maneuver against the platoon, or in open terrain provided a means of keeping the enemy in the kill zone can be effected.

(b) *L-Shaped Ambush*. In an L-shaped ambush, the assault element forms the long leg parallel to the enemy's direction of movement along the kill zone. The support element forms the short leg at one end of and at right angles to the assault element. This provides both flanking (long leg) and enfilading fires (short leg) against the enemy. The L-shaped ambush can be used at a sharp bend in a trail, road, or stream. It should not be used where the short leg would have to cross a straight road or trail.

**Figure 5-3. AMBUSH FORMATIONS**

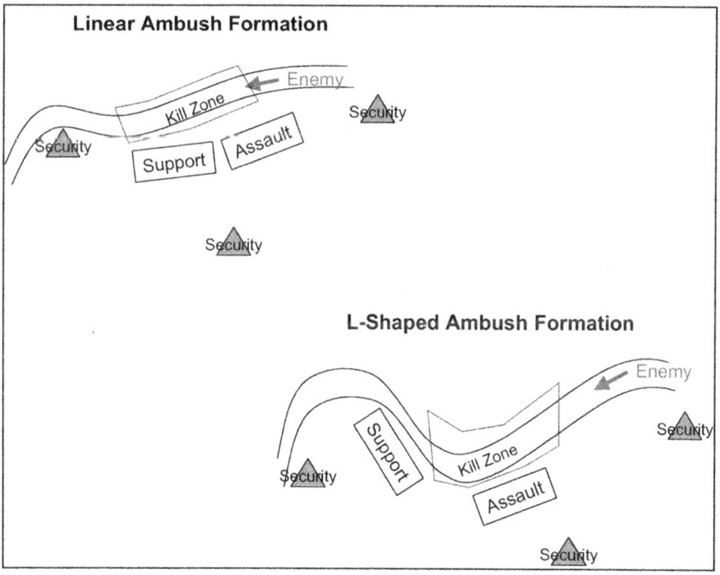

c. **Hasty Ambush**.

(1) ***Task Standards***. The platoon moves quickly to concealed positions. The ambush is not initiated until the majority of the enemy is in the kill zone. The unit does not become decisively engaged. The platoon surprises the enemy. The patrol captures, kills, or forces the withdrawal of all of the enemy within the kill zone. On order, the patrol withdraws all personnel and equipment in the kill zone from observation and direct fire. The unit does not become decisively engaged by follow-on elements. The platoon continues follow-on operations.

(2) ***Actions on the Objective (Hasty Ambush)***. See Figure 5-4.

(a) Using visual signals, any Soldier alerts the unit that an enemy force is in sight. The Soldier continues to monitor the location and activities of the enemy force until his team or squad leader relieves him, and gives the enemy location and direction of movement.

(b) The platoon or squad halts and remains motionless.

- The PL gives the signal to conduct a hasty ambush, taking care not to alert the enemy of the patrol's presence.
- The leader determines the best nearby location for a hasty ambush. He uses arm-and-hand signals to direct the unit members to covered and concealed positions.

(c) The leader designates the location and extent of the kill zone.

d) Teams and squads move silently to covered and concealed positions, ensuring positions are undetected and have good observation and fields of fire into the kill zone.

(e) Security elements move out to cover each flank and the rear of the unit. The leader directs the security elements to move a given distance, set up, and then rejoin the unit on order or, after the ambush (the sound of firing ceases). At squad level, the two outside buddy teams normally provide flank security as well as fires into the kill zone. At platoon level, fire teams make up the security elements.

(f) The PL assigns sectors of fire and issues any other commands necessary such as control measures.

(g) The PL initiates the ambush, using the greatest casualty-producing weapon available, when the largest percentage of enemy is in the kill zone. The PL--

- Controls the rate and distribution of fire.
- Employs indirect fire to support the ambush.
- Orders cease fire.
- (If the situation dictates) Orders the patrol to assault through the kill zone.

(h) The PL designates personnel to conduct a hasty search of enemy personnel and process enemy prisoners and equipment.

(i) The PL orders the platoon to withdraw from the ambush site along a covered and concealed route.

(j) The PL gains accountability, reorganizes as necessary, disseminates information, reports the situation, and continues the mission as directed.

d. **Deliberate (Point/Area) Ambush**.

(1) ***Task Standards***. The ambush is emplaced NLT the time specified in the order. The patrol surprises the enemy and engages the enemy main body. The patrol kills or captures all enemy in the kill zone and destroys equipment based on the commander's intent. The patrol withdraws all personnel and equipment from the objective, on order, within the time specified in the order. The patrol obtains all available PIR from the ambush and continues follow-on operations.

(2) ***Actions on the Objective (Deliberate Ambush).*** See Figure 5-5.

(a) The PL prepares the patrol for the ambush in the ORP.

(b) The PL prepares to conduct a leader's reconnaissance. He--

- Designates the members of the leader's recon party (typically includes squad leaders, surveillance team, FO, and possibly the security element.
- Issues a contingency plan to the PSG.

Figure 5-4. ACTIONS ON THE OBJECTIVE -- HASTY AMBUSH

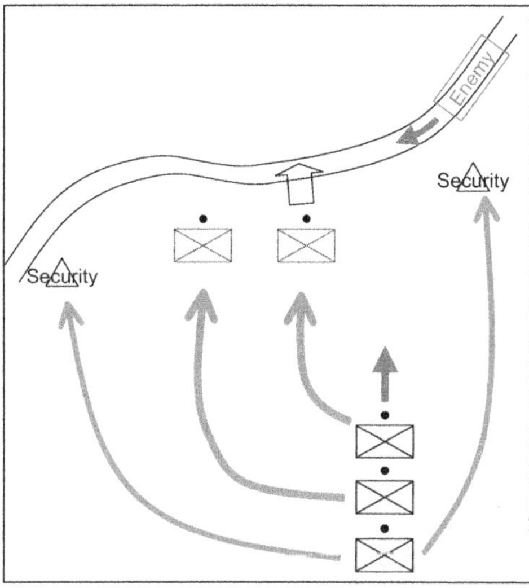


**Critical Tasks**

- Patrol detects an enemy unit; PL is notified
- Patrol halts and remains motionless
- PL gives signal for Hasty Ambush
- PL directs elements to covered and concealed positions
- Security Elements move to flanks of patrol
- PL establishes control measures
- PL initiates and controls ambush
- PL directs a hasty search
- Patrol consolidates, reorganizes, withdraws, reports, and continues mission

(c) The PL conducts his leader's reconnaissance. He--

- Ensures the leader's recon party moves undetected.
- Confirms the objective location and suitability for the ambush.
- Selects a kill zone.
- Posts the surveillance team at the site and issues a contingency plan.
- Confirms suitability of assault and support positions, and routes from them to the ORP.
- Selects the position of each weapon system in the support-by-fire position, and then designates sectors of fire.
- Identifies all offensive control measures to be used. Identifies the PLD, the assault position, LOA, any boundaries or other control measures. If available, the PL can use infrared aiming devices to identify these positions on the ground.

(d) The PL adjusts his plan based on info from the reconnaissance. He--

- Assigns positions.
- Designates withdrawal routes.

(e) The PL confirms the ambush formation.

(f) The security team(s) occupy first, securing the flanks of the ambush site, and providing early warning. The security element must be in position before the support and assault elements move forward of the release point. A security team remains in the ORP if the patrol plans to return to the ORP after actions on the objective. If the ORP is abandoned, a rear security team should be emplaced.

(g) Support element leader assigns sectors of fire. He--

- Emplaces claymores and obstacles as designated.
- Identifies sectors of fire and emplaces limiting stakes to prevent friendly fires from hitting other elements.
- Overwatches the movement of the assault element into position.

(h) Once the support element is in position, or on the PLs order, the assault element--

- Departs the ORP and moves into position.
- Upon reaching the PLD, the assault element transitions from the movement formation to the battle formation.
- Identifies individual sectors of fire as assigned by the PL. Emplaces aiming stakes or uses metal-to-metal contact with the machine gun tripods to prevent fratricide on the objective.
- Emplaces claymores to help destroy the enemy in the kill zone.
- Camouflages positions.

(i) The security element spots the enemy and notifies the PL, and reports the direction of movement, size of the target, and any special weapons or equipment carried. The security element also keeps the platoon leader informed if any enemy forces are following the lead force.

(j) The PL alerts other elements, and determines if the enemy force is too large, or if the ambush can engage the enemy successfully.

(k) The PL initiates the ambush using the highest casualty-producing device. He may use a command-detonated claymore. He must also plan a backup method for initiating the ambush, in case his primary means fails. This should also be a casualty-producing device such as his individual weapon. He passes this information to all Rangers, and practices it during rehearsals.

(l) The PL ensures that the assault and support elements deliver fire with the heaviest, most accurate volume possible on the enemy in the kill zone. In limited visibility, the PL may use infrared lasers to further define specific targets in the kill zone.

(m) Before assaulting the target, the PL gives the signal to lift or shift fires.

(n) The assault element--

- Assaults before the remaining enemy can react.
- Kills or captures enemy in the kill zone.
- Uses individual movement techniques or bounds by fire teams to move.
- Upon reaching the limit of advance, halts and establishes security. If needed, it reestablishes the chain of command and remains key weapon systems. All Soldiers will load a fresh magazine or drum of ammunition using the buddy system. ACE reports will be submitted through the chain of command. The PL will submit an initial contact report to higher.

(o) The PL directs special teams (EPW search, aid and litter, demo) to accomplish their assigned task once the assault element has established its LOA.

- Once the kill zone had been cleared, collect and secure all EPWs and move them out of the kill zone before searching bodies. Coordinate for an EPW exchange point to link up with higher to extract all EPWs and treat them IAW the five S's.
- Search from one side to the other and mark bodies that have been searched to ensure the area is thoroughly covered. Units should use the "*clear out, search in*" technique, clear from the center of the objective out ensuring the area is clear of all enemy combatants; then search all enemy personnel towards the center of the objective. Search all dead enemy personnel using two-man search techniques.
  -- As the search team approaches a dead enemy Soldier, one-man guards while the other man searches. First, he kicks the enemy weapon away.
  -- Second, he rolls the body over (if on the stomach) by lying on top and when given the go ahead by the guard (who is positioned at the enemy's head), the searcher rolls the body over on him. This is done for protection in case the enemy Soldier has a grenade with the pin pulled underneath him.
  -- The searchers then conduct a systematic search of the dead Soldier from head to toe removing all papers and anything new (different type rank, shoulder boards, different unit patch, pistol, weapon, or NVD). They note if the enemy has a fresh or shabby haircut and the condition of his uniform and boots. They note the radio frequency, and then they secure the SOI, maps, documents, and overlays.
  -- Once the body has been thoroughly searched, the search team will continue in this manner until all enemy personnel in and near the kill zone have been searched.
- Identify, collect, and prepare all equipment to be carried back or destroyed.
- Evacuate and treat friendly wounded first, then enemy wounded, time permitting.

- The demolition team prepares dual-primed explosives or incendiary grenades and awaits the signal to initiate. This is normally the last action performed before the unit departs the objective and may signal the security elements to return to the ORP.
- Actions on the objective with stationary assault line; all actions are the same with the exception of the search teams. To provide security within the teams to the far side of the kill zone during the search, they work in three-Ranger teams. Before the search begins, the Rangers move all KIAs to the near side of the kill zone.

(p) If enemy reinforcements try to penetrate the kill zone, the flank security will engage to prevent the assault element from being compromised.

(q) The platoon leader directs the unit's withdrawal from the ambush site:

- Elements normally withdraw in the reverse order that they established their positions.
- The elements may return to the RP or directly to the ORP, depending on the distance between elements.
- The security element of the ORP must be alert to assist the platoon's return to the ORP. It maintains security for the ORP while the rest of the platoon prepares to leave.
- If possible, all elements should return to the location at which they separated from the main body. This location should usually be the RP.

(r) The PL and PSG direct actions at the ORP, to include accountability of personnel and equipment and recovery of rucksacks and other equipment left at the ORP during the ambush.

(s) The platoon leader disseminates information, or moves the platoon to a safe location (no less than one kilometer or one terrain feature away from the objective) and disseminates information.

(t) As required, the PL and FO execute indirect fires to cover the platoon's withdrawal.

**Figure 5-5. ACTIONS ON THE OBJECTIVE -- DELIBERATE AMBUSH**

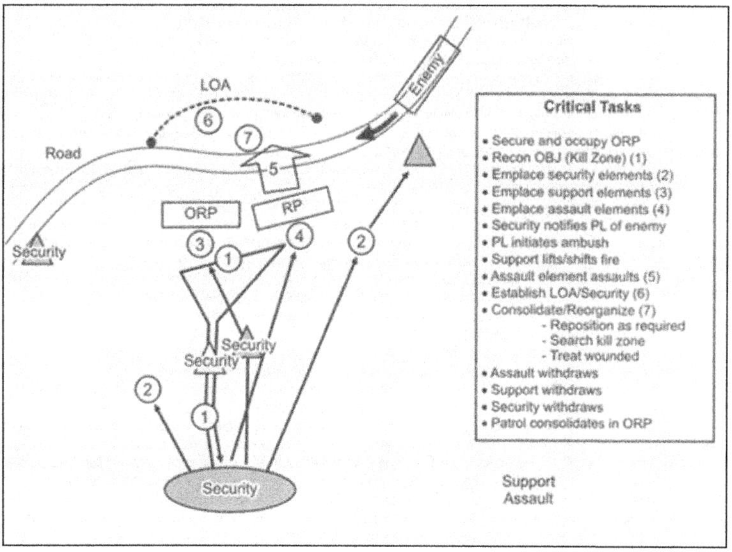

e. **Perform Raid**. The patrol initiates the raid NLT the time specified in the order, surprises the enemy, assaults the objective, and accomplishes its assigned mission within the commander's intent. The patrol does not become decisively engaged en route to the objective. The patrol obtains all available PIR from the raid objective and continues follow-on operations.

(1) ***Planning Considerations***. A Raid is a form of attack, usually small scale, involving a swift entry into hostile territory to secure information, confuse the enemy, or destroy installations followed by a planned withdrawal. Squads do not conduct raids. The sequence of platoon actions for a raid is similar to those for an ambush. Additionally, the assault element of the platoon may have to conduct a breach of an obstacle. It may have additional tasks to perform on the objective such as demolition of fixed facilities. Fundamentals of the raid include--

- Surprise and speed. Infiltrate and surprise the enemy without being detected.
- Coordinated fires. Seal off the objective with well-synchronized direct and indirect fires.
- Violence of action. Overwhelm the enemy with fire and maneuver.
- Planned withdrawal. Withdraw from the objective in an organized manner, maintaining security.

(2) ***Actions on the Objective (Raid)***. See Figure 5-6.

(a) The patrol moves to and occupies the ORP IAW the patrol SOP. The patrol prepares for the leader's recon.

(b) The PL, squad leaders, and selected personnel conduct a leader's recon.

- PL leaves a five point contingency plan with the PSG.
- PL establishes the RP, pinpoints the objective, contacts the PSG to prep men, weapons, and equipment, emplaces the surveillance team to observe the objective, and verifies and updates intelligence information.
- Leader's recon verifies location of and routes to security, support, and assault positions.
- Security teams are brought forward on the leader's reconnaissance and emplaced before the leader's recon leaves the RP
- Leaders conduct the recon without compromising the patrol.
- Leaders normally recon support by fire position first, then the assault position.

(c) The PL confirms, denies, or modifies his plan and issues instructions to his squad leaders.

- Assigns positions and withdrawal routes to all elements.
- Designates control measures on the objective (element objectives, lanes, limits of advance, target reference points, and assault line).
- Allows SLs time to disseminate information, and confirm that their elements are ready.

(d) Security elements occupy designated positions, moving undetected into positions that provide early warning and can seal off the objective from outside support or reinforcement.

(e) The support element leader moves the support element to designated positions. The support element leader ensures his element can place well-aimed fire on the objective.

(f) The PL moves with the assault element into the assault position. The assault position is normally the last covered and concealed position before reaching the objective. As it passes through the assault position the platoon deploys into its assault formation; that is, its squads and fire teams deploy to place the bulk of their firepower to the front as they assault the objective.

- Makes contact with the surveillance team to confirm any enemy activity on the objective.
- Ensures that the assault position is close enough for immediate assault if the assault element is detected early.
- Moves into position undetected, and establish local security and fire control measures.

(g) Element leaders inform the PL when their elements are in position and ready.

(h) The PL directs the support element to fire.

(i) Upon gaining fire superiority, the PL directs the assault element to move towards the objective.

- Assault element holds fire until engaged, or until ready to penetrate the objective.
- PL signals the support element to lift or shift fires. The support element lifts or shifts fires as directed, shifting fire to the flanks of targets or areas as directed in the FRAGO.

(j) The assault element attacks and secures the objective. The assault element may be required to breech a wire obstacle. As the platoon, or its assault element, moves onto the objective, it must increase the volume and accuracy of fires. Squad leaders assign specific targets or objectives for their fire teams. Only when these direct fires keep the enemy suppressed can the rest of the unit maneuver. As the assault element gets closer to the enemy, there is more emphasis on suppression and less on maneuver. Ultimately, all but one fire team may be suppressing to allow that one fire team to break into the enemy position. Throughout the assault, Soldiers use proper individual movement techniques, and fire teams retain their basic shallow wedge formation. The platoon does not get "*on-line*" to sweep across the objective.

- Assault element assaults through the objective to the designated LOA.
- Assault element leaders establish local security along the LOA, and consolidate and reorganize as necessary. They provide ACE reports to the PL and PSG. The platoon establishes security, operates key weapons, provides first aid, and prepares wounded Soldiers for MEDEVAC. They redistribute ammunition and supplies, and they relocate selected weapons to alternate positions if leaders believe that the enemy may have pinpointed them during the attack. They adjust other positions for mutual support. The squad and team leader provide ammunition, casualty, and equipment (ACE) reports to the platoon leader. The PL/PSG reorganizes the patrol based on the contact.
  -- On order, special teams accomplish all assigned tasks under the supervision of the PL, who positions himself where he can control the patrol.
  -- Special team leaders report to PL when assigned tasks are complete.

(k) On order or signal of the PL, the assault element withdraws from the objective. Using prearranged signals, the assault line begins an organized withdrawal from the objective site, maintaining control and security throughout the withdrawal. The assault element bounds back near the original assault line, and begins a single file withdrawal through the APL's choke point. All Rangers must move through the choke point for an accurate count. Once the assault element is a safe distance from the objective and the headcount is confirmed, the platoon can withdraw the support element. If the support elements were a part of the assault line, they withdraw together, and security is signaled to withdraw. Once the support is a safe distance off the objective, they notify the platoon leader, who contacts the security element and signals them to withdraw. All security teams link up at the release point and notify the platoon leader before moving to the ORP. Personnel returning to the ORP immediately secure their equipment and establish all-round security. Once the security element returns, the platoon moves out of the objective area as soon as possible, normally in two to three minutes.

- Before withdrawing, the demo team activates demo devices and charges.
- Support element or designated personnel in the assault element maintain local security during the withdrawal.
- Leaders report updated accountability and status (ACE report) to the PL and PSG.

(l) Squads withdraw from the objective in the order designated in the FRAGO to the ORP.

- Account for personnel and equipment.
- Disseminate information.
- Redistribute ammunition and equipment as required.

(m) The PL reports mission accomplishment to higher and continues the mission.

- Reports raid assessment to higher.
- Informs higher of any IR/PIR gathered.

NOTES

Figure 5-6. ACTIONS ON THE OBJECTIVE -- RAID

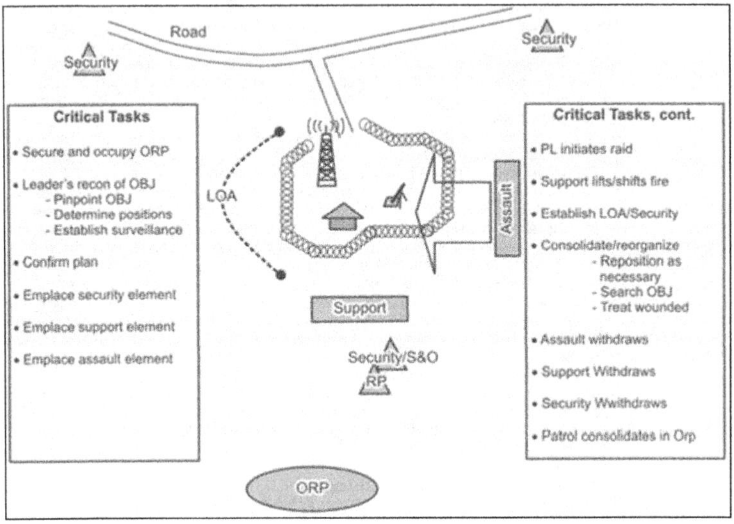


5-5. **SUPPORTING TASKS**. This section covers Linkup, Patrol Debriefing, and Occupation of an ORP.

a. **Linkup**. A linkup is a meeting of friendly ground forces. Linkups depend on control, detailed planning, communications, and stealth.

(1) ***Task Standard***. The units link up at the time and place specified in the order. The enemy does not surprise the main bodies. The linkup units establish a consolidated chain of command.

(2) ***Linkup Site Selection***. The leader identifies a tentative linkup site by map reconnaissance, other imagery, or higher headquarters designates a linkup site. The linkup site should have the following characteristics:

- Ease of recognition.
- Cover and concealment.
- No tactical value to the enemy.
- Location away from natural lines of drift.
- Defendable for a short period of time.
- N=Multiple access and escape routes.

(3) ***Execution***. Linkup procedure begins as the unit moves to the linkup point. The steps of this procedure are:

(a) The stationary unit performs linkup actions.

- Occupies the linkup rally point (LRP) NLT the time specified in the order.
- Establishes all-around security, establishes commo, and prepares to accept the moving unit.
- The security team clears the immediate area around the linkup point. It then marks the linkup point with the coordinated recognition signal. The security team moves to a covered and concealed position and observes the linkup point and immediate area around it.

(b) The moving unit--

- Performs linkup actions.
- The unit reports its location using phase lines, checkpoints, or other control measures.
- Halts at a safe distance from the linkup point in a covered and concealed position (the linkup rally point).

(c) The PL and a contact team--

- Prepare to make physical contact with the stationary unit.
- Issue a contingency plan to the PSG.
- Maintain commo with the platoon; verify near and far recognition signals for linkup (good visibility and limited visibility).
- Exchange far and near recognition signals with the linkup unit; conduct final coordination with the linkup unit.

(d) The stationary unit--

- Guides the patrol from its linkup rally point to the stationary unit linkup rally point.
- Linkup is complete by the time specified in the order.
- The main body of the stationary unit is alerted before the moving unit is brought forward.

(e) The patrol continues its mission IAW the order.

(4) ***Coordination Checklist***. The PL coordinates or obtains the following information from the unit that his patrol will link up with:

- Exchange frequencies, call signs, codes, and other communication information.
- Verify near and far recognition signals.
- Exchange fire coordination measures.
- Determine command relationship with the linkup unit; plan for consolidation of chain of command.
- Plan actions following linkup.
- Exchange control measures such as contact points, phase lines, contact points, as appropriate.

b. **Debrief**. Immediately after the platoon or squad returns, personnel from higher headquarters conduct a thorough debrief. This may include all members of the platoon or the leaders, RTOs, and any attached personnel. Normally the debriefing is oral. Sometimes a written report is required. Information on the written report should include--

- Size and composition of the unit conducting the patrol.
- Mission of the platoon such as type of patrol, location, and purpose.
- Departure and return times.
- Routes. Use checkpoints, grid coordinates for each leg or include an overlay.
- Detailed description of terrain and enemy positions that were identified.
- Results of any contact with the enemy.
- Unit status at the conclusion of the patrol mission, including the disposition of dead or wounded Soldiers.
- Conclusions or recommendations.

c. **Objective Rally Point**. The ORP is a point out of sight, sound, and small arms range of the objective area. It is normally located in the direction that the platoon plans to move after completion of actions on the objective. The ORP is tentative until the objective is pinpointed.

(1) ***Occupation of the ORP***. See Figure 5-7.

(a) The patrol halts beyond sight and sound of the tentative ORP (200-400m in good visibility, 100-200m in limited visibility).

(b) The patrol establishes a security halt IAW the unit SOP.

(c) After issuing a five-point contingency plan to the PSG, the PL moves forward with a recon element to conduct a leader's recon of the ORP.

(d) For a squad-sized patrol, the PL moves forward with a compass man and one member of each fire team to confirm the ORP.

- After physically clearing the ORP location, the PL leaves two Rangers at the 6 o'clock position facing in opposite directions.
- The PL issues a contingency plan and returns with the compass man to guide the patrol forward.
- The PL guides the patrol forward into the ORP, with one team occupying from 3 o'clock through 12 o'clock to 9 o'clock, and the other occupying from 9 o'clock through 6 o'clock to 3 o'clock.

(e) For a platoon-size patrol, the PL, RTO, WSL, three ammo bearers, a team leader, a SAW gunner, and riflemen go on the leaders recon for the ORP and position themselves at 10, 2, and 6 o'clock.

- The first squad in the order of march is the base squad, occupying from 10 to 2 o'clock.
- The trail squads occupy from 2 to 6 o'clock and 6 to 10 o'clock, respectively.
- The patrol headquarters element occupies the center of the triangle.

(2) ***Actions in the ORP***. The unit prepares for the mission in the ORP. Once the leader's recon pinpoints the objective, the PSG generally lines up rucks IAW unit SOP in the center of the ORP.

**Figure 5-7. OCCUPATION OF THE ORP**

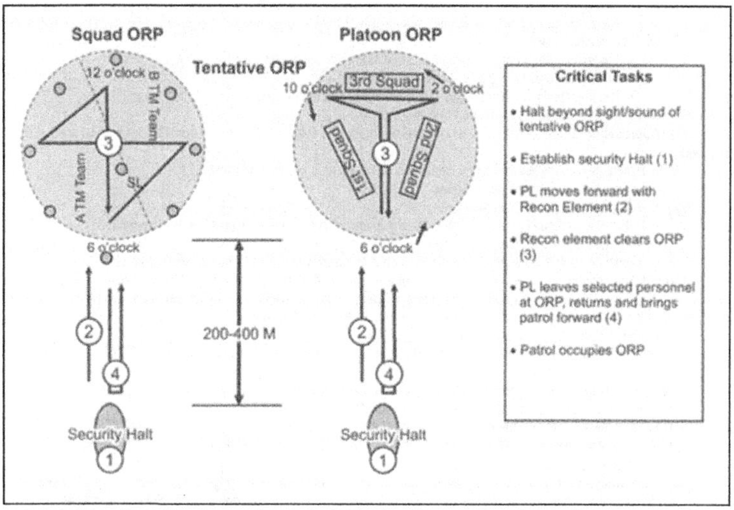


d. **Patrol Base**. A patrol base is a security perimeter that is set up when a squad or platoon conducting a patrol halts for an extended period. Patrol bases should not be occupied for more than a 24-hour period (except in emergency). A patrol never uses the same patrol base twice.

(1) ***Use***. Patrol bases are typically used--

- To avoid detection by eliminating movement.
- To hide a unit during a long detailed reconnaissance.
- To perform maintenance on weapons, equipment, eat and rest.
- To plan and issue orders.
- To reorganize after infiltrating on an enemy area.
- To establish a base from which to execute several consecutive or concurrent operations.

(2) ***Site Selection***. The leader selects the tentative site from a map or by aerial reconnaissance. The site's suitability must be confirmed and secured before the unit moves into it. Plans to establish a patrol base must include selecting an alternate patrol base site. The alternate site is used if the first site is unsuitable or if the patrol must unexpectedly evacuate the first patrol base.

(3) ***Planning Considerations***. Leaders planning for a patrol base must consider the mission and passive and active security measures. A patrol base (PB) must be located so it allows the unit to accomplish its mission.

- Observation posts and communication with observation posts.
- Patrol or platoon fire plan.
- Alert plan.
- Withdrawal plan from the patrol base to include withdrawal routes and a rally point, rendezvous point, or alternate patrol base.
- A security system to make sure that specific Soldiers are awake at all times.
- Enforcement of camouflage, noise, and light discipline.
- The conduct of required activities with minimum movement and noise.
- Priorities of Work.

(4) ***Security Measures***.

- Select terrain the enemy would probably consider of little tactical value.
- Select terrain that is off main lines of drift.
- Select difficult terrain that would impede foot movement, such as an area of dense vegetation, preferably bushes and trees that spread close to the ground.
- Select terrain near a source of water.
- Select terrain that can be defended for a short period and that offers good cover and concealment.
- Avoid known or suspected enemy positions.
- Avoid built-up areas.
- Avoid ridges and hilltops, except as needed for maintaining communications.
- Avoid small valleys.
- Avoid roads and trails.

(5) ***Occupation***. See Figure 5-8.

(a) A PB is reconned and occupied in the same manner as an ORP, with the exception that the platoon will typically plan to enter at a 90 degree turn. The PL leaves a two-man OP at the turn, and the patrol covers any tracks from the turn to the PB.

(b) The platoon moves into the PB. Squad-sized patrols will generally occupy a cigar-shaped perimeter; platoon-sized patrols will generally occupy a triangle-shaped perimeter.

(c) The PL and another designated leader inspect and adjust the entire perimeter as necessary.

(d) After the PL has checked each squad sector, each SL sends a two-man R&S team to the PL at the CP. The PL issues the three R&S teams a contingency plan, reconnaissance method, and detailed guidance on what to look for (enemy, water, built-up areas or human habitat, roads, trails, or possible rally points).

(e) Where each R&S team departs is based on the PLs guidance. The R&S team moves a prescribed distance and direction, and reenters where the PL dictates.

- Squad-sized patrols do not normally send out an R&S team at night.
- R&S teams will prepare a sketch of the area to the squad front if possible.
- The patrol remains at 100 % alert during this recon.
- If the PL feels the patrol was tracked or followed, he may elect to wait in silence at 100 % alert before sending out R&S teams.
- The R&S teams may use methods such as the "*I*," the "*Box*," or the "*T*." Regardless of the method chosen, the R&S team must be able to provide the PL with the same information.
- Upon completion of R&S, the PL confirms or denies the patrol base location, and either moves the patrol or begins priorities of work.

(6) ***Passive (Clandestine) Patrol Base (Squad)***.

- The purpose of a passive patrol base is for the rest of a squad or smaller size element.
- Unit moves as a whole and occupies in force.
- Squad leader ensures that the unit moves in at a 90-degree angle to the order of movement.
- A claymore mine is emplaced on route entering patrol base.
- Alpha and Bravo teams sit back-to-back facing outward, ensuring that at least one individual per team is alert and providing security.

(7) ***Priorities of Work (Platoon and Squad)***. Once the PL is briefed by the R&S teams and determines the area is suitable for a patrol base, the leader establishes or modifies defensive work priorities in order to establish the defense for the patrol base. Priorities of work are not a laundry list of tasks to be completed; to be effective, priorities of work must consist of a task, a given time, and a measurable performance standard. For each priority of work, a clear standard must be issued to guide the element in the successful accomplishment of each task. It must also be designated whether the work will be controlled in a centralized or decentralized manner. Priorities of work are determined IAW METT-TC. Priorities of Work may include, but are not limited to the following tasks:

(a) *Security (Continuous)*.

- Prepare to use all passive and active measures to cover all of the perimeter all of the time, regardless of the percentage of weapons used to cover that all of the terrain.
- Readjust after R&S teams return, or based on current priority of work (such as weapons maintenance).
- Employ all elements, weapons, and personnel to meet conditions of the terrain, enemy, or situation.
- Assign sectors of fire to all personnel and weapons. Develop squad sector sketches and platoon fire plan.
- Confirm location of fighting positions for cover, concealment, and observation and fields of fire. SLs supervise placement of aiming stakes and claymores.
- Only use one point of entry and exit, and count personnel in and out. Everyone is challenged IAW the unit SOP.
- Hasty fighting positions are prepared at least 18 inches deep (at the front), and sloping gently from front to rear, with a grenade sump if possible.

(b) *Withdrawal Plan*. The PL designates the signal for withdrawal, order of withdrawal, and the platoon rendezvous point and/or alternate patrol base.

(c) *Communication (Continuous)*. Commo must be maintained with higher headquarters, OP's, and within the unit. May be rotated between the patrol's RTOs to allow accomplishment of continuous radio monitoring, radio maintenance, act as runners for PL, or conduct other priorities of work.

(d) *Mission Preparation and Planning*. The PL uses the patrol base to plan, issue orders, rehearse, inspect, and prepare for future missions.

(e) *Weapons and Equipment Maintenance*. The PL ensures that machine guns, weapon systems, commo equipment, and night vision devices (as well as other equipment) are maintained. These items are not disassembled at the same time for maintenance (no more than 33 percent at a time), and weapons are not disassembled at night. If one machine gun is down, then security for all remaining systems is raised.

(f) *Water Resupply*. The PSG organizes watering parties as necessary. The watering party carries canteens in an empty rucksack or duffel bag, and must have commo and a contingency plan prior to departure.

(g) *Mess Plan*. At a minimum, security and weapons maintenance are performed prior to mess. Normally no more than half the platoon eats at one time. Rangers typically eat 1 to 3 meters behind their fighting positions.

- Rest/sleep plan management. The patrol conducts rest as necessary to prepare for future operations.
- Alert Plan and Stand-to. The PL states the alert posture and the stand-to time. He develops the plan to ensure all positions are checked periodically, OP's are relieved periodically, and at least one leader is always alert. The patrol typically conducts stand-to at a time specified by unit SOP (such as 30 minutes before and after BMNT or EENT).
- Resupply. Distribute or cross-load ammunition, meals, equipment, etc.
- Sanitation and personal hygiene. The PSG and medic ensure a slit trench is prepared and marked. All Soldiers will brush teeth, wash face, shave, wash hands, armpits, groin, feet, and darken (brush shine) boots daily. The patrol will not leave trash behind.

**Figure 5-8. PATROL BASE**

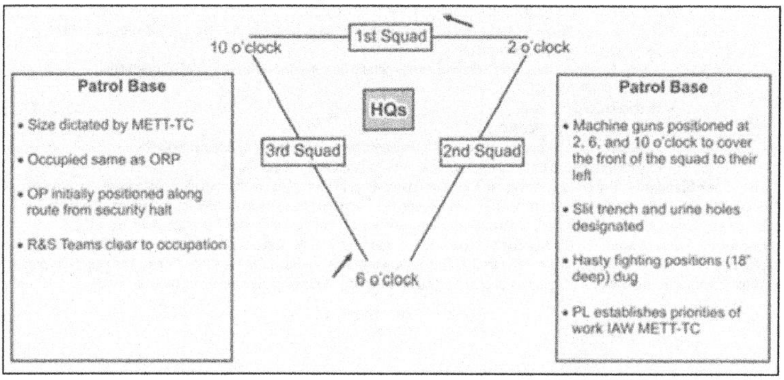


5-6. **MOVEMENT TO CONTACT**. The MTC is one of the five types of offensive operations. An MTC gains or regains contact with the enemy. Once contact is made, the unit develops the situation. Normally a platoon conducts an MTC as part of a larger force.

a. **Techniques**. The two techniques of conducting a movement to contact are search and attack and approach march.

(1) ***Search and Attack (S&A)***. This technique is used when the enemy is dispersed, is expected to avoid contact, disengage or withdraw, or you have to deny his movement in an area. The search and attack technique involves the use of multiple platoons, squads, and fire teams coordinating their actions to make contact with the enemy. Platoons typically try to find the enemy and then fix and finish him. They combine patrolling techniques with the requirement to conduct hasty or deliberate attacks once the enemy has been found.

(a) *Planning Considerations*.

- The factors of METT-TC.
- The requirement for decentralized execution.
- The requirement for mutual support.
- The length of operations.
- Minimize Soldier's load to improve stealth and speed.
- Resupply and MEDEVAC.
- Positioning key leaders and equipment.
- Employment of key weapons.
- Requirement for patrol bases.
- Concept for entering the zone of action.
- The concept for linkups while in contact.

(b) *Critical Performance Measures*.

- The platoon locates the enemy without being detected.
- Once engaged, fixes the enemy in position and maneuvers against the enemy.
- Maintains security throughout actions to avoid being flanked.

(2) ***Approach March***. The concept of the approach march is to make contact with the smallest element, allowing the commander the flexibility of destroying or bypassing the enemy. A platoon uses the approach march method as part of a larger unit. It can be tasked as the advance guard, move as part of the main body, or provide flank or rear security for the company or battalion. They may also receive on-order missions as part of the main body.

(a) *Fundamentals*. These basics are common to all movements to contact.

- Make enemy contact with smallest element possible.
- Rapidly develop combat power upon enemy contact.

- Provide all-round security for the unit.
- Support higher unit's concept.
- Reports all information rapidly and accurately and strives to gain and maintain contact with the enemy.
- Requires decentralized execution.

(b) *Planning Considerations*. The following issues should be considered heavily for MTC operations:

- Factors of METT-TC.
- Reduced Soldier's load.

(c) *Critical Performance Measures*.

- PL selects the appropriate movement formation based on likelihood of enemy contact.
- Maintains contact, once contact is made, until ordered to do otherwise.

b. **Task Standards**. The platoon moves NLT the time specified in the order, the platoon makes contact with the smallest element possible, and the main body is not surprised by the enemy. Once the platoon makes contact, it maintains contact. The platoon destroys squad and smaller-sized elements, and fixes elements larger than a squad. The platoon maintains sufficient fighting force capable of conducting further combat operations. Reports of enemy locations and contact are forwarded. If not detected by the enemy, the PL initiates a hasty attack. The platoon sustains no casualties from friendly fire. The platoon is prepared to initiate further movement within 25 minutes of contact, and all personnel and equipment are accounted for.

NOTES

# Chapter 6
# BATTLE DRILLS

Infantry battle drills describe how platoons and squads apply fire and maneuver to commonly encountered situations. They require leaders to make decisions rapidly and to issue brief oral orders quickly.

## Section I. INTRODUCTION

This section defines and describes the format for battle drills.

6-1. **DEFINITION.** FM 7-1 defines a battle drill as "*a collective action rapidly executed without applying a deliberate decision-making process.*"

a. Characteristics of a battle drill are—

- They require minimal leader orders to accomplish and are standard throughout the Army.
- Sequential actions are vital to success in combat or critical to preserving life.
- They apply to platoon or smaller units.
- They are trained responses to enemy actions or leader's orders.
- They represent mental steps followed for offensive and defensive actions in training and combat.

b. A unit's ability to accomplish its mission often depends on Soldiers and leaders to execute key actions quickly. All Soldiers and their leaders must know their immediate reaction to enemy contact, as well as follow-up actions. Drills are limited to situations requiring instantaneous response; therefore, Soldiers must execute drills instinctively. This results from continual practice. Drills provide small units with standard procedures essential for building strength and aggressiveness.

(1) They identify key actions that leaders and Soldiers must perform quickly.

(2) They provide for a smooth transition from one activity to another; for example, from movement to offensive action to defensive action.

(3) They provide standardized actions that link Soldier and collective tasks at platoon level and below. (Soldiers perform individual tasks to CTT or SDT standard.)

(4) They require the full understanding of each individual and leader, and continual practice.

6-2. **FORMAT.** Drills include a ***title***, a ***situation*** to cue the unit or leader to start the drill, the ***required actions*** in sequence, and supporting illustrations. Where applicable, drills are cross-referenced with material in other chapters, or other drills, or both. Training standards for battle drills are in the mission training plan (MTP).

## Section II. DRILLS

This section provides the battle drills for small units.

### BATTLE DRILL 1
### REACT TO CONTACT

**SITUATION:** A squad or platoon receives fires from enemy individual or crew-served weapons.

**REQUIRED ACTIONS:** See Figure 6-1.

1. Soldiers immediately assume the nearest covered positions.
2. Soldiers return fire immediately on reaching the covered positions.
3. Squad/team leaders locate and engage known or suspected enemy positions with well-aimed fire, and they pass information to the platoon/squad leader.
4. Fire team leaders control the fire of their Soldiers by using standard fire commands (initial and supplemental) containing the following elements:
   - Alert
   - Direction
   - Description of target
   - Range
   - Method of fire (manipulation and rate of fire)
   - Command to commence firing

5. Soldiers maintain contact (visual or oral) with the Soldiers on their left and right.

6. Soldiers maintain contact with their team leaders and indicate the location of enemy positions.

7. The leaders (visual or oral) check the status of their personnel.

8. The squad/team leaders maintain visual contact with the platoon/squad leader.

9. The platoon/squad leader moves up to the squad/fire team in contact and links up with its leader.
   a. The platoon leader brings his RTO, platoon FO, the nearest squad's squad leader, and one machine gun team.
   b. The squad leader of the trail squad moves to the front of his lead fire team.
   c. The PSG also moves forward with the second machine gun team and links up with the platoon leader, ready to assume control of the base-of-fire element.

10. The platoon/squad leader determines whether or not his unit must move out of an EA.

11. The platoon/squad leader determines whether his unit can gain and maintain suppressive fires with the element already in contact (based on the volume and accuracy of enemy fires against the element in contact). The platoon/squad leader assesses the situation. He identifies—
- The location of the enemy position and obstacles.
- The size of the enemy force engaging the unit in contact. The number of enemy automatic weapons, the presence of any vehicles, and the employment of indirect fires are indicators of the enemy strength.
- Vulnerable flanks.
- Covered and concealed flanking routes to the enemy position.

12. The platoon/squad leader determines the next COA, such as fire and movement, assault, breach, knock out bunker, enter and clear a building or trench.

13. The platoon/squad leader reports the situation to the company commander/platoon leader and begins to maneuver his unit.

14. Platoon leader directs platoon FO to call for and adjust indirect fires (mortars or artillery). Squad leaders relay requests through the platoon leader. The platoon/squad leader in conjunction with the platoon FO maintains accurate battle tracking of all friendly elements to facilitate quick clearance of fires.

15. Leaders relay all commands and signals from the platoon chain of command.

16. The PSG positions the base of fire element to observe and to provide supporting fires.

***Note:*** Once the platoon has executed the React to Contact drill, the platoon leader quickly assesses the situation, for example, enemy size and location. He picks a COA. The platoon leader reports the situation to the company commander.

Figure 6-1. REACT TO CONTACT

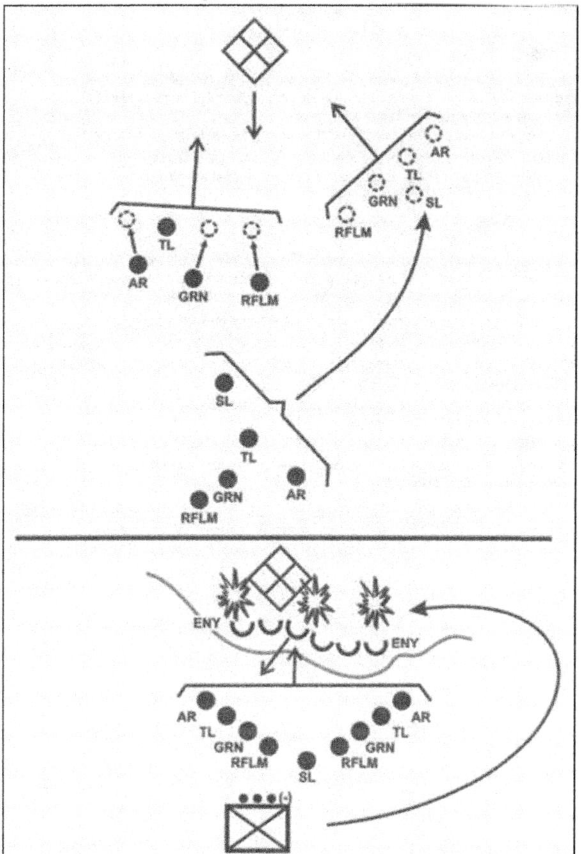

## BATTLE DRILL 2
## BREAK CONTACT

**SITUATION:** The squad or platoon is under enemy fire and must break contact.

**REQUIRED ACTIONS:** See Figure 6-2.

1. The platoon/squad leader directs fire support for the disengagement. He accomplishes this by—
   - Directing one squad/fire team in contact to support the disengagement of the remainder of the unit.
   - Considering the use of indirect fires for breaking contact.
   - Clearing the location, task, purpose, and method of conducting the fire mission with the platoon FO.

2. The platoon/squad leader orders a distance and direction, a terrain feature, or last ORP for the movement of the first squad/fire team. In conjunction with the platoon FO, the leader maintains accurate battle tracking of all friendly elements to facilitate quick clearance of fires.

3. The base of fire squad/team continues to suppress the enemy. The platoon/squad leader directs the platoon FO to execute the fire mission, if needed.

4. The moving squad/team assumes the overwatch position. The squad/team should use M203 grenade launchers, throw fragmentation and concussion grenades, and use smoke grenades to mask movement. The platoon/squad leader directs the platoon FO to execute smoke mission to screen friendly elements movement, if needed.

5. The moving squad/team takes up the designated position and engages the enemy position.

6. The platoon leader directs the base-of-fire element to move to its next location. Based on the terrain and the volume and accuracy of the enemy's fire, the moving fire team or squad may need to use fire and movement techniques.

7. The platoon/squad continues to bound away from the enemy until—
   - It breaks contact. It must continue to suppress the enemy as it does this.
   - It passes through a higher-level SBF position.
   - Its squad/fire teams are in the assigned position to conduct the next mission.

8. The leader should consider changing his unit's direction of movement once contact is broken. This will reduce the ability of the enemy to place effective indirect fires on the unit.

9. If the platoon/squad becomes disrupted, Soldiers stay together and move to the last designated rally point.

10. The platoon/squad leaders account for Soldiers, report, reorganize as necessary, and continue the mission. With the platoon FO, the platoon/squad develops a quick fire plan to support the hasty defense or new route of march.

Figure 6-2. BREAK CONTACT

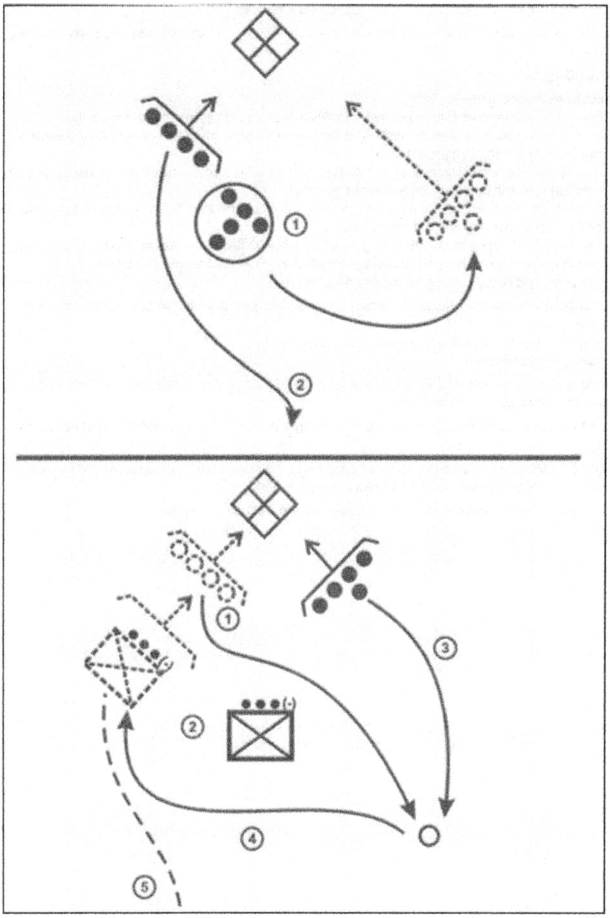

## BATTLE DRILL 3
## REACT TO AMBUSH

**SITUATION:** The squad or platoon enters a kill zone and the enemy initiates an ambush with a casualty-producing device and a high volume of fire.

**REQUIRED ACTIONS:** See Figure 6-3.

1. Near ambush (within hand-grenade range).
   a. Soldiers in the kill zone return fire immediately. How they go about this depends on the terrain. If—
      - Cover is not available, Soldiers immediately, without order or signal, assume prone position and throw concussion or fragmentation and smoke grenades.
      - Cover is available, without order or signal, Soldiers seek the nearest covered position, assume the prone position, and throw fragmentation or concussion and smoke grenades.

   b. Immediately after the explosion of the concussion or fragmentation grenades, Soldiers in the kill zone return fire and assault through the ambush position using fire and movement.

   c. Soldiers not in the kill zone identify the enemy's position. Fire is shifted as the personnel in the kill zone begin to assault.

   d. Soldiers in the kill zone continue the assault to eliminate the ambush or until contact is broken.

   e. The platoon/squad conducts consolidation and reorganization.

2. Far ambush (out of hand-grenade range). Soldiers receiving fire immediately return fire, take up covered positions, and suppress the enemy by—
   - Destroying or suppressing enemy crew-served weapons.
   - Sustaining suppressive fires.

3. Soldiers (squad/teams) not receiving fires move by a covered and concealed route to a vulnerable flank of the enemy position and assault using fire and movement techniques.

4. Soldiers in the kill zone continue suppressive fires and shift fires as the assaulting squad/team fights through the enemy position.

5. The platoon FO calls for and adjusts indirect fires as directed by the platoon leader. On order, he shifts or ceases fires to isolate the enemy position or to attack them with indirect fires as they retreat.

6. The platoon/squad leader reports, reorganizes as necessary, and continues the mission.

Figure 6-3. REACT TO AMBUSH

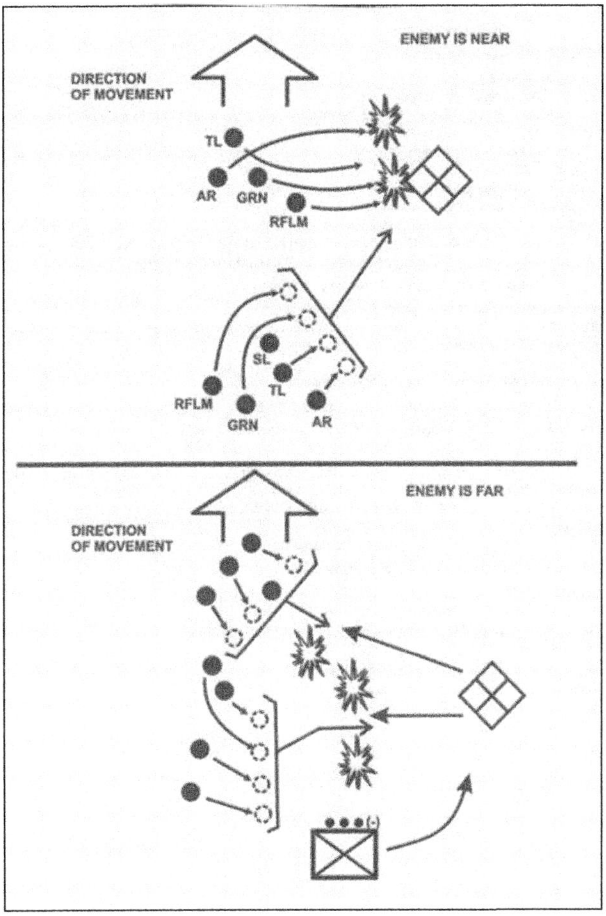

## BATTLE DRILL 4
## KNOCK OUT BUNKERS

**SITUATION:** While moving as part of a larger force, the platoon identifies the enemy in bunkers.

**REQUIRED ACTIONS:** See Figures 6-4 and 6-5.

1. The platoon initiates contact.
   a. The squad in contact establishes a base of fire.
   b. The platoon leader, his RTO, platoon FO, and one machine gun team move forward to link up with the squad leader of the squad in contact.
   c. The PSG moves forward with the second machine gun team and assumes control of the base-of-fire squad.
   d. The base-of-fire squad—
      - Destroys or suppresses enemy crew-served weapons first.
      - Obscures the enemy position with smoke (M203).
      - Sustains suppressive fires at the lowest possible level.

   e. The platoon leader directs platoon FO to call for and adjust indirect fires. The platoon leader in conjunction with the platoon FO maintains accurate battle tracking of all friendly elements to facilitate quick clearance of fires.
2. The platoon leader determines that he can maneuver by identifying—
   - Enemy bunkers, other supporting positions, and any obstacles.
   - Size of the enemy force engaging the platoon. The number of enemy automatic weapons, the presence of any vehicles, and the employment of indirect fires are indicators of enemy strength.
   - A vulnerable flank of at least one bunker.
   - A covered and concealed flanking route to the flank of the bunker.
3. The platoon leader determines which bunker is to be assaulted first and directs one squad (not in contact) to knock it out.
4. If necessary, the PSG repositions a squad, fire team, or machine gun team to isolate the bunker as well as to continue suppressive fires.
5. The assaulting squad, with the platoon leader and his RTO, move along the covered and concealed route and take action to knock out the bunker. The following occurs.
   a. On the platoon leader's signal, the support squad ceases or shifts fires to the opposite side of the bunker from which the squad is assaulting.
   b. At the same time, the platoon FO shifts indirect fires to isolate enemy positions.
6. The assaulting squad leader reports to the platoon leader and reorganizes his squad.
7. The platoon leader—
   - Directs the supporting squad to move up and knock out the next bunker.
   - Directs the assaulting squad to continue and knock out the next bunker.
   - Rotates squads as necessary.
8. The platoon leader reports, reorganizes as necessary, and continues the mission. The company follows up the success of the platoon attack and continues to assault enemy positions.

Figure 6-4. KNOCK OUT A BUNKER (SQUAD)

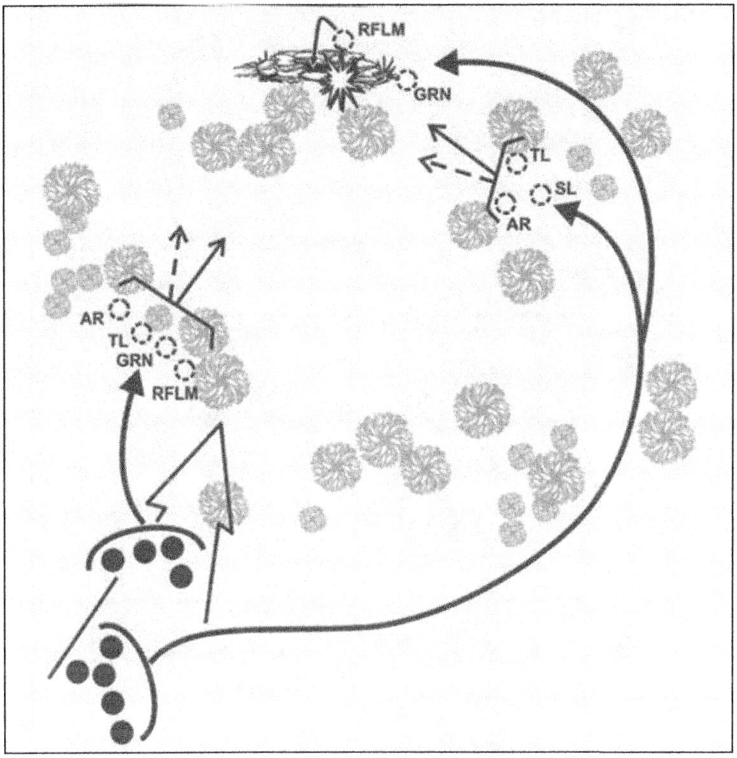

Figure 6-5. KNOCK OUT BUNKERS (PLATOON)

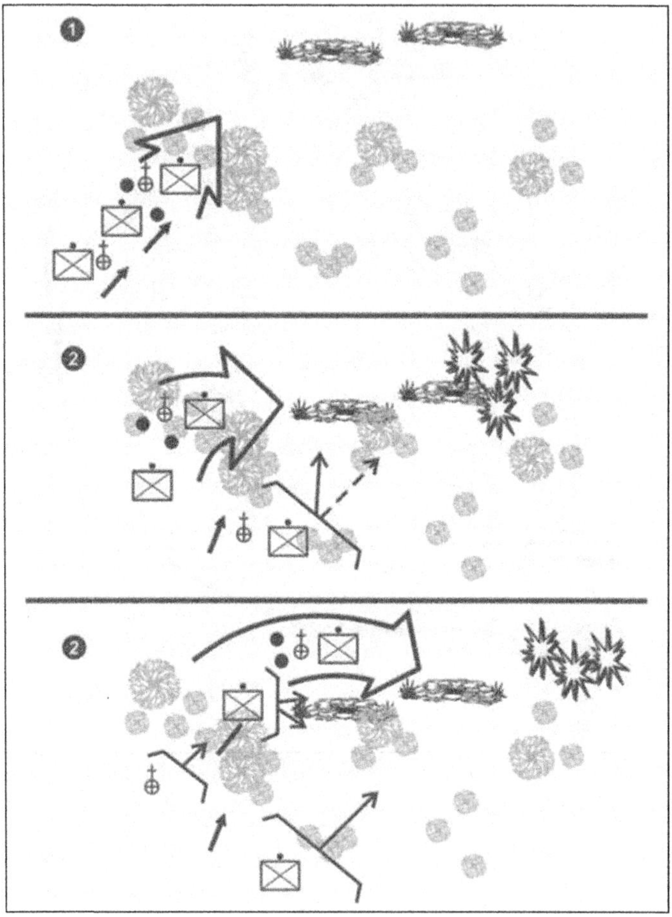

## BATTLE DRILL 5
## ENTER BUILDING/CLEAR ROOM

**SITUATION:** Operating as part of a larger force (in daylight or darkness), the squad is tasked to participate in clearing a building. The platoon leader directs the squad to enter the building or to clear a room. An entry point or breach has already been identified or will be created before initiating the entry. For a detailed discussion of urban entry breaching techniques, see FM 3.06-11. Enemy forces and noncombatants may or may not be present in the room and or building to be cleared.

**REQUIRED ACTIONS:** See Figures 6-6 through 6-14.

1. Platoon and squad leaders must consider the task and purpose they have been given and the method they are to use to achieve the desired results. They must operate IAW the ROE and must be aware of the affects that platoon weapons will have on the type and composition of the buildings.

2. To seize or gain control of a building may not always require committing troops into the structure or closing with the enemy. Before initiating this action and exposing members of the clearing squad to direct enemy contact and risking casualties, the platoon leader should consider/direct employment of all organic, crew-served, and supporting weapon systems onto the objective area to suppress and neutralize the threat, providing the mission, purpose, building composition, and ROE permit.

3. When conducting urban operations, Soldiers must be equipped at all times with a properly-mounted and immediately-useable NVD or light source to illuminate the immediate area.

***Note:*** The following discussion assumes that only the platoon's organic weapons are supporting the infantry squad. Urban situations may require precise application of firepower. This situation is especially true of an urban environment in which the enemy is mixed with noncombatants. Noncombatants may be found in the room, which can restrict the use of fires and reduce the combat power available to a squad leader. His squad may have to operate in no-fire areas. ROE can prohibit the use of certain weapons until a specific hostile action takes place. All Soldiers must be aware of the ROE. Leaders must include the specific use of weapons in their planning for precision operations in urban terrain. Leaders should always consider the use of snipers or designated marksmen to apply precise fires to the objective.

4. Clearing team members must approach the entry point quickly, quietly, and in standard order. The squad leader must ensure he is in a position to control the actions of both fire teams. This approach preserves the element of surprise and allows for quick entry and domination of the room. If a breach is required, the order may be slightly modified based on the breach technique (FM 3-06.11). The members of the fire team are assigned numbers one through four. The rifleman is # 1 and the grenadier is # 3. If one member of the clearing team is armed with the SAW rather than an M16 rifle or carbine, he should be designated # 4. The team leader is normally the # 2 man, because he will have the most immediate decision to make as he enters the room.

5. The entire team enters the room as quickly and smoothly as possible and clears the doorway immediately. If possible, the team moves from a covered or concealed position already in their entry order. Ideally, the team arrives and passes through the entry point without having to stop. If the team must stop to effectively "*stack*" outside the entry point, it must do so only momentarily, and it must provide cover.

6. The door is the focal point of anyone in the room. It is known as the "*fatal funnel*," because it focuses attention at the precise point where the individual team members are the most vulnerable. Moving into the room quickly reduces the chance of anyone being hit by enemy fire directed at the doorway.

7. For this battle to be effectively employed, each member of the team must know his sector of fire and how his sector overlaps and links with the sectors of the other team members. Team members do not move to the points of domination and then engage their targets. Rather, they engage targets as they calmly and quickly move to their designated points. Engagements must not slow movement to their points of domination. Team members may shoot from as short a range as 1 to 2 inches. They engage the most immediate threat first and then the less immediate threats in sector. Immediate threats are personnel who—

- Are armed and prepared to return fire immediately.
- Block movement to the position of domination.
- Are within arm's reach of a clearing team member.
- Are within 3 to 5 feet of the breach point.

8. The squad leader designates the assault team and identifies the location of the entry point for them.

9. The squad leader positions the follow-on assault team to provide overwatch and supporting fires for the initial assault team.

10. Assault team members use available cover and concealment, and move as close to the entry point as possible.

a. If an explosive breach or a ballistic breach is to be performed by a supporting element, the assault team remains in a covered position until the breach is made. They may provide overwatch and fire support for the breaching element if necessary.

b. All team members must signal one another that they are ready before the team moves to the entry point.

c If stealth is a consideration, team members avoid the use of verbal signals, which may alert the enemy and remove the element of surprise.

d. Assault team members must move quickly from the covered position to the entry point, minimizing the time they are exposed to enemy fire.

11. The assault team enters through the entry point or breach. Unless a grenade will be thrown prior to entry, the team should avoid stopping outside the point of entry.

a. If required, the # 2 man throws a grenade of some type (fragmentary, concussion, stun) into the room before entry.

b. The use of grenades should be consistent with the ROE and building structure. The grenade should be cooked off before being thrown, if applicable to the type of grenade used.

c. If stealth is not a factor, the thrower should sound off with a verbal indication ("*Frag out*," "*Concussion out*," "*Stun out*") that a grenade of some type is being thrown. If stealth is a factor, only visual signals are given as the grenade is thrown.

> **CAUTION**
>
> If walls and floors are thin, fragments from fragmentation grenades and debris created by concussion grenades can injure Soldiers outside the room. If the structure has been stressed by previous explosive engagements, the use of these grenades could cause it to collapse. Leaders must determine the effectiveness of these types of grenades compared to possibilities of harm to friendly troops.

12. On the signal to go, or after the grenade detonates, the assault team moves through the entry point (Figure 6-6 through Figure 6-9) and quickly takes up positions inside the room that allow it to completely dominate the room and eliminate the threat. Unless restricted or impeded, team members stop movement only after they have cleared the door and reached their designated point of domination. In addition to dominating the room, all team members are responsible for identifying possible loopholes and mouseholes in the ceiling, walls, and floor.

***Note:*** Where enemy forces may be concentrated and the presence of noncombatants is highly unlikely, the assault team can precede their entry by throwing a fragmentation or concussion grenade (structure dependent) into the room, followed by bursts of automatic small-arms fire by the # 1 man as he enters. Carefully consider the ROE and building composition before employing this method.

13. The # 1 and # 2 men are initially concerned with the area directly to their front, then along the wall on either side of the door or entry point. This area is in their path of movement, and it is their primary sector of fire. Their alternate sector of fire is from the wall they are moving toward, back to the opposite far corner.

14. The # 3 and # 4 men start at the center of the wall opposite their point of entry and clear to the left if moving toward the left, or to the right if moving toward the right. They stop short of their respective team member (either the # 1 man or the # 2 man).

15. The team members move toward their points of domination, engaging all threat or hostile targets in sequence in their sector. Team members must exercise fire control and discriminate between hostile and noncombatant room occupants. (The most practical way to do this is to identify whether or not the target has a weapon in his or her hands.) Shooting is done without stopping, using reflexive shooting techniques. Because the Soldiers are moving and shooting at the same time, they must move using the careful hurry. Figure 6-10, shows all four team members at their points of domination for a room with a center door and their overlapping sectors of fire.

16. The first man (rifleman) enters the room and eliminates the immediate threat. He has the option of going left or right, normally moving along the path of least resistance to one of two corners. When using a doorway as the point of entry, the team uses the path of least resistance, which they determine initially based on the way the door opens. If the door opens inward, the first man plans to move away from the hinges. If the door opens outward, he plans to move toward the hinged side. Upon entering, his direction is influenced by the size of the room, the enemy situation, and furniture or other obstacles that hinder or channel his movement.

17. The direction each man moves in should not be preplanned unless the exact room layout is known. Each man should go in a direction opposite the man in front of him (Figure 6-6). Every team member must know the sectors and duties of each position.

18. As the first man goes through the entry point, he can usually see into the far corner of the room. He eliminates any immediate threat and continues to move along the wall if possible and to the first corner, where he assumes a position of domination facing into the room.

**Figure 6-6. FIRST MAN ENTERS A ROOM, FOLLOWED BY TEAM LEADER**

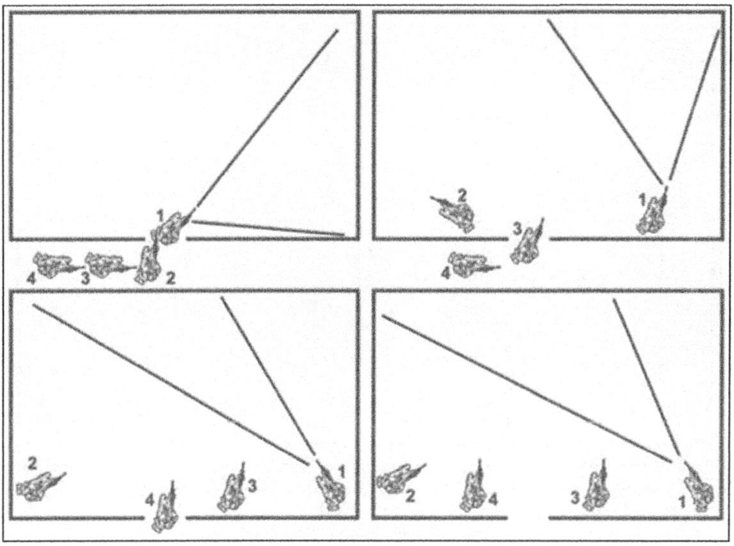


***Note:*** Team members must always stay within 1 meter of the wall. If a team member finds his progress blocked by some object that will force him more than 1 meter from the wall, he must either step over it (if able) or stop where he is and clear the rest of his sector from where he is. If this action creates dead space in the room, the team leader directs which clearing actions to take once other members of the team have reached their points of domination.

19. The second man (normally the team leader), entering almost simultaneously with the first, moves in the opposite direction, following the wall (Figure 6-7). The second man must clear the entry point, clear the immediate threat area, clear his corner, and move to a dominating position on his side of the room. The second man must also immediately determine if he is entering a center door or corner door and act accordingly (Figure 6-7 and Figure 6-10).

**Figure 6-7. SECOND MAN (TEAM LEADER) ENTERS A ROOM**

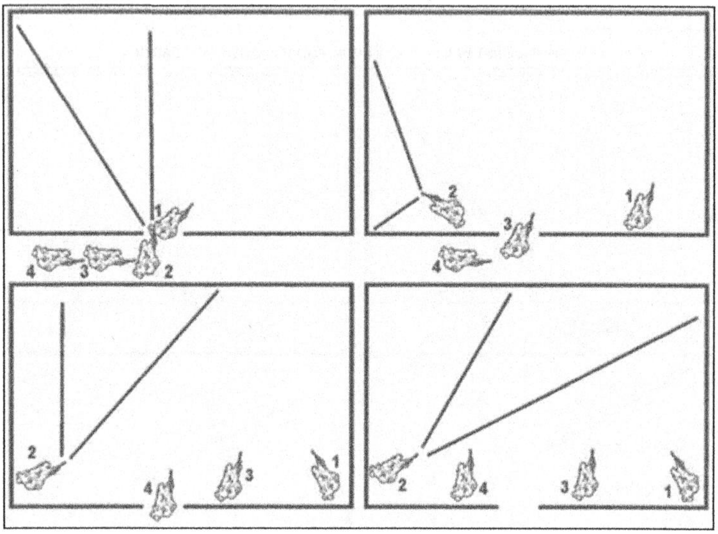


20. The third man (normally the grenadier) simply goes opposite of the second man inside the room, moves at least 1 meter from the entry point, and takes a position that dominates his sector (Figure 6-8).

**Figure 6-8. THIRD MAN ENTERS A ROOM**

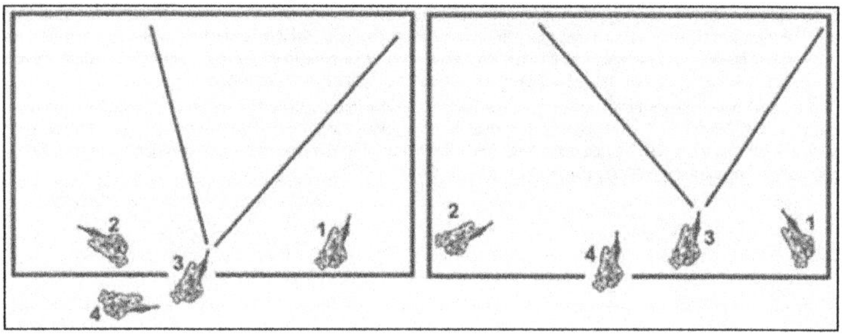

21. The fourth man (normally the SAW gunner) moves opposite of the third man, clears the doorway by at least 1 meter, and moves to a position that dominates his sector (Figure 6-9).

**Figure 6-9. FOURTH MAN IN A ROOM**

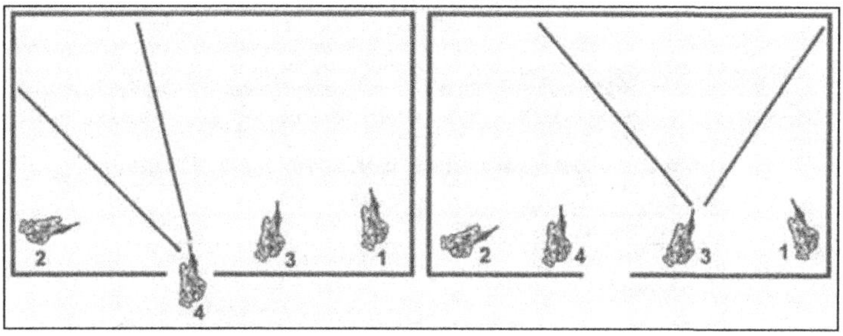


***Note:*** If the path of least resistance takes the first man to the left, then all points of domination are the mirror image of those shown in the diagrams.

22. Points of domination should not be in front of doors or windows so team members are not silhouetted to the outside of the room (Figure 6-10). No movement should mask the fire of any of the other team members.

**Figure 6-10. POINTS OF DOMINATION AND SECTORS OF FIRE, CENTER DOOR VERSUS CORNER DOOR**

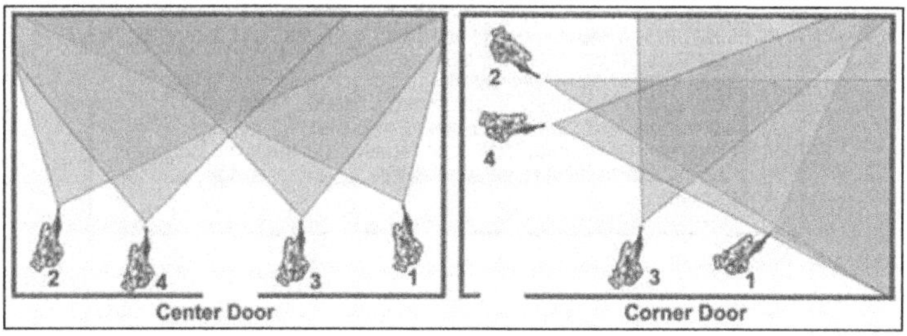


23. On order, any member of the assault team may move deeper into the room overwatched by the other team members. The team leader must control this action.

24. Once the room is cleared, the team leader signals to the squad leader that the room has been cleared. The squad leader marks the room IAW unit SOP (see FM 3.06-11). The squad leader determines whether or not his squad can continue to clear through the building.

25. The squad reorganizes as necessary. Leaders redistribute the ammunition. The squad leader reports to the platoon leader when the room is clear.

***Note:*** If at any point a team member experiences a weapon malfunction in the presence of enemy combatants, he has to make an immediate decision. If the enemy is outside immediate danger distance from the Soldier, then the Soldier should drop to one knee, which indicates that he has experienced a weapon malfunction and prevents fratricide by ensuring that the next Soldier's fields of fire are clear. Once on a knee, the Soldier must remain there until the team leader directs him to stand up. If the kneeling Soldier corrects his weapon malfunction, he can continue to engage targets from his kneeling position. If, however, the enemy Soldier is so close that he presents an immediate threat, and if dropping to a knee would expose the US Soldier to immediate harm, then the US Soldier must try to subdue or disable the enemy Soldier. He can strike the enemy with the muzzle of his weapon or a bayonet to the face, throat, or chest. Or, he could grapple with the enemy soldier and take him to the ground immediately to clear the other team members' fields of fire. Once they have cleared their sectors and eliminated any other threats in the room, the other members of the clearing team assist.

26. Although this battle drill is effective, leaders might have to modify it for the situation. Some reasons and methods for modifying the technique are shown in Table 6-1.

**Table 6-1. REASONS FOR AND METHODS OF MODIFYING ENTRY TECHNIQUES**

| REASON | METHOD |
|---|---|
| Objective rooms are consistently small | Clear in teams of 2 or 3 |
| Shortage of personnel | Clear in teams of 2 or 3 |
| Enemy poses no immediate threat | 2 or 3 men search rooms to ensure no enemy or noncombatants are present |
| No immediate threat, speed is critical | 1 to 3 men visually search each room |

27. When full four-man teams are not available for room clearing, two- and three-man teams can be used. If the # 1 or # 2 man discovers that the room is very small, he can yell, "*Short room*" or "*Short*," which tells the # 3 or # 4 man (whoever following the # 1 or # 2 man) to stay outside the room. Figures 6-11 and 6-12 show the points of domination and sectors of fire for a two-man clearing team. Figures 6-13 and 6-14 show the actions for a three-man team.

**DANGER**

**RICOCHETS POSE A HAZARD. ALL SOLDIERS MUST BE AWARE OF THE TYPE OF WALL CONSTRUCTION OF THE ROOM BEING CLEARED. THE WALLS OF AN ENCLOSED ROOM PRESENT MANY RIGHT ANGLES. COMBINED WITH HARD SURFACES SUCH AS CONCRETE, A BULLET MAY CONTINUE TO RICOCHET AROUND A ROOM UNTIL ITS ENERGY IS SPENT. AFTER HITTING THREAT PERSONNEL, BALL AMMUNITION MAY PASS THROUGH THE BODY AND RICOCHET. BODY ARMOR AND THE KEVLAR HELMET PROVIDE SOME PROTECTION FROM THIS HAZARD.**

Figure 6-11. POINTS OF DOMINATION AND SECTORS OF FIRE (TWO-MAN TEAM, CENTER DOOR)

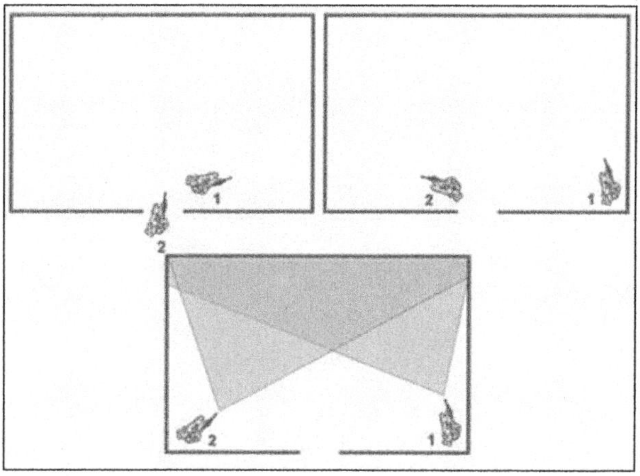

Figure 6-12. POINTS OF DOMINATION AND SECTORS OF FIRE (TWO-MAN TEAM, CORNER DOOR)

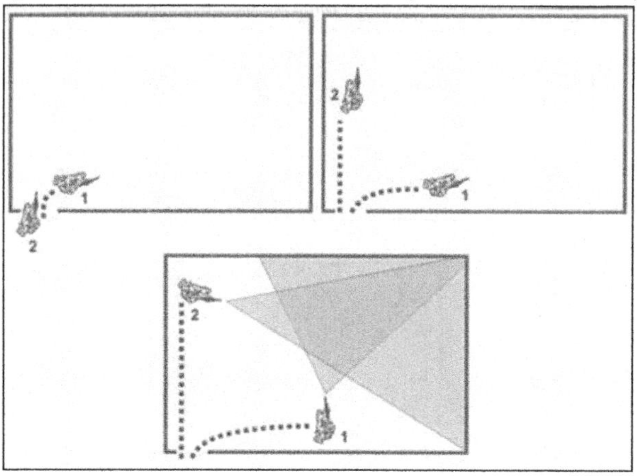

Figure 6-13. POINTS OF DOMINATION AND SECTORS OF FIRE (THREE-MAN TEAM, CENTER DOOR)

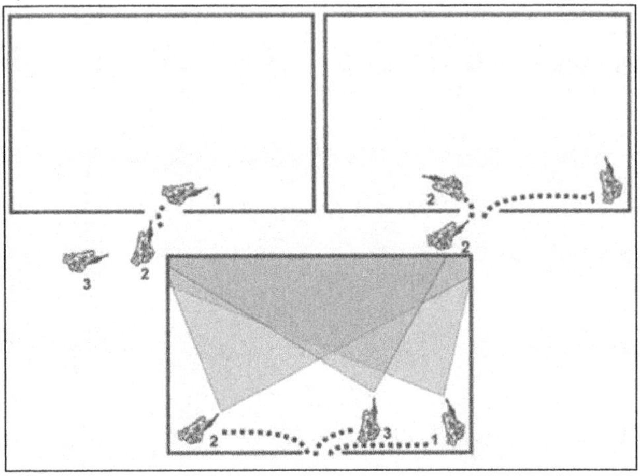

Figure 6-14. POINTS OF DOMINATION AND SECTORS OF FIRE (THREE-MAN TEAM, CORNER DOOR)

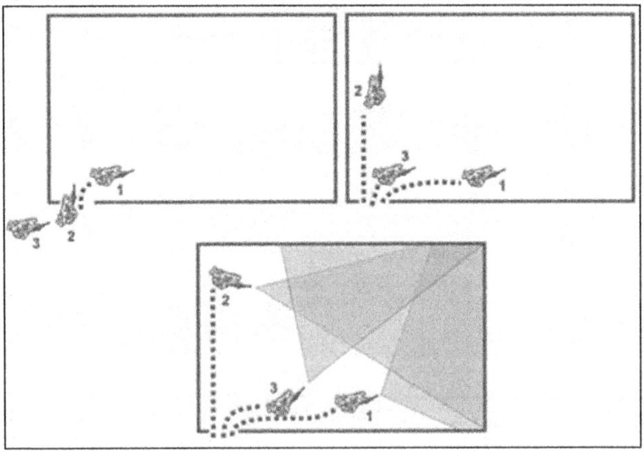

28. While moving through a building Soldiers may encounter the following architectural building features.

a. **Multiple Team/Multiple Rooms**. Figure 6-15 shows this.

**STEP 1:** First team enters and clears Room 1.

**STEP 2:** Squad leader determines the direction from which the second clearing team must enter Room 1, based on the location of the Room 2 entry point.

**STEP 3:** First team collapses inward to allow the second team to move into the room.

**STEP 4:** Second team 'stacks left' and prepares to enter Room 2.

**Figure 6-15. MULTIPLE TEAMS / MULTIPLE ROOMS**

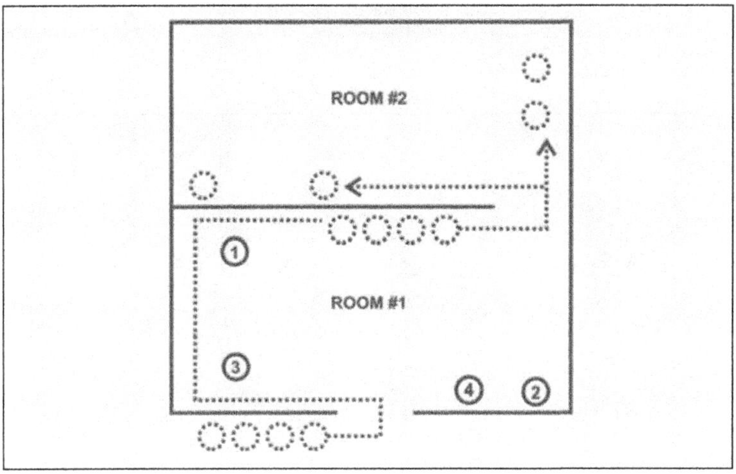

b. **Open Stairwell**. This is a gap between flights of stairs that allows a person to look up and down between the flights. Figure 6-16 shows an open stairwell.

**STEP 1:** The # 1 man pulls security on the highest point he can see / engage.

**STEP 2:** The # 2 man moves up the stairs on the inside with the # 3 man to a point that he can see / engage the next landing, where he turns around and continues to move up to the next landing.

**STEP 3:** The # 3 man moves up the stairs with the # 2 man on the outside and engages the threat on the immediate landing.

**STEP 4:** The # 4 man moves up the stairs with the # 1 man, on the squeeze, the # 2 man turns around to engage the next landing.

**STEP 5:** The flow continues with the # 2 man picking up the sector of the # 1 man had. The # 3 man picks up where the # 2 man was. The # 4 man picks up where the # 3 man was. The # 1 man picks up where the # 4 man was.

***Note:*** Most stairwells require a second team.

**Figure 6-16. OPEN STAIRWELL**

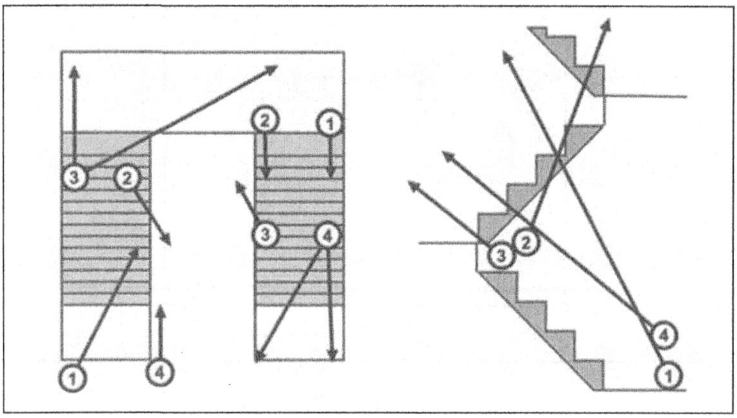

c. **Closed Stairwell**. This is any stairwell with walls separating the flights of stairs. Figure 6-17 shows a closed stairwell.

**STEP 1:** The # 1 man checks high to ensure there is no opening on the landing or between the stairs.

**STEP 2:** The # 2 man pulls long security to the next bend or landing.

**STEP 3:** The # 1 or # 3 man moves up the steps with the # 2 man. As they approach the corner, the # 2 man signals his presence by tapping the # 1 man on the shoulder.

**STEP 4:** Keying off the # 1 man's movement, the other men simultaneously break around the corner.

**STEP 5:** If no fire is received, the # 2 man moves to the opposite wall. Both men continue to move up until they reach their objective.

**STEP 6:** The # 3 and # 4 men will continue to move 3 to 4 steps behind.

***Note:*** Avoid getting locked into a security position such as inside a stairwell. Also, avoid spreading yourselves thin or getting separated by more than one floor of stairs.

**Figure 6-17. CLOSED STAIRWELL**

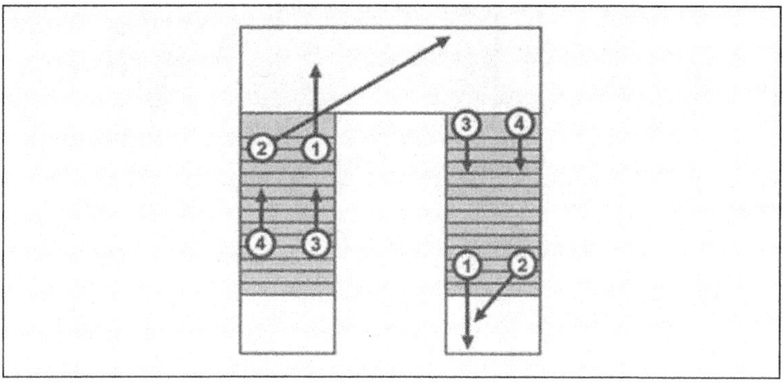

d. **Hallway Movement**. Clearing team(s) move down the hallway using the frontal security (cross cover) technique (Figure 6-18).

**Figure 6-18. HALLWAY MOVEMENT**

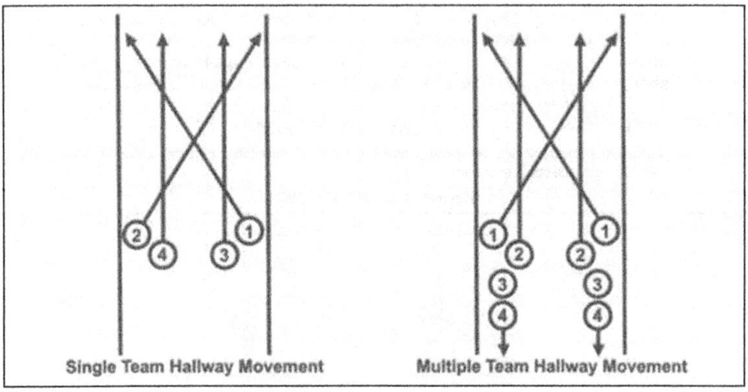

e. **T-Shaped Intersection**. See Figure 6-19. This technique can incorporate the dynamic corner clear (Figure 6-20).

**STEP 1:** Each # 1 man goes to a knee covering his sector.

**STEP 2:** On a predetermined signal, each two-man team will break the corner, picking up their sectors of fire.

**Figure 6-19. T-SHAPED INTERSECTION**

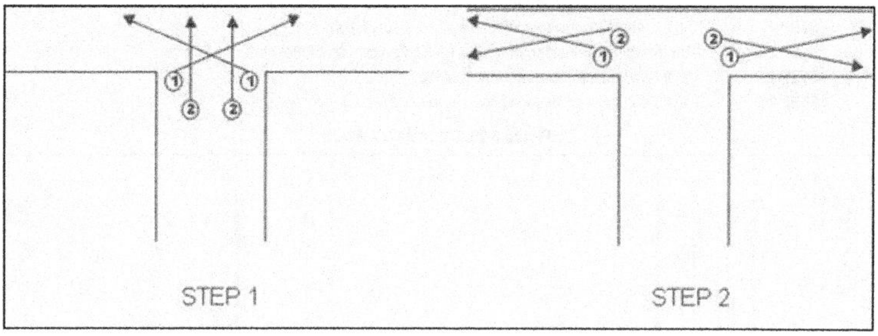

f. **Dynamic Corner**. (See Figure 6-20.)

**STEP 1:** The #1 and #2 man, as they approach the corner, have to clear. Do not slow down.

**STEP 2:** The #2 man will tap the #1 man on the shoulder about 2-3 meters away from the corner letting the #1 man know the #2 man is with him.

**STEP 3:** Keying off of the #1 man's movement, they both break the corner simultaneously.

**STEP 4:** The #1 man goes low to a knee, the #2 man stays high.

**STEP 5:** If the Rangers are not receiving fire the #2 man rabbits, or moves, to the far side.

**STEP 6:** The #1 and #2 man take up sectors of fire.

**STEP 7:** The #3 and #4 man take long security in the direction of movement.

**Figure 6-20. DYNAMIC CORNER**

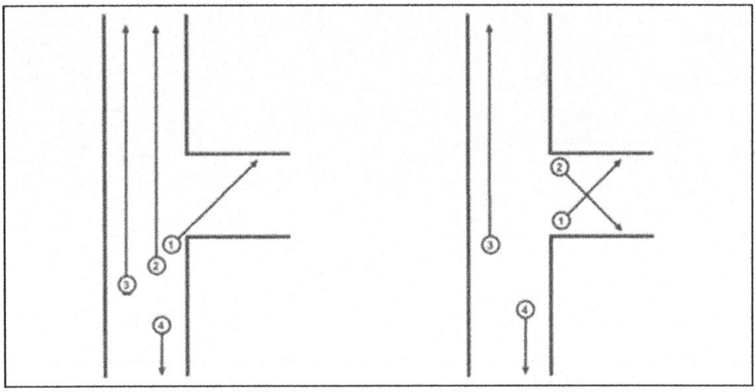

g. **Three-Way Intersection**. (See Figure 6-21.)

**STEP 1:** The #1 through #4 man will use one of the corner clearing techniques to clear the intersection.

**STEP 2:** When the intersection is secure, the #5 and #6 man move through the intersection and enter either more hallway or a room. If they are entering more hallway, the #1 through #4 man provide cross coverage as the #5 and #6 man move through the intersection into more hallway. If they are moving into a room, the #5 and #6 man move to one side of the door and signal the #1 and #3 or #2 and #4 man to enter the stack depending on which side of the door they stack on. The remaining members of the team, not in the stack, will continue to provide security down the hallway.

**STEP 3:** If the #5 and #6 man move into more hallway the #1 through #4 man will enter the stack and proceed down the hallway with the #3 and #4 man providing rear security.

**Figure 6-21. THREE WAY INTERSECTION**

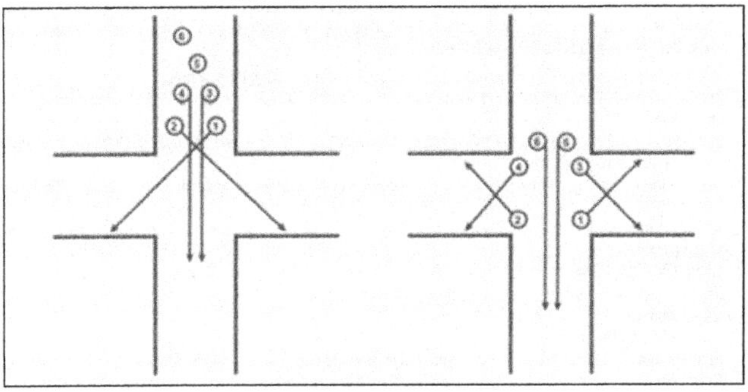

**BATTLE DRILL 6**
**ENTER/CLEAR A TRENCH**

**SITUATION:** The platoon is attacking as part of a larger force and identifies the enemy in a trench line. The platoon deploys and establishes a base of fire. The platoon leader determines that he has sufficient combat power to maneuver and assault the trench line.

**REQUIRED ACTIONS:** See Figure 6-22 and Figure 6-23.

1. The platoon leader directs one squad to enter the trench and secure a foothold.
2. The platoon leader designates the entry point of the trench line and the direction of movement once the platoon begins clearing.
3. The PSG positions Soldiers and machine guns to suppress the trench and isolate the entry point.
4. Platoon leader directs platoon FO to initiate fire mission, if necessary, in support of assault. They maintain accurate battle tracking of all friendly elements to facilitate quick clearance of fires. The platoon FO ceases or shifts fires to isolate the OBJ as assault team advances.
5. The assaulting squad executes actions to enter the trench and establish a foothold. The squad leader signals to the platoon leader that the foothold is secure, and the follow on elements can move into the trench. The squad leader remains at the entry point and marks it. The platoon follows the success of the seizure of the foothold with the remainder of the platoon as part of the platoon actions to clear a trench line.
6. The platoon leader moves into the trench with the assaulting squad.

7 The platoon leader directs one of the base-of-fire squads to move into the trench and begin clearing it in the direction of movement from the foothold.

8. The base-of-fire element repositions as necessary to continue suppressive fires.

9. The assaulting squad passes the squad that has secured the foothold and executes actions to take the lead and clear the trench.

a. The squad leader designates a lead fire team and a trail fire team.

b. The lead fire team and the squad leader move to the forward most secure corner or intersection. The squad leader tells the team securing that corner or intersection that his squad is ready to continue clearing the trench. The trail fire team follows, maintaining visual contact with the last Soldier of the lead team.

***Notes:*** 1. The fire support element must be able to identify the location of the lead fire team in the trench at all times.

2. Throughout this battle drill, the team leader positions himself at the rear of the fire team to have direct control (physically, if necessary) of his Soldiers. Other Soldiers in the fire team rotate the lead to change magazines and prepare grenades. Rotating the lead provides constant suppressive fires down the trench and maintains the momentum of the attack as the squad clears the trench.

c. The lead fire team passes the element securing the foothold. The following then occurs.

(1) The lead Soldier of the fire team moves abreast of the Soldier securing the corner or intersection, taps him, and announces, "*Taking the lead.*"

(2) The Soldier securing the corner or intersection acknowledges that he is handing over the lead by shouting, "*Okay!*" He allows the fire team to pass him.

d. The lead fire team starts clearing in the direction of movement. They arrive at a corner or intersection. The following then occurs.

(1) Allowing for cook-off (2 seconds maximum) and shouting, "*Frag out,*" the second Soldier prepares and throws a grenade around the corner.

(2) Upon detonation of the grenade, the lead Soldier moves around the corner firing three-round bursts and advancing as he fires. The entire fire team follows him to the next corner or intersection.

e. The squad leader—

- Follows behind the team.
- Ensures that the trailing fire team moves up and is ready to pass the lead at his direction.
- Rotates fire teams as necessary to keep his Soldiers fresh and to maintain the momentum of the attack.
- Requests indirect fires, if required, through the platoon leader. The squad leader also directs the employment of the M203 to provide immediate suppression against positions along the trench line.
- Ensures fire teams maintain sufficient interval to prevent them from being engaged by the same enemy fires.

f. At each corner or intersection, the lead fire team performs the same actions previously described.

g. If the lead Soldier finds that he is nearly out of ammunition before reaching a corner or intersection, he announces, "*Ammo,*" and then--

(1) The lead Soldier stops and moves against one side of the trench, ready to let the rest of the team pass. He continues to aim his weapon down the trench in the direction of movement.

(2) The next Soldier ensures that he has a full magazine, moves abreast of the lead Soldier, taps him, and announces, "*Taking the lead.*"

(3) The lead Soldier acknowledges that he is handing over the lead by shouting, "*Okay.*" Positions rotate and the squad continues forward.

h. The trailing fire team secures intersections and marks the route within the trench as the squad moves forward. The trailing fire team leader ensures that follow-on squads relieve his buddy teams to maintain security.

i. The squad leader reports the progress of the clearing operation. The base-of-fire element must be able to identify the location of the lead fire team in the trench at all times.

10. The platoon leader rotates squads to keep the Soldiers fresh and to maintain the momentum of the assault.

11. The PSG calls forward ammunition resupply and organizes teams to move it forward into the trench.

12. The base-of-fire element ensures that all friendly forces move into the trench only through the designated entry point to avoid fratricide.

13. The platoon leader reports to the company commander that the trench line is secured, or he is no longer able to continue clearing. If trench line is secured, then platoon leader directs platoon FO to develop a fire plan to support the defense of the platoon position.

**DANGER**

**FIRE TEAMS: MAINTAIN SUFFICIENT INTERVALS TO AVOID ENGAGEMENT BY THE SAME ENEMY FIRES.**

**Figure 6-22. ENTER A TRENCH (SQUAD)**

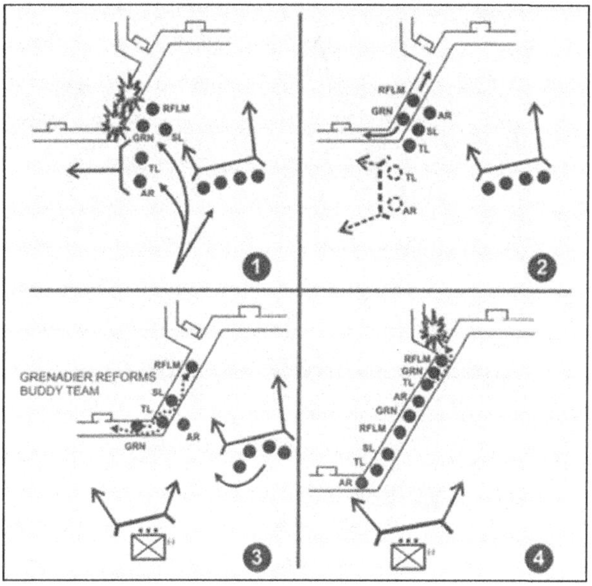

Figure 6-23. CLEAR A TRENCH LINE (PLATOON)

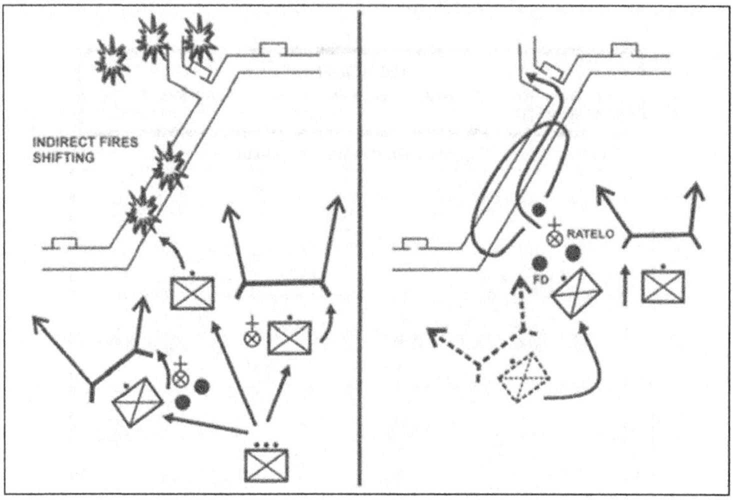


**BATTLE DRILL 7**
**CONDUCT INITIAL BREACH OF A MINED WIRE OBSTACLE (PLATOON)**

**SITUATION:** The platoon is operating as part of a larger force. The lead squad identifies a wire obstacle reinforced with mines that cannot be bypassed and enemy positions on the far side of the obstacle.

**REQUIRED ACTIONS:** See Figure 6-24 and Figure 6-25.

1. The squad in contact reacts to contact.
2. The platoon gains suppressive fires. The following then occurs.
   a. The squad in contact establishes a base-of-fire position.
   b. The platoon leader, his RTO, platoon FO, and the squad leader of the next squad with one machine gun team move forward to link up with the squad leader of the squad in contact.
3. The platoon leader determines that he can maneuver by identifying—
   - The obstacle and enemy positions.
   - The size of the enemy force engaging the squad, for example, the number of enemy automatic weapons, the presence of any vehicles, and the employment of indirect fires are indicators of enemy strength.
   - A breach point.
   - A covered and concealed route to the breach point.
4. The platoon leader directs the squad in contact to support the movement of another squad to the breach point. He indicates the base-of-fire position and the route to it, the enemy position to be suppressed, and the breach point and route the rest of the platoon will take to it. The platoon leader also clears the location, task, purpose, and method of conducting the fire mission with the platoon FO.

5. On the platoon leader's signal, the base-of-fire squad—
- Destroys or suppresses enemy weapons that are firing effectively against the platoon.
- Obscures the enemy position with smoke.
- Continues suppressive fires at the lowest possible level.

6. The platoon leader designates one squad as the breach squad and the remaining squad as the assault squad once the breach has been made. The assault squad may add its fires to the base-of-fire squad. Normally, it follows the covered and concealed route of the breach squad and assaults through immediately after the breach is made.

7. The base-of-fire squad moves to the breach point and establishes a base of fire.

8. The PSG moves forward to the base-of-fire squad with the second machine gun and assumes control of the squad.

9. The platoon leader leads the breach and assault squads along the covered and concealed route.

10. The platoon FO calls for and adjusts indirect fires as directed by the platoon leader. The platoon leader in conjunction with the platoon FO maintains accurate battle tracking of all friendly elements to facilitate quick clearance of fires.

11. The breach squad executes actions to breach the obstacle.

a. The squad leader directs one fire team to support the movement of the other fire team to the breach point.

b. The squad leader identifies the breach point.

c. The base-of-fire element continues to provide suppressive fires and isolates the breach point.

d. The breaching fire team, with the squad leader, moves to the breach point using the covered and concealed route. From there, the following takes place:

(1) The squad leader and breaching fire team leader employ smoke grenades to obscure the breach point. The platoon base-of-fire element shifts direct fires away from the breach point and continues to suppress key enemy positions. The platoon FO ceases indirect fires or shifts them beyond the obstacle.

(2) The breaching fire team leader positions himself and the automatic rifleman on one flank of the breach point to provide close-in security.

(3) The grenadier and rifleman of the breaching fire team probe for mines and cut the wire obstacle, marking their path as they proceed. If available, a bangalore is preferred.

(4) Once the obstacle has been breached, the breaching fire team leader and the automatic rifleman move to the far side of the obstacle and take up covered and concealed positions with the rifleman and grenadier. The team leader signals to the squad leader when they are in position and ready to support.

(5) The squad leader signals the base-of-fire team leader to move his fire team up and through the breach. He then moves through the obstacle and joins the breaching fire team, leaving the grenadier (or anti-armor specialist) and rifleman of the supporting fire team on the near side of the breach to guide the rest of the platoon through.

(6) Using the same covered and concealed route as the breaching fire team, the base-of-fire team moves through the breach and takes up covered and concealed positions on the far side.

12. The squad leader reports the situation to the platoon leader and posts guides at the breach point.

13. The platoon leader leads the assault squad through the breach in the obstacle and positions them beyond the breach to support the movement of the remainder of the platoon or assaults the enemy position covering the obstacle. Platoon leader directs the platoon FO to shift indirect fires off the enemy position.

14. The breaching squad continues to widen the breach to allow vehicles to pass through.

15. The platoon leader reports the situation to the company commander, and then he directs his base-of-fire squad to move up and through the obstacle. The platoon leader appoints Soldiers to guide the company through the breach point.

16. The PSG brings the remaining elements forward and through the breach on the platoon leader's command.

17. The company follows up the success of the platoon as it conducts the breach and continues the assault against the enemy positions.

Figure 6-24. CONDUCT INITIAL BREACH OF A MINED WIRE OBSTACLE (SQUAD)

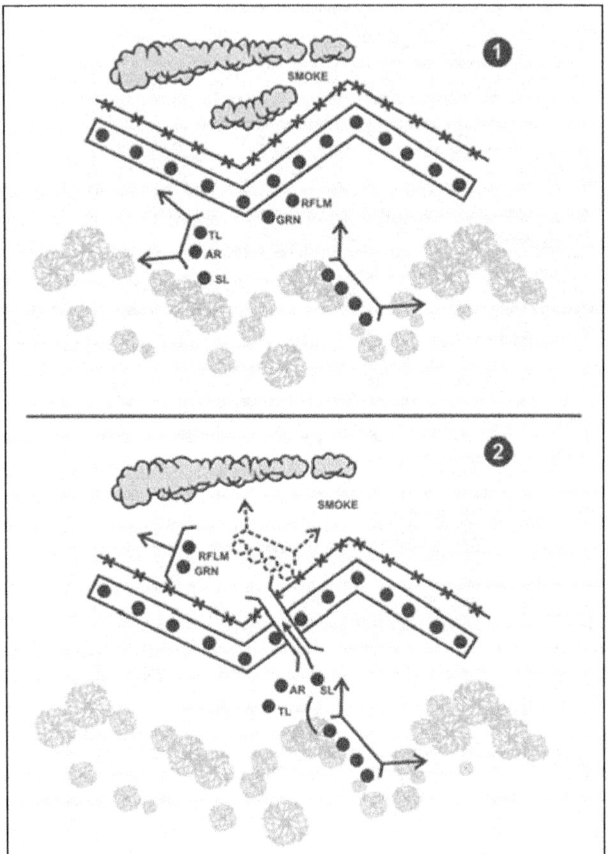

Figure 6-25. CONDUCT INITIAL BREACH OF A MINED WIRE OBSTACLE (PLATOON)

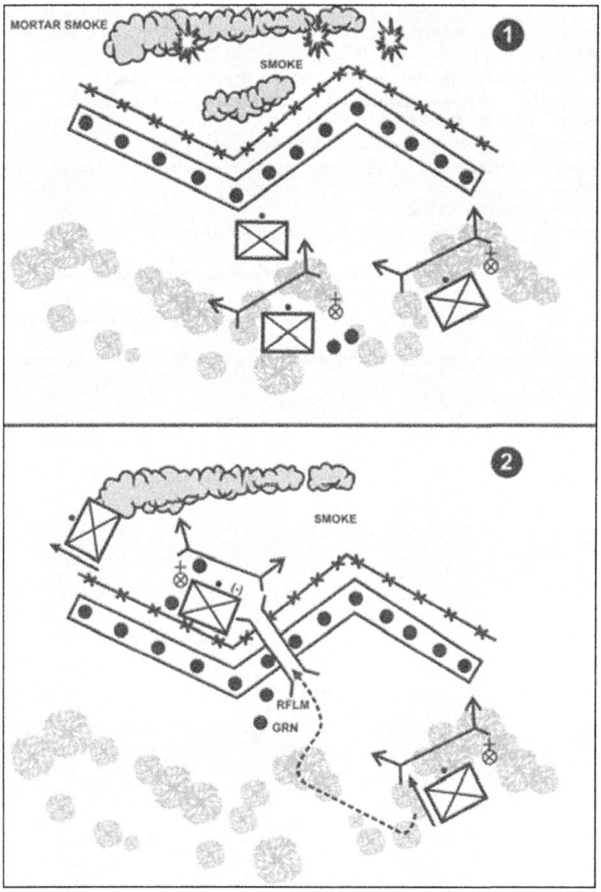

## BATTLE DRILL 8
## REACT TO INDIRECT FIRE

**SITUATION:** A platoon or squad while moving or at the halt (not dug in) receives indirect fire.

**REQUIRED ACTIONS:** NA

Any Soldier announces, "*Incoming.*" Then, if the platoon/squad is moving, the following takes place:

1. The platoon/squad leader gives direction and distance for the platoon/squad to move to a rally point by ordering direction and distance, for example, "*Four o'clock, three hundred meters.*"

2. Platoon/squad members move rapidly along the direction and distance to the rally point.

3. At the rally point, the leader immediately accounts for personnel and equipment, and forms the platoon/squad for a move to an alternate position.

4. The senior leader submits a SITREP to higher headquarters.

5. If the platoon/squad is halted (not dug in) or is preparing to move because they hear incoming artillery, the following takes place:

   a. The platoon/squad leader gives direction and distance for the platoon/squad to move to a rally point by ordering direction and distance, for example, "*Four o'clock, three hundred meters.*"

   b. Platoon/squad members secure all mission-essential equipment and ammunition and move rapidly along the direction and distance to the rally point.

   c. At the rally point, the leader immediately accounts for personnel and equipment, and forms the platoon/squad for a move to an alternate position.

   d. The senior leader submits a SITREP to higher headquarters.

***Note:*** If platoon/squad members are in defensive (dug in) positions, then members will remain in those positions if appropriate. Senior leader submits SITREP.

NOTES

# Chapter 7
# COMMUNICATIONS

The basic requirement of combat communications is to provide rapid, reliable, and secure interchange of information. Communications are a vital aspect in successful mission accomplishment. The information in this chapter helps the Ranger squad/platoon maintain effective communications and correct any radio antenna problems.

## Section I. INTRODUCTION

This section discusses military frequency modulated radios and automated net control devices (ANCDs).

7-1. **MILITARY FREQUENCY MODULATED RADIOS**. Military frequency modulated (FM) radios are composed of a receiver and a transmitter. Together, the receiver-transmitter has many capabilities and features that enable you to perform your mission more effectively. The radio can operate in Single Channel mode or frequency hopping mode. The radio has about 2,320 SC Channels and includes voice and digital communication. The operating voltage for the manpack radio is 13.5 volts from the primary battery. The range of the manpack radio is 5-10 KM on Hi power. This range is based upon line of sight and is derived from averages achieved under normal atmospheric and weather conditions. Ranges depend upon location, sighting, weather, and surrounding noise level, among other factors. the automated net control device (ANCD) is an essential tool. It replaces the signal operating instructions booklet. The ANCD has many capabilities. The main purpose of the ANCD for your mission is to obtain signal operating instructions electronically. The following discusses assembly, operation, and troubleshooting for the AN/PRC-119; provides step-by-step instructions to obtain SOI from an ANCD; and explains how to repair broken antennas:

a. **Manpack Radio Assembly**. To assemble a manpack radio you must first check and install a battery.
   (1) Inspect the battery box for dirt or damage.
   (2) Stand RT on front panel guards.
   (3) Check battery life condition (you will be using the rechargeable BB-390 batteries).
   (4) Place battery in box.
   (5) Close battery cover, and secure using latches.
   (6) Return radio to upright position.
   (7) If used battery was installed, enter the battery life condition into radio by performing the following:
      (a) Set **FCTN** to **LD**.
      (b) Press **BAT**, then **CLR**.
      (c) Enter number recorded on side of battery.
      (d) Press **STO**.
      (e) Set **FCTN** to **SQ ON**.

b. **Antenna**.
   (1) Inspect whip antenna connector on antenna and on radio for damage.
   (2) Screw whip antenna into base.
   (3) Hand tighten.
   (4) Carefully mate antenna base with RT ANT connector.
   (5) Hand tighten.
   (6) Position antenna as needed by bending goose neck.

***Note:*** Keep antenna straight, if possible. If the antenna is bent to a horizontal position, it may be necessary to turn the radio in order to receive and transmit messages.

c. **Handset**.
   (1) Inspect the handset for damage.
   (2) Push handset on **AUD/DATA** and twist clockwise to lock in place.

d. **Field Pack**.
   (1) Place radio transmitter (RT) in field pack with antenna on the left shoulder.
   (2) Fold top flap of field over RT and secure flap to field pack using straps and buckles.

e. **Presets**.
   (1) Set **CHAN** to **1**.
   (2) Set **MODE** to **SC**.
   (3) Set **RF PWR** to **HI**.

(4) Set **VOL** to mid range.
(5) Set **DIM** full clockwise.
(6) Set **FCTN** to **LD**.
(7) Set **DATA RATE** to **OFF**.

f. **Single Channel Loading Frequencies**.
(1) Obtain Ranger SOI.
(2) Set **FCTN** to **LD**.
(3) Set **MODE** to **SC**.
(4) Set **CHAN** to **MAN**, **Cue**, or desired channel (**1** to **6**) where frequency is to be stored.
(5) Press **FREQ** (display will show "**00000**," or the frequency the RT is currently set to).
(6) Press **CLR** (display will show five lines).
(7) Enter the number of the new frequency.
(8) If you make a mistake with a number, press **CLR**.
(9) Press **STO** (display will blink).
(10) Set **FCTN** to **SQ ON**.

g. **Clearing of Frequencies**.
(1) Set **MODE** to **SC**.
(2) Set **CHAN** to **MAN**, **Cue** or the channel whose frequency is to be cleared.
(3) Press **FREQ**.
(4) Press **CLR**.
(5) Press **Load**; then press **STO**.
(6) Set **FCTN** to **SQ ON**.

h. **Scanning of Multiple Frequencies**.
(1) Load all desired frequencies using "Single Channel Loading Frequencies" instructions (see 7-2h).
(2) Set **CHAN** to **CUE**.
(3) Set **SC** to **FH**.
(4) Set **FCTN** to **SQ ON**.
(5) Press **STO** (display will say **SCAN**).
(6) Press **8**. You will now be able to scan more than one frequency.

7-2. **AUTOMATED NET CONTROL DEVICE**. The RTO retrieves all necessary COMSEC and information from the ANCD. To retrieve the SOI from an ANCD:

a. Press the **ON** button on the ANCD keypad.
b. Press the letter lock button to unlock the keys on the ANCD.
c. Press **Main Menu** key.
d. Once you press the main menu key, this is what should show on the screen.

**Appl Date Time Setup**
**Util Bit.**

e. Using the arrow keys, scroll over to **Appl** (as shown above) and press the enter button on the keypad.
f. Once you have entered into the **Appl** (applications) menu using the arrow keys scroll over to **SOI** and press the enter button on the key pad. For example--

**RDS SOI Radio.**
**(Appl)** <--*(Indicates that you are in the applications menu).*

**SOI menu will look like this:**
**Qref Group Net Sufx Pyro**
**Tmpd Set C/S Find Memo**

g. Using the arrow keys scroll over to **Set** and press the enter button on the keypad.
h. Press enter on choose once in the **Set** menu.

> *EXAMPLE:*
> SELECT:
> **Choose Send Receive.**

i. Once in the **Choose** menu press the enter key on set **1-5** or **6-10**.

j. After you choose your selection out of the **Set** menu, the ANCD will automatically return back to the **SOI** menu.

k. Using the keypad scroll over to **Tmpd** and enter the number out of the set you need for that day. After this operation, the ANCD will automatically return back to the **SOI** menu.

l. Using the arrow keys on the keypad scroll over to **Net** and press the enter button on the keypad. The Net menu should look like the following example:

> *EXAMPLE:*
> T05 <--(time period) 1 <--(platoon) / A <--(company) /A3Q <--(prefix).
> C 60.000.

***Note:*** When in **Net** menu, this is where you will obtain your frequency and the first part of your call sign.

m. After you obtain the information you need out of **Net**, press the **Abort** button on the keypad. This will return you to the **SOI** menu.

n. Using the keypad, scroll over to **Sufx** and press the **Enter** button.

o. Once you are in the **Sufx** menu, use the up and down arrow keys on the ANCD to locate your two-digit designator. These two digits will go at the end of your prefix; your prefix plus the two digits equal your call sign.

p. Once you are done, press the **OFF** button on the ANCD to end operation.

## Section II. ANTENNAS

This section discusses repair techniques, construction and adjustment, field-expedient antennas, antenna length and orientation, and improvement of marginal communications.

7-3. **REPAIR**. Antennas are sometimes broken or damaged, causing either a communications failure or poor communications. If a spare antenna is available, the damaged antenna is replaced. When there is no spare, the Ranger squad/platoon may have to construct an emergency antenna. The following paragraphs contain suggestions for repairing antennas and antenna supports and the construction and adjustment of emergency antennas.

> **DANGER**
>
> **SERIOUS INJURY OR DEATH CAN RESULT FROM CONTACT WITH THE RADIATING ANTENNA OF A MEDIUM-POWER OR HIGH-POWER TRANSMITTER. TURN THE TRANSMITTER OFF WHILE ADJUSTING THE ANTENNA.**

a. **Whip Antennas**. When a whip antenna is broken into two sections, the part of the antenna that is broken off can be connected to the part attached to the base by joining the sections.

(1) Use the method shown in A, Figure 7-1 when both parts of the broken whip are available and usable.

(2) Use the method in B, Figure 7-1, when the part of the whip that was broken off is lost, or when the whip is so badly damaged that it cannot be used.

(3) To restore the antenna to its original length, a piece of wire is added that is nearly the same length as the missing part of the whip. The pole support is then lashed securely to both sections of the antenna. The two antenna sections are cleaned thoroughly to ensure good contact before connecting them to the pole support. If possible, the connections are soldered.

b. **Wire Antennas**. Emergency repair of a wire antenna may involve the repair or replacement of the wire used as the antenna or transmission line; or the repair or replacement of the assembly used to support the antenna.

(1) When one or more wires of an antenna are broken, the antenna can be repaired by reconnecting the broken wires. To do this, lower the antenna to the ground, clean the ends of the wires, and twist the wires together. Whenever possible, solder the connection.

(2) If the antenna is damaged beyond repair, construct a new one. Make sure that the lengths of the wires on the substitute antenna match the length of the original.

(3) Antenna supports may also require repair or replacement. A substitute item may be used in place of a damaged support and, if properly insulated, can be of any material of adequate strength. If the radiating element is not properly insulated, field antennas may be shorted to ground and be ineffective. Many commonly found items can be used as field-expedient insulators.

(4) The best of these items are plastic or glass to include plastic spoons, buttons, bottle necks, and plastic bags. Though less effective than plastic or glass but still better than no insulator at all are wood and rope. The radiating element--the actual antenna wire-should touch only the antenna terminal and should be physically separated from all other objects, other than the supporting insulator. (Figure 7-2 shows how to make emergency insulators.)

**Figure 7-1. USE OF BROKEN ANTENNA**

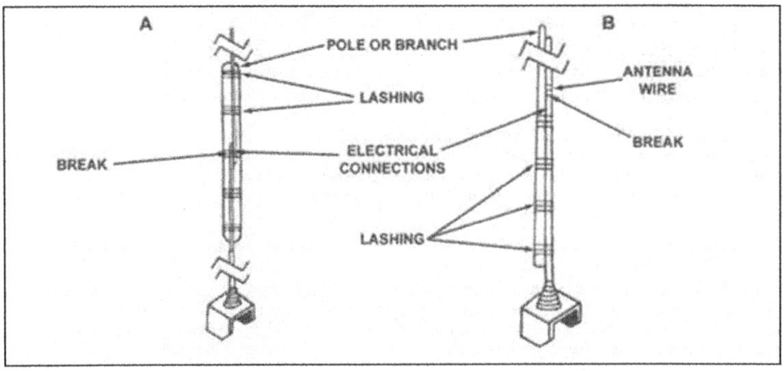


**Figure 7-2. METHODS OF CONSTRUCTING EMERGENCY INSULATORS**

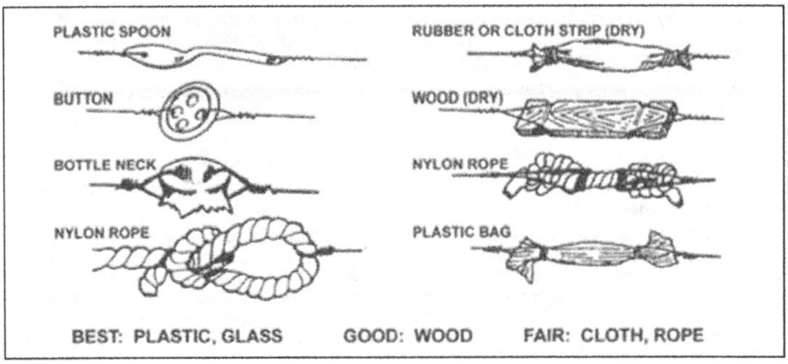

7-4. **CONSTRUCTION AND ADJUSTMENT**. Ranger squad/platoons may use the following methods to construct and adjust antennas:

a. **Construction**. The best kinds of wire for antennas are copper and aluminum. In an emergency, however, Rangers use any type of wire that is available.

(1) The exact length of most antennas is critical. The emergency antenna should be the same length as the antenna it replaces.

(2) Antennas supported by trees can usually survive heavy wind storms if the trunk of a tree or a strong branch is used as a support. To keep the antenna taut and to prevent it from breaking or stretching as the trees sway, the Ranger attaches a spring or old inner tube to one end of the antenna. Another technique is to pass a rope through a pulley or eyehook. The rope is attached to the end of the antenna and loaded with a heavyweight to keep the antenna tightly drawn.

(3) Guidelines used to hold antenna supports are made of rope or wire. To ensure the guidelines will not affect the operation of the antenna, the Ranger cuts the wire into several short lengths and connects the pieces with insulators.

b. **Adjustment**. An improvised antenna may change the performance of a radio set. The following methods can be used to determine if the antenna is operating properly:

(1) A distant station may be used to test the antenna. If the signal received from this station is strong, the antenna is operating satisfactorily. If the signal is weak, the Ranger adjusts the height and length of the antenna and the transmission line to receive the strongest signal at a given setting on the volume control of the receiver. This is the best method of tuning an antenna when transmission is dangerous or forbidden.

(2) In some radio sets, the Ranger uses the transmitter to adjust the antenna. First, he sets the controls of the transmitter to normal; then, he tunes the system by adjusting the antenna height, the antenna length, and the transmission line length to obtain the best transmission output.

7-5. **FIELD-EXPEDIENT OMNIDIRECTIONAL ANTENNAS**. Vertical antennas are omnidirectional. The omnidirectional antenna transmits and receives equally well in all directions.

a. **Vertical Antennas**. Most tactical antennas are vertical. For example, the man-pack portable radio uses a vertical whip and so do the vehicular radios in tactical vehicles. A vertical antenna can be made by using a metal pipe or rod of the correct length, held erect by means of guidelines. The lower end of the antenna should be insulated from the ground by placing it on a large block of wood or other insulating material. A vertical antenna may also be a wire supported by a tree or a wooden pole (Figure 7-3).

**Figure 7-3. VERTICAL ANTENNA MADE FROM A WIRE SUPPORTED BY A TREE OR WOODEN POLE**

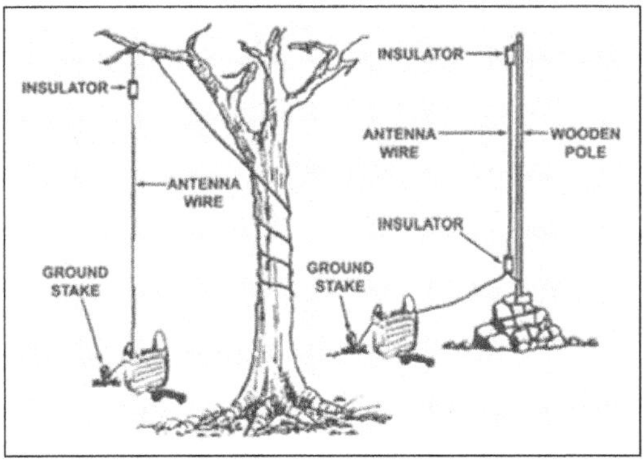

b. **Modification of Connection on Shorter Vertical Antennas**. For short vertical antennas, the pole may be used without guidelines (if properly supported at the base). If the length of the vertical mast is not long enough to support the wire upright, it may be necessary to modify the connection at the top of the antenna (Figure 7-4; see FM 24-18).

**Figure 7-4. MODIFIED CONNECTION AT TOP OF ANTENNA**

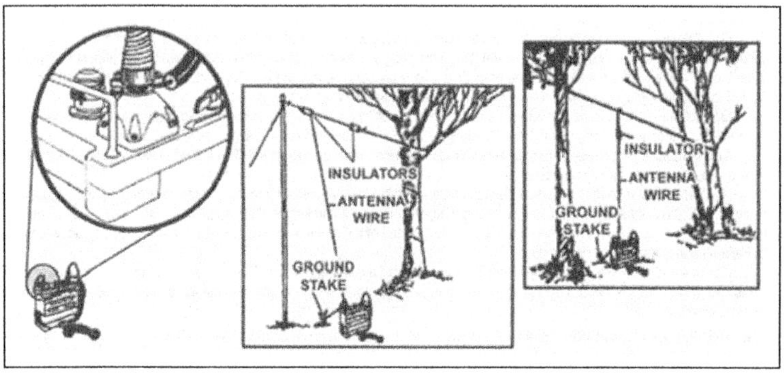

c. **End-Fed, Half-Wave Antenna**. An emergency, end-fed half-wave antenna (Figure 7-5) can be constructed from available materials such as field wire, rope, and wooden insulators. The electrical length of this antenna is measured from the antenna terminal on the radio set to the far end of the antenna. The best performance can be obtained by constructing the antenna longer than necessary and then shortening it, as required, until the best results are obtained. The ground terminal of the radio set should be connected to a good earth ground for this antenna to function efficiently.

**Figure 7-5. END-FED, HALF-WAVE ANTENNA**

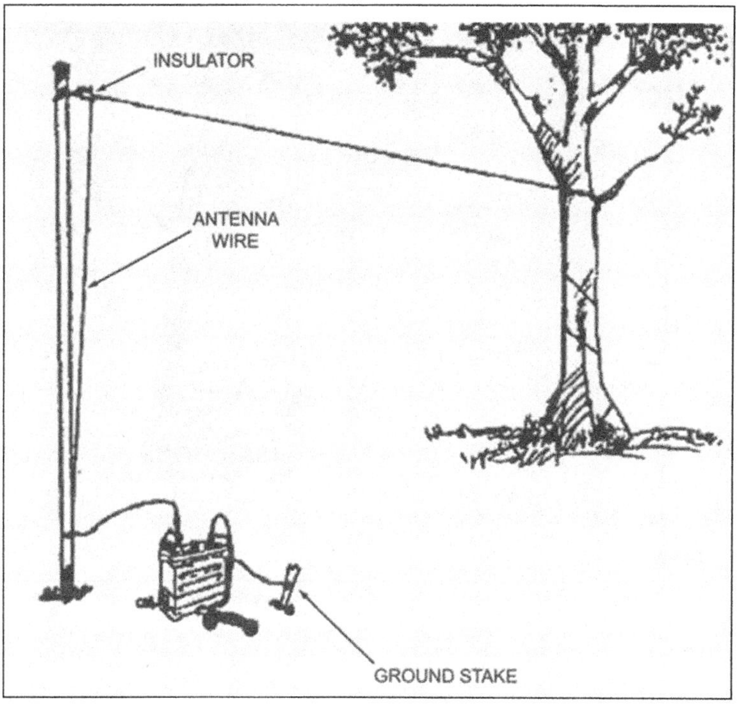

d. **Center-Fed, Doublet Antenna**. The center-fed doublet is a half-wave antenna consisting of two quarter wavelength sections on each side of the center (Figure 7-6). Doublet antennas are directional broadside to their length, which makes the vertical doublet antenna omnidirectional. This is because the radiation pattern is doughnut-shaped and bidirectional.

**Figure 7-6. CENTER-FED, DOUBLET ANTENNA**

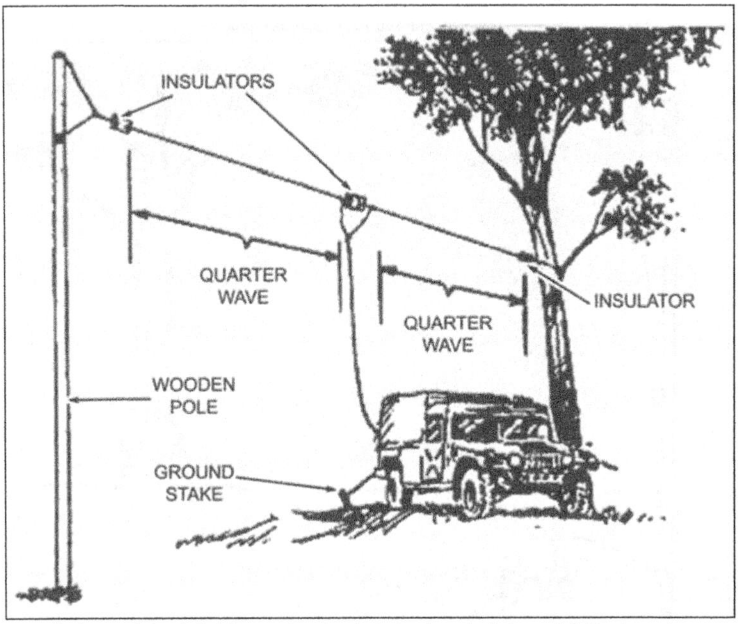


(1) Compute the length of a half-wave antenna by using the formula previously discussed. Cut the wires as close as possible to the correct length; this is very important.

(2) Uses transmission line for conducting electrical energy from one point to another and for transferring the output of a transmitter to an antenna. Although it is possible to connect an antenna directly to a transmitter, the antenna is usually located some distance away.

(3) Support center-fed half-wave FM antennas entirely with pieces of wood. (A horizontal antenna of this type is shown in A, Figure 7-7 and a vertical antenna in B, Figure 7-7.) Rotate these antennas to any position to obtain the best performance.

(a) If the antenna is erected vertically, bring out the transmission line horizontally from the antenna for a distance equal to at least one-half of the antenna's length before it is dropped down to the radio set.

**Figure 7-7. CENTER-FED ANTENNAS SUPPORTED WITH WOOD**

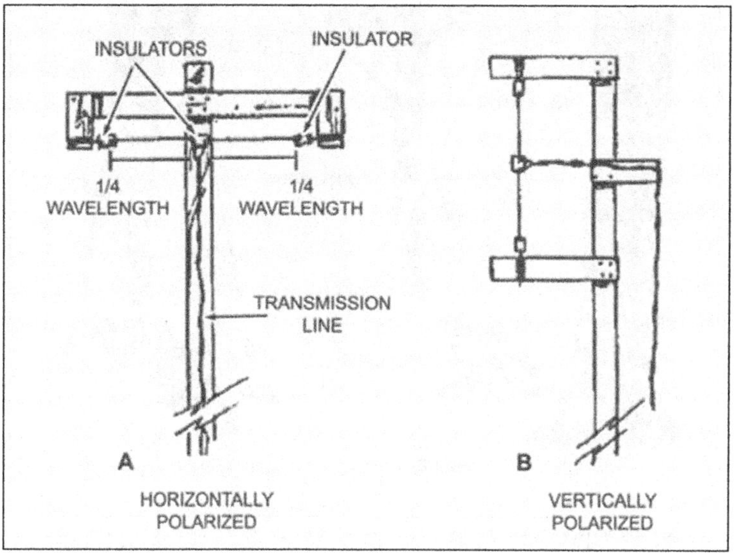

(b) The half-wave antenna is used with FM radios (Figure 7-8). It is effective in heavily wooded areas to increase the range of portable radios. Connect the top guidelines to a limb or pass it over the limb and connect it to the tree trunk or a stake.

**Figure 7-8. HALF-WAVE ANTENNA USED WITH FM RADIO**

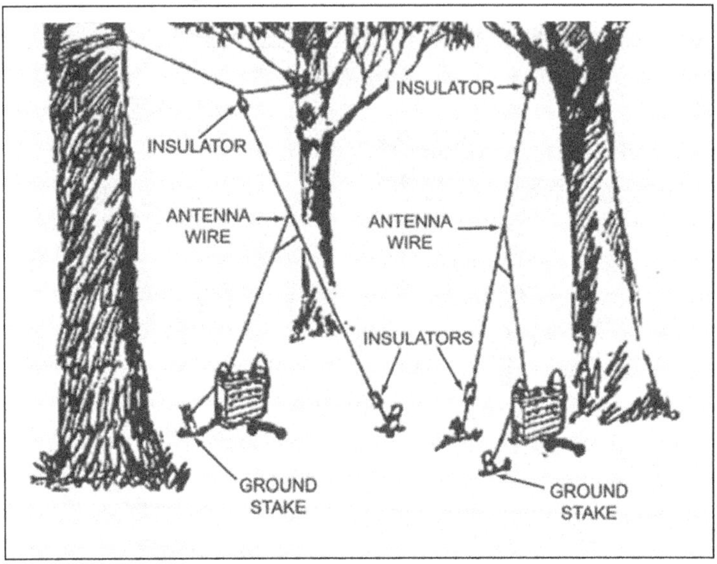


c. **Expedient 292-Type Antenna**. The expedient 292-type antenna was developed for use in the jungle and, if used properly, can increase the team's ability to communicate. In its entirety, the antenna is bulky, heavy, and not acceptable for sniper team operations. The team can, however, carry only the mast head and antenna sections, mounting these on wood poles or hanging them from trees; or, the team can make a complete expedient 292-type antenna (Figure 7-9), using WD-1, wire, and other readily available material. The team can also use almost any plastic, glass, or rubber objects for insulators. Dry wood is acceptable when nothing else is available. (See Figure 7-2 for types of insulators that may be used.) The following describes how to make this antenna:

(1) Use the quick-reference table (Table 7-2) to determine the length of the elements (one radiating and three ground planes) for the frequency that will be used. Cut these elements (A, Figure 7-9) from WD-1 field wire (or similar wire). Cut spacing sticks (B, Figure 7-9) the same length. Place the ends of the sticks together to form a triangle and tie the ends with wire, tape, or rope. Attach an insulator to each corner. Attach a ground-plane wire to each insulator. Bring the other ends of the ground-plane wires together, attach them to an insulator (C, Figure 7-9), and tie securely. Strip about 3 inches of insulation from each wire and twist them together (D, Figure 7-9).

(2) Tie one end of the radiating element wire to the other side of insulator C and the other end to another insulator (D, Figure 7-9). Strip about 3 inches of insulation from the radiating element at insulator C.

(3) Cut enough WD-1 field wire to reach from the proposed location of the antenna to the radio set. Keep this line as short as possible, because excess length reduces the efficiency of the system. Tie a knot at each end to identify it as the *hot* lead. Remove insulation from the *hot* wire and tie it to the radiating element wire at insulator C. Remove insulation from the other wire and attach it to the bare ground-plane element wires at insulator C. Tape all connections and do not allow the radiating element wire to touch the ground-plane wires.

(4) Attach a rope to the insulator on the free end of the radiating element and toss the rope over the branches of a tree. Pull the antenna as high as possible, keeping the lead-in routed down through the triangle. Secure the rope to hold the antenna in place.

(5) At the radio set, remove about 1 inch of insulation from the *hot* lead and about 3 inches of insulation from the other wire. Attach the *hot* line to the antenna terminal (doublet connector, if so labeled). Attach the other wire to the metal case-the handle, for example. Be sure both connections are tight or secure.

(6) Set up correct frequency, turn on the set, and proceed with communications.

**Figure 7-9. COMPLETE EXPEDIENT 292-TYPE ANTENNA**

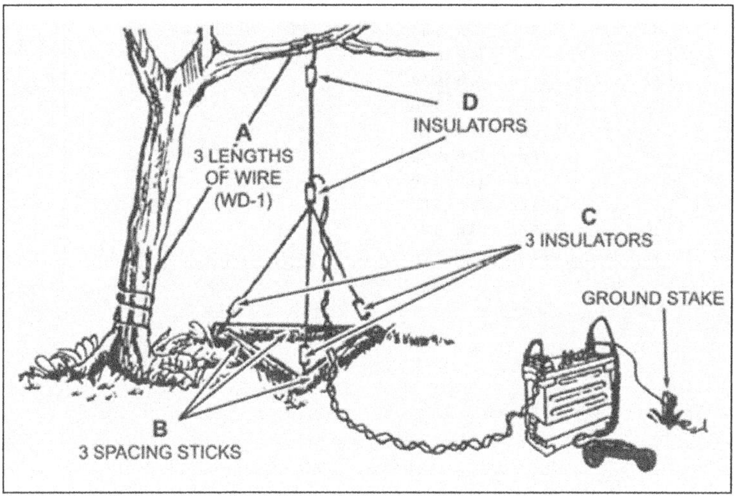

Table 7-2. QUICK REFERENCE TABLE

| OPERATING FREQUENCY IN MHz | ELEMENT LENGTH (RADIATING ELEMENTS AND GROUND-PLANE ELEMENTS) | |
|---|---|---|
| 30 | 2.38 m | (7 ft 10 in) |
| 32 | 2.23 m | (7 ft 4 in) |
| 34 | 2.1 m | (6 ft 11 in) |
| 36 | 1.98 m | (6 ft 6 in) |
| 38 | 1.87 m | (6 ft 2 in) |
| 40 | 1.78 m | (5 ft 10 in) |
| 43 | 1.66 m | (5 ft 5 in) |
| 46 | 1.55 m | (5 ft 1 in) |
| 49 | 1.46 m | (4 ft 9 in) |
| 52 | 1.37 m | (4 ft 6 in) |
| 55 | 1.3 m | (4 ft 3 in) |
| 58 | 1.23 m | (4 ft 0 in) |
| 61 | 1.17 m | (3 ft 10 in) |
| 64 | 1.12 m | (3 ft 8 in) |
| 68 | 1.05 m | (3 ft 5 in) |
| 72 | 0.99 m | (3 ft 3 in) |
| 76 | 0.94 m | (3 ft 1 in) |

7-6. **FIELD-EXPEDIENT DIRECTIONAL ANTENNAS**. The vertical half-rhombic antenna (Figure 7-10) and the long-wire antenna (Figure 7-11) are two field-expedient directional antennas. These antennas consist of a single wire, preferably two or more wavelengths long, supported on poles at a height of 3 to 7 meters (10 to 20 feet) above the ground. The antennas will, however, operate satisfactorily as low as 1 meter (about 3 feet) above the ground--the radiation pattern is directional. The antennas are used mainly for either transmitting or receiving high-frequency signals.

**Figure 7-10. VERTICAL HALF-RHOMBIC ANTENNA**

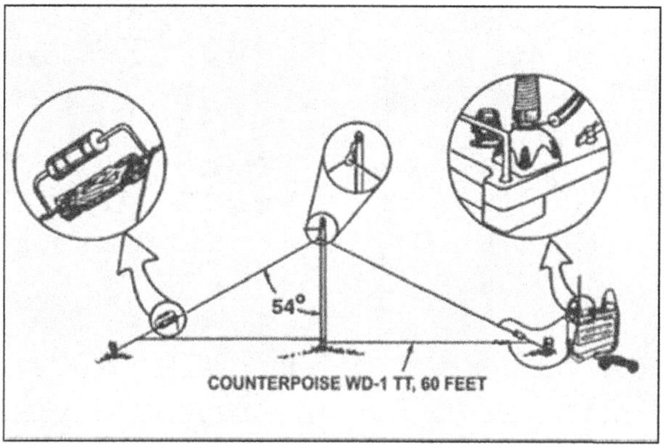


**Figure 7-11. LONG-WIRE ANTENNA**

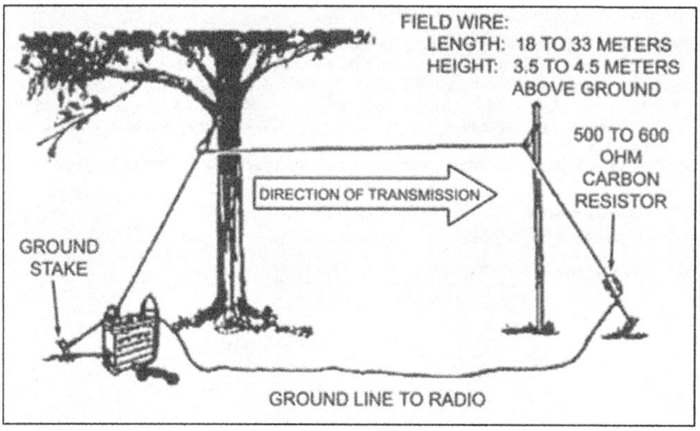

a. The vee antenna (Figure 7-12) is another field-expedient directional antenna. It consists of two wires forming a V with the open area of the "V" pointing toward the desired direction of transmission or reception. To make construction easier, the legs should slope downward from the apex of the "V." This is called a sloping vee antenna (Figure 7-13). The angle between the legs varies with the length of the legs to achieve maximum performance. (Use Table 7-1 to determine the angle and the length of the legs.)

**Figure 7-12. VEE ANTENNA**

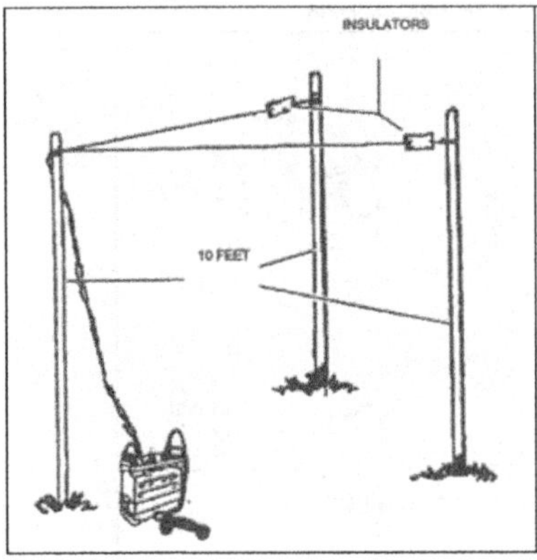


**Figure 7-13. SLOPING VEE ANTENNA**

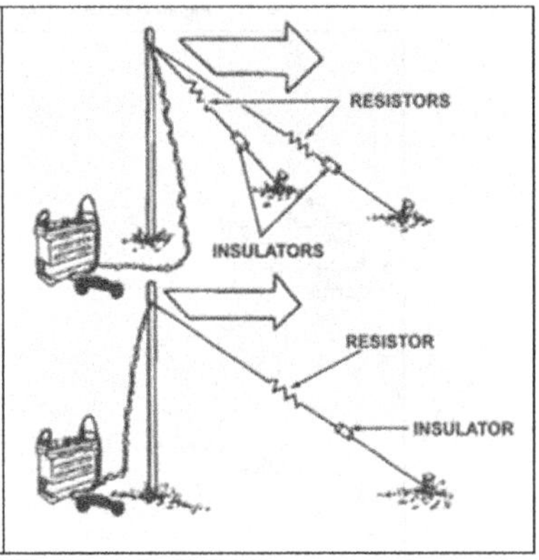


b. When the antenna is used with more than one frequency or wavelength, use an apex angle that is midway between the extreme angles determined by the chart. To make the antenna radiate in only one direction, add noninductive terminating resistors from the end of each leg (not at the apex) to ground. (See TM 11-666.)

7-7. **ANTENNA LENGTH**. The length of an antenna must be considered in two ways: both a physical and an electrical length. These two lengths are never the same. The reduced velocity of the wave on the antenna and a capacitive effect (known as end *effect*) make the antenna seem longer electrically than it is physically. The contributing factors are the ratio of the diameter of the antenna to its length and the capacitive effect of terminal equipment, such as insulators and clamps, used to support the antenna.

a. To calculate the physical length of an antenna, use a correction of 0.95 for frequencies between 3.0 and 50.0 MHz The figures given below are for a half-wave antenna.

b. Use the following formula to calculate the length of a long-wire antenna (one wavelength or longer) for harmonic operation:

***EXAMPLE:***

If N = the number of half-wavelengths in the total length of the antenna = 3,
and if the frequency in MHz = 7, then--

$$\text{Length (meters)} = \frac{150(N-0.05)}{\text{Frequency in MHz}} = \frac{150(3-.05)}{7} =$$

$$\frac{150 \times 2.95}{7} = \frac{442.50}{7} = 63.2 \text{ meters}$$

7-8. **ANTENNA ORIENTATION**. If the azimuth of the radio path is not provided, the azimuth should be determined by the best available means. The accuracy required in determining the azimuth of the path depends on the radiation pattern of the directional antenna. In transportable operation, the rhombic and V antennas may have such a narrow beam as to require great accuracy in azimuth determination. The antenna should be erected for the correct azimuth. Great accuracy is not required in erecting broad-beam antennas. Unless a line of known azimuth is available at the site, the direction of the path is best determined by a magnetic compass.

7-9. **IMPROVEMENT OF MARGINAL COMMUNICATIONS**. Under certain situations, it may not be feasible to orient directional antennas to the correct azimuth of the desired radio path. As a result, marginal communications may suffer. To improve marginal communications, the following procedure can be used:

a. Check, tighten, and tape cable couplings and connections.
b. Return all transmitters and receivers in the circuit.
c. Ensure antennas are adjusted for the proper operating frequency.
d. Change the heights of antennas.
e. Move the antenna a short distance away and in different locations from its original location.

NOTES

## Chapter 8

# ARMY AVIATION

Army aviation and infantry units can be fully integrated with other members of the combined arms team to form powerful and flexible air assault task forces that can project combat power throughout the entire depth, width, and breadth of the modern battlefield with little regard for terrain barriers. These combat operations are deliberate, precisely planned, and vigorously executed. They are assigned to strike the enemy when and where he is most vulnerable (See page 2-28, *Air Movement Annex* and page 2-37, *Army Aviation Coordination Checklist*).

8-1. **REVERSE PLANNING SEQUENCE.** Successful air assault execution is based on a careful analysis of METT-TC and detailed, precise reverse planning. Five basic plans that comprise the reverse planning sequence are developed for each air assault operation. The battalion is the lowest level that has sufficient personnel to plan, coordinate, and control air assault operations. When company size or lower operations are conducted, the bulk of the planning takes place at battalion or higher headquarters. They are:

a. **Ground Tactical Plan.** The foundation of a successful air assault operation is the commander's ground tactical plan. All additional plans must support this plan. The plan specifies actions in the objective area to ultimately accomplish the mission and address subsequent operations.

b. **Landing Plan.** The landing plan must support the ground tactical plan. This plan outlines a sequence of events that allows elements to move into the area of operations, and ensures that units arrive at designated locations at prescribed times prepared to execute the ground tactical plan.

c. **Air Movement Plan.** The air movement plan is based on the ground tactical and landing plans. It specifies the schedule and provides instructions for air movement of troops, equipment, and supplies from PZs to LZs.

d. **Loading Plan.** The loading plan is based on the air movement plan. It ensures that troops, equipment, and supplies are loaded on the correct aircraft. Unit integrity is maintained when aircraft loads are planned. Cross-loading may be necessary to ensure survivability of command and control assets, and that the mix of weapons arriving at the LZ is ready to fight.

e. **Staging Plan.** The staging plan is based on the loading plan and prescribes the arrival time of ground units (troops, equipment and supplies) at the PZ in the order of movement

8-2. **SELECTION AND MARKING OF PICKUP AND LANDING ZONES**

a. Small unit leaders should consider the following when selecting a PZ/LZ:

(1) ***Size***. Minimal circular landing point separation from other aircraft and obstacles is needed:

- OH-58D – 25 meters.
- UH-1, AH-1 – 35 meters.
- UH-60, AH-64 – 50 meters.
- Cargo helicopters – 80 meters.

(2) ***Surface Conditions***. Avoid potential hazards such as sand, blowing dust, snow, tree stumps, or large rocks.

(3) ***Ground Slope***.

- 0% - 6 % -- land upslope.
- 7% - 15% -- land sideslope.
- over 15% -- no touchdown (aircraft may hover).

(4) ***Obstacles***. An obstacle clearance ratio of 10 to 1 is used in planning approach and departure of the PZ and LZ, for example, a 10-foot tall tree requires 100 feet of horizontal distance for approach or departure. Obstacles will be marked with a red chemlight at night, or red panels in daytime. Avoid using markings if the enemy would see them.

(5) ***Approach/Departure***. Approach and departure are made into the wind and along the long axis of the PZ/LZ.

(6) ***Loads***. The greater the load, the larger the PZ/LZ must be to accommodate the insertion or extraction.

b. **Marking of PZs and LZs.**

(1) ***Day***. A ground guide will mark the PZ or LZ for the lead aircraft by holding an M16/M4 rifle over his head, by displaying a folded VS-17 panel chest high, or by other coordinated and identifiable means.

(2) ***Night***. The code letter "Y" (inverted "Y") is used to mark the landing point of the lead aircraft at night (Figure 8-1). Chemical lights or "beanbag" lights are used to maintain light discipline. A swinging chemlight may also be used to mark the landing point.

Figure 8-1. INVERTED "Y"

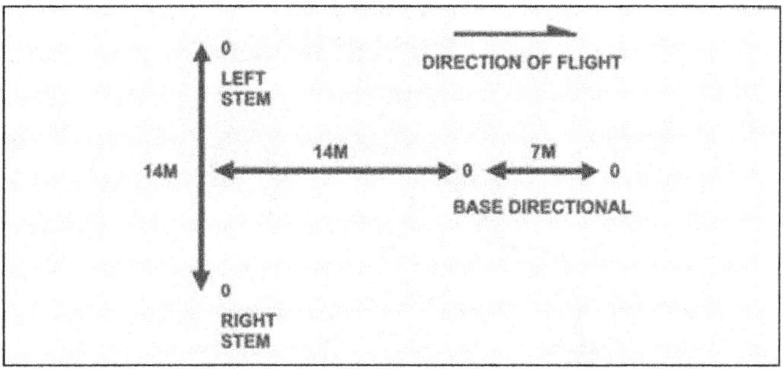


8-3. **AIR ASSAULT FORMATIONS**. Aircraft supporting an operation may use any of the following PZ/LZ configurations which are prescribed by the air assault task force (AATF) commander working with the air mission commander (AMC):

a. ***Heavy Left or Right***. Requires a relatively long, wide landing area; presents difficulty in pre-positioning loads; restricts suppressive fire by inboard gunners; provides firepower to front and flank (Figure 8-2).

Figure 8-2. HEAVY LEFT/HEAVY RIGHT

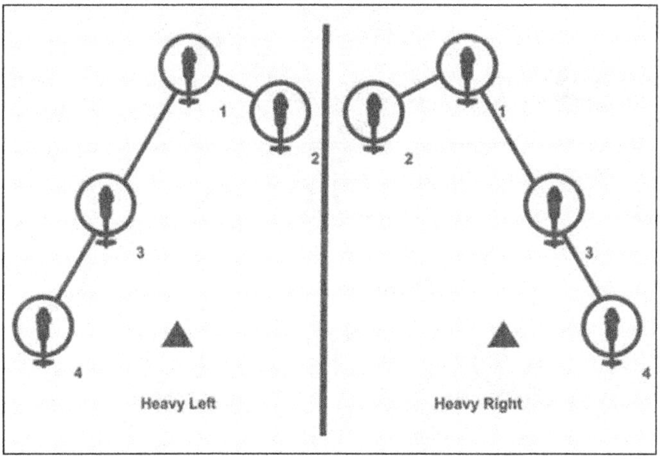

b. **Diamond**. Allows rapid deployment for all-round security; requires small landing area; presents some difficulty in pre-positioning loads; restricts suppressive fire of inboard gunners (Figure 8-3).

**Figure 8-3. DIAMOND**

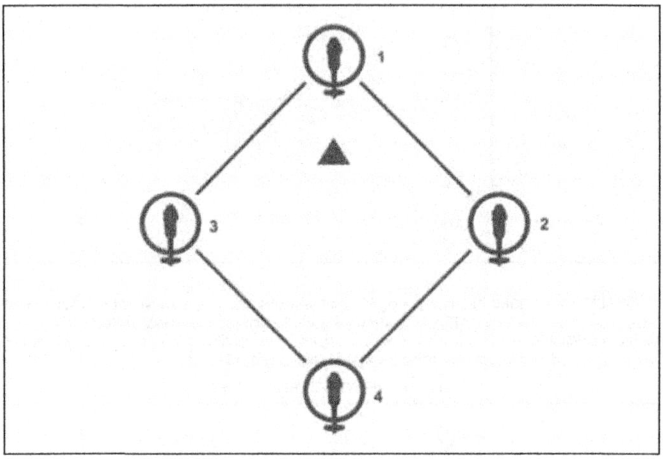


c. **Vee**. Requires a relatively small landing area; allows rapid deployment of forces to the front; restricts suppressive fire of inboard gunners; presents some difficulty in pre-positioning loads (Figure 8-4).

**Figure 8-4. VEE**

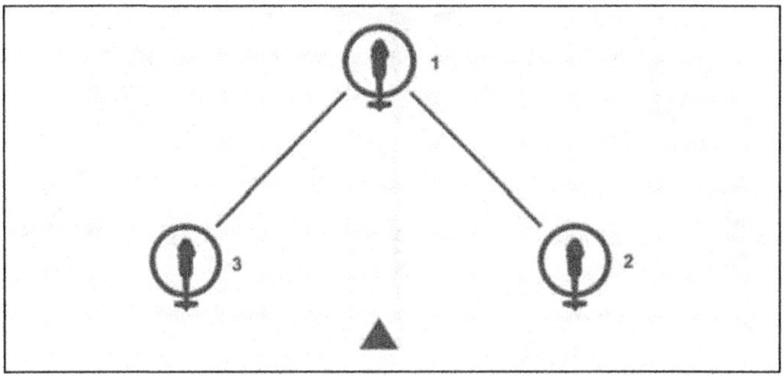

d. **Echelon Left or Right**. Requires a relatively long, wide landing area; presents some difficulty in pre-positioning loads; allows rapid deployment of forces to the flank; allows unrestricted suppressive fire by gunners (Figure 8-5).

**Figure 8-5. ECHELON LEFT/ECHELON RIGHT**

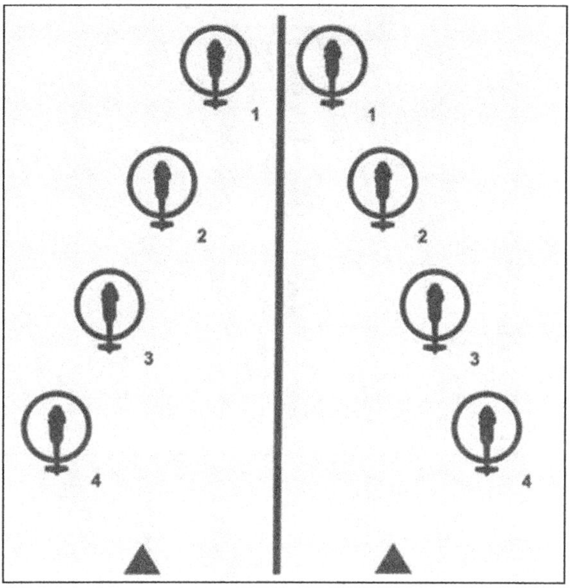

e. **Trail**. Requires a relatively small landing area; allows rapid deployment of forces to the flank; simplifies pre-positioning loads; allows unrestricted suppressive fire by gunners (Figure 8-6).

**Figure 8-6. TRAIL**

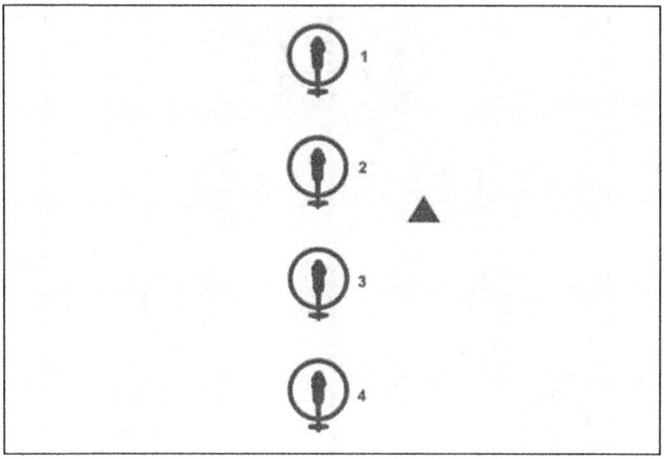

f. **Staggered Trail Left or Right**. Requires a relatively long, wide landing area; simplifies pre-positioning loads; allows rapid deployment for all round security; gunners' suppressive fire restricted somewhat (Figure 8-7).

**Figure 8-7. STAGGERED TRAIL LEFT/STAGGERED TRAIL RIGHT**

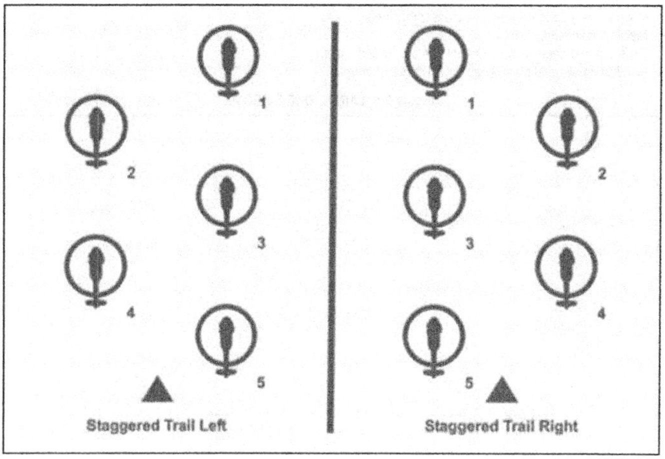

8-4. **PICKUP ZONE OPERATIONS**. Prior to arrival of aircraft, the PZ is secured, PZ control party is positioned, and the troops and equipment are positioned in platoon/squad assembly areas. In occupying a patrol/squad assembly area, the patrol/squad leader does the following. Figure 8-8 shows an example of a large, one-sided PZ. Figures 8-8 through 8-12 show loading/unloading procedures and techniques:

- Maintain all-around security of the assembly area.
- Maintain communications.
- Organize personnel and equipment into chalks and loads.
- Conduct safety briefing and equipment check of troops.

**Figure 8-8. LARGE, ONE SIDED PZ**

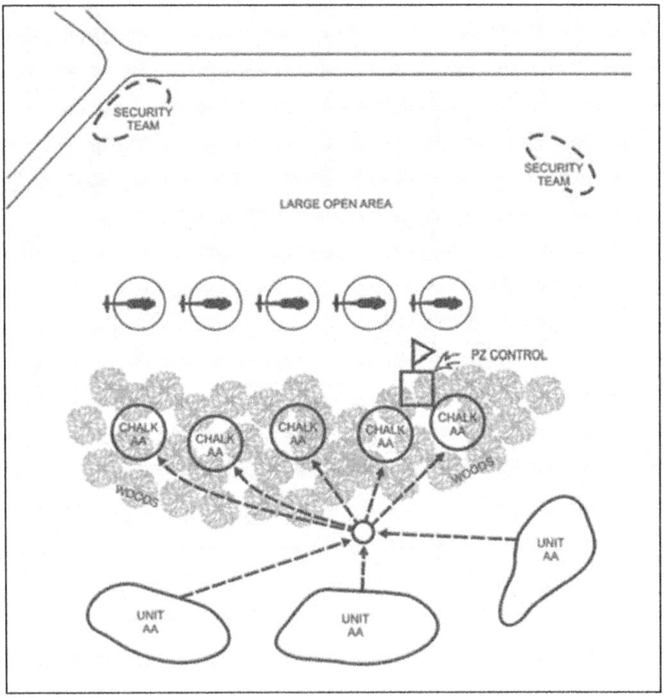

Figure 8-9. UH-60 LOADING SEQUENCE

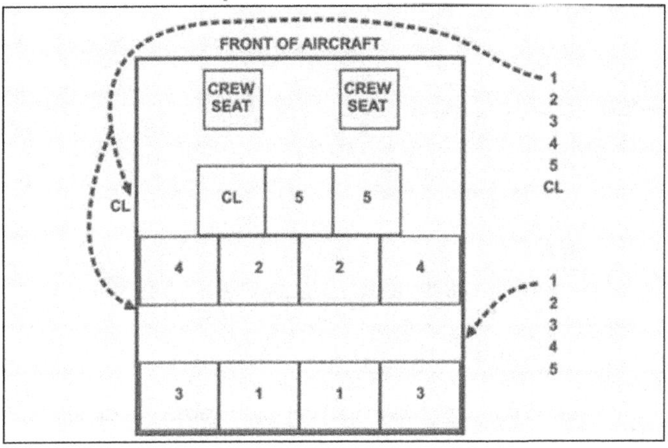


Figure 8-10. UNLOADING SEQUENCE

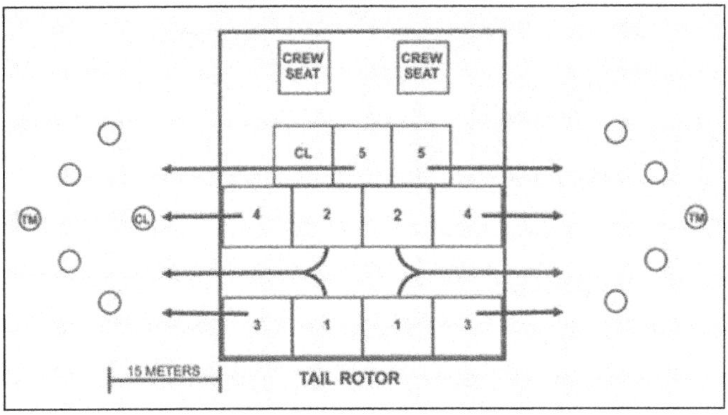

Figure 8-11. TACTICAL LOADING SEQUENCE

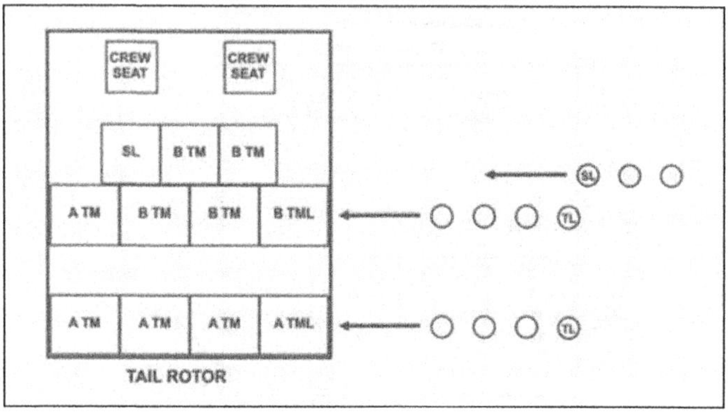


Figure 8-12. TACTICAL UNLOADING USING DOOR NEAREST COVER, CONCEALMENT

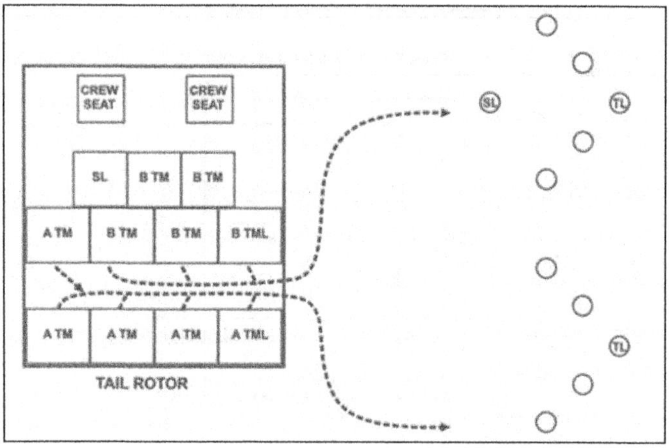

8-5. **SAFETY**. Safety is the primary concern of all leaders when operating in/around aircraft. The inclusion of aircraft into Ranger operations carries an inherent high-risk factor. The following guidelines are to be considered.

a. Approach the aircraft from 90 degrees to 45 degrees off the nose.

b. Weapons with blank firing adapters are pointed muzzles up.

c. Weapons loaded with live ammunition are muzzles down.

d. Wear the ballistic helmet.

e. When possible, conduct an air crew safety brief with all personnel. At a minimum, cover loading/off-loading, emergencies, and egress procedures.

f. Leaders need to carry a manifest and turn a copy into higher.

8-6. **CLOSE COMBAT ATTACK AVIATION**. Close combat attack is defined as a hasty or deliberate attack in support of units engaged in close combat. During CCA, armed helicopters engage enemy units with direct fire that impacts near friendly forces. Targets may range from a few hundred meters to a few thousand meters. Close combat attack is coordinated and directed by a team, platoon or company level ground unit soldiers using standardized CCA procedures in unit SOP's.

a. **Aircraft Capabilities and Limitations**.

(1) AH-64D (Apache):

- Air speed (knots): 164 max, 120 Cruise combat radius: 200 KM
- 30 mm chain gun: 1200 rds, 3,500 m range
- 2.75" rockets: 19 per pod (76), 3-5 km range
- Hellfire Missile: 8 per side (16), 5-8 km range

(2) Capabilities.

- MOBILITY - AO can cover corps or division area
- SPEED - 100-120 knots day/80-100 Knots night
- VERSATILITY - scout-weapons teams vs. pure attack
- LETHALITY - Attack bn can engage 288 targets
- VIDEO RECONNAISSANCE - Provide near real time intelligence

(3) ***Limitations***.

- Throat ID with FLIR
- Low ceilings (Clouds) <500' AGL degrades hellfire capability
- Combat service support consume large amounts of class III, V, and IX

(4) **OH-58D (Kiowa)**.

- Air speed (knots): 125 max, 100 cruise combat radius: 120 KM
- .50 cal MG: 500 rds, 2,000 m range
- 2.75" rockets: 7 per pod (14), 3-5 km range
- Hellfire: 2 per side (4), 5-8 km range
- Stinger: 2 per side (4), 4 km range

(5) ***Capabilities***.

- Mast-mounted sight
- Ability to designate targets while remaining masked
- Thermal imaging system (day/night)
- Laser designator / Aim1 laser
- Video image crosslink (VIXL)
- Moving map display
- Video

(6) ***Limitations***.

- Power limited
- Infrared crossover
- Battlefield obscurants
- Low ceilings - Hellfire
- Remote designation constraints
- Instrument meteorological conditions

c. **CCA Call-for-Fire Format**.

(1) IP/BP/ABF or friendly location________
- Grid/Lat Long

(2) Known point
- Terrain feature

(3) HDG to TGT ________ (magnetic)
- Specify from IP/BP/ABF or friendly location

(4) DST to TGT __________ (m) (Specify from IP/BP/ABF or friendly location.)

(5) TGT elevation_________ (feet, mean sea level)

(6) TGT Description ________________

(7) TGT location ____________
- Grid
- Known point
- Terrain Feature

(8) Type TGT Mark _________ (day/night)

(9) location of friendly _________
- Omit if previously given
- Grid/Lat Long
- Known point
- Terrain feature

(10) Egress direction__________
- Cardinal direction not to over fly threats

d. **Marking Method Chart**. Table 8-1 shows this chart.

**Table 8-1. MARKING METHOD CHART**

| Method | D | N | NVG | NVS | Friendly | Target |
|---|---|---|---|---|---|---|
| Smoke | X | | | | X | X |
| Smoke IR | X | | X | | X | X |
| Signal Mirror | X | | | | X | |
| IR Laser | | | X | | X | X |
| Tracer | X | X | X | | | X |
| Glint Tape | | | | X | X | |
| CID | X | | | | X | |
| Strobe | | X | X | | X | |
| IR Strobe | | | X | | X | |
| IR Panel | | | | X | X | |
| Chem Light | | X | X | | X | |
| AN/PAQ-4 | | | X | | | X |
| VS-17 Panel | X | | | | X | |
| Spot Light | | X | X | | X | |
| MRE Heater | | | | X | X | |
| AN/PEQ-2 | | | X | | | X |
| Hydra 70 Illum | X | X | X | X | | X |
| Ref FM 3-04.111 Appendix Q for more detail | | | | | | |

e. **Risk-Estimate Distances**. REDs for CCA munitions are shown in Table 8-2.

**Table 8-2. RISK ESTIMATE DISTANCES FOR CCA MUNITIONS**

| *WEAPON* | *DANGER CLOSE* | *MAX EFF RANGE)* |
|---|---|---|
| 50 cal | 150 meters | 2,000 meters |
| 30 mm | 150 meters | 3,500 meters |
| 2.75 rocket | 175 meters | 8,000 meters |
| Hellfire Missile | 500 meters | 8,000 meters |
| Danger Close = Risk-Estimate Distances for 0.1% (1/10th of a percent) probability of incapacitation (PI) per FM 3-09.32. | | |

## Chapter 9
# WATERBORNE OPERATIONS

This chapter discusses rope bridges, poncho rafts, and watercraft.

9-1. **ROPE BRIDGE TEAM.** The Ranger patrol seldom has ready-made bridges, so they must know how to employ field-expedient stream crossing techniques.

a. **Personnel.**

(1) # 1 man................Lead safety swimmer and far side lifeguard.

(2) # 2 man................Rope puller, swims water obstacle pulling 150-foot rope, ties off rope on far side anchor point.

(3) # 3 man................Near side lifeguard is the last man to cross water obstacle.

(4) # 4 man................Bridge team commander (BTC; most knowledgeable person on team).

(5) # 5/6 men............Rope tighteners.

b. **One-Rope Bridge (Wet Crossing).** Special equipment follows:

- Two snaplinks per piece of heavy equipment.
- Two snaplinks for every 120 feet of rope.
- One 14-foot utility rope per person (Swimmer safety line).
- Two snaplinks per person.
- One waterproof bag per RTO.
- Two B-7 life preservers.
- Three flotation work vests.
- Two 150-foot nylon ropes.

c. **Planning.** A stream crossing annex is prepared in conjunction with the unit's operation order. Special organization is accomplished at this time. For a platoon size patrol, a squad is normally given the task of providing the bridge team. The squad leader designates the most technically proficient soldier in the squad as the bridge team commander (BTC).

d. **Rehearsals and Inspections.**

(1) Rehearse the entire stream crossing, emphasizing--

- Security and actions on enemy contact.
- Actual construction of the rope bridge within the 8-minute time standard on dry land.
- Individual preparation.
- Order of crossing.
- All signals and control measures.
- Reorganization.

(2) Conduct rehearsals as realistically as possible.

(3) Ensure personnel are proficient in the mechanics of a stream crossing operation.

(4) Inspect for equipment completeness, correct rigging and preparation, personnel knowledge and understanding of the operation.

(5) Actions of the # 4 man (BTC) during the preparation phase.

(a) Rehearse the bridge team.

(b) Account for all equipment in the bridge kit.

(c) Ensure 150-foot rope is coiled.

e. **Execution Phase.**

(1) Steps for the establishment and conduct of bridge stream crossing:

(a) Leader halts short of the river, local security is established, and a recon is conducted of the area for the presence of the enemy and for crossing site suitability/ necessity. He directs the BTC to construct the bridge.

(b) Security is established up and downstream while unit leader briefs BTC on anchor points. The unit leader counts individuals across.

(c) The bridge team begins to establish the rope bridge while unit members begin individual preparation.

(d) The BTC is responsible for construction of one-rope bridge and selection of the near side anchor point as well as the far side anchor point if visibility permits. He will tie a swimmer's safety line to anchor himself into the bridge. Tying an end of the line bowline around the waist secured with an overhand knot, and on the free running end, an end of line bowline with an overhand knot. A snaplink will be placed on the loop portion of the end of line bowline, which will extend no further than arms length away from the soldier (standard waterborne uniform). This will ensure that the soldier is never more than one arms length away from the rope bridge should he lose his grip.

(e) Noise and light discipline are enforced and security is maintained.

(f) Individual soldiers put a snaplink in their end of the line bowline and the front sight assembly of every M-4/M16 or M203. M240 gunners put a snaplink through the front sight assembly and rear swivel of their M240 MG. RTOs (and others with heavy rucksacks) place an additional snaplink on their rucksack frame, top center.

(2) The responsibilities of the bridge team while constructing the rope bridge are as follows:

(a) # 1 man (Lead safety swimmer and far side lifeguard). He grounds his rucksack (with snaplink through top of frame) to the rear of the near side anchor point. He wears equipment in the following order (body out). He carries a safety line to assume duties of far side lifeguard:

-- Waterborne uniform (top button buttoned, pants unbloused)
-- B7 life preserver or engineer work vest
-- LBV
-- Weapon (across the back)

(b) # 1 man enters the water upstream from the # 2 man and stays an arms length away from the # 2 man on the upstream side. He identifies the far side anchor point upon exiting the water and once the # 2 man has exited the water moves to his far side lifeguard position downstream of the rope bridge with knotted safety line on wrist, LBV LBV/weapon grounded, and work vest held in throwing hand. He continues to wear the B7.

(c) # 2 man (rope puller) in waterborne uniform (same as # 1 man) wears his equipment in the following order. He grounds his rucksack (with snaplink through top of frame) to the rear of the near side anchor point. His duties are to swim across the water obstacle pulling the rope. He ties off the rope on the anchor point identified by the # 1 man with a round turn and two half hitches with a quick release. The direction of the round turn is the same direction as the flow of water (current) to facilitate exit off the rope bridge:

- Work vest.
- LBV.
- Weapon (across the back).
- Swimmer safety line.

(d) # 3 man (near side lifeguard) in the same waterborne uniform as the far side lifeguard. The # 3 man positions himself on the downstream side of the bridge before the # 1/2 men enter the water. He (# 1 man) grounds his rucksack (with snaplink through top of frame) on rear of near-side anchor point. After the PSG crosses and verifies the headcount, the # 1 man unties the quick release at the near side anchor point. The # 3 man reties his safety line into an Australian rappel seat, hooks the end of line bowline into his snaplink, and connects his snaplink to the one on the end-of-line bowline . # 3 man is the last pulled across the water obstacle. Before crossing the water obstacle, he dons his equipment in the following order:

- B7.
- Work vest.
- LBV.
- Weapon.

(e) # 4 man (BTC). He is in the standard waterborne uniform with LBV and sling rope tied in safety line (round the waist bowline with end of line bowline no more than one arms length). He is responsible for construction of rope bridge and organization of bridge team. He is also responsible for back feeding the rope and tying end of line bowlines. He designates the near side anchor point, ties the Figure Eight slip of the transport tightening system, and hooking all personnel to the rope bridge. He ensures that the transport tightening knot is on the upstream side of the rope bridge. He ensures that all individuals are in the waterborne uniform, hooked into the rope facing the current with the safety line routed through the trailing shoulder of the individual's LBV and rucksack. He ensures that the weapon is hooked onto the rope. He controls the flow of traffic on the bridge. He is responsible for crossing with the # 1 man's rucksack. He is generally the next to the last man to cross (follows PSG who is keeping head count).

(f) # 5 and # 6 men (rope tighteners) wear the waterborne uniform with LBV and safety line. They tighten the transport tightening knot. They also take the rucksacks of the # 2 and # 3 men across. Once they reach the far side, # 5 and # 6 men pull the last man (# 3 man) across.

(g) The rucksacks of # 1/2/3 men are transported across by # 4/5/6 men. The rucksacks of the # 1/2/3 men are hooked into the rope by the snaplink through the top of the frames and the # 4/5/6 men pull them across. The weapons of # 4/5/6 men are attached between the # 4/5/6 men and the rucksack that they are pulling across the bridge.

(h) BTC rehearses the bridge team during the planning sessions and directs construction and emplacement. The unit leader selects the crossing site which complements the tactical plan.

- # 3 man positions himself downstream of crossing site.
- # 1 man enters water upstream of # 2. He stays one arm's length from the # 2 man and is prepared to render any assistance to the # 2 man. Both swim in conjunction upstream to compensate for the current. BTC feeds rope out of rucksack positioned on the downstream side of near side anchor point.

(i) The # 1 man exits and identifies the far side anchor point (if BTC cannot identify it for the # 2 man). # 2 man exits on the upstream side of the far side anchor point. The rope is now routed to facilitate movement onto and off the bridge.

(j) Radios and heavy equipment are waterproofed and rigged. All individuals don waterborne uniform and tie safety lines. PSG moves to anchor point and maintains accountability through headcount.

(k) The # 2 man signals the BTC that the rope is temporarily attached to the far side anchor point, and the BTC pulls out excess slack and ties the transport tightening system using a Figure Eight slip. The BTC signals the # 2 man to pull the knot twelve to fifteen feet from the near side anchor point. After this is done the # 2 two man ties round turns 18 to 24 inches off the water with the remaining rope and secures the rope, to itself, with a snaplink. The # 2 man signals the BTC and the pulling team (# 4/5/6) tightens the bridge, pulling the transport tightening system as close as possible to the near side anchor point.

(l) The # 1 man moves downstream and assumes his duties as the far side lifeguard. The bridge team commander will tie off the rope with a round turn and two half hitches around the near side anchor point. The BTC will place himself on the upstream side of the bridge (facing downstream). He begins to hook individuals into the rope inspecting them for safety.

***Note:*** Any Ranger identified as a weak swimmer will cross with no other personnel on the rope bridge. The weak swimmer crosses individually to allow the near and far side lifeguards to focus their attention exclusively on the weak swimmer and not be distracted by other personnel crossing the bridge.

(m) The # 2 man moves to the upstream side of the rope bridge, assists personnel off the rope on the far side, and maintains head count. The # 5 and # 6 men cross with the # 1 and # 2 men's rucksacks..

(n) The BTC maintains the flow of traffic ensuring that no more than three men are on the bridge at any one time (one hooking up, one near the center, and one being unhooked). Once the PSG has accounted for everyone on the near side, he withdraws left and right (L/R) security and sends them across. PSG follows security across. The # 3 man hooks the BTC (with # 3's ruck) onto the rope. Once the BTC has crossed, the # 3 man unhooks near side anchor point and the BTC unties far side anchor point. The # 3 man ties an Australian rappel seat with snaplink to the front, hooks onto the snaplink that is in the end of the line bowline on the 120-foot rope, and then signals the # 4/5/6 men to take in slack. The # 3 man extends arms in front of his head, slightly upstream to fend off debris and is pulled across by the # 4/5/6. All individuals (except # 1/2/3 and RTOs) wear rucksacks across. The # 4/5/6 men hook the rucksacks of the # 1/2/3 men onto the bridge by the snaplink. All the men cross facing upstream.

(o) Once the far side headcount, weapons and equipment are verified (between PSG and # 5 man), personnel reorganize, and then they continue the mission.

(p) Personnel with heavy equipment:

**M240** All major groups are tied together with 1/4-inch cord. An anchor line bowline runs through the rear swivel, down the left side of gun. Tie a round turn through the trigger guard. Route the cord down the right side and tie off two half hitches around the forearm assembly with a round turn and two half hitches through the front sight posts. The remainder of the working end is tied off with an end of the rope bowline approximately one foot from the front sight post large enough to place leading hand through. The M240 is secured to the bridge by snaplinks on the front sight post and rear swivel. The M240 is pulled across by the trailing arm of the M240 gunner.

**AN-PRC/II9s** These are waterproofed before crossing a one-rope bridge. Once far side FM commo is established, the near side RTO breaks down and waterproofs his radio, and prepares to cross the bridge. He puts a snaplink in the top center of the rucksack frame (same as for # 1/2/3 men). The BTC will hook the rucksack to the rope.

***Note:*** Using two snaplinks binds the load on the rope. Adjust arm straps all the way out. RTO pulls radio across the rope bridge.

9-2. **PONCHO RAFT.** Normally a poncho raft is constructed to cross rivers and streams when the current is not swift. A poncho raft is especially useful when the unit is still dry and the platoon leader desires to keep the individuals equipment dry.

a. **Equipment Requirements**.

(1) Two serviceable ponchos.

(2) Two weapons (poles can be used in lieu of weapons).

(3) Two rucksacks per team.

(4) 10 feet of utility cord per team.

(5) One sling rope per team.

b. **Conditions**. Poncho rafts are used to cross water obstacles when any or all of the following conditions are found:

(1) The water obstacle is too wide for 150-foot rope.

(2) No sufficient near or far shore anchor points are available to allow rope bridge construction.

(3) Under no circumstances will poncho rafts be used as a means to cross a water obstacle if an unusually swift current is present.

c. **Choosing a Crossing Site**. Before a crossing site is used, a thorough reconnaissance of the immediate area is made. Analyzing the situation using METT-TC, the patrol leader chooses a crossing site that offers as much cover and concealment as possible and has entrance and exit points that are as shallow as possible. For speed of movement, it is best to choose a crossing site that has near and far shore banks that are easily traversed by an individual Ranger.

d. **Execution Phase**. Steps for the construction of a poncho raft:

(1) Pair off the unit/patrol in order to have the necessary equipment.

(2) Tie off the hood of one poncho and lay out on the ground with the hood up.

(3) Weapons are then placed in the center of the poncho, approximately 18 inches apart, muzzle to butt.

(4) Next, rucksacks and LBV are placed between the weapons with the two individuals placing their rucksacks as far apart as possible.

(5) The two will then start to undress (bottom to top), first with their boots, taking the laces completely out for subsequent use as tie downs if necessary).

(6) The boots are then placed over muzzle/butt of weapon toe in.

(7) Members continue to undress, folding each item neatly and placing it on top of their boots.

(8) Once all of the equipment is placed between the two weapons or poles, the poncho is snapped together. The snapped portion of the poncho is then lifted into the air and tightly rolled down to the equipment. Start at the center and work out to the end of the raft creating pigtails at the end. This is accomplished much easier if done by both soldiers together. The pigtailed ends are then folded in toward the center top of the raft and tied off with a single boot lace.

(9) The other poncho is then laid out on the ground with the hood up and the first poncho with equipment is placed in its center. The second poncho is then snapped, rolled and tied in the same manner as the first poncho. The third and fourth boot laces (or utility cord) are then tied around the raft approximately one foot from each end for added security. The poncho raft is now complete.

***Note:*** The patrol leader must analyze the situation using METT-TC and make a decision on the uniform for crossing the water obstacle such as weapons inside the poncho raft or slung across the back, remaining dressed or stripped down with clothes inside raft.

9-3. **WATERCRAFT.** Use of inland and coastal waterways may add flexibility, surprise, and speed to tactical operations. Use of these waterways will also increase the load carrying capacity of normal dismounted units. Watercraft are employed in reconnaissance and assault operations.

a. **Inflation Method**. Inflatable with foot pumps using four separate valves located on the inside of the buoyancy tubes. Each of the four valves are used to section off the assault boat into eight separate airtight compartments. To pump air into the boat, turn all valves into the "orange" or "inflate" section of the valve. Once the assault boat is filled with air, turn all valves onto the "green" or "navigation" section. This will section the assault boat into eight separate compartments.

b. **Characteristics**.

- Overall length: 15 feet, 5 inches.
- Overall width: 6 feet, 3 inches.
- Weight: 265 pounds.
- Maximum payload: 2,756 pounds.
- Crew : 1 coxswain + either 10 paddlers or a 65 HP short-shaft outboard motor.

c. **Preparation**.

(1) ***Rubber Boat***.

- All rubber boats will have a twelve-foot bowline secured to the front starboard D-Ring. This rope will be tied with an anchor-line bowline, and the knot will be covered with 100-MPH tape.
- All rubber boats will have a fifteen foot center line tied to the rear floor D-Ring. The same procedure for securing the bowline will be used for the centerline.
- All rubber boats will be filled to 240 millibars of air, and checked to ensure that all valve caps are tight, and set in the NAVIGATE position.
- All rubber boats will be equipped with one foot pumps. These will be placed in the front pouches of the rubber boat. If no pouches are present, the foot pumps will be placed on the floor.
- All rubber boats will be inspected with the maintenance chart.

(2) ***Personnel and Equipment***.

- All personnel will wear work vest or kapoks (or another suitable positive flotation device).
- LBV is worn over the work vest, unbuckled at the waist.
- Individual weapon is slung across the back, muzzle pointed down and facing toward the inside of the boat.
- Crew served weapons, radios, ammunition and other bulky equipment is lashed securely to the boat to prevent loss if the boat should overturn. Machine guns with hot barrels are cooled prior to being lashed inside the boats.
- Radios and batteries are waterproofed.
- Pointed objects are padded to prevent puncturing the boat.

NOTES

d. **Positions.** Assign each individual a specific boat position (Figure 9-1).

e. **Duties**.

(1) Designate a commander for each boat, (normally coxswain)

(2) Designate a navigator (normally a leader within the platoon) - observer team as necessary.

(3) Crew is positioned as shown in Figure 9-2.

(4) Duties of the Coxswain.

(a) Responsible for control of the boat and actions of the crew.

(b) Supervises the loading, lashing and distribution of equipment.

(c) Maintains the course and speed of the boat.

(d) Gives all commands.

(5) # 2 paddler (long count) is responsible for setting the pace.

(6) # 1 paddler is the observer, stowing and using the bowline unless another observer is assigned.

f. **Embarkation and Debarkation Procedures**.

(1) When launching, the crew will maintain a firm grip on the boat until they are inside it: similarly, when beaching or debarking, they hold on to the boat until it is completely out of the water. Loading and unloading is done using the bow as the entrance and exit point.

(2) Keep a low center of mass when entering and existing the boat to avoid capsizing. Maintain 3 points of contact at all times.

(3) The long count is a method of loading and unloading by which the boat crew embarks or debarks individually over the bow of the boat. It is used at river banks, on loading ramps, and when deep water prohibits the use of the short count method.

(4) The short count is a method of loading or unloading by which the boat crew embarks or debarks in pairs over the sides of boat while the boat is in the water. It is used in shallow water allowing the boat to be quickly carried out of the water. The short count method of organization is primarily used during surf operations.

(5) Beaching the boat is a method of debarking the entire crew at once into shallow water and quickly carrying the boat out of the water.

g. **Commands**. Commands are issued by the coxswain to ensure the boat is transported over land and controlled in the water. All crew members learn and react immediately to all commands issued by the coxswain. T he various commands are as follows:

(1) "Short Count...count off," Crew counts off their position by pairs, such as 1,2,3,4,5 (passenger # 1, # 2, if applicable) coxswain.

(2) "Long Count...count off," Crew counts off the position by individual, such as 1,2,3,4,5,6,7,8,9,10, (Passenger # 1, # 2, if applicable), coxswain.

(3) "Boat Stations," Crew takes positions along side the boat.

(4) "High Carry...Move," (used for long distance move overland).

(a) On the preparatory command of "High Carry," the crew faces the rear of the boat and squats down grasping carrying handles with the inboard hand.

(b) On the command "Move," the crew swivels around, lifting the boat to the shoulders so that the crew is standing and facing to the front with the boat on their inboard shoulders.

(c) Coxswain guides the crew during movement.

(5) "Low Carry...Move," (used for short distance moves overland).

(a) On preparatory command of "Low Carry," the crew faces the front of the boat, bent at the waist, and grasps the carrying handles with their inboard hands.

(b) On the command of "Move," the crew stands up straight raising the boat approximately six to eight inches off the ground.

(c) Coxswain guides the crew during movement.

(6) "Lower the Boat----Move," Crew lowers the boat gently to the ground using the carrying handles.

(7) "Give Way Together," Crew paddles to front with # 2 setting the pace.

(8) "Hold," Entire crew keeps paddles straight downward motionless in the water stopping the boat.

(9) "Left side hold (Right)," Left crew holds, right continues with previous command.

(10) "Back paddle," Entire crew paddles backward propelling the boat to the rear.

**Figure 9-1. BOAT POSITIONS**

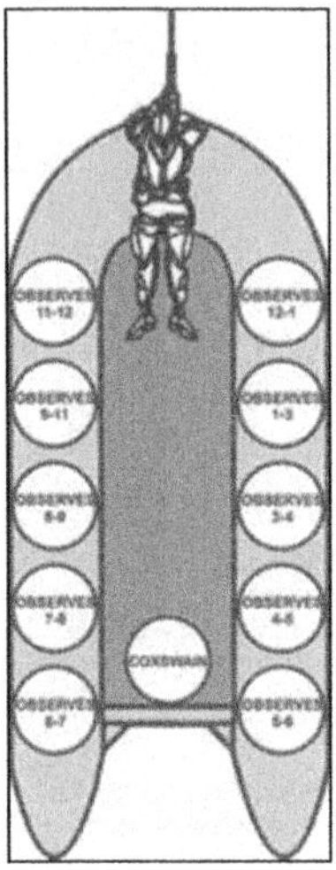

**Figure 9-2. CREW POSITIONS, LONG COUNT AND SHORT COUNT**

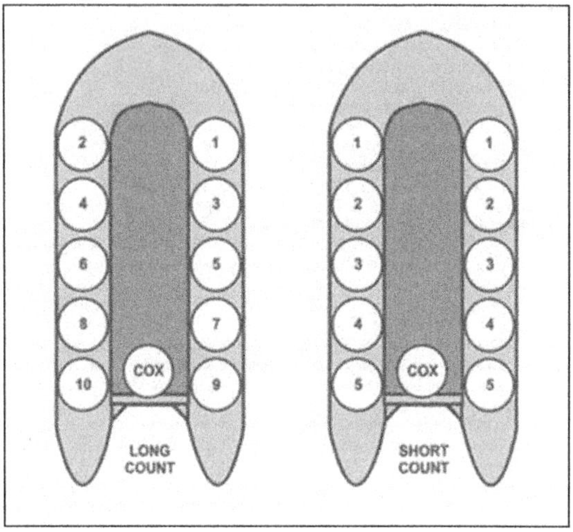


(11) "Back Paddle Left" (Right), Left crew back paddles causing the boat to turn left, right crew continues with previous command.

(12) "Rest Paddles," Crew members place paddles on their laps with blades outboard. This command may be given in pairs such as "# 1s, rest paddles").

h. **Securing of Landing Site.**

(1) If the landing site cannot be secured prior to the waterborne force landing, some form of early warning, such as scout swimmers, is considered. These personnel swim to shore from the assault boats and signal the boats to land. All signals and actions are rehearsed prior to the actual operation.

(2) If the patrol is going into an unsecured landing site, it can provide security by having a security boat land, reconnoiter the landing site and then signal to the remaining boats to land. This is the preferred technique.

(3) The landing site can be secured by force with all the assault boats landing simultaneously in a line formation. While this is the least desirable method of securing a landing site, it is rehearsed in the event the tactical situation requires its use.

(4) Arrival at the debarkation point.

(a) Unit members disembark according to leaders order (Figure 9-3).

(b) Local security is established.

(c) Leaders account for personnel and equipment.

(d) Unit continues movement.

- Soldiers pull security initially with work vest on.
- Coxswains and two men unlash and de-rig rucksacks.
- Soldiers return in buddy teams to secure rucksack and drop off work vest.
- Boats are camouflaged/cached if necessary prior to movement.

**Figure 9-3. DEBARKATION**

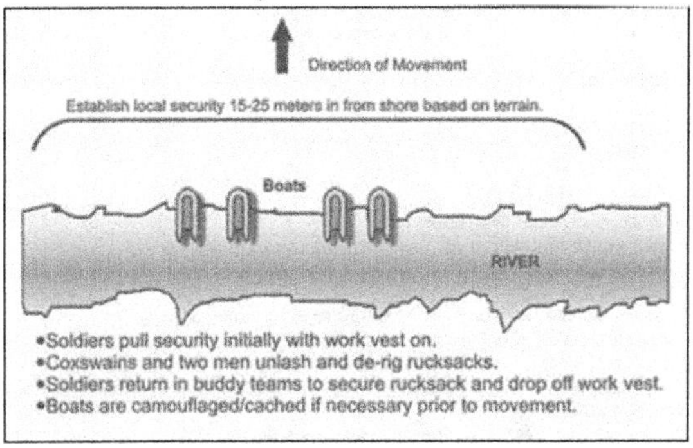


i. **Capsize Drill.** The following commands and procedures are used for capsize drills or to right an over-turned boat,

(1) ***PREPARE TO CAPSIZE.*** This command alerts the crew and they raise paddles above their heads, with the blade pointed outward. Before capsizing, the coxswain will conduct a long count.

(2) ***PASS PADDLES.*** All paddles are passed back and collected by the # 9/10 men.

(3) ***CAPSIZE THE BOAT.*** All personnel slide into the water except the # 3/5/7 men. The # 1 man secures the bowline. The three men in the water grasp the capsize lines (ensuring the lines are routed under the safety lines) and stand on the buoyancy tubes opposite the capsize lines anchor points. The boat is then turned over by the # 3/5/7 men, who lean back and straighten their legs while pulling back on the capsize lines. As the boat lifts off the water, the # 4 man grasps the center carrying handle and rides the boat over. Once the boat is over, the # 4 man helps the # 3/7 men back onto the boat. The # 5 man holds onto the center carrying handle and turns the boat over the same way. The # 5 man rides the boat back over and helps the rest of the crew into the boat. As soon as the boat is capsized, the coxswain commands a long count to ensure that no one sank or was trapped under the boat. Every time the boat is turned over, he conducts another long count.

j. **River Movement.**

(1) ***Characteristics of River.***

(a) Know local conditions prior to embarking on river movement.

(b) A bend is a turn in the river course.

(c) A reach is a straight portion of river between two curves.

(d) A slough is a dead end branch from a river. They are normally quite deep and can be distinguished from the true river by their lack of current

(e) Dead water is a part of the river, due to erosion and changes in the river course that has no current. Dead water is characterized by excessive snags and debris.

(f) An island is usually a pear-shaped mass of land in the main current of the river. Upstream portions of islands usually catch debris and are avoided.

(g) The current in a narrow part of a reach is normally greater than in the wide portion.

(h) The current is greatest on the outside of a curve; sandbars and shallow water are found on the inside of the curve.

(i) Sandbars are located at those points where a tributary feeds into the main body of a river or stream.

(j) Because the # 1 and # 2 men are sitting on the front left and right sides of the boat, they observe for obstacles as the boat moves downriver. If either notices an obstacle on either side of the boat, he notifies the coxswain. The coxswain then adjusts steering to avoid the obstacle.

(2) ***Navigation***. The patrol leader is responsible for navigation. The three acceptable methods of river navigation are--

(a) Checkpoint and general route. These methods are used when the drop site is marked by a well-defined checkpoint and the waterway is not confused by a lot of branches and tributaries. They are best used during daylight hours and for short distances.

(b) Navigator-observer method. This method is the most accurate means of river navigation and is used effectively in all light conditions.

(c) *Equipment Needed*.

- Compass
- GPS
- Photo map (1st choice)
- Topographical map (2nd choice)
- Poncho (for night use)
- Pencil/Grease pencil
- Flashlight (for night use)

(d) *Procedure*. Navigator is positioned in center of boat and does not paddle. During hours of darkness, he uses his flashlight under the poncho to check his map. The observer (or **# 1** man) is at the front of the boat.

(a) The navigator keeps his map and compass oriented at all time.

(b) The navigator keeps the observer informed of the configuration of the river by announcing bends, sloughs, reaches, and stream junctions as shown on his map.

(c) The observer compares this information with the bends, sloughs, reaches, and stream junctions he actually sees. When these are confirmed, the navigator confirms the boat's location on his map.

(d) The navigator also keeps the observer informed of the general azimuths of reaches as shown on his map and the observer confirms these with actual compass readings of the river.

(e) The navigator announces only one configuration at a time to the observer and does not announce another until it is confirmed and completed.

(f) A strip-map drawn on clear acetate backed by luminous tape may be used. The drawing is to scale or a schematic. It should show all curves and the azimuth and distance of all reaches. It may also show terrain features, stream junctions, and sloughs.

k. **Formations.** Various boat formations are used (day and night) for control, speed, and security (Figure 9-4). The choice of which is used depends on the tactical situation and the discretion of the patrol leader. He should use hand and arm signals to control his assault boats. The formations are:

- Wedge
- Line
- File
- Echelon
- Vee

**Figure 9-4. FORMATIONS**

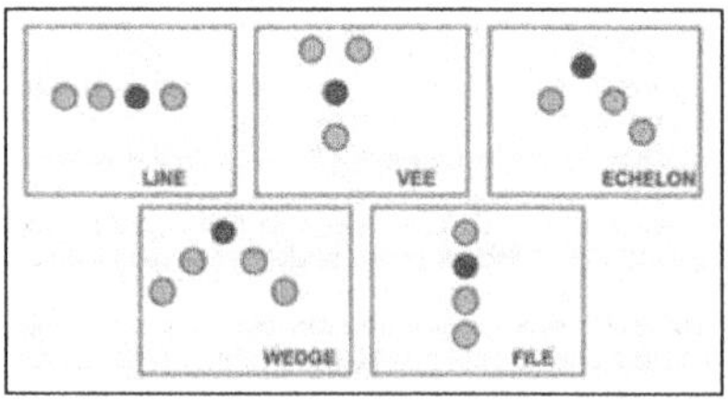

## Chapter 10
# MILITARY MOUNTAINEERING

In the mountains, commanders face the challenge of maintaining their units' combat effectiveness and efficiency. To meet this challenge, commanders conduct training that provides Rangers with the mountaineering skills necessary to apply combat power in a rugged mountain environment, and they develop leaders capable of applying doctrine to the distinct characteristics of mountain warfare.

10-1. **TRAINING**. Military mountaineering training provides tactical and fundamental mobility skills needed to move units safely and efficiently in mountainous terrain. Rangers should receive extensive training prior to executing combat operations in mountainous environments. Some of the areas to train are as follows:

- Characteristics of the mountain environment.
- Care and use of basic mountaineering equipment.
- Mountain bivouac techniques.
- Mountain communications.
- Mountain travel and walking techniques.
- Mountain navigation, hazard recognition and route selection.
- Rope management and knots.
- Natural and artificial anchors.
- Belay and rappel techniques.
- Installation construction and use such as rope bridges.
- Rock climbing fundamentals.
- Rope bridges and lowering systems.
- Individual movement on snow and ice.
- Mountain stream crossings (to include water survival techniques).

10-2. **DISMOUNTED MOBILITY**. Movement in more technical terrain demands specialized skills and equipment. Before Rangers can move in such terrain, a technical mountaineering team might have to secure the high ground.

10-3. **MOUNTAINEERING EQUIPMENT**. Mountaineering equipment is everything that allows the trained Ranger to accomplish many tasks in the mountains. The importance of this gear to the mountaineer is no less than that of the rifle to the Infantry Soldier.

a. **Ropes and Cords**. Ropes and cords are the most important pieces of mountaineering equipment. They secure climbers and equipment during steep ascents and descents. They are also used to install other ropes and to haul equipment. From WWII until the 1980's, US armed forces mainly used 7/16 *nylon laid rope*, often referred to as *green line* for all mountaineering operations. Since kernmantle ropes were introduced, other, more specialized ropes have in many cases replaced the all-purpose green line. Kernmantle ropes are made much like parachute cord, with a smooth sheath around a braided or woven core. Laid ropes are still used, but should be avoided whenever rope failure could cause injury or loss of equipment.

(1) ***Types***. Kernmantle rope is classified as *dynamic* or *static*:

(a) *Dynamic Ropes*. Ropes used for climbing are classified as *dynamic ropes*. These rope stretch or elongate 8 to 12 percent once subjected to weight or impact. This stretching is critical in reducing the impact force on the climber, anchors, and belayer during a fall by softening the catch. An 11-mm by 150-meter rope is standard for military use. Other sizes are available.

(b) *Static Ropes*. Static ropes are used in situations where rope stretch is undesired, and when the rope is subjected to heavy static weight. Static ropes should never be used while climbing, since even a fall of a few feet could generate enough impact force to injure climber and belayer, and/or cause anchor failure. Static Ropes are usually used when constructing rope bridges, fixed rope installations, vertical haul lines, and so on.

(2) ***Sling Ropes and Cordelettes***. A short section of static rope or static cord is referred to as a sling rope or cordelette. These are a critical piece of personal equipment during mountaineering operations. They are usually 7 to 8mm diameter and up to 21 feet long. The minimum Ranger standard is 8mm by 15 feet.

(3) ***Rope Care***. Ropes used daily should be used no longer that one year. An occasionally used rope can be used generally up to 4 to 5 years if properly cared for.

- Inspect ropes thoroughly before, during, and after use for cuts, frays, abrasion, mildew, and soft or worn spots.
- Never step on a rope or drag it on the ground unnecessarily.
- Avoid running rope over sharp or rough edges (pad if necessary).

- Keep ropes away from oil, acids, and other corrosive substances.
- Avoid running ropes across one another under tension (nylon to nylon contact will damage ropes).
- Do not leave ropes knotted or under tension longer than necessary.
- Clean in cool water, loosely coil, and hang to dry out of direct sunlight. Ultraviolet light rays harm synthetic fibers. When wet, hang rope to drip-dry on a rounded wooden peg, at room temperature (do not apply heat).

(4) ***Webbing and Slings***. Loops of tubular webbing or cord, called slings or runners, are the simplest pieces of equipment and some of the most useful. The uses for these simple pieces are endless, and they are a critical link between the climber, the rope, carabiners, and anchors. Runners are usually made from either 9/16- or 1-inch tubular webbing and are either tied or sewn by a manufacturer.

b. **Carabiners**. The carabiner is one of the most versatile pieces of equipment available in the mountains. This simple piece of gear is the critical connection between the climber, his rope, and the protection attaching him to the mountain. New metal alloys are stronger to hold hard falls; and, they are lighter than steel, which allows the Ranger to carry many at once. Consequently, steel carabiners are used less and less often. Basic carabiners come in several different shapes.

c. **Protection**. This is the generic term used to describe a piece of equipment, natural or artificial, that is used to construct an anchor. Protection, along with a climber, belayer, and climbing rope, forms the lifeline of the climbing team. The rope connects two climbers, and the protection connects them to the rock or ice. The two types of artificial protection are traditional (removable) and fixed (usually permanent) (Figures 10-1 and 10-2).

**Figure 10-1. EXAMPLES OF TRADITIONAL (REMOVABLE) PROTECTION USED ON ROCK**

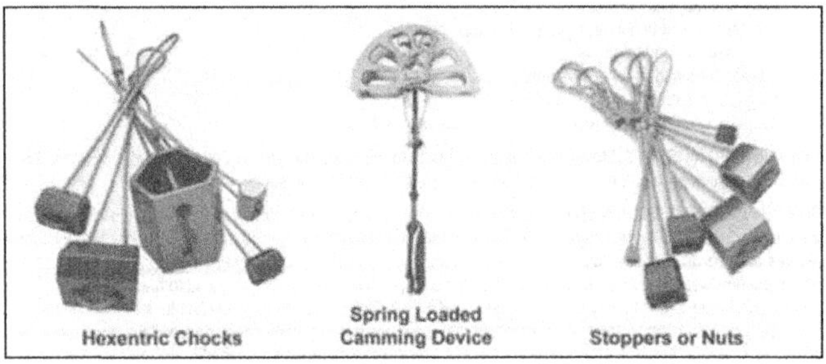


**Figure 10-2. EXAMPLES OF FIXED (PERMANENT OR SEMI) PROTECTION USED ON ROCK**

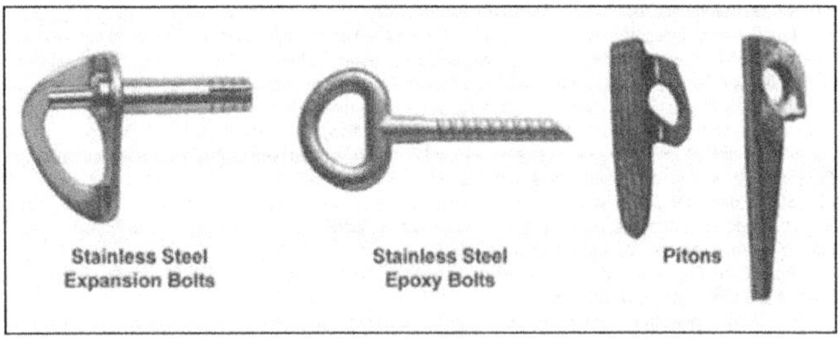

10-4. **ANCHORS**. Anchors form the base for all installations and roped mountaineering techniques. Anchors must be strong enough to support the entire weight of the load or of the impact placed upon them. Several pieces of artificial or natural protection may be incorporated together to make one multipoint anchor. Anchors are classified as artificial or natural:

a. **Artificial Anchors**. Artificial anchors are made from man-made materials. The most common ones use traditional or fixed protection (Figure 10-3).

**Figure 10-3. CONSTRUCTION OF THREE-POINT, PRE-EQUALIZED ANCHOR WITH FIXED ARTIFICIAL PROTECTION**

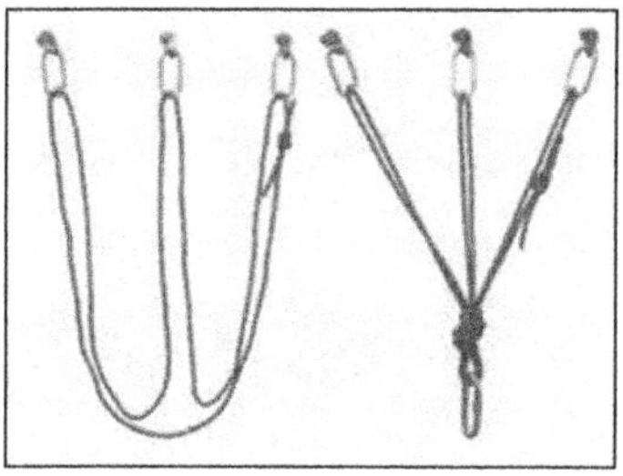

b. **Natural Anchors**. Natural anchors are usually very strong and often simple to construct with little equipment. Trees, shrubs, and boulders are the most common anchors. All natural anchors require is a method of attaching a rope. Regardless of the type of natural anchor used, it must support the weight of the load.

(1) ***Trees***. These are probably the most widely used of all anchors. In rocky terrain, trees usually have a very shallow root system. Check this by pushing or tugging on the tree to see how well it is rooted. Anchor as low as possible to keep the pull on the base of the tree, where it is nearer its own anchor and therefore stronger. Use padding on soft, sap-producing trees to keep sap off ropes and slings.

(2) ***Rock Projections and Boulders***. You can use these, but they must be heavy enough, and they must have a stable enough base to support the load.

(3) ***Bushes and Shrubs***. If no other suitable anchor is available, route a rope around the bases of several bushes. As with trees, place the anchoring rope as low as possible. Make sure all vegetation is healthy and well rooted to the ground.

(4) ***Tensionless Anchor***. Use this to anchor rope on high-load installations such as bridging. The wraps of the rope around the anchor (Figure 10-4) absorb the tension, keeping it off the knot and carabiner. Tie this anchor with at least four wraps around the anchor. A smooth anchor, such as a pipe or rail, might require several more wraps. Wrap the rope from top to bottom. Place a fixed loop into the end of the rope, and attach it loosely back onto the rope with a carabiner.

**Figure 10-4. TENSIONLESS NATURAL ANCHOR**

10-5. **KNOTS**.

a. **Square Knot**. This joins two ropes of equal diameter (Figure 10-5). It has two interlocking bights; the running ends exit on the same side of the standing portion of the rope. Each tail is secured with an overhand knot on the standing end. When you dress the knot, leave at least a 4-inch tail on the working end.

**Figure 10-5. SQUARE KNOT**

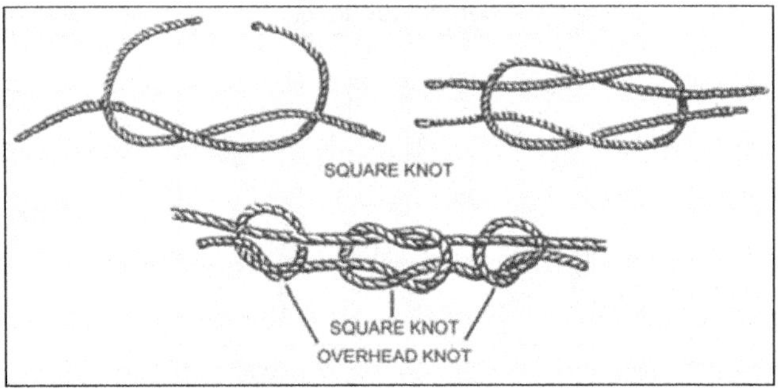


b. **Round-Turn with Two Half Hitches**. This is a constant tension anchor knot (Figure 10-6). The rope forms a complete turn around the anchor point (thus the name round turn), with both ropes parallel and touching, but not crossing. Both half hitches are tightly dressed against the round turn, with the locking bar on top. When you dress the knot, leave at least a 4-inch tail on the working end.

**Figure 10-6. ROUND-TURN WITH TWO HALF-HITCHES**

c. **End-of-the-Rope Clove Hitch**. The end-of-the-rope clove hitch (Figure 10-7) is an intermediate anchor knot that requires constant tension. Make two turns around the anchor. A locking bar runs diagonally from one side to the other. Leave no more than one rope width between turns of rope. Locking bar is opposite direction of pull. When you dress the knot, leave at least a 4-inch tail on the working end.

**Figure 10-7. END-OF-THE-ROPE CLOVE HITCH**

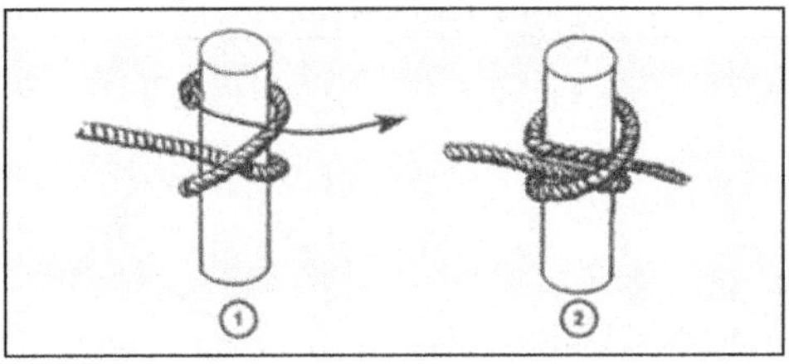


d. **Middle of the Rope Clove Hitch**. This knot (Figure 10-8) secures the middle of a rope to an anchor. The knot forms two turns around the anchor. A locking bar runs diagonally from one side to the other. Leave no more than one rope width between turns. Ensure the locking bar is opposite the direction of pull.

**Figure 10-8. MIDDLE-OF-THE-ROPE CLOVE HITCH**

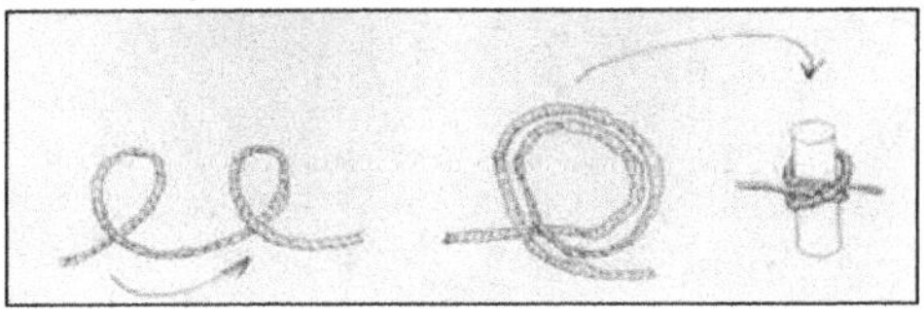

e. **Rappel Seat**. The rappel seat (Figure 10-9) is a rope harness used in rappelling and climbing. It can be tied for use with the left or right hand. Leg straps do not cross, and are centered on buttocks and tight. Leg straps form locking half hitches on rope around waist. A square knot is tied on the right hip and finished with two overhand knots. Tails must be even, no longer than 6 inches. A carabiner is inserted around all ropes, with the gate opening up and away. The carabiner must not contact the square or overhand knot. The rappel seat must be tight enough to prevent a fist from fitting between the rappeller's body and the harness.

**Figure 10-9. RAPPEL SEAT**

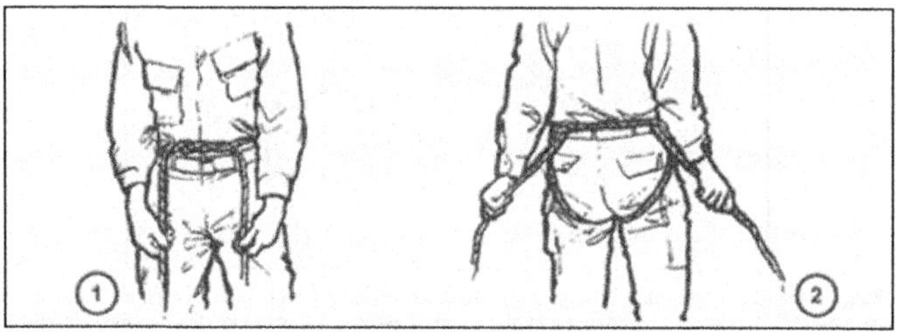


**Figure 10-9. RAPPEL SEAT (CONTINUED)**

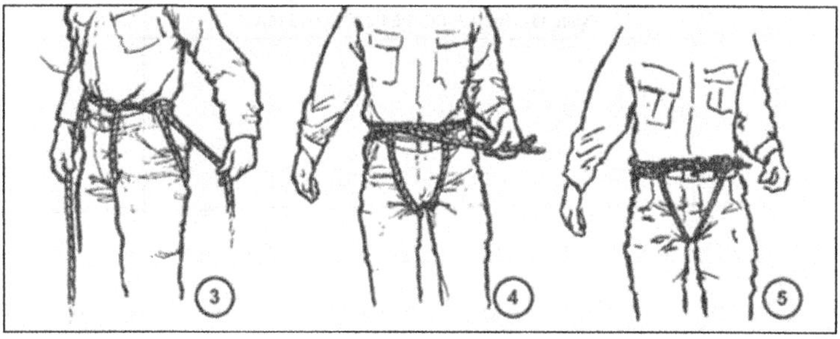

f. **Double Figure Eight.** Use a Figure Eight loop knot (Figure 10-10) to form a fixed loop in the end of the rope. It can be tied at the end of the rope or anywhere along the length of the rope. Figure Eight loop knots are formed by two ropes parallel to each other in the shape of a Figure Eight, no twists are in the Figure Eight. Fixed loops are large enough to insert a carabiner. When you dress the knot, leave at least a 4-inch tail on the working end.

**Figure 10-10. DOUBLE FIGURE EIGHT-LOOP KNOT**

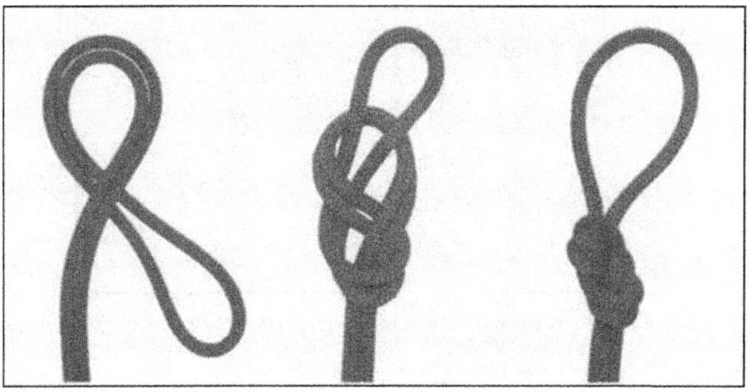

g. **Rerouted Figure Eight Knot.** This anchor knot also attaches a climber to a climbing rope. Form a Figure Eight in the rope, and pass the working end around an anchor. Reroute the end back through to form a double Figure Eight (Figure 10-11). Tie the knot with no twists. When you dress the knot, leave at least a 4-inch tail on the working end.

**Figure 10-11. REROUTED FIGURE EIGHT KNOT**

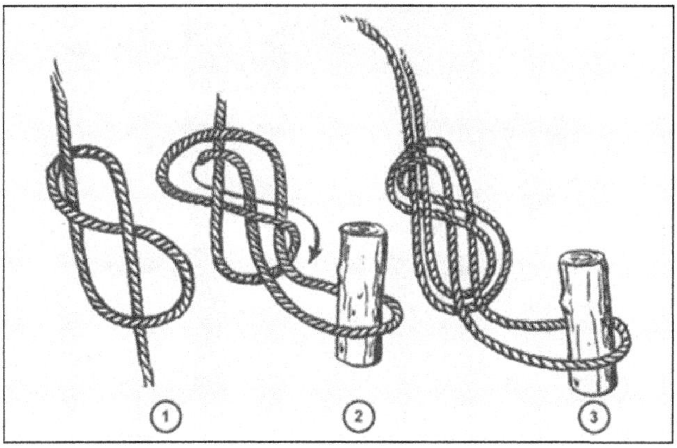

h. **Figure Eight Slip Knot**. The Figure Eight slip knot is used to form an adjustable bight in the middle of a rope. Knot is in the shape of a Figure Eight. Both ropes of the bight pass through the same loop of the Figure Eight. The bight is adjustable by means of a sliding section (Figure 10-12).

**Figure 10-12. FIGURE EIGHT SLIP KNOT**

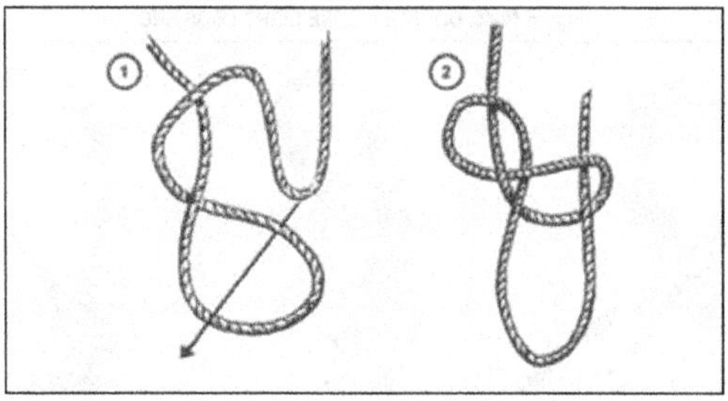


i. **End-of-the-Rope Prusik**. The End-of-the-Rope Prusik (Figure 10-13) attaches a movable rope to a fixed rope. The knot has two round turns, with a locking bar perpendicular to the standing end of the rope. Tie a bowline within 6 inches of the locking bar. When you dress the knot, leave at least a 4-inch tail on the working end.

**Figure 10-13. END-OF-THE-ROPE PRUSIK**

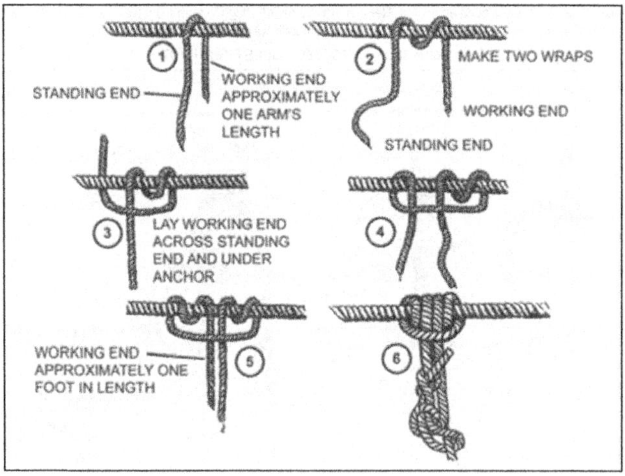

j. **Middle-of-the-Rope Prusik**. The middle-of-the-rope Prusik (Figure 10-14) attaches a movable rope to a fixed rope, anywhere along the length of the fixed rope. Make two round turns with a locking bar perpendicular to the standing end. Ensure the wraps do not cross and that the overhand knot is within 6 inches from the horizontal locking bar. Ensure the knot does not move freely on fixed rope.

**Figure 10-14. MIDDLE-OF-THE-ROPE PRUSIK**

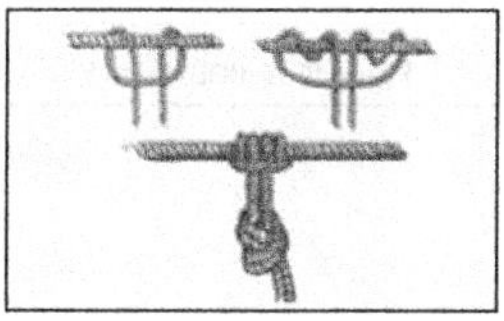

k. **Bowline on a Coil**. This knot (Figure 10-15) secures a climber to the end of a climbing rope. Make at least four parallel wraps around the body, between the hipbone and lower set of ribs. Keep all coils free of clothing, touch and are tight enough to ensure that a fist cannot be inserted between the wraps and the body. Three distinct coils show through the bight of the bowline. The rope coming off the bottom of the coils is on the right side, forward of the hip and forms the bight and the overhand knot. The rope coming off the top of the coils is on the left side, forward of the hips and forms the third and final coil showing through the bight of the bowline. Bowline is centered on the gig line and secured with an overhand knot with a minimum of 4 inch tail remaining after the knot is dressed.

**Figure 10-15. BOWLINE ON A COIL**

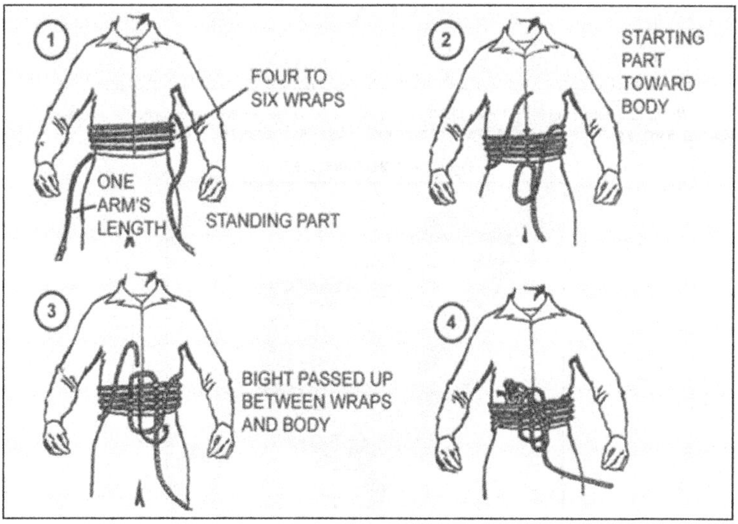

10-6. **BELAYS**. Belaying is any action taken to stop a climber's fall or to control the rate a load descends. The belayer also helps manage the rope for a climber by controlling the amount of rope that is taken out or in, or by controlling a rappeller's rate of descent. The belayer must be stably anchored to prevent him from being pulled out of his position, and losing control of the rope. The two types of belays are body and mechanical belays.

a. **Body Belay**. This belay (Figure 10-16) uses the belayer's body to apply friction by routing the rope around the belayer's body. Caution must be used when belaying from the body since the entire weight of the load may be placed on the belayer's body.

**Figure 10-16. BODY BELAY**

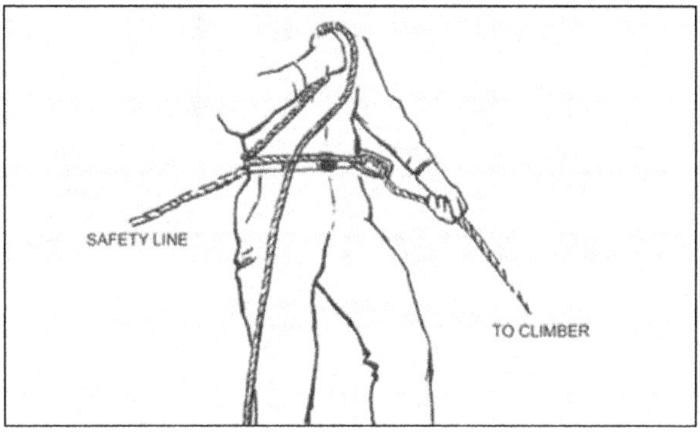


b. **Mechanical Belay**. This belay (Figure 10-17) uses mechanical devices to help the belayer control the rope, as in rappelling. A variety of devices are used to construct a mechanical belay for mountaineering.

**Figure 10-17. MECHANICAL BELAY**

(1) ***Munter Hitch***. One of the most often used belays, the munter hitch (Figure 10-18), requires very little equipment. The rope is routed through a locking, pear-shaped carabiner, then back on itself. The belayer controls the rate of descent by manipulating the working end back on itself with the brake hand.

**Figure 10-18. MUNTER HITCH**

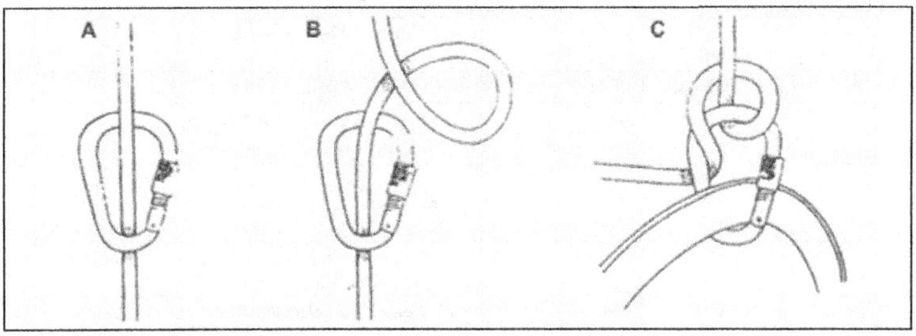


(2) ***Air Traffic Controller***. The ATC (air traffic controller) is a mechanical belay device (Figure 10-19) that locks down on itself when tension is applied in opposite directions. The belayer need apply very little force with his brake hand to control descent or arrest a fall.

**Figure 10-19. AIR TRAFFIC CONTROLLER**

10-7. **CLIMBING COMMANDS**. Table 10-1 shows the sequence of commands used by climber and belayer.

**Table 10-1. SEQUENCE OF CLIMBING COMMANDS**

| *SEQUENCE* | *COMMAND* | *GIVEN BY* | *MEANING* |
|---|---|---|---|
| 1. | BELAY ON, CLIMB | Belayer | Alerts climber that belay is on and climber may climb. |
| 2. | CLIMBING | Climber | Alerts belayer that climber is climbing. |
| 3. | UP-ROPE | Climber | Belayer, remove excess slack in the rope. |
| 4. | BRAKE | Climber | Alerts belayer to immediately apply brake. |
| 5. | FALLING | Climber | Alerts belayer that climber is falling and that the belayer should immediately apply the brake and prepare to arrest the fall. |
| 6. | TENSION | Climber | Alerts belayer to remove all slack from climbing rope until rope is tight, then apply brake and hold position. |
| 7. | SLACK | Climber | Alerts belayer to pull slack into the climbing rope (Belayer may have to assist) |
| 8. | ROCK | Anyone | Alerts everyone that an object is about to fall near them. Belayer immediately applies brake. |
| 9. | POINT | Climber | Alerts belayer that the direction of pull on the climbing rope has changed in the event of a fall. |
| 10. | STAND-BY | Climber or Belayer | Alerts the other to hold position, stand by, I am not ready. |
| 11. | DO YOU HAVE ME? | Climber | Alerts belayer to prepare for a fall or to prepare to lower the climber. |
| 12. | I HAVE YOU | Belay | Alerts climber that the brake is on and the belayer is prepared for the climber to fall, or to lower him. |
| 13. | OFF-BELAY | Climber | Alerts belayer that climber is safetied in or it is safe to come off belay. |
| 14. | 3-METERS | Belayer | Alerts climber to the amount of rope between climber and belayer (May be given in feet or meters) |
| 15. | BELAY-OFF | Belayer | Alerts climber that belayer is off belay |

10-8. **ROPE INSTALLATIONS**. Rope installations may be constructed by teams to assist units in negotiating natural and manmade obstacles. installation teams consist of a squad sized element, with 2 to 4 trained mountaineers. Installation teams deploy early and prepare the AO for safe, rapid movement by constructing various types of mountaineering installations. Following construction of an installation, the squad, or part of it, remains on site to secure and monitor the system, assist with the control of forces across it, and make adjustments or repairs during its use. After passage of the unit, the installation team may then disassemble the system and deploy to another area as needed.

a. **Fixed Rope Installations**. A fixed rope is anchored in place to help Rangers move over difficult terrain. Its simplest form is a rope tied off on the top of steep terrain. As terrain becomes steeper or more difficult, fixed rope systems may require intermediate anchors along the route. Planning considerations are as follows:

- Does the installation allow you to bypass the obstacle?
- (Tactical considerations) Can obstacle be secured from construction through negotiation, to disassembly?
- Is it in a safe and suitable location? Is it easy to negotiate? Does it avoid obstacles?
- Are natural and artificial anchors available?
- Is the area safe from falling rock and ice?

h **Vertical Hauling Line.** This installation (Figure 10-20) is used to haul men and equipment up vertical or near vertical slopes. It is often used in conjunction with the fixed rope.

**Figure 10-20. VERTICAL HAULING LINE**

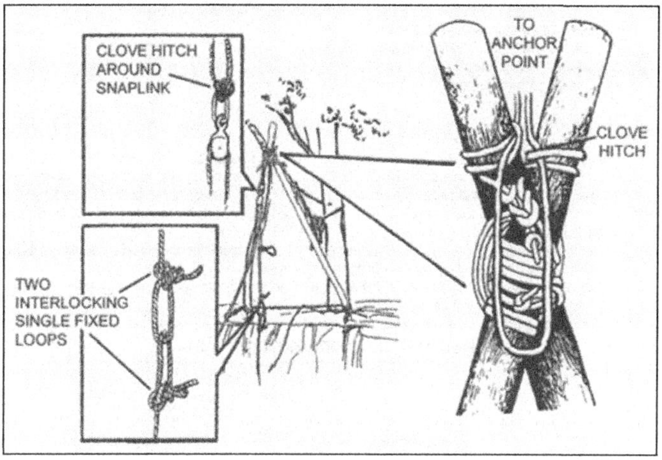


(1) ***Planning Considerations.***

- Does the installation allow you to bypass the obstacle?
- (Tactical considerations) Can you secure the installation from construction through negotiation, to disassembly?
- Does it have good loading and off-loading platforms? Are the platforms natural and easily accessible? Do they provide a safe working area?
- Does it allow sufficient clearance for load? Is there enough space between the slope and the apex of the A-frame to allow easy on-loading and off-loading of troops and equipment?
- Does it have an A-Frame for artificial height?
- Does it allow you to haul line in order to move personnel and equipment up and down slope?
- Does the A frame have a pulley or locking carabiner to ease friction on hauling line?
- Does it have a knotted hand line to help Rangers up the installation?
- Does it allow for Rangers top and bottom to monitor safe operation?

(2) ***Equipment.***

- Three 120-foot (50m) static ropes.
- Three 15-foot sling ropes for constructing A-frame.
- Two A-frame poles, 7 to 9 feet long, 4 to 6 inches in diameter (load dependent).
- Nine carabiners.
- One pulley with steel locking carabiner.

c. **Bridges**. Rope bridges are employed in mountainous terrain to bridge linear obstacles such as streams or rivers where the force of flowing water may be too great or temperatures are too cold to conduct a wet crossing.

(1) ***Construction.*** The rope bridge is constructed using a static ropes. The max span that can be bridged is 50 percent of the ropes entire length for a dry crossing, 75 percent for a wet crossing. The ropes are anchored with an anchor knot on the far side

of the obstacle, and are tied off at the near end with a transport tightening system. Rope Bridge Planning Considerations are as follows:

- Does the installation allow you to bypass the obstacle?
- (Tactical considerations) Can you secure the installation from construction through negotiation, to disassembly?
- Is it in the most suitable location such as a bend in the river? Is it easily secured?
- Does it have near and far-side anchors?
- Does it have good loading and off-loading platforms?

(2) ***Equipment (1-Rope Bridge)***.

- One sling rope per-man.
- One steel locking carabineer.
- Two steel ovals.
- Two 120-foot static ropes.

(3) ***Construction Steps***. The first Ranger swims the rope to the far side and ties a tensionless anchor (Fig 10-4) between knee and chest level, with at least 6 to 8 wraps. The BTC ties a transport tightening system (Fig 10-21) to the near side anchor point. Then, he ties a Figure Eight slipknot and incorporates a locking half hitch around the adjustable bight. Insert 2 steel oval carabineers into the bight so the gates are opposite and opposed. The rope is then routed around the near side anchor point at waist level and dropped into the steel oval carabineers.

- A three-Ranger pulling team moves forward from the PLT. No more than three men should tighten the rope. Using more can overtighten the rope, bringing it near failure.
- Once the rope bridge is tight enough, the bridge team secures the transport-tightening system using two half hitches, without losing more than 4 inches of tension.
- Personnel cross using the commando crawl, rappel seat, or monkey crawl method (Figures 10-22 thru 24).

**Figure 10-21. TRANSPORT TIGHTENING SYSTEM**

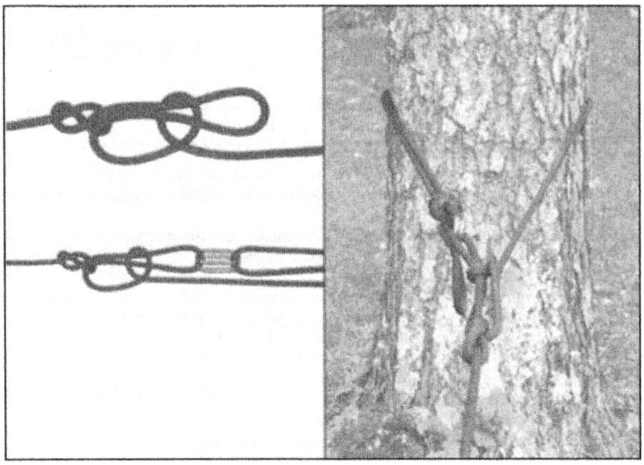

Figure 10-22. COMMANDO CRAWL METHOD

Figure 10-23. RAPPEL SEAT (TYROLEAN TRAVERSE) METHOD

Figure 10-24. MONKEY CRAWL METHOD

(4) ***Bridge Recovery***. Once all but two Rangers have crossed the rope bridge, the bridge team commander (BTC) chooses either the wet or dry method to dismantle it. If he chooses the dry method, he should he should first anchor his tightening system with the transport knot.

- The BTC back-stacks all of the slack coming out of the transport knot, then ties a fixed loop and places a carabiner into the fixed loop.
- The next to last Ranger to cross attaches the carabiner to his rappel seat or harness, and then moves across the bridge using the Tyrolean traverse method.
- The BTC then removes all knots from the system. The far side remains anchored. The rope should now only pass around the near side anchor.
- A three-man pull team, assembled on the far side, takes the end brought across by the next to last Ranger and pulls the rope tight again and holds it.
- The BTC then attaches himself to the rope bridge and moves across.
- Once across, the BTC breaks down the far side anchor, removes the knots, and then pulls the rope across. If the BTC chooses a wet crossing, any method can be used to anchor the tightening system.
- All personnel cross except the BTC or the strongest swimmer.
- The BTC then removes all knots from the system.
- The BTC ties a fixed loop, inserts a carabiner, and attaches it to his rappel seat or harness. He then manages the rope as the slack is pulled to the far side.
- The BTC then moves across the obstacle while being belayed from the far side.

d. **Suspension Traverse**. The suspension traverse is used to move personnel and equipment over rivers, ravines, chasms, and up or down a vertical obstacle. By combining the transport tightening system used during the Rope bridge, an A-Frame used for the Vertical haul Line (Figure 10-25), and belay techniques device, Units can construct a suspension traverse (Figures 10-26 and 10-27). Installation of a suspension traverse can be time-consuming and equipment-intensive. All personnel must be well trained and rehearsed in the procedures.

(1) ***Construction***. The suspension traverse must be constructed using a static ropes. The max span that can be bridged is generally 75 percent length of the shortest rope. Planning considerations are the same as rope bridge and vertical haul line combined.

(2) ***Equipment***.

- Three static installation ropes.
- Seven sling Ropes.
- Nine carabiners.
- One heavy duty double pulley.
- One locking carabiner.
- One canvas pad.

**Figure 10-25. ANCHORING THE TRAVERSE ROPE TO THE A-FRAME**

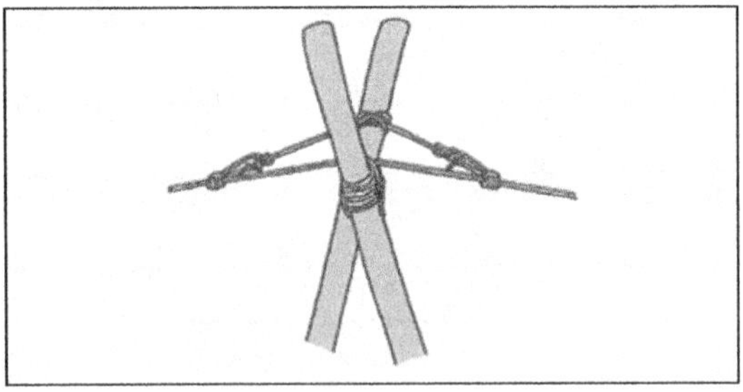

Figure 10-26. CARRYING ROPE FOR USE ON A SUSPENSION TRAVERSE

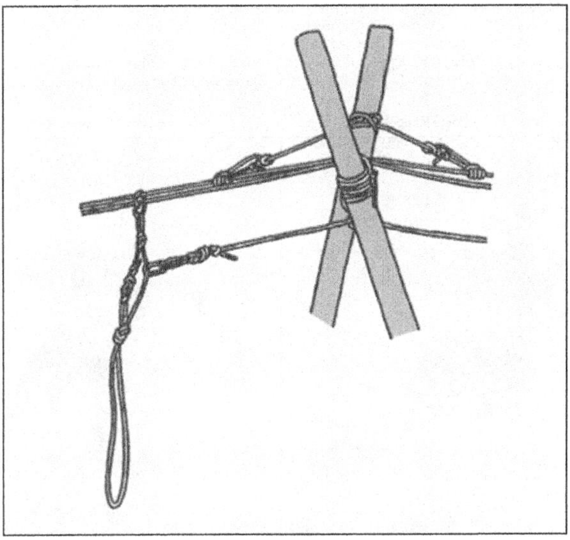

Figure 10-27. SUSPENSION TRAVERSE

10-9. **RAPPELLING**. Rappelling is a quick method of descent but can be extremely dangerous. These dangers include anchor failure, equipment failure, and individual error. Anchors in a mountainous environment should be selected carefully. Great care must be taken to load the anchor slowly and to ensure that no excessive stress is placed on the anchor. The best way to ensure this is to prohibit bounding rappels, and to use only walk-down rappels.

a. **Hasty and Body Rappels**. Hasty (Figure 10-28) and body rappels (Figure 10-29) are easier and faster than other methods, but should only be used on moderate pitches--never on vertical or overhanging terrain. Wear gloves to prevent rope burns.

**Figure 10-28. Hasty Rappel**

**Figure 10-29. Body Rappel**

b. **Seat-Hip Rappel**. The seat-hip rappel uses a mechanical rappel device that is inserted in a sling rope seat and is fastened to the rappeller to provide the necessary friction to remain in control. This method provides a faster and more controlled descent than any other method (Figure 10-30). There are several mechanical rappel devices (Figure 10-31 and 10-32) that can be used for the seat-hip rappel.

**Figure 10-30. FIGURE EIGHT DESCENDER**

**Figure 10-30. FIGURE EIGHT DESCENDER (CONTINUED)**

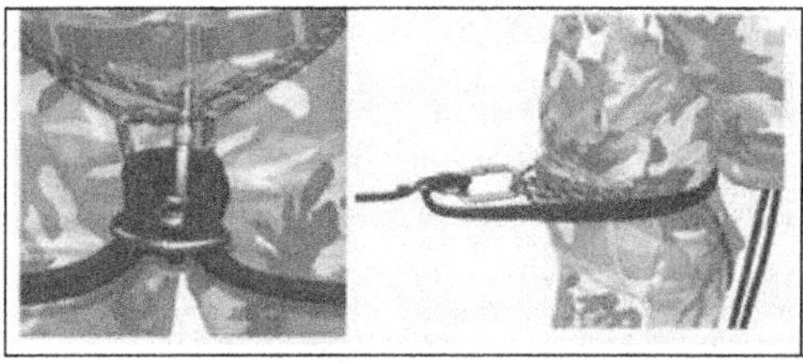

**Figure 10-31. CARABINER WRAP**

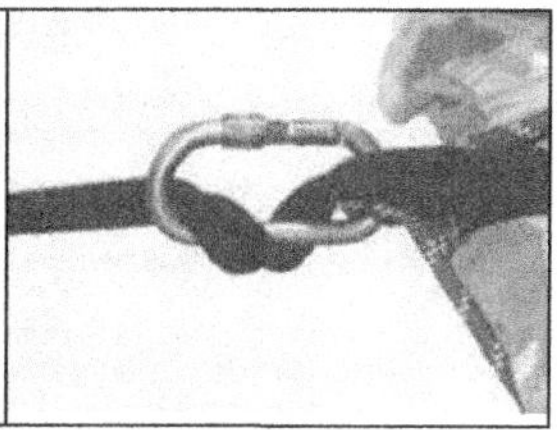

**Figure 10-32. SEAT HIP RAPPEL WITH CARABINER WRAP DESCENDER**

c. **Site Selection**. The selection of the rappel point depends on factors such as mission, cover, route, anchor points, and edge composition (loose or jagged rocks). There must be good anchors (primary and secondary). The anchor point should be above the rappeller's departure point. Suitable loading and off-loading platforms should be available.

- Each rappel point has primary and secondary anchors.
- Rappel point has equal tension between all anchor points.
- Double rope is used when possible.
- Ropes must reach the off loading platform.
- Site has suitable on and off loading platforms.
- Personnel working near the edge are tied in.
- Select a smooth route free of loose rock and debris.

# Chapter 11
# EVASION AND SURVIVAL

This chapter will cover basic material dealing with evading the enemy and survival techniques. Field expedient methods of acquiring food/ water, land navigation, constructing shelters, and making fires will be presented.

11-1. **EVASION**. When you become isolated or separated in a hostile area, either as an individual or as a group, your evasion and survival skills will determine whether or not you return to friendly lines.

a. When unable to continue the mission or unable to rejoin your unit, leave the immediate area, and move to your last rally point.

b. Observe activity in the area and form a plan.

c. Traveling alone offers the least possibility of detection, but traveling in groups of two to three is more desirable.

d. Plan a primary and alternate route. Consider distance, cover, food, and water. The easiest and shortest route may not be the best.

e. Food and water are daily requirements. You can do without food for several days; water, however, is essential.

f. Move at night. Use the daylight to observe, plan, and rest in a hide position.

g. Linkup only during daylight hours. Place friendly lines under observation.

h. Attempt to identify the unit you approach, note their movements and routine.

i. After carefully considering your approach route, make voice contact with the unit as soon as possible.

11-2. **SURVIVAL**.
With training, equipment, and the ***will to survive***, you can overcome *any* obstacle you may face. ***You will survive.*** Understand the emotional states associated with survival. "Knowing thyself" is extremely important in a survival situation. It bears directly on how well you cope with serious stresses, anxiety, pain, injury, illness; cold, heat, thirst, hunger, fatigue, sleep deprivation, boredom, loneliness and isolation.

a. **Memory Aid**. You can overcome and reduce the shock of being isolated behind enemy lines if you keep the key word S-U-R-V-I-V-A-L foremost in your mind. Its letters can help guide you in your actions.

(1) S - Size up the situation, the surroundings, your physical condition, and your equipment.

(2) U - Undue haste makes waste; don't be too eager to move. Plan your moves.

(3) R - Remember where you are relative to friendly and enemy units and controlled areas; water sources (most important in the desert); and good cover and concealment. This information will help you make intelligent decisions.

(4) V - Vanquish fear and panic.

(5) I – Improvise/Imagine. You can improve your situation. Learn to adapt what is available for different uses. Use your imagination.

(6) V - Value living. Remember your goal - to get out alive. Remain stubborn. Refuse to give in to problems and obstacles. This will give you the mental and physical strength to endure.

(7) A - Act like the natives; watch their daily routines and determine when, where, and how they get their food and water.

(8) L - Live by your wits. Learn basic skills.

b. **Navigation**. In a survival situation, you might find yourself without a compass. The ability to determine direction can enable you to navigate back to your unit or to a friendly sanctuary. In sunlight, there are two simple ways to determine direction: the shadow-tip and the watch.

(1) ***Shadow Tip***. Use the sun to find approximate true North. Use this in light bright enough to cast shadows. Find a fairly straight stick about 3 feet long, and follow the diagram below (Figure 11-l).

**Figure 11-1. SHADOW-TIP METHOD**

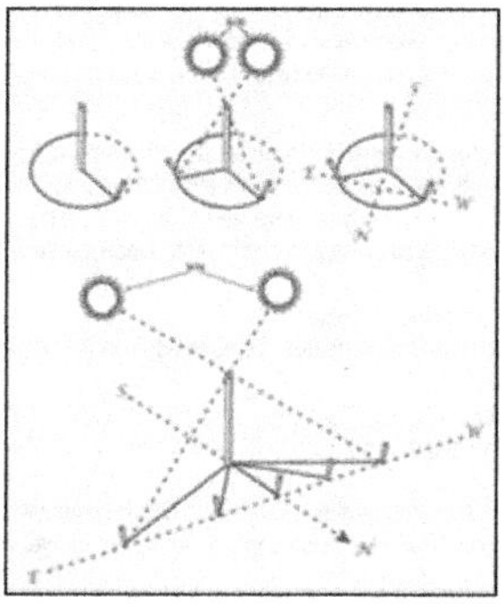

(2) ***Watch Method***. You can also determine direction using a watch (Figure 11-2). The steps you take will depend on whether you are in the northern Temperate Zone or in the southern Temperate Zone. The northern temperate zone is located between 23.4 north and 26.6 north. The southern Temperate Zone is located between 23.4 south and 66.6 south.

**Figure 11-2. WATCH METHOD**

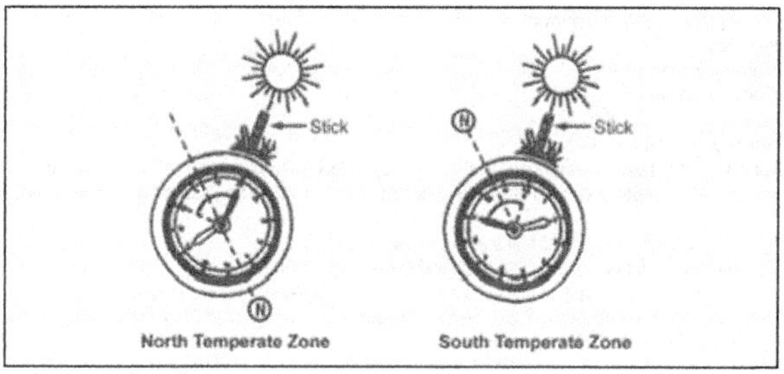


c. **Northern Temperate Zone**. Procedures in the northern temperate zone using a conventional watch are as follows:

(1) Place a small stick in the ground so that it casts a definite shadow.

(2) Place your watch on the ground so that the hour hand points toward and along the shadow of the stick.

(3) Find the point on the watch midway between the hour hand and 12 o'clock and draw an imaginary line from that point through and beyond the center of the watch. This imaginary line is a north-south line. You can then tell the other directions.

***Note:*** If your watch is set on daylight savings time, then use the midway point between the hour hand and 1 o'clock to draw your imaginary line.

d. **Southern Temperate Zone**. Procedures in the southern temperate zone using a conventional watch are as follows:

(1) Place a small stick in the ground so that it casts a definite shadow.

(2) Place your watch on the ground so that 2 o'clock points to and along the shadow.

(3) Find the midway point between the hour and 12 o'clock and draw an imaginary line from the point through and beyond the center of the watch. This is a north-south line.

e. A hasty shortcut using a conventional watch is simply to point the hour hand at the sun in the northern temperate zone (or point the 12 at the sun in the southern temperate zone) and then follow the last step of the watch method above to find your directions. This shortcut, of course, is not as accurate as the regular method but quicker. Your situation will dictate which method to use.

11-3. **WATER**. Water is one of your most urgently needed resources in a survival situation. You can't live long without it, especially in hot areas where you lose so much through sweating. Even in cold areas, you need a minimum of 2 quarts of water a day to maintain efficiency. More than three-fourths of your body is composed of fluids. Your body loses fluid as a result of heat, cold, stress, and exertion. The fluid your body loses must be replaced for you to function effectively. So, one of your first objectives is to obtain an adequate supply of water.

a. **Purification**. Purify all water before drinking. Either--

(1) Boil it for at least one minute (plus 1 more minute for each additional 1,000 feet above sea level) or for a maximum of 10 minutes anywhere.

(2) Use water purification tablets.

(3) Add eight drops of 2-1/2 percent iodine solution to a quart (canteen full) of water. Let it stand for 10 minutes before drinking.

(4) Collect rain water directly in clean containers or on plants. This is generally safe to drink without purifying. Never drink urine or sea water; the salt content is too high. Use old, bluish sea ice, but newer, grayer ice may be salty. Glacier ice is safe to melt and drink.

b. **Desert Environment**. In a desert environment, water has a huge effect on Rangers. If a unit fails to plan properly for water, and resupply is unavailable, then they can run out of water. In the desert, look for four signs of water: animal trails, vegetation, birds, and civilization. Adequate water is critical in a hot environment if a unit is to survive and maintain the physical condition necessary to accomplish the mission. Unit leaders must enforce water discipline and plan for water resupply. The leader can use the following planning considerations for water resupply:

(1) Units average water consumption.

(2) Drop sites.

(3) Aviation support.

(4) DZ and LZ parties.

(5) Caches.

(6) Targets of opportunity (enemy).

c. **Survival Water Still**. Dig a belowground still (Figure 11-3). Select a site where you believe the soil will contain moisture (such as a dry stream bed or a spot where rain water has collected). The soil should be easy to dig, and be in sunlight most of the day:

(1) Dig a bowl-shaped hole about 3 feet across and 2 feet deep.

(2) Dig a sump in center of the hole. The depth and the perimeter of the sump will depend on the size of the container that you have to set inside of it. The bottom of the sump should allow the container to stand upright.

(3) Anchor the tubing to the bottom of the container by forming a loose overhand knot in the tubing.

(4) Place the container upright in the sump.

(5) Extend the unanchored end of the tubing up, over, and beyond the lip of the hole.

(6) Place plastic sheeting over the hole covering the edge with soil to hold it in place.

(7) Place a rock in the center of the plastic.

(8) Allow the plastic to lower into the hole until it is about 15 inches below ground level. The plastic now forms an inverted cone with the rock at its apex. Make sure that the apex of the cone is directly over your container. Also, make sure the plastic cone does not touch the sides of the hole, because the earth will absorb the condensed water.

(9) Put more soil on the edges of the plastic to hold it securely in place and to prevent loss of moisture.

(10) Plug the tube when not being used so that moisture will not evaporate.

d. You can drink water without disturbing the still by using the tube as a straw. You may want to use plants in the hole as a moisture source. If so, when you dig the hole you should dig out additional soil from the sides of the hole to form a slope on which to place the plants. Then proceed as above.

Figure 11-3. SURVIVAL WATER STILL

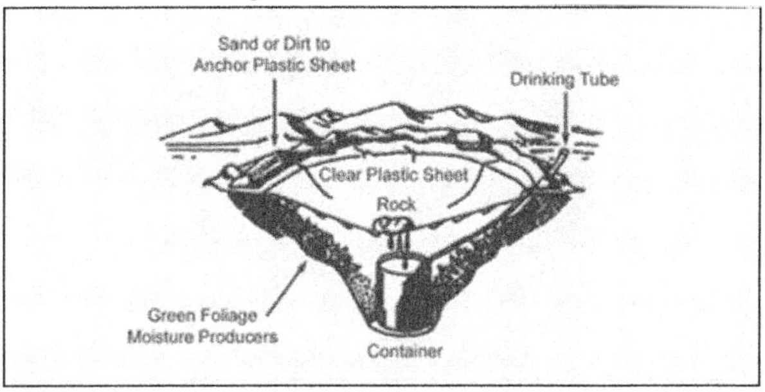


11-4. **PLANT FOOD**. There are many plants throughout the world. Tasting or swallowing even a small portion of some can cause severe discomfort, extreme internal disorders, or death. Therefore, if you have the slightest doubt as to the edibility of a plant, apply the universal edibility test described below before eating any part of it.

a. **Universal Edibility Test**. Before testing a plant for edibility, make sure there are a sufficient number of plants to make testing worth your time and effort. You need more than 24 hours to apply the edibility test outlined below:

(1) Test only one part of a potential food plant at a time.

(2) Break the plant into its basic components, leaves, stems, roots, buds, and flowers.

(3) Smell the food for strong or acid odors. Keep in mind that smell alone does not indicate a plant is edible.

(4) Do not eat for 8 hours before starting the test.

(5) During the 8 hours you are abstaining from eating, test for contact poisoning by placing a piece of the plant you are testing on the inside of your elbow or wrist. Usually 15 minutes is enough time to allow for reaction.

(6) During the test period, take nothing by mouth except purified water and the plant part being tested.

(7) Select a small portion and prepare it the way you plan to eat it.

(8) Before putting the prepared plant part in your mouth, touch a small portion (a pinch) to the outer surface of the lip to test for burning or itching.

(9) If after 3 minutes there is no reaction on your lip, place the plant part on your tongue, holding there for 15 minutes.

(10) If there is no reaction, thoroughly chew a pinch and hold it in your mouth for 15 minutes. *Do not swallow.*

(11) If no burning, itching, numbing, stinging, or other irritation occurs during the 15 minutes, swallow the food.

(12) Wait 8 hours. If any ill effects occur during this period, induce vomiting and drink a lot of water.

(13) If no ill effects occur, eat 1/2 cup of the same plant part prepared the same way. Wait another 8 hours. If no ill effects occur, the plant part as prepared is safe for eating.

b. **Poisonous Plants**. ***Do not*** eat unknown plants that have the below characteristics:

(1) Have a milky sap or a sap that turns black when exposed to air.

(2) Are mushroom like.

(3) Resemble onion or garlic.

(4) Resemble parsley, parsnip, or dill.

(5) Have carrot-like leaves, roots, or tubers.

11-5. **ANIMAL FOOD**. Animal food contains the most food value per pound. Anything that creeps, crawls, swims, or flies is a possible source of food, however you must first catch, kill and butcher it before this is possible. There are numerous methods for catching fish and animals in a survival situation. You can catch fish by using a net across a small stream, (Figure 11-4) or by making fish traps and baskets. Improvise fish hooks and spears as indicated in Figure 11-5, and use them for conventional fishing, spearing and digging.

Figure 11-4. SETTING A GILL NET IN THE STREAM

Figure 11-5. SPEARS AND FISH HOOKS

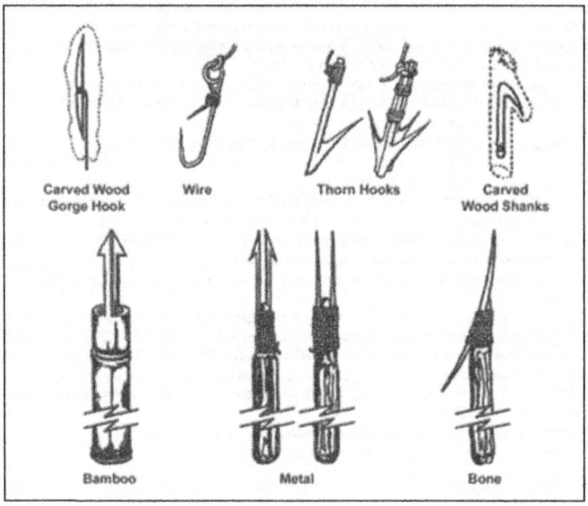


11-6. **TRAPS AND SNARES**. For an unarmed survivor or evader, or when the sound of a rifle shot could be a problem, trapping or snaring wild game is a good alternative. Several well-placed traps have the potential to catch much more game than a Ranger with a rifle is likely to shoot.

a. To be effective with any type of trap or snare, you must---

(1) Know the species of animal you intend to catch.

(2) Know how to construct a proper trap.

(3) Avoid alarming the prey with signs of your presence.

b. There are no catchall traps you can set for all animals. You must determine what species are in a given area and set your traps specifically with those animals in mind. Look for the following:

(1) Runs and trails.
(2) Tracks.
(3) Droppings.
(4) Chewed or rubbed vegetation.
(5) Nesting or roosting sites.
(6) Feeding and watering areas.

c. Position your traps and snares where there is proof that animals pass through. You must determine if it is a "run" or a "trail." A trail will show signs of use by several species and will be rather distinct. A run is usually smaller and less distinct and will only contain signs of one species. You may construct a perfect snare, but it will not catch anything if haphazardly placed in the woods. Animals have bedding areas, waterholes, and feeding areas with trails leading from one to another. You must place snares and traps around these areas to be effective.

d. An evader in a hostile environment must conceal traps and snares. It is equally important, however, to avoid making a disturbance that will alarm the animal and cause it to avoid the trap. Therefore, if you must dig, remove all fresh dirt from the area. Most animals will instinctively avoid a pitfall-type trap. Prepare the various parts of a trap or snare away from the site, carry them in, and set them up. Such actions make it easier to avoid disturbing the local vegetation, thereby alerting the prey. Do not use freshly cut, live vegetation to construct a trap or snare. Freshly cut vegetation will "bleed" sap that has an odor the prey will be able to smell. It is an alarm signal to the animal.

e. You must remove or mask the human scent on and around the trap you set. Although birds do not have a developed sense of smell, nearly all mammals depend on smell even more than on sight. Even the slightest human scent on a trap will alarm the prey and cause it to avoid the area. Removing the scent from a trap is difficult but masking it is relatively easy. Use the fluid from the gall and urine bladders of previous kills. Do not use human urine. Mud, particularly from an area with plenty of rotting vegetation, is also good. Use it to coat your hands when handling the trap and to coat the trap when setting it. In nearly all parts of the world, animals know the smell of burned vegetation and smoke. It is only when a fire is actually burning that they become alarmed. Therefore, smoking the trap parts is an effective means to mask your scent. If one of the above techniques is not practical, and if time permits, allow a trap to weather for a few days and then set it. Do not handle a trap while it is weathering. When you position the trap, camouflage it as naturally as possible to prevent detection by the enemy and to avoid alarming the prey.

f. Traps or snares placed on a trail or run should use canalization. To build a channel, construct a funnel-shaped barrier extending from the sides of the trail toward the trap, with the narrowest part nearest the trap. Canalization should be inconspicuous to avoid alerting the prey. As the animal gets to the trap, it cannot turn left or right and continues into the trap. Few wild animals will back up, preferring to face the direction of travel. Canalization does not have to be an impassable barrier. You only have to make it inconvenient for the animal to go over or through the barrier. For best effect, the canalization should reduce the trail's width to just slightly wider than the targeted animal's body. Maintain this constriction at least as far back from the trap as the animal's body length, then begin the widening toward the mouth of the funnel.

(1) Use a treadle snare against small game on a trail (Figure 11-6). Dig a shallow hole in the trail. Then drive a forked stick (fork down) into the ground on each side of the hole on the same side of the trail. Select two fairly straight sticks that span the two forks. Position these two sticks so that their ends engage the forks. Place several sticks over the hole in the trail by positioning one end over the lower horizontal stick and the other on the ground on the other side of the hole. Cover the hole with enough sticks so that the prey must step on at least one of them to set off the snare. Tie one end of a piece of cordage to a twitch-up or to a weight suspended over a tree limb. Bend the twitch-up or raise the suspended weight to determine where you will tie a 5 centimeter or so long trigger. Form a noose with the other end of the cordage.

(2) Route and spread the noose over the top of the sticks over the hole. Place the trigger stick against the horizontal sticks and route the cordage behind the sticks so that the tension of the power source will hold it in place. Adjust the bottom horizontal stick so that it will barely hold against the trigger. As the animal places its foot on a stick across the hole, the bottom horizontal stick moves down, releasing the trigger and allowing the noose to catch the animal by the foot. Because of the disturbance on the trail, an animal will be wary. You must therefore use canalization.

Figure 11-6. TREADLE SNARE

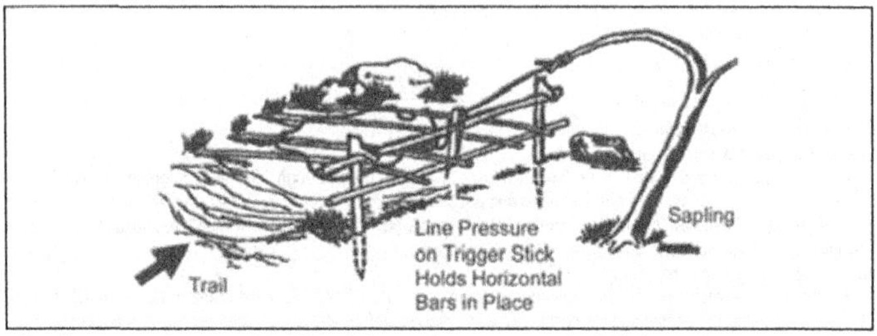


g. Trapping game can be accomplished through the use of snares, traps, or deadfalls. A snare is a noose that will slip and strangle or hold any animal caught in it. You can use inner core strands of parachute suspension lines, wire, bark of small hardwood saplings as well as hide strips from previously caught animals to make snares.

(1) The drag noose snare, Figure 11-7, is usually the most desirable in that it allows you to move away from the site, plus it is one of the easiest to make and fastest to set. It is especially suitable for catching rabbits. To make the drag noose snare, make a loop in the string using a bowline or wireman's knot. (When using wire, secure the loop by intertwining the end of the wire with the wire at the top of the loop). Pull the other end of the string (or wire) through the loop to form a noose that is large enough for the animal's head but too small for its body; tie the string (or attach the wire) to a sturdy branch. The branch should be large enough to span the trail and rest on the bush or support (two short forked sticks) you have selected. A snared animal will dislodge the drag stick, pulling it until it becomes entangled in the brush. Any attempt to escape tightens the noose, strangling or holding the animal.

Figure 11-7. DRAG NOOSE SNARE

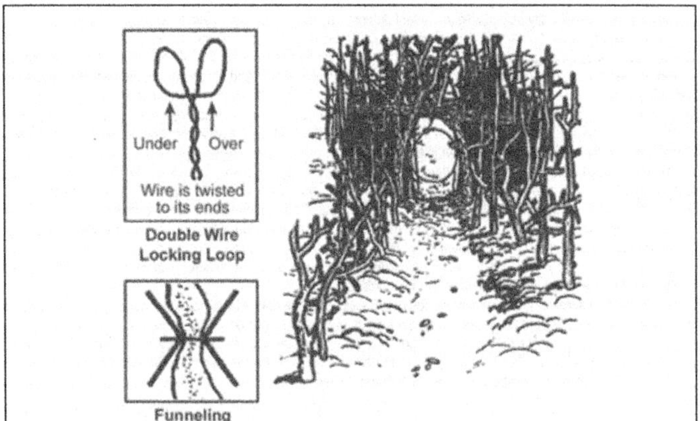

(2) Another type snare is the locking type snare loop (Figure 11 8) that will lock when pulled tight, ensuring the snared animal cannot escape. Use lightweight wire to make this snare, such as, trip wire or the wire from a vehicle or aircraft electrical system. To construct this snare, cut a piece of wire twice the length of the desired snare wire. Double the wire and attach the running ends to a securely placed object, such as the branch of a tree. Place a stick about 1/2 inch in diameter through the loop end of the wire; holding the wire taut, turn the stick in a winding motion so that the wire is twisted together. You should have four to five twists per inch. Detach the wire from the branch and then remove the loop from the stick; make a Figure Eight in the l/2-inch loop by twisting the loop over itself then fold the Figure Eight so the small loops are almost overlapping; run the loose wire ends through these loops. This forms a stiff noose that is strong. Tie the loose end to the stick (for a drag noose square) or branch you are using to complete the snare. This is an excellent snare for catching large animals.

**Figure 11-8. FORMING A LOCKING-TYPE SNARE LOOP**

(3) Another means of obtaining game is the use of the deadfall trap (Figure 11-9 and Figure 11-10).

**Figure 11-9. TRIGGER WITH DEADFALL**

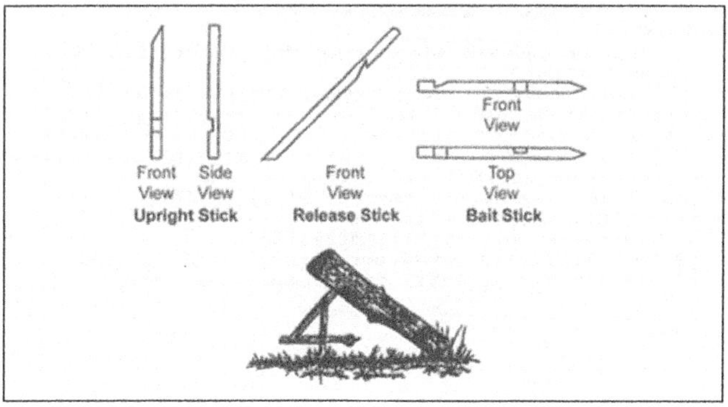

**Figure 11-10. TRIP-STRING DEADFALL TRAP**

h. Once you have obtained your fish or game, you must clean/butcher and cook/store it. Improper cleaning storing can result in inedible fish and game.

(1) ***Fish***. You must know how to tell if fish are free of bacterial decomposition that makes the fish dangerous to eat. Although cooking may destroy the toxin from bacterial decomposition, do not eat fish that appear spoiled.

(a) Signs of spoilage are:

- A peculiar odor.
- A suspicious color. (Gills should be red or pink. Scales should be a pronounced-not faded shade of gray).
- A dent remaining after pressing the thumb against the flesh.
- A slimy rather than moist or wet body.
- A sharp or peppery taste.

(b) Eating spoiled or poisoned fish may cause diarrhea, nausea, cramps, vomiting, itching; paralysis, or a metallic taste in the mouth. These symptoms appear suddenly 1 to 6 hours after eating. If you are near the sea, drink sea water immediately upon on set of such symptoms and force yourself to vomit.

(c) Fish spoil quickly after death, especially on a hot day, so prepare fish for eating as soon as possible after you catch them.

(d) Cut out the gills and large blood vessels that lie next to the backbone. (You can leave the head if you plan to cook the fish on a spit).

(e) Gut fish that are more than 4 inches long cut along the abdomen and scrape out the intestines.

(f) Scale or skin the fish.

(g) You can impale a whole fish on a stick and cook it over an "open fire." However, boiling the fish with the skin on is the best way to get the most food value. The fats and oil are under the skin, and by boiling the fish, you can save the juices for broth. Any of the methods used for cooking plant food can be used for cooking fish. Fish is done when the meat flakes off.

(h) To dry fish in the sun, hang them from branches or spread them on hot rocks. When the meat has dried, splash it with sea water, if available, to salt the outside. Keep seafood only if it is well dried or salted.

(2) ***Snakes***. All poisonous and nonpoisonous fresh water and land snakes are edible. To prepare snakes for eating use the following steps (Figure 11-11):

(a) Grip the snake firmly behind the head and cut off the head with a knife.

(b) Slit the belly and remove the innards. (You can use the innards for baiting traps and snares).

(c) Skin the snake. (You can use the skin for improvising, belts, straps, or similar items).

**CAUTION**

Take extreme care in securing snakes--the bite of some poisonous snakes can be fatal. Even after a snake's head is cut off, its reflex action can cause it to bite, injecting poison. The best time to capture snakes is in the early morning or late evening when temperatures are low and they move slow. Kill or use a long stick to pin down its head and capture it. To pick up a snake, place the index finger on the top rear of its head with your thumb and middle finger on either side of the head behind the jaws. Keep your index finger on top of snake's head to prevent it from turning inside its skin and biting you.

**Figure 11-11. CLEANING A SNAKE**

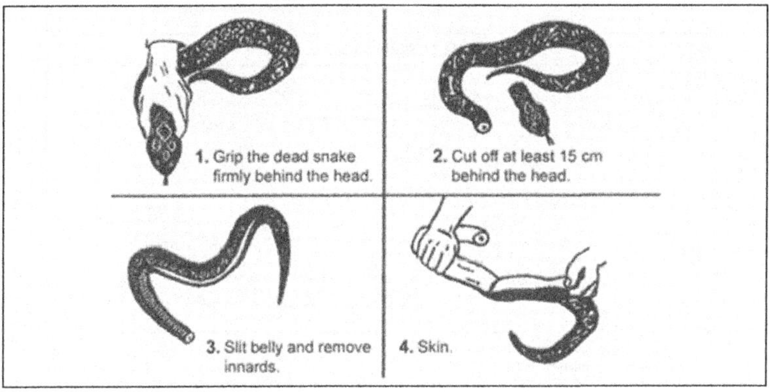


(3) ***Fowl***. Your first step after killing a fowl for eating or preserving is to pluck its feathers. If plucking is impractical, you can skin the fowl. Keep in mind, however, that a fowl cooked with the skin on retains more food value. Waterfowl are easier to pluck while dry, but other fowl are easier to pluck after scalding. After you pluck the fowl--

(a) Cut off its neck close to the body.

(b) Cut an incision in the abdominal cavity and clean out the insides. Save the neck, liver, and heart for stew. Thoroughly clean and dry the entrails to use for cordage.

(c) Wash out the abdominal cavity with fresh clean water. You can boil fowl or cook it on a spit over a fire. You should boil scavenger birds such as vultures and buzzards for at least 20 minutes to kill any parasites. Use the feathers from fowl for insulating your shoes clothing, or bedding. You can also use feathers for fish lures.

(4) **Medium-Sized Mammals**. The game you trap or snare will generally be alive when you find it and therefore dangerous. Be careful when you approach a trapped animal. Use a spear or club to kill it so you can keep a safe distance from it. After you kill an animal, immediately bleed it by cutting its throat. If you must drag the carcass any distance, do so before you cut off the hide so that the carcass is protected from dirt and debris that might contaminate it. Clean the animal near a stream if possible so that you can wash and cool the carcass and edible parts. Fleas and parasites will leave a cooled body so if the situation allows, wait until the animal cools before cleaning and dressing the carcass. To skin and dress the animal (Figure 11-12 and 11-13).

Figure 11-12. SKINNING AND BUTCHERING LARGE GAME

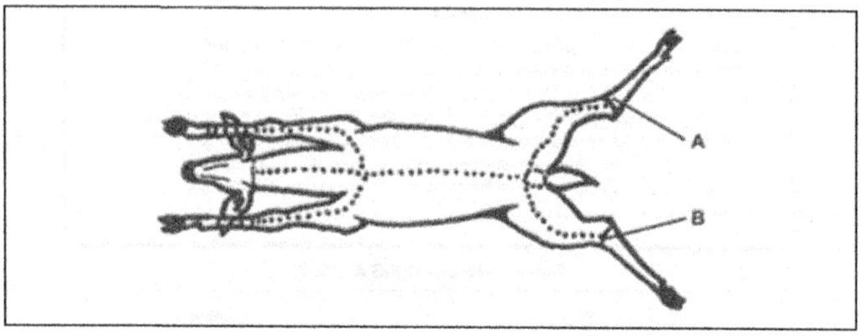


Figure 11-13. SKINNING SMALL GAME

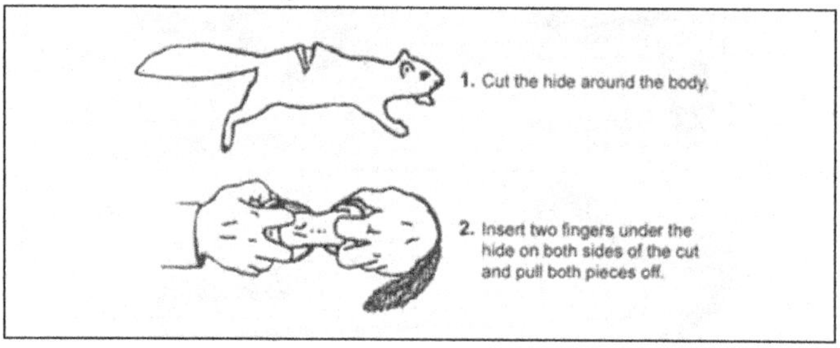


(a) Place carcass, belly up, on a slope if available. You can use rocks or brush to support it.
(b) Remove genitals or udder.
(c) Remove musk glands to avoid tainting meat.
(d) Split hide from tail to throat. Make the cut shallow so that you do not pierce the stomach.
(e) Insert your knife under the skin, taking care not to cut into the body cavity. Peel the hide back several inches on each side to keep hair out of the meat.
(f) Open the chest cavity by splitting the sternum. You can do this by cutting to one side of the sternum where the ribs join.
(g) Reach inside and cut the windpipe and gullet as close to the base of the skull as possible.
(h) With the forward end of the intestinal tract free, work your way to the rear lifting out internal organs and intestines. Cut only where necessary to free them.
(i) Carefully cut the bladder away from the carcass so that you do not puncture the bladder (urine can contaminate meat). Pinch the urethra tightly and cut it beyond the point you are pinching.
(j) Remove the bladder.
(k) From the outside of the carcass, cut a circle around the anus.
(l) Pull the anus into the body cavity and out of the carcass.
(m) Lift or roll the carcass to drain all blood.

***Note:*** Try to save as much blood as you can as it is a vital source of food and salt. Boil the blood.

(n) Remove the hide, and make cuts along the inside of the legs to just above the hoof or paw. Peel the skin back, using your knife in a slicing motion to cut the membrane between the skin and meat. Continue this until the entire skin is removed.

(o) Most of the entrails are usable. The heart, liver, and kidneys are edible. Cut open the heart and remove the blood from its chambers. Slice the kidneys and if enough water is available, soak or rinse them. In all animals except those of the deer family, the gall bladder (a small, dark-colored, clear-textured sac) is attached to the liver.

(p) Sometimes, the sac looks like a blister on the liver. To remove the sac, hold the top portion of it and cut the liver around and behind the sac. If the gall bladder breaks and gall gets on the meat, wash it off immediately so the meat will not become tainted. Dispose of the gall.

(q) Clean blood splattered on the meat will glaze over and help preserve the meat for a short time. However, if an animal is not bled properly, the blood will settle in the lowest part of its body and will spoil in a short time. Cut out any meat that becomes contaminated.

(r) When temperatures are below 40 degrees, you can leave meat hanging for several days without danger of spoilage. If maggots get on the meat, remove the maggots and cut out the discolored meat. The remaining meat is edible. Maggots, which are the larvae of insects, are also edible.

(s) Blood, which contains salts and nutrients, is a good base for soups.

(t) Thoroughly clean the intestines and use them for storing or smoking food or lashings for general use. Make sure they are completely dry to preclude rotting.

(u) The head of most animals contains a lot of meat, which is relatively easy to get. Skin the head, saving the skin for leather. Clean the mouth thoroughly and cut out the tongue. Remove the outer skin from the tongue after cooking. Cut or scrape the meat from the head. If you prefer, you can roast the head over an open fire before cutting off the meat. Eyes are edible. Cook them but discard the retina (this is a plastic like disc). The brain is also edible; in fact, some people consider it a delicacy. The brain is also used to tan leather, the theory being that the brain of an animal is adequate to tan its hide.

(v) Use the tendons and ligaments of the body of large animals for lashings.

(w) The marrow in bones is a rich-food source. Crack the bones and scrap out the marrow, and use bones to make weapons.

(x) If the situation and time allow, you should preserve the extra meat for later use. If the air is cold enough, you can freeze the meat. In warmer climates however, you will need to use a drying or smoking process to preserve it. One night of heavy smoking will make meat edible for about 1 week. Two nights will make it remain edible for 2 to 4 weeks. To prepare meat for drying or smoking, cut it with the grain in quarter inch strips. To air dry the meat, hang it in the wind and hot sun out the reach of animals; cover it so that blow flies cannot land on it.

(y) To smoke meat, you will need an enclosed area – for instance, a teepee (Figure 11-14) or a pit. You will also need wood from deciduous trees, preferably green. Do not use conifer trees such as pines, firs, spruces, or cedars, as the smoke from these trees give the meat a disagreeable taste.

(z) When using the para-teepee or other enclosed area with a vent at the top, set the fire in the center and let it burn down to coals, then stoke it with green wood. Place the strips of meat on a grate or hang them from the top of the enclosure so that they are about 2 feet above the smoking coals. To use the pit method of smoking meat dig, a hole about 3 feet (1 meter) deep and 1 1/2 feet (1/2 meter) in diameter. Make a fire at the bottom of the hole. After it starts burning well, add chipped green wood or small branches of green wood to make it smoke. Place a wooden grate about 1 1/2 feet (1/2 meter) above the fire and lay the strips of meat on the grate. Cover the pit with poles, boughs, leaves, or other material.

11-7. **SHELTERS**. A shelter can protect you from the sun, insects, wind, rain, snow, hot or cold temperatures, and enemy observation. In some areas your need for shelter may take precedence over your need for food, possibly even your need for water. After determining your shelter site, you should keep in mind the type of shelter (protection) you need. You need to know how to make different types of shelters. Only two are described in this handbook. Additional information is available in FM 21-76.

a. **Planning Considerations**.

- How much time and effort are needed to build the shelter?
- Will the shelter adequately protect you from the elements (rain, snow, wind, sun, and so on)?
- Do you have tools to build it? If not, can you improvise tools from materials in the area?
- Do you have the type and amount of manmade materials needed to build it? If not, are there sufficient natural materials in the area?

Figure 11-14. SMOKING MEAT

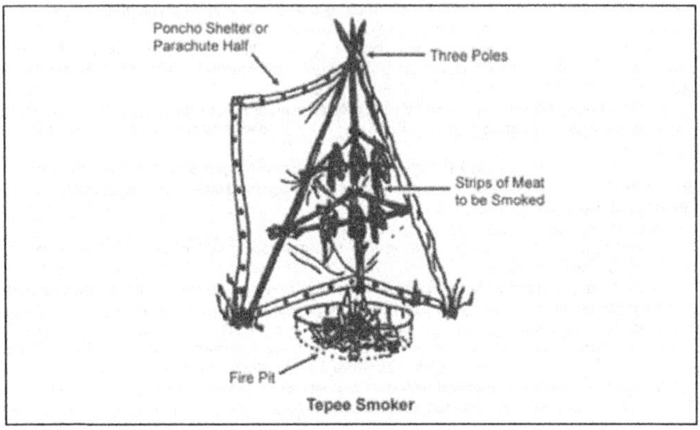


b. **Types.**

(1) ***Poncho Lean-To.*** It takes only a short time and minimal equipment to build this lean-to (Figure 11-15). You need a poncho, 6 to 10 feet of rope, three stakes about 6 inches long, and two trees (or two poles) 7 to 9 feet apart. Before you select the trees you will use (or decide where to place the poles), check the wind direction. Make sure the back of your lean-to will be into the wind. To make the lean-to:

(a) Tie off the hood of the poncho. To do this, pull the draw cord tight; roll the hood long ways, fold it into thirds, and tie it with the draw cord.

(b) Cut the rope in half. On one long side of the poncho, tie half of the rope to one corner grommet, and the other half to the other corner grommet.

(c) Attach a drip stick (about a 4-inch stick) to each rope 1/4 to 3/4 inches away from the grommet. These drip sticks will keep rainwater from running down the ropes into the lean-to. Using drip lines is another way to prevent dripping inside the shelter. Tie lines or string about 4 inches long to each grommet along the top edge of the shelter. This allows water to run to and down the line without dripping into the shelter.

(d) Tie the ropes about waist high on the trees (uprights). Use a round turn and two half hitches with quick-release knot.

(e) Spread the poncho into the wind and anchor to the ground. To do this, put three sharpened sticks through the grommets and into the ground.

Figure 11-15. PONCHO LEAN-TO

(f) If you plan to use the lean-to for more than one night, or if you expect rain, make a center support to the lean-to. You can do this by stretching a rope between two upright poles or trees that are in line with the center of the poncho.

(g) Tie another rope to the poncho hood; pull it upward so that it lifts the center of the poncho, and tie it firmly to the rope stretched between the two uprights.

(h) Another method is to cut a stick to place upright under the center of the lean-to. This method, however, will restrict your space and movements in the shelter.

(i) To give additional protection from wind and rain, place boughs, brush, your rucksack, or other equipment at the sides of the lean-to.

(j) To reduce heat loss to the ground, place some type of insulating material, such as leaves or pine needles, inside your lean-to.

***Note:*** When at rest, as much as 80 percent of your body heat can be lost to the ground.

(l) To increase your security from enemy observation, lower the silhouette of the lean-to by making two modifications.

- Secure the support lines to the trees knee-high rather than waist-high.
- Use two knee-high sticks in the two center grommets (sides of lean-to), and angle the poncho to the ground, securing it with sharpened sticks as above.

(2) ***Field Expedient Lean-To***. If you are in a wooded area and have sufficient natural materials, you can make an expedient lean-to (Figure 11-16) without the aid of tools or with only a knife. You need more time to make it than the shelter previously mentioned, but it will protect you from most environmental elements. You will need two trees, (or two upright poles), about 6 feet apart; one pole about 7 feet long and 1 inch in diameter. Five to eight poles about 10 feet long and 1 inch in diameter for beams, cord or vines for securing, the horizontal support to the trees and other poles, saplings, or vines to crisscross the beams. To make this lean-to:

**Figure 11-16. FIELD-EXPEDIENT LEAN-TO**

(a) Tie the 7-foot pole to the two trees at point about waist to chest high. This is your horizontal support. If there is a fork in the tree, you can rest the pole in it instead of tying the pole in place. If a standing tree is not available, construct a bipod using an Y-shaped sticks or two tripods.

(b) Place one end of the beams (10-foot poles) one side of the horizontal support. As with all lean-to type shelters, make sure the backside of the lean-to is placed into the wind.

(c) Crisscross sapling or vines on the beams.

(d) Cover the framework with brush, leaves, pine needles, or grass, starting at the bottom and working your way up like shingling.

(e) Place straw, leaves, pine needles, or grass inside the shelter for bedding.

(f) In cold weather, you can add to the comfort of your lean-to by building a fire-reflector wall (Figure 11-16). Drive four stakes about 4 feet long into the ground to support the wall. Stack green logs on top of one another between the support stales. Bind the top of the support stakes so the green logs will stay in place. Fill in the spaces between the logs with twigs or small branches. With just a little more effort, you can have a drying rack. Cut a few 3/4-inch diameter poles. The length depends on distance between the lean-to support and the top of the fire-reflector wall. Lay one end of the poles on the lean-to horizontal support and the other ends on top of the reflector wall. Place and tie into place smaller sticks across these poles. You now have a place to dry clothes, meat, or fish.

11-8. **FIRES**. A fire can full fill several needs. It can keep you warm, it can keep you dry: you can use it to cook food, to purify water, and to signal. It can also cause you problems when you are in enemy territory: it creates smoke, which can be smelled and seen from a long distance: It causes light which can be seen day or night and it leaves signs of your presence. Remember you should always weigh your need for a fire against your need to avoid enemy protection. When operating in remote areas you should always take a supply of matches in a waterproof case and keep them on your person.

a. **Selection**. When selecting a site to build a fire, you should consider the following:

(1) The area (terrain and climate) in which you are operating.
(2) The material and tools available.
(3) How much time you have.
(4) Why you need a fire.
(5) The nearness of the enemy.

b. **Preparation**. If you are in a wooded or brush-covered area, clear brush away, and scrape the surface soil from the spot you selected. The cleared circle should be at least 3 feet (1 meter) in diameter so that there is little chance of the fire spreading. To prepare the site for a fire, ensure that it is dry and that it--look for a dry spot that--

(1) Offers protection from the wind.
(2) Is suitably placed in relation to your shelter (if any).
(3) Concentrates the heat in the direction you desire.
(4) Has a supply of wood or other fire burning material.

c. **Dakota Fire Hole**. In some situations, an underground fireplace will best meet your needs. It conceals the fire to some extent and serves well for cooking food. To make an underground fireplace or Dakota fire hole (Figure 11-17):

**Figure 11-17. DAKOTA FIRE HOLE**

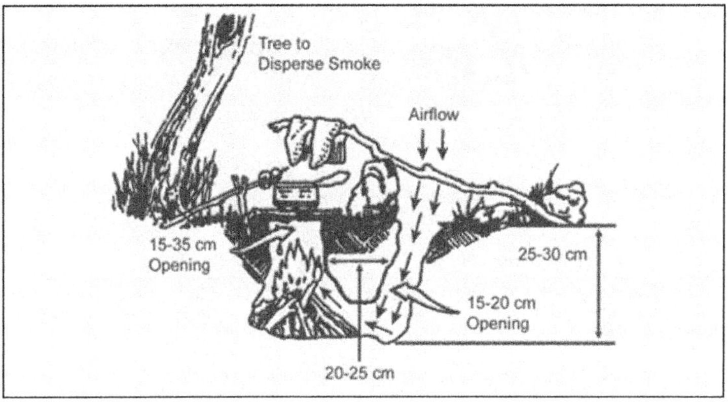


(1) Dig a hole in the ground.

(2) On the upwind side of this hole, poke one large connecting hole for ventilation.

d. **Aboveground fire**. If you are in a snow covered or wet area, you can use green logs to make a dry base for your fire (Figure 11-18). Trees with wrist-size trunks are easily broken in extreme cold. Cut or break several green logs and lay them side by side on top of the snow. Add one or two more layers, laying the top layer logs in a direction opposite those of the layer below it.

**Figure 11-18. BASE FOR FIRE IN SNOW COVERED AREA**

e. **Methods**. There are several methods for laying a fire for quick fire making. Three easy methods are Tepee, lean-to, and cross-ditch. Tepee (Figure 11-19). Arrange tinder and a few sticks of kindling in the shape of a cone. Fire the center. As the cone burns away, the outside logs will fall inward, feeding the heart of the fire. This type of fire burns well even with wet wood.

(1) ***Lean-To*** (Figure 11-19). Push a green stick into the ground at a 30 degree angle. Point the end of the stick in the direction of the wind. Place some tinder (at least a handful) deep inside this lean-to stick. Light the tinder. As the kindling catches fire from the tinder, add more kindling.

Figure 11-19. METHODS FOR LAYING A FIRE

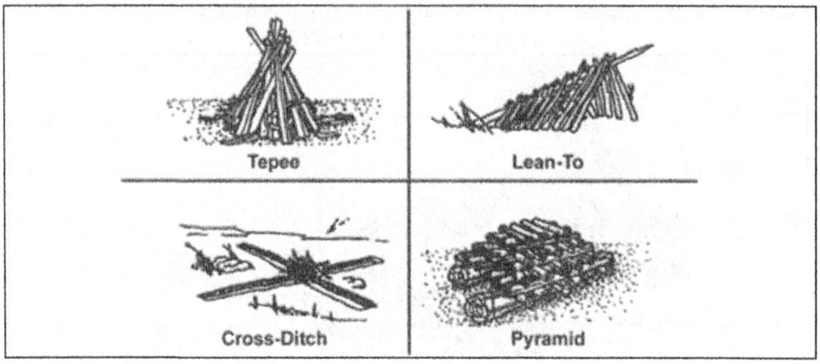


(2) ***Cross-Ditch*** (Figure 11-19). Scratch a cross about 1 foot in size in the ground. Dig the cross 3 inches deep. Put a large wad of tinder in the middle of the cross. Build a kindling pyramid above the tinder. The shallow ditch allows air to sweep under the fire to provide a draft.

NOTES

# Chapter 12
# FIRST AID

Patrolling, more than some other types of missions, puts Rangers in harms way. CASEVAC planning is vital. Also, because trained medical personnel might be unavailable at the initial point of injury, everyone must know how to diagnose and treat injuries, wounds, and common illnesses. The unit should also have a plan for handling KIAs.

12-1. **LIFESAVING STEPS**. Whatever the injury, (1) open the airway and restore breathing; (2) stop the bleeding and protect the wound; (3) check, treat and monitor for shock; and (4) MEDEVAC the casualty.

12-2. **CARE UNDER FIRE**. When still under fire, (1) maintain situational awareness; (2) return fire; (3) protect the casualty; (4) Identify and control severe bleeding with bandage or tourniquet; (5) and move the casualty to cover.

12-3. **PRIMARY SURVEY**. Use the ABC's to help you remember how to identify and deal with life-threatening injuries, airway complications, breathing problems, or uncontrolled hemorrhaging (bleeding).

| | |
|---|---|
| A | Airway. Open airway by patient position or with airway adjuncts. |
| B | Breathing. Seal open chest wounds with occlusive dressing. |
| C | Circulation. Identify uncontrolled bleeding and control with pressure or tourniquet. Start IV if needed. |
| D | Disability. Determine Level of consciousness. |
| E | Exposure. Fully expose patient. (Environment dependent) |

12-4. **AIRWAY MANAGEMENT**. The *airway* is most often obstructed at the base of the tongue.

a. If this happens, open the airway using the chin lift (nontrauma, Figure 12-1) or jaw thrust (trauma, Figure 12-2).

**Figure 12-1. CHIN LIFT**

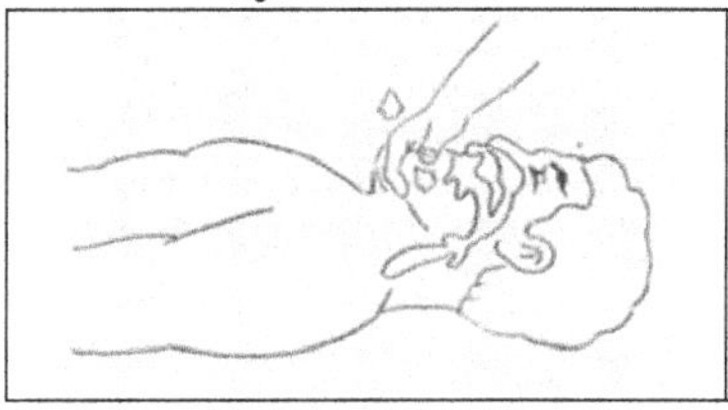

**Figure 12-2. JAW THRUST**

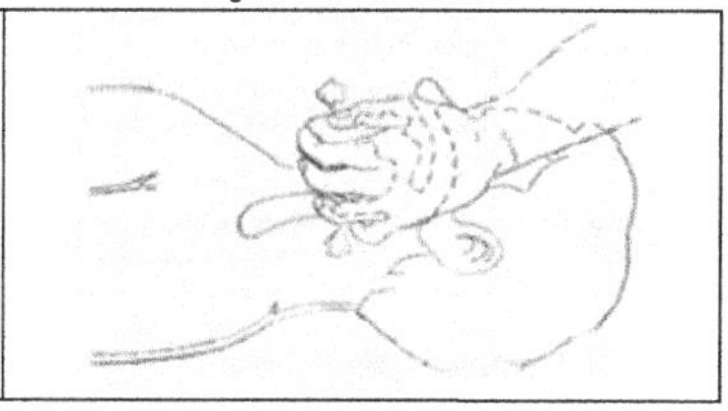

b. **Remove** debris (teeth, blood clots, bone) from the oral cavity; use suction if you have it; and place airway adjuncts to allow the victim to breathe through their nose (Figure 12-3) or mouth (Figure 12-4).

**Figure 12-3. NASAL AIRWAY**

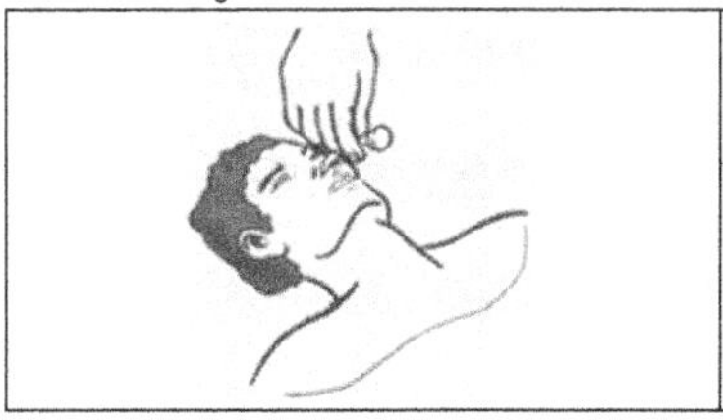

**Figure 12-4. MOUTH AIRWAY**

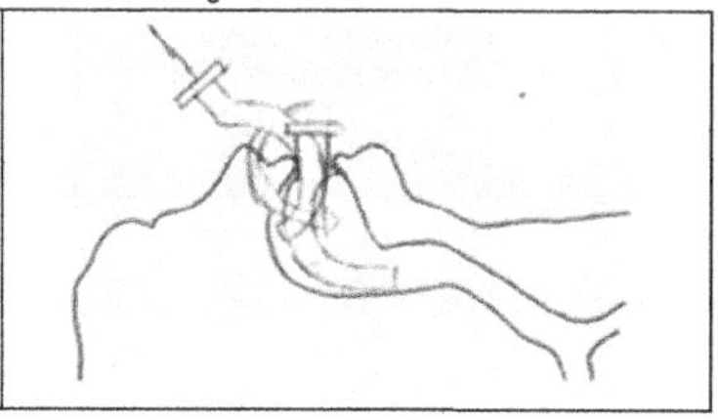

12-5. **BREATHING**. If the patient is having difficulty breathing, (1) expose the chest and identify open chest injuries; (2) place occlusive dressing (plastic and tape, or a COTS chest dressing) over open entry and exit chest wounds; and (3) place the patient on his injured side, or position him where he can breathe most comfortably.

12-6. **BLEEDING**. Quickly identify and control bleeding. (1) Apply a standard or pressure dressing. (2) If this does not control the bleeding, immediately apply a tourniquet. (3) Reassess dressings frequently to ensure bleeding has been controlled.

12-7. **SHOCK**. Shock causes an inadequate flow of oxygen to the tissues. Shock can cause bleeding (hemorrhagic) or not (nonhemorrhagic). Signs and symptoms of shock can include increased pulse, increased respiration; and decreased level of consciousness. (1) Open the victim's airway; (2) restore his breathing; (3) control the bleeding; (4) initiate an IV or a saline lock; and (5) monitor his condition.

12-8. **EXTREMITY INJURIES**. Identify and control bleeding. If you suspect a fracture, splint it as it lies. Do not reposition injured extremity.

12-9. **ABDOMINAL INJURIES**. Identify and control bleeding. Treat for shock. (1) If internal organs are exposed, cover them with dry, sterile dressing. Do not place organs back in wound area. Do not handle internal organs. (2) Place patient in position of comfort. Flex knees to relax abdomen. (3) Do not give anything by mouth to the patient.

12-10. **BURNS**. Remove patient from burn source. Then (1) remove all clothing and jewelry from the area; (2) cover with dry, sterile dressings, and ensure fingers and toes have dressings between them before covering entire area; and (3) Evacuate immediately any casualties with burns of the face, neck, hands, genitalia, or over 20 percent of his body surface.

12-11. **WEATHER (HEAT AND COLD) INJURIES**. Tables 12-1, 12-2, and 12-3 show first aid for (respectively) heat injuries, cold injuries, and environmental injuries.

**Table 12-1. HEAT INJURIES**

| *INJURY* | *SIGNS/SYMPTOMS* | *FIRST AID* |
|---|---|---|
| Heat Cramps | Casualty experiences muscle cramps in arms, legs and/or stomach, may also have wet skin and extreme thirst. | 1. Move the casualty to a shaded area and loosen clothing.<br>2. Allow casualty to drink 1 quart of cool water slowly per hour.<br>3. Monitor casualty and provide water as needed.<br>4. Seek medical attention if cramps persist. |
| Heat Exhaustion | Casualty experiences loss of appetite, headache, excessive sweating, weakness or faintness, dizziness, nausea, muscle cramps. The skin is moist, pale, and clammy. | 1. Move the casualty to a cool, shaded area and loosen clothing.<br>2. Pour water on casualty and fan to increase cooling effect of evaporation.<br>3. Provide at least one quart of water to replace lost fluids.<br>4. Elevate legs.<br>5. Seek medical aid if symptoms continue. |
| Heatstroke (Sunstroke) | Casualty stops sweating (hot, dry skin), may experience headache, dizziness, nausea, vomiting, rapid pulse and respiration, seizures, mental confusion. Casualty may suddenly collapse and lose consciousness. THIS IS A MEDICAL EMERGENCY! | 1. Move casualty to a cool, shaded area, loosen clothing, and remove outer clothing if the situation permits.<br>2. Immerse in cool water. If cool bath is not available, massage arms and legs with cool water. Fan casualty to increase the cooling effect of evaporation.<br>3. If conscious, slowly consume one quart of water.<br>4. SEEK MEDICAL AID AND EVACUATE AS SOON AS POSSIBLE. Perform any lifesaving measures. |

Table 12-2. COLD INJURIES

| *INJURY* | *SIGNS/SYMPTOMS* | *FIRST AID* |
|---|---|---|
| Chilblain | Red, swollen, hot, tender, itching skin. Continued exposure may lead to infected (ulcerated bleeding) skin lesions. | 1. Area usually responds to locally applied warming (body heat).<br>2. Do Not rub or massage area.<br>3. Seek medical treatment. |
| Immersion (trench) foot | Affected parts are cold, numb, and painless. As parts warm, they may be hot, with burning and shooting pains. Advanced stage: skin pale with bluish cast: pulse decreases, blistering, swelling, heat hemorrhages, and gangrene may follow. | 1. Gradual warming by exposure to warm air.<br>2. DO NOT massage or moisten skin.<br>3. Protect affected parts from trauma.<br>4. Dry feet thoroughly: avoid walking.<br>5. Seek medical treatment. |
| Frostbite | ***SUPERFICIAL***<br>Redness, blisters in 24-36 hours and sloughing of the skin.<br><br>***DEEP***<br>Preceded by superficial frostbite; skin is painless, pale- yellowish, waxy, "wooden" or solid to touch, blisters form in 12-36 hours | ***SUPERFICIAL***<br>1. Keep casualty warm; gently warm affected parts.<br>2. Decrease constricting clothing, increase exercise and insulation.<br>***DEEP***<br>1. Protect the part from additional injury.<br>2. Seek medical treatment as fast as possible. |
| Snow Blindness | Red, scratchy, or watery eyes; headache; increased pain in eyes with exposure to light. | 1. Cover the eyes with a dark cloth.<br>2. Seek medical treatment. |
| Dehydration | Similar to heat exhaustion. | 1. Keep warm, loosen clothes.<br>2. Replace lost fluids, rest, and additional medical treatment. |
| Hypothermia | Casualty is cold, shivers uncontrollably until shivering stops. Rectal (core) temp less 95° F can affect consciousness. Uncoordinated movements, shock, and coma may occur as body temperature drops. | ***MILD HYPOTHERMIA***<br>1. Warm body evenly and without delay. (Heat source must be provided.)<br>2. Keep dry, protect from elements.<br>3. Warm liquids may be given to conscious casualty only.<br>4. Be prepared to start CPR.<br>5. Seek medical treatment immediately.<br>***SEVERE HYPOTHERMIA***<br>1. Quickly stabilize body temperature.<br>2. Attempt to prevent further heat loss.<br>3. Handle the casualty gently.<br>4. Evacuate to nearest medical treatment facility as soon as possible. |

**Table 12-3. ENVIRONMENTAL INJURIES**

| *TYPE* | *FIRST AID* |
|---|---|
| Snake bite | 1. Get the casualty away from the snake.<br>2. Remove all rings and bracelets from the affected extremity.<br>3. Reassure the casualty and keep him quiet.<br>4. Apply constricting band(s) 1 to 2 finger widths close to the bite. You should be able to slip 1 finger between the band and skin.<br>***ARM OR LEG BITE*** - Place one band above and one band below the bite site.<br>***HAND OR FOOT BITE*** - Place one band above the wrist or ankle.<br>5. Immobilize the affected limb below the level of the heart.<br>6. Kill the snake, if possible, (without damaging its head or endangering yourself) and send it with the casualty.<br>7. Seek medical treatment immediately. |
| Brown recluse or black widow, spider bite | 1. Keep the casualty calm.<br>2. Wash the area.<br>3. Apply ice or a freeze pack, if available.<br>4. Seek medical treatment. |
| Tarantula bite, scorpion sting, ant bite | 1. Wash the area.<br>2. Apply ice or a freeze pack, if available.<br>3. Apply baking soda, calamine lotion, or meat tenderizer to the bite site to relieve pain and itching.<br>4. If site of bite(s) or sting(s) is on the face, neck (possible airway blockage), or genital area, or if reaction is severe, or if the sting is by the dangerous Southwestern scorpion, keep the casualty as quiet as possible, administer the epinephrine auto-injector (EAI) if needed, and then seek immediate medical aid. |
| Wasp or bee sting | 1. If the stinger is present, remove by scraping with a knife or finger nail. DO NOT squeeze venom sack on stinger, more venom may be injected.<br>2. Wash the area.<br>3. Apply ice or freeze pack, if available.<br>4. If allergic signs or symptoms appear, be prepared to administer the EAI, and then seek medical assistance. |
| Human and other animal bites | 1. Cleanse the wound thoroughly with soap or detergent solution.<br>2. Flush bite well with water.<br>3. Cover bite with a sterile dressing.<br>4. Immobilize injured extremity.<br>5. Transport casualty to a medical treatment facility.<br>6. For animal bites, without endangering yourself or damaging the animal's head, kill the animal and send its head with the casualty.<br>7. For human bites, try to extract some of the attacker's saliva from the wound and send that in a sealed, identified container with the casualty. |

Table 12-3. ENVIRONMENTAL INJURIES (CONTINUED)

| | |
|---|---|
| Poison Ivy, Oak, Sumac | 1. Gently clean affected area. Clean areas 2-3 times daily. Wash clothing.<br>2. Apply topical anti-itch lotion or ointment as needed, and cover.<br>3. Avoid scratching or itching the area.<br>4. Observe for signs of infection (increasing redness, tenderness, warm to touch).<br>5. Seek medical attention if rash persists or signs of infection develop. |

## 12-12. POISONOUS PLANT IDENTIFICATION

Poison plants include, among others, poison ivy, oak, and sumac, as well as a few more such as stinging nettles which are not discussed here.

a. **Poison Ivy**. Poison ivy is grows in two forms, a vine or shrub. The compound leaves of poison ivy consist of three pointed leaflets; the middle leaflet has a much longer stalk than the two side ones. The leaflet edges can be smooth or toothed but are rarely lobed. The leaves vary greatly in size, from 8 to 55 mm (0.31" to 2.16") in length. They are reddish when they emerge in the spring, turn green during the summer, and become various shades of yellow, orange, or red in the autumn. Small greenish flowers grow in bunches attached to the main stem close to where each leaf joins it. Later in the season, clusters of poisonous, berrylike drupes form. They are whitish, with a waxy look.

b. **Poison Oak**. Poison oak is a widespread deciduous shrub throughout mountains and valleys of North America, generally below 5,000 feet elevation. It commonly grows as a climbing vine with aerial (adventitious) roots that adhere to the trunks of oaks and sycamores. Poison oak also forms dense thickets in chaparral and coastal sage scrub. Poison oak leave typically have three leaflets (sometimes five), the terminal one on a slender rachis (also called a stalk or petiolule). Eastern poison ivy often has a longer rachis and the leaflet margins tend to be less lobed and serrated (less "oak-like"). Like many members of the sumac family (Anacardiaceae) new foliage and autumn leaves often turn brilliant shades of pink and red.

c. **Poison Sumac**. Poison Sumac is a woody perennial shrub or small tree growing from 5 - 25 feet tall that grows in peat bogs and swamps. To identify Poison Sumac, look for the fruit that grows between the leaf and the branch. Leaves are arranged in an alternate pattern on the vine. There are about 7-13 leaflets forming a feather-like appearance. The foliage has brilliant orange or scarlet coloring in the fall.

Figure 12-5. POISONOUS PLANTS

12-13. **FOOT CARE.** Use moleskin to prevent blisters prior to movement or foot march. Drain large blisters. Clean area, puncture with needle, drain blister. Place moleskin over area. Observe for signs of infection. Keep feet as clean and dry as possible. Use foot powder and change socks. Let feet air dry as mission permits.

12-14. **CASUALTY RESCUE AND TRANSPORT SYSTEM.** Remove the hoistable CRTS litter from its pack, and place the litter on the ground. Unfasten the retainer strap, step on the foot end of the litter, and unroll the litter completely to the opposite end. Then--

a. Bend the CRTS litter in half and backroll it. Repeat with the opposite end of the litter. The litter will now lay flat. Lift the sides of the litter, and fasten the four cross-straps to the buckles directly opposite the straps.

b. Place the CRTS litter next to the casualty. Insure the head end of the stretcher is adjacent to the head of casualty. Place the cross-straps under the litter. Logroll the casualty and slide the litter as far under the casualty as possible. Gently roll the casualty down on the litter. Slide the casualty to the center of the litter. Be sure to keep his spinal column as straight as possible.

c. Pull the straps out from under the CRTS litter and fasten it to the buckles. Position the foot end of the litter at the head of the casualty. Have one rescuer straddle the litter and support the casualty's head, neck, and shoulders.

d. Grasp the foot straps of the litter, and slide the litter under the casualty. Center the casualty on the litter, and feed foot straps thru the unused grommets at the foot end of the litter. Fasten the straps to the buckles.

12-15. **HYDRATION AND ACCLIMATIZATION**

a. Table 12-4 shows strategies for minimizing dehydration and increasing acclimatization and good hydration practices.

**Table 12-4. HYDRATION MANAGEMENT AND ACCLIMATIZATION**

| ***STRATEGY*** | ***SUGGESTIONS FOR IMPLEMENTATION*** |
|---|---|
| Start early | 1. Start at least 1 month prior to School<br>2. Be flexible and patient: performance benefits take longer than the physiological benefits |
| Mimic the training environment climate | 1. In warm climates, acclimatize in the heat of day.<br>2. In temperate climates workout in a warm room wearing sweats. |
| Ensure adequate heat stress | 1. Induce sweating.<br>2. Work up to 100 minutes of continuous physical exercise in the heat. Be patient. The first few days, you may not be able to go 100 minutes without resting.<br>3. Once you can comfortably exercise for 100 minutes in the heat, then continue for at least 7-14 days with added exercise intensity (loads or training runs). |
| Teach yourself to drink and eat | 1. Your thirst mechanism will improve as you become heat acclimatized, but you will still under-drink if relying on thirst sensation.<br>2. Heat acclimatization will increase your water requirements.<br>3. Dehydration will negate most benefits of physical fitness and heat acclimatization.<br>4. You will sweat out more electrolytes when not acclimatized, so add salt to your food, or drink electrolyte solutions during the first week of heat acclimatization.<br>5. A convenient way to learn how much water your body needs to replace is to weigh yourself before and after the 100 minutes of exercise in the heat. For each pound lost, you should drink about one-half quart of fluid.<br>6. Do not skip meals, as this is when your body replaces most of its water and salt losses. |

12-16 **WORK, REST, AND WATER CONSUMPTION.** Table 12-5 shows a work, rest, and water consumption table. The guidance applies to the average size, heat acclimated Soldier wearing ACU, not hot weather gear, except as specified otherwise. The work and rest times and fluid replacement volumes shown will help the Soldier sustain his performance and hydration for at least 4 hours of work in the specified heat category. Fluid needs can vary based on individual differences (±1 quart per hour).

a. "NL" means that there is no limit to work time per hour. "Rest" means minimal physical activity such as sitting or standing, preferably in the shade.

b. Consume no more than 1.5 quarts of fluid per hour, and no more than 12 quarts per day.

c. If you are wearing body armor in a humid climate, then add 5° F to the WBGT. if wearing MOPP 4 clothing, add 10° F to the WBGT.

d. Work categories include easy, moderate, and hard.

1) ***Easy Work***. This includes, for example, maintaining weapons; walking on hard surfaces at 2.5 mph with a load weighing no more than 30 pounds; participating in marksmanship training; and participating in drills or ceremonies.

2) ***Moderate Work***. This includes, for example, walking in loose sand at 2.5 mph (maximum) or under with no load; walking on a hard surface at 3.5 mph (maximum) with a load weighing no more than 40 pounds; performing calisthenics; patrolling; or conducting individual movement techniques such as the low or high crawl.

3) ***Hard Work***. This includes, for example, walking on a hard surface at 3.5 mph with a load weighing 40 or more pounds; walking in loose sand at 2.5 mph while carrying a load; and conducting field assaults.

**Table 12-5. WORK, REST, AND WATER CONSUMPTION TABLE**

| HEAT CATEGORY | WBGT INDEX, F° | ***EASY WORK*** | | ***MODERATE WORK*** | | ***HARD WORK*** | |
|---|---|---|---|---|---|---|---|
| | | WORK/ REST | WATER INTAKE (QT/H) | WORK/ REST | WATER INTAKE (QT/H) | WORK/ REST | WATER INTAKE (QT/H) |
| 1 | 78° to 81.9° | NL | 0.50 | NL | 0.75 | 40/20 | 0.75 |
| 2 (Green) | 82° to 84.9° | NL | 0.50 | 50/10 | 0.75 | 30/30 | 1.00 |
| 3 (Yellow) | 95° to 87.9° | NL | 0.75 | 40/20 | 0.75 | 30/30 | 1.00 |
| 4 (Red) | 88° to 89.9° | NL | 0.75 | 30/30 | 0.75 | 20/40 | 1.00 |
| 5 (Black) | 90° or more | 50/10 min | 1.00 | 20/40 | 1.00 | 10/50 | 1.00 |

NOTES

# Chapter 13
# DEMOLITIONS

This chapter introduces Rangers to the characteristics of explosives, to initiation systems, MDI components, detonation systems, safety considerations, expedient explosives, breaching charges, and timber cutting charges (FM 5-250). The two categories of explosives are low and high (Table 13-1).

- *Low explosives* have a detonating velocity up to 1,300 feet per second, which produces a pushing or shoving effect.
- *High explosives* have a detonating velocity of 3,280 to 27,888 feet per second, which produces a shattering effect.

**Table 13-1. CHARACTERISTICS OF US DEMOLITIONS EXPLOSIVES**

| NAME | APPLICATIONS | DETONATION VELOCITY | | RE FACTOR* | FUME TOXICITY | WATER RESISTANCE |
|---|---|---|---|---|---|---|
| | | MIN/SEC | FT/SEC | | | |
| AMMONIUM NITRATE | CRATERING CHG | 2,700 | 8.800 | 0.42 | DANGEROUS | POOR |
| PETN | DET CORD<br>BLASTING CAPS<br>DEMOLITION CHGS | 8,300 | 27,200 | 1.66 | SLIGHT | EXCELLENT |
| RDX | BLASTING CAPS<br>COMPOSITION EXPLOSIVE | 8,350 | 27,400 | 1.60 | DANGEROUS | EXCELLENT |
| TRINITROTOLUENE (TNT) | DEMOLITION CHG<br>COMPOSITION EXPLOSIVE | 6,900 | 22,600 | 1.00 | DANGEROUS | EXCELLENT |
| TETRYL | BOOSTER CHG<br>COMPOSITION EXPLOSIVE | 7,100 | 23,300 | 1.25 | DANGEROUS | EXCELLENT |
| NITROGLYCERIN | COMMERCIAL DYNAMITE | 7,700 | 25,200 | 1.50 | DANGEROUS | GOOD |
| BLACK POWDER | TIME FUSE | 400 | 1,300 | 0.55 | DANGEROUS | POOR |
| AMATOL 80/20 | BURSTING CHG | 4,900 | 16,000 | 1.17 | DANGEROUS | POOR |
| COMPOSITION A3 | BOOSTER CHG<br>BURSTING CHG | 8,100 | 26,500 | --- | DANGEROUS | GOOD |
| COMPOSITION B | BURSTING CHG | 7,800 | 25,600 | 1.35 | DANGEROUS | EXCELLENT |
| COMPOSITION C4 (M112) | CUTTING CHG<br>BREACHING CHG | 8.040 | 26,400 | 1.34 | SLIGHT | EXCELLENT |
| COMPOSITION H6 | CRATERING CHG | 7,190 | 23,600 | 1.33 | DANGEROUS | EXCELLENT |
| TETRYTOL 75/25 | DEMOLITION CHG | 7,000 | 23,000 | 1.20 | DANGEROUS | EXCELLENT |
| PENTOLITE 50/50 | BOOSTER CHG<br>BURSTING CHG | 7,450 | 24,400 | --- | DANGEROUS | EXCELLENT |

| NAME | APPLICATIONS | DETONATION VELOCITY MIN/SEC | FT/SEC | RE FACTOR* | FUME TOXICITY | WATER RESISTANCE |
|---|---|---|---|---|---|---|
| M1 DYNAMITE | DEMOLITION CHG | 6,100 | 20,000 | 0.92 | DANGEROUS | FAIR |
| DET CORD | PRIMING DEMOLITION CHG | 6,100 TO 7,300 | 20,000 TO 24,000 | --- | SLIGHT | EXCELLENT |
| SHEET EXPLOSIVE M118 AND M186 | CUTTING CHG | 7,300 | 24,000 | 1.14 | DANGEROUS | EXCELLENT |
| BANGALORE TORPEDO M1A2 | DEMOLITION CHG | 7,800 | 25,600 | 1.17 | DANGEROUS | EXCELLENT |
| SHAPED CHARGES M2A3, M2A4, AND M3A1 | CUTTING CHG | 7,800 | 25,600 | 1.17 | DANGEROUS | EXCELLENT |
| ** TNT = 1.00 RELATIVE EFFECTIVENESS* | | | | | | |

13-1. **INITIATING (PRIMING) SYSTEMS**. The best way to prime demolition systems is with modernized demolition initiators (MDI). These consist of blasting caps attached to various lengths of time fuse or shock tube. MDI can be used with a fuse igniter and detonating cord to create many firing systems. In the absence of MDI, field expedient methods may be used.

a. **Shock Tube**.

(1) Thin, plastic tube of extruded polymer with a layer of special explosive material on the interior surface.

(2) Explosive material propagates a detonation wave which moves along the shock tube to a factory crimped and sealed blasting cap.

(3) Detonation is normally contained within the plastic tubing. However, burns may occur if the shock tube is held.

(4) Advantages of shock tube.

(a) The shock tube offers the instantaneous action of electric initiation without the risk of accidental initiation caused by radio transmitters, static electricity, and so on.

(b) Extremely reliable.

(c) May be extended using left over pieces from previous operations.

b. **Blasting Caps**. Five MDI blasting caps are available which replace the M6 electric and M7 Non-electric blasting cap. Three of these are high-strength caps and two are low-strength. High-strength blasting caps can be used to prime all standard military explosives (including detonating cord) or to initiate the shock tube of other MDI blasting caps.

(1) ***M11***.

- Factory crimped to 30 feet of shock tube.
- A movable "J" hook is attached for quick and easy attachment to detonating cord.
- A red flag is attached 1 meter from the blasting cap and a yellow flag 2 meters from the blasting cap.

(2) ***M14***.

- Factory crimped to 7.5 feet of time fuse.
- May be initiated using a fuse igniter or match.
- Burn-time for total length is about five minutes.
- Yellow bands indicate calibrated one-minute time intervals.

(3) ***M15***.

- Two blasting caps factory crimped to 70 feet of shock tube.
- Each blasting cap has delay elements to allow for staged detonations.
- Low-strength blasting caps. Used as a relay device to transmit a shock tube detonation impulse from an initiator to a high strength-blasting cap.

(4) ***M12***. Factory crimped to 500 feet of shock tube on a cardboard spool.

(5) ***M13***. Factory crimped to 1,000 feet of shock tube.

c. **Matches**. If fuse igniter is unavailable, light the time blasting fuse with a match. Split the fuse at the end (Figure 13-1), and place the head of an unlit match in the powder train. Light the inserted match head with a flaming match, or rub the abrasive on the match box against it. It may be necessary to use two match heads during windy conditions. Burn time will increase with altitude and colder temperatures.

d. **M81 Fuse Igniter**. This is used to ignite time blasting fuse or to initiate the shock tube of MDI blasting caps.

**Figure 13-1. LIGHTING TIME FUSE WITH A MATCH**

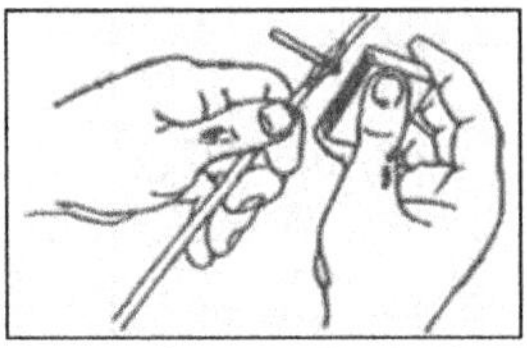

***Notes:*** Low strength blasting caps cannot reliably set off explosives. They should only be used to set off additional shock tubes. The M60 fuse igniter may still be used to ignite time blasting fuse. However, it will not reliably initiate the shock tube.

13-2. **DETONATION (FIRING) SYSTEMS**. The two types of firing systems are MDI alone, or MDI plus detonating cord.

a. **MDI Alone**. An MDI firing system is one in which the initiation set, transmission and branch lines are constructed using MDI components and the explosive charges are primed with MDI blasting caps. Construct the charge in the following manner.

(1) Emplace and secure explosive charge (C4, TNT, cratering charge, and so on.) on target.

(2) Place a sandbag or other easily identifiable marker over the M11, M14, or M15 blasting cap to be used.

(3) Connect to an M12 or M13 transmission line if desired.

(4) Connect blasting cap with shock tube to an M14 cap with time fuse. Cut the time blasting fuse to desired delay time.

(5) Prime the explosive charge by inserting the blasting cap into the charge.

(6) Visually inspect firing system for possible misfire indicators such as cracks, bulges, or corrosion.

(7) Return to the firing point and secure a fuse igniter to the cut end of the time fuse.

(8) Remove the safety cotter pin from the igniter's body.

(9) Actuate the charge by grasping the igniter body with one hand while sharply pulling the pull ring.

b. **MDI and Detonating Cord**. Construct the charge using the above steps for MDI stand-alone system. Incorporate detonating cord branch lines into the system using the "J" hooks of the M11 shock tube. Taping the ends of the Detonation Cord reduces the effect of moisture on the system.

13-3. **SAFETY**. MDI is not recommended for below ground use, except in quarry operations with water-gel or slurry explosives. Use detonating cord when it is necessary to bury primed charges.

a. Do not handle misfires downrange until the required 30 minute waiting period for both primary and secondary initiation systems has elapsed and other safety precautions have been accomplished.

b. Never yank or pull hard on the shock tube. This may actuate the blasting cap.

c. Do not dispose of used shock tubes by burning because of potentially toxic fumes given off from the burning plastic.

d. Do not use M1 dynamite with the M15 blasting cap. The M15 delay-blasting cap should be used only with water-gel or slurry explosives.

e. Always use protective equipment when handling demolitions. Minimum protection consists of leather gloves, ballistic eye protection, and helmet.

13-4. **EXPEDIENT EXPLOSIVES--IMPROVISED SHAPED CHARGE**. An improvised shaped charge (Figure 13-2) concentrates the energy of the explosion released on a small area, making a tubular or linear fracture in the target.

a. The versatility and simplicity of these charges make them effective against targets, especially those made of concrete or those with armor plating.

(1) Bowls, funnels, cone shaped glasses, (champagne glasses with stem removed) used as cones. Champagne or cognac bottles are excellent.

(2) Charge characteristics.

(a) *Cavity Liners.* These are made of copper, tin, or zinc. If none is available, cut a cavity out of the plastic explosive.

(b) *Cavity Angle.* This will work with 30 to 60 degree angles. The cavity angle in most high-explosive antitank (HEAT) ammunition is 42 to 45 degrees.

(c) *Explosive Height (In Container).* 2 x the height of the cone measured from the base of the cone to the top of the explosive.

(d) *Standoff.* Normal standoff is one and one half the cone's diameter.

(e) *Detonation Point.* The exact top center of the charge is the detonation point. Cover the blasting cap with a small amount of C4 if any part of the blasting cap is exposed.

b. Remove the narrow neck of a bottle or the stem of a glass by wrapping it with a piece of soft, absorbent twine or by soaking the string in gasoline and lighting it. Place two bands of adhesive tape, one on each side of the twine, to hold the twine firmly in place. The bottle or stem must be turned continuously with the neck up, to heat the glass uniformly.

c. A narrow band of plastic explosive placed around the neck and burned gives the same result. After the twine or plastic has burned, submerge the neck of the bottle in water and tap it against some object to break it off. Tape the sharp edge of the bottle to prevent cutting hands while tamping the explosive in place.

d. Do not immerse the bottle in water before the plastic has been completely burned, or it could detonate.

**Figure 13-2. IMPROVISED SHAPED CHARGE**

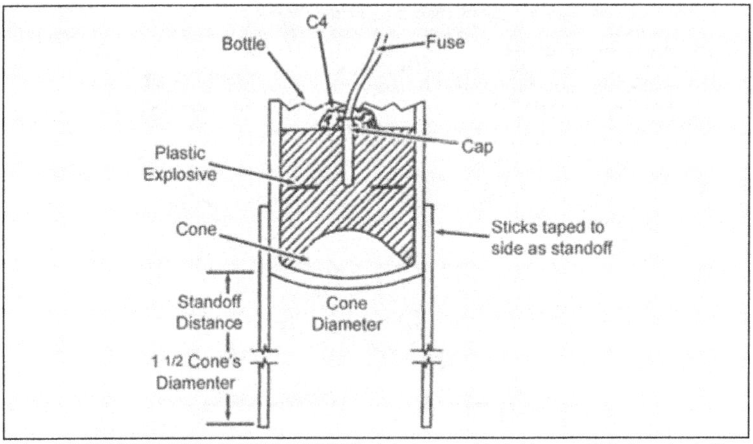


13-5. **EXPEDIENT EXPLOSIVES--PLATTER CHARGE**. This device turns a metal plate into a powerful blunt-nosed projectile (Figure 13-3). The plate should be steel, preferably round but square will work, and should weigh from 2 to 6 pounds.

a. The weight of the explosive should equal the weight of the platter.

b. Uniformly pack the explosive behind the platter. You will only need a container if the explosives fail to remain firmly against the platter. You can use tape to anchor the explosives, if needed.

c. Prime the charge at the exact, rear center of the charge. If any part of the blasting cap is exposed, cover the blasting cap with a small quantity of C4.

d. Aim charge at the direct center of the target, and ensure that the charge is on the opposite side of the platter from the target. Effective range is 35 yards for a small target. With practice, you might hit a 55-gallon drum at 25 yards 90 percent of the time. A gutted fuse igniter can serve as an expedient aiming device.

**Figure 13-3. PLATTER CHARGE**

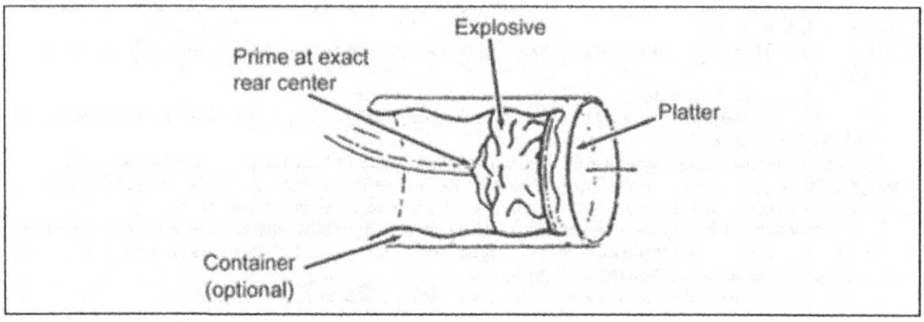


13-6. **EXPEDIENT EXPLOSIVES--GRAPESHOT CHARGE**. To use this antipersonnel fragmentation mine (Figure 13-4)--

a. **Hole**. Create a hole in the center, bottom of the container, for the blasting cap.

b. **Explosives**. Place explosives evenly on the bottom of the container. Remove all voids and air pockets by pressing the C4 into place using a non-sparking instrument.

c. **Buffer**. Place buffer material directly over top of the explosives.

d. **Projectiles**. Place projectiles over top of the buffer materials, then cover to prevent spilling from movement.

e. **Aim**. Aim at target from about 100 feet. Use a small amount of C4 on any exposed portion of the blasting cap.

**Figure 13-4. GRAPESHOT CHARGE**

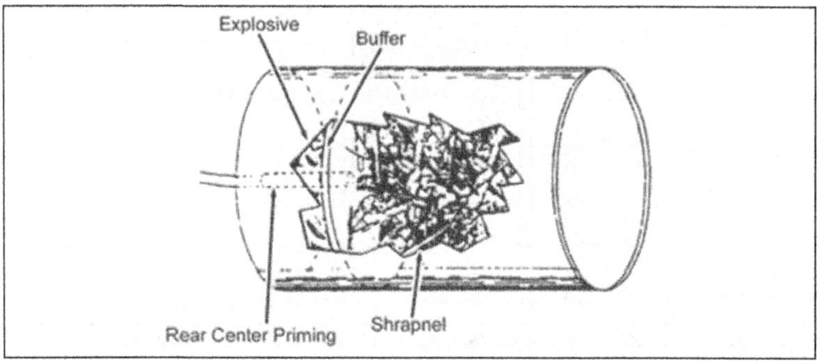

13-7. **DEMOLITION KNOTS.** Several knots are used in demolitions. Figures 13-5 and 13-6 show a few simple knots that can join demolitions to detonation cord.

**Figure 13-5. VARIOUS JOINING KNOTS USED IN DEMOLITIONS**

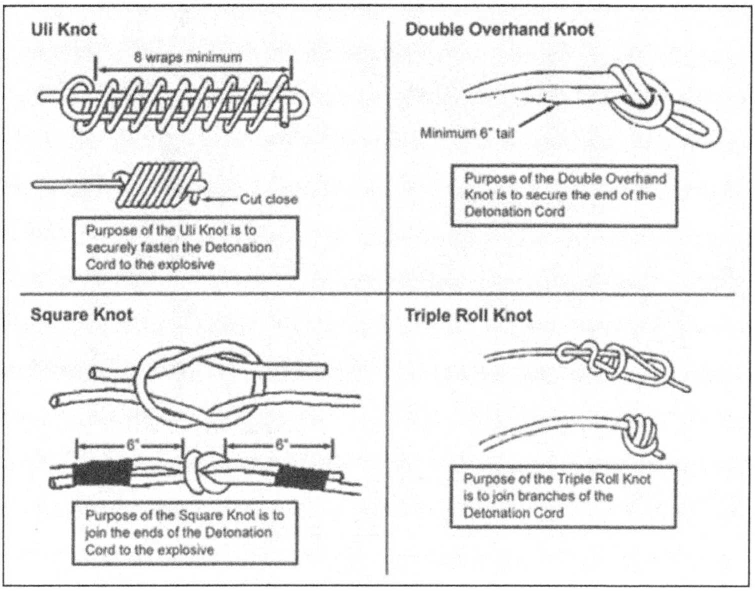

Figure 13-6. BRITISH JUNCTION

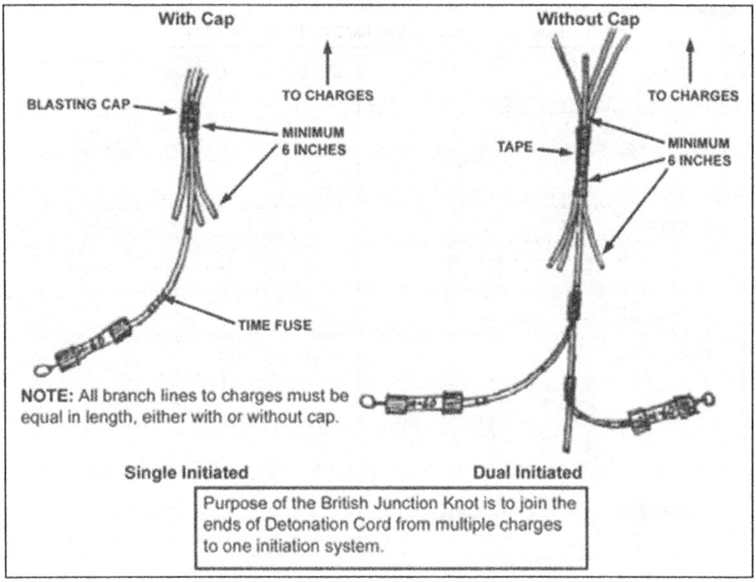

13-8. **MINIMUM SAFE DISTANCES**. Rangers must remain especially aware of their situation when using demolitions. (Table 13-2 shows minimum safe distances for employing demolitions.) For charges over 500 pounds, use the formulas shown in Figure 13-7.

**Table 13-2. MINIMUM SAFE DISTANCE FOR PERSONNEL IN OPEN (BARE CHARGE)**

| EXPLOSIVE WEIGHT (LB) | SAFE DISTANCE | | EXPLOSIVE WEIGHT (LB) | SAFE DISTANCE | |
|---|---|---|---|---|---|
| | FEET | METERS | | FEET | METERS |
| 27 OR LESS | 985 | 300 | 175 | 1,838 | 560 |
| 30 | 1,021 | 311 | 200 | 1,920 | 585 |
| 35 | 1,073 | 327 | 225 | 1,999 | 609 |
| 40 | 1,123 | 342 | 250 | 2,067 | 630 |
| 45 | 1,168 | 356 | 275 | 2,136 | 651 |
| 50 | 1,211 | 369 | 300 | 2,199 | 670 |
| 60 | 1,287 | 392 | 325 | 2,258 | 688 |
| 70 | 1,355 | 413 | 350 | 2,313 | 705 |
| 80 | 1,415 | 431 | 375 | 2,369 | 722 |
| 90 | 1,474 | 449 | 400 | 2,418 | 737 |
| 100 | 1,526 | 465 | 425 | 2,461 | 750 |
| 125 | 1,641 | 500 | 500 | 2,625 | 800 |
| 150 | 1,752 | 534 | --- | --- | --- |

**Figure 13-7. MINIMUM SAFE DISTANCES FOR CHARGES OVER 500 POUNDS**

$$\text{safe distance (meters)} = 100\sqrt[3]{\text{pounds of explosive}}$$

$$\text{safe distance (feet)} = 300\sqrt[3]{\text{pounds of explosive}}$$

***Notes:*** For charges on targets, the minimum radius of danger is 1,000 meters. Minimum safe distance when in a missile-proof shelter from the point of detonation is 100 meters.

13-9. **BREACHING CHARGES**. For Table 13-3, the left column represents the thickness of reinforced concrete wall. The remaining 7 columns show the number of packages of C4 required to breach the wall using the charge placements shown in the drawings above the columns.

a. Use Table 13-3, 13-4 and 13-5 for breaching charges.
b. Use the formula in Figure 13-8 to calculate the charges.

c. Multiply number of packages of C4 from Table 13-3 by conversion factor from Table 13-4 for materials other than reinforced concrete.

**Table 13-3. BREACHING CHARGES FOR REINFORCED CONCRETE**

| Reinforced-Concrete Thickness (ft) | Placement Methods: Placed in center of mass | Tamped or stemmed (Fill) | Deep water | Elevated untamped | Shallow water | Earth tamping | Ground placed, untamped |
|---|---|---|---|---|---|---|---|
| | C = 1.0 | C = 1.0 | C = 1.0 | C = 1.8 | C = 2.0 | C = 2.0 | C = 3.6 |
| | Packages of M112 (C4) | | | | | | |
| 2.0 | 1 | 5 | 5 | 9 | 10 | 10 | 17 |
| 2.5 | 2 | 9 | 9 | 17 | 18 | 18 | 33 |
| 3.0 | 2 | 13 | 13 | 24 | 26 | 26 | 47 |
| 3.5 | 4 | 21 | 21 | 37 | 41 | 41 | 74 |
| 4.0 | 5 | 31 | 31 | 56 | 62 | 62 | 111 |
| 4.5 | 7 | 44 | 44 | 79 | 88 | 88 | 157 |
| 5.0 | 9 | 48 | 48 | 85 | 95 | 95 | 170 |
| 5.5 | 12 | 63 | 63 | 113 | 126 | 126 | 226 |
| 6.0 | 13 | 82 | 82 | 147 | 163 | 163 | 293 |
| 6.5 | 17 | 104 | 104 | 186 | 207 | 207 | 372 |
| 7.0 | 21 | 111 | 111 | 200 | 222 | 222 | 399 |
| 7.5 | 26 | 137 | 137 | 245 | 273 | 273 | 490 |
| 8.0 | 31 | 166 | 166 | 298 | 331 | 331 | 595 |

NOTE: The results of all calculations for this table have been rounded UP to the next whole package.

**Table 13-4. CONVERSION FACTORS FOR MATERIALS OTHER THAN REINFORCED CONCRETE**

| *MATERIAL* | *CONVERSION FACTOR* |
|---|---|
| • EARTH | 0.1 |
| • ORDINARY MASONRY<br>• HARDPAN<br>• SHALE<br>• ORDINARY CONCRETE<br>• ROCK<br>• GOOD TIMBER<br>• EARTH CONSTRUCTION | 0.5 |
| • DENSE CONCRETE<br>• FIRST-CLASS MASONRY | 0.7 |

**Table 13-5. MATERIAL FACTOR (K) FOR BREACHING CHARGES**

| Material | R | K |
|---|---|---|
| Earth | All values | 0.07 |
| Poor masonry, shale, hardpan, good timber, and earth construction | Less than 1.5 m (5 ft) | 0.32 |
| | 1.5 m (5 ft) or more | 0.29 |
| Good masonry, concrete block, and rock | 0.3 m (1 ft) or less | 0.88 |
| | Over 0.3 m (1 ft) to less than 0.9 m (3 ft) | 0.48 |
| | 0.9 m (3 ft) to less than 1.5 m (5 ft) | 0.40 |
| | 1.5 m (5 ft) to less than 2.1 m (7 ft) | 0.32 |
| | 2.1 m (7 ft) or more | 0.27 |
| Dense concrete and first-class masonry | 0.3 m (1 ft) or less | 1.14 |
| | Over 0.3 m (1 ft) to less than 0.9 m (3 ft) | 0.62 |
| | 0.9 m (3 ft) to less than 1.5 m (5 ft) | 0.52 |
| | 1.5 m (5 ft) to less than 2.1 m (7 ft) | 0.41 |
| | 2.1 m (7 ft) or more | 0.35 |
| Reinforced concrete (factor does not consider cutting steel) | 0.3 m (1 ft) or less | 1.76 |
| | Over 0.3 m (1 ft) to less than 0.9 m (3 ft) | 0.96 |
| | 0.9 m (3 ft) to less than 1.5 m (5 ft) | 0.80 |
| | 1.5 m (5 ft) to less than 2.1 m (7 ft) | 0.63 |
| | 2.1 m (7 ft) or more | 0.54 |

Figure 13-8. FORMULA FOR COMPUTING SIZES OF CHARGES TO BREACH CONCRETE, MASONRY, AND ROCK

$$P = R^3KC$$

where—

*P* = *TNT required, in pounds*

*R* = *breaching radius, in feet*

*K* = *material factor, which reflects the strength, hardness, and mass of the material to be demolished (Table 3-5)*

*C* = *tamping factor, which depends on the location and tamping of the charge (Figure 3-15)*

13-10. **TIMBER CUTTING CHARGES**. Table 13-6 shows timber-cutting charge sizes. Figures 13-9 through 13-15 show the types of charges and the formulas to use with each.

Table 13-6. TIMBER-CUTTING CHARGE SIZE

| | Packages of C4 Required (1.25-lb Packages) by Timber Diameter (in) | | | | | | | | | | | |
|---|---|---|---|---|---|---|---|---|---|---|---|---|
| Charge Type | 6 | 8 | 10 | 12 | 15 | 18 | 21 | 24 | 27 | 30 | 33 | 36 |
| Internal | 1 | 1 | 1 | 1 | 1 | 1 | 2 | 2 | 2 | 3 | 3 | 4 |
| External | 1 | 1 | 2 | 3 | 4 | 5 | 7 | 9 | 11 | 14 | 17 | 20 |
| Abatis | --- | --- | --- | --- | --- | --- | --- | 7 | 9 | 11 | 14 | 16 |
| NOTE: Packages required are rounded UP to the next whole package. | | | | | | | | | | | | |

Figure 13-9. ABATIS

Figure 13-10. FORMULA FOR FALLEN TREE OBSTACLES OR TEST SHOT

$$P = \frac{D^2}{50} = P = 0.02D^2$$

where—

*P* = *TNT required per tree, in pounds*

*D* = *diameter or least dimension of dimensioned timber, in inches*

Figure 13-11. TIMBER-CUTTING RING CHARGE

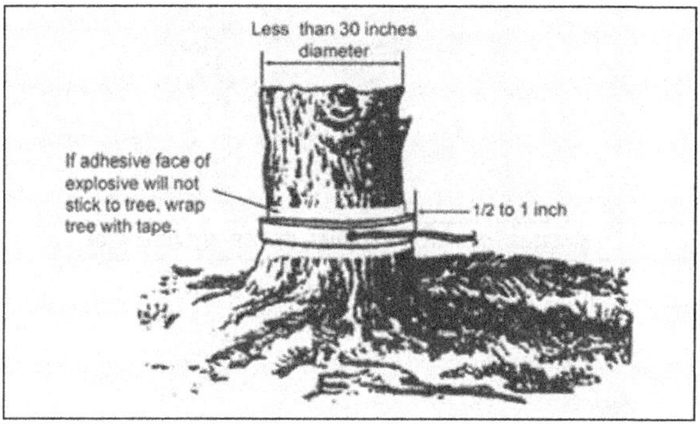

Figure 13-12. TIMBER-CUTTING CHARGE (EXTERNAL)

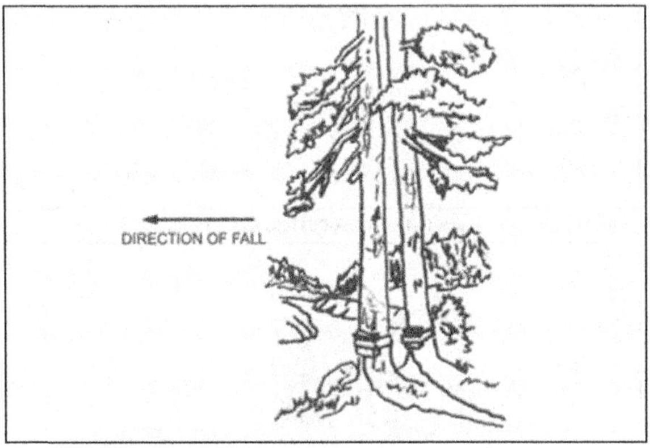


Figure 13-13. FORMULA FOR EXTERNAL TIMBER-CUTTING CHARGE

$$P = \frac{D^2}{40} \text{ or } P = 0.025D^2$$

where—

*P* = *TNT required per target, in pounds*

*D* = *diameter or least dimension of dimensioned timber, in inches*

Figure 13-14. TIMBER-CUTTING CHARGE (INTERNAL)

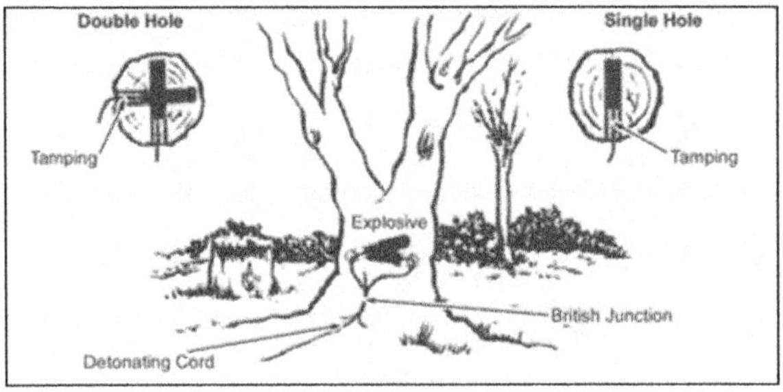

**Figure 13-15. FORMULA FOR INTERNAL TIMBER-CUTTING CHARGE**

$$P = \frac{D^2}{250} \text{ or } P = 0.004D^2$$

where—

$P$ = *TNT required per tree, in pounds*

$D$ = *diameter or least dimension of dressed timber, in inches*

NOTES

## Chapter 14

# RANGER URBAN OPERATIONS

Urban operations are planned and conducted in an area of operations (AO) that includes one or more urban areas. An urban area is a topographical complex where man-made construction or high population density is the dominant feature (FM 30-06.11). Urban terrain is likely to be one of the most significant future areas of operations for American forces throughout the world. Expanding urban development affects military operations as the terrain is altered. The increasing focus on stability and support operations, urban terrorism, and civil disorder emphasizes that combat in urbanized areas is unavoidable. Urban areas are the power centers, the centers of gravity, and thus the future battlefield. For further study, see FM 30-06.11, FM 90-10, FM 90-1, FM 7-8, 75th Ranger Regiment Advanced UO SOP, and Ranger Training Circular 350-1-2.

14-1. **TYPES**. Infantry units must be trained to conduct urban combat under high-intensity conditions. High-intensity urban combat requires the employment of combat power of the joint combined arms team. An Infantry unit's mission is normally to recon, isolate, penetrate, systematically clear, defend the urban area, and engage and defeat the enemy with decisive combat power. Although the changing world situation may have made urban combat under high-intensity conditions less likely for US forces, it represents the high end of the combat spectrum, and units must be trained for it. High-intensity urban operations can be casualty-intensive for both sides. With the integrated firepower of the joint, combined arms team, leaders must make every attempt to limit unnecessary destruction of critical infrastructure and casualties among noncombatants.

a. **Precision Conditions**. Infantry units train to defeat an enemy that is mixed with non-combatants in precision urban combat. Leaders plan to limit civilian casualties and collateral damage through the establishment of strict rules of engagement (ROE) and the employment of precision weapons and munitions. The ROE provides the focus for the use and restraint of combat power. The ROE may be significantly more restrictive than under high-intensity conditions.

b. **Surgical Conditions**. Operations conducted under surgical conditions include special-purpose raids, small precision strikes, or small-scale personnel seizure or recovery operations in an urban environment (for example, hostage rescue). Joint special operation forces usually conduct these operations. They may closely resemble US police operations performed by Special Weapons and Tactics (SWAT) teams. They may even involve cooperation between US forces and host nation police. Though regular units may not usually be involved in the actual surgical operation, they may support it by isolating the area, by providing security or crowd control, or providing search and rescue teams.

c. **Transitions**. Leaders must always be prepared to transition rapidly from one type of urban combat to another, and back. Real-world combat shows us that urban operations can deteriorate rapidly and without warning. A force involved in stability and support operations can suddenly find itself in a high-intensity combat situation.

14-2. **PRINCIPLES**

a. **Surprise**. Strike the enemy at a time or place or in a manner for which he is unprepared. Key to success: gives the assaulting element the advantage.

b. **Security**. Never permit the enemy to acquire unexpected advantage.

(1) Maintain during all phases of the operation.

(2) Four-dimensional battlefield (height, depth, width, subterranean).

(3) Always maintain 360 degree security (include elevated and subterranean areas).

(4) Mission is never complete as long as you remain in the urban environment. The status of actors in the urban environment does not afford the sense of security offered by "open" terrain. The key to survivability is a constant state of situational awareness.

c. **Simplicity**. Prepare clear, uncomplicated plans, and provide subordinates with concise orders to ensure thorough understanding.

(1) Always keep plans simple.

(2) Ensure everyone understands the mission and the commander's intent.

(3) Plan and prepare for the worst.

d. **Speed**. Rate of military action.

(1) Act as security.

(2) Move in a careful hurry.

(3) Smooth is fast and fast is smooth.

(4) Never move faster than you can accurately engage targets.

(5) Exercise tactical patience.

e. **Violence of Action**. Eliminate the enemy with sudden, explosive force.

(1) Combined with speed gives surprise.

(2) Prevents enemy reaction.

(3) Both physical and mental.

14-3. **METT-TC**. To effectively plan combat operations in urban environments, leaders must use Troop Leading Procedures and conduct a thorough analysis utilizing METT-TC factors. The following lists specific planning guidance that must be incorporated when planning for urban operations. For more specifics on mission planning, refer to Chapter 2, Ranger Handbook.

a. **Mission**. Know correct Task Organization to accomplish the mission (Offense, Defense, or Stability and Support Operations).

b. **Enemy**.

(1) ***Disposition***. Analyze the arrayal of enemy forces in and around your objective, known and suspected. Example: Known or suspected locations of minefields, obstacles, and strong points.

(2) ***Composition and Strength***. Analyze the enemy's task organization, troops available, suspected strength, and amount of support from local civilian populace based on intelligence estimates. Is the enemy a conventional or unconventional force?

(3) ***Morale***. Analyze the enemy's current operational status based on friendly intelligence estimates. For example, is the enemy well supplied; have they had recent success against friendly forces, taken many casualties; and what is the current weather?

(4) ***Capabilities***. Determine what the enemy can employ against your forces. Example: Enemy's weapons, artillery assets, engineer assets, air defense assets, NBC threats, thermal/NVG capabilities, close air support, armor threat, and so on.

(5) ***Probable Course(s) of Action***. Based on friendly intelligence estimates, determine how the enemy will fight within his area of operation (in and around your area of operation).

c. **Terrain**. Leaders conduct a detailed terrain analysis of each urban setting, considering the types of built-up areas and composition of existing structures. They use OCOKA when analyzing terrain, in and around the area of operation.

(1) ***Observation and Fields of Fire***. Always be prepared to conduct urban operations under limited visibility conditions. This includes the effects of reduced illumination, as well as natural and manmade obscuration. Leaders should ensure that Rangers are equipped with adequate resources, which allow them to successfully operate in the urban environment under these types of conditions. (2) ***Cover and Concealment***. Leaders must perform a thorough analysis of peripheral as well as intra-urban areas. Leaders should identify routes to objectives, which afford assault forces with the best possible cover and concealment. Additionally, leaders should take advantage of limited visibility conditions, which would allow forces to move undetected to their final assault / breaching positions. When in the final assault position, forces should move as rapidly as tactically possible to access structures, which afford additional cover and concealment. Leaders must learn to properly employ obscurants and exercise "tactical patience" to fully take advantage of these effects. Finally, all members of the urban force must practice noise and light discipline. Soldiers must avoid unnecessary voice communications, learn the proper use of white light, and limit contact with surfaces that may alert the enemy of their presence.

(3) ***Obstacles***. There are many manmade and natural obstacles on the periphery, as well as within the urban environment. Leaders should conduct a detailed reconnaissance of routes and objectives (this must include subterranean complexes), taking into consideration route adjustments and special equipment needs.

(4) ***Key Terrain***. Analyze which buildings, intersections, bridges, LZ/PZ, airports, and elevated areas that provide a tactical advantage to you or the enemy. Additionally, the leader must identify critical infrastructure within his area of operations, which would provide the enemy with a tactical advantage on the battlefield. These may include, but are not limited to, communication centers, medical facilities, governmental facilities, and facilities that are of psychological significance.

(5) ***Avenues of Approach***. Consider roads, intersections, inland waterways, and subterranean constructions (subways, sewers, and basements). Leaders should classify areas as go, slow go, or no-go based on the navigability of the approach.

***Note:*** Military maps may not provide enough detail for urban terrain analysis or reflect the underground sewer systems, subways, water systems, or mass transit routes.

(6) ***Troops***. Analyze your forces utilizing their disposition, composition, strength, morale, capabilities, and so on. Leaders must also consider the type and size of the objective to plan effective use of troops available.

(7) ***Time***. Operations in an urban environment have a slower pace and tempo. Leaders must consider the amount of time required to secure, clear, or seize the urban objective and stress and fatigue Rangers will encounter. Additional time must also be allowed for area analysis efforts; these may include, but are not limited to--

- Maps and urban plans Recon and analysis.
- Hydrological data analysis.
- Line-of-sight surveys.
- Long Range Surveillance and Scout reconnaissance.
- Artillery.
- Armor.
- Aviation.
- Armor.
- Engineers.

(8) ***Civilians***. Authorities such as the National Command will establish the Rules of Engagement. Commanders at all levels may provide further guidance regarding civilians occupying the area of operations (AO). Leaders must daily reiterate the ROE to subordinates, and immediately inform them of any changes to the ROE. Rangers must have the discipline to identify the enemy from noncombatants and ensure civilians understand and follow all directed commands.

***Note:*** Civilians may not speak English, may be hiding (especially small children), or dazed from a breach. Civilians must not be given the means to resist. Rehearse how clearing/search teams will react to these variables. Never compromise the safety of your Rangers.

14-4. **CLOSE QUARTERS COMBAT.** Due to the very nature of a CQC encounter, engagements will be very close (within 10 meters) and very fast (targets exposed for only a few seconds). Most close quarter's engagements are won by who hits first and puts the enemy down. It is more important to knock a Ranger down as soon as possible than it is to kill him. In order to win a close quarters engagement, Rangers must make quick, accurate shots by mere reflex. This is accomplished by reflexive fire training. Remember, no matter how proficient you are, always fire until the enemy goes down. All reflexive fire training is conducted with the eyes open.

***Note:*** Research has determined that, on average, only three out of ten people actually fire their weapons when confronted by an enemy during room clearing operations. Close quarters combat success for the Ranger begins with the Ranger being psychologically prepared for the close quarter's battle. The foundation for this preparedness begins with the Ranger's proficiency in basic rifle marksmanship. Survival in the urban environment does not depend on advanced skills and technologies. Rangers must be proficient in the basics.

14-5. **REHEARSALS**. Similar to the conduct of other military operations, leaders need to designate time for rehearsals. Urban operations require a variety of individual, collective, and special tasks, which are not associated with operations on less complex terrain. These task require additional rehearsal time for clearing, breaching, obstacle reduction, casualty evacuation, and support teams. Additionally, rehearsal time must be identified for rehearsals with combined arms elements. These may include, but are not limited to--

a. **Stance**. Feet are shoulder width apart, toes pointed straight to the front (direction of movement). The firing side foot is slightly staggered to the rear of the non-firing foot. Knees are slightly bent and the upper body leans slightly forward. Shoulders are not rolled or slouched. Weapon is held with the butt stock in the pocket of the shoulder. The firing side elbow is kept in against the body. The stance should be modified to ensure the Ranger maintains a comfortable boxer stance.

(1) ***Low-Carry Technique***. The butt stock of the weapon is placed in the pocket of the shoulder. The barrel is pointed down so the front sight post and day optic is just out of the field of vision. The head is always up identifying targets. This technique is safest and is recommended for use by the clearing team once inside the room.

(2) ***High-Carry Technique***. The butt stock of the weapon is held in the armpit. The barrel is pointed slightly up with the front sight post in the peripheral vision of the individual. Push out on the pistol grip and thrust the weapon forward and pull straight back into the pocket of the shoulder to assume the proper firing position. This technique is best suited for the line-up outside the door. Exercise caution with this technique always maintaining situational awareness, particularly in a multi-floored building.

***Note:*** Muzzle awareness is critical to the successful execution of close quarter's operations. Rangers must never point their weapons or cross the bodies of their fellow Rangers at any time. Additionally Rangers should always avoid exposing the muzzle of their weapons around corners; this is referred to as "flagging."

b. **Malfunction**. If a Ranger has a malfunction with his weapon during any CQC training, he will take a knee to conduct immediate action. Once the malfunction is cleared, there is no need to immediately stand up to engage targets. Rangers can save precious seconds by continuing to engage from one knee. Whenever other members of the team see a Ranger down, they must automatically clear his sector of fire. Before rising to his feet, the Ranger warns his team members of his movement and only rises after they acknowledge him. If a malfunction occurs once committed to a doorway, the Ranger must enter the room far enough to allow those following him to enter and move away from the door. This drill must be continually practiced until it is second nature.

c. **Approach to a Building or Breach Point**. One of the trademarks of ranger operations is the use of limited visibility conditions. Whenever possible, breaching and entry operations should be executed during hours and conditions of limited visibility. Rangers should always take advantage of all available cover and concealment when approaching breach and entry points. When natural or manmade cover and concealment is not available, Rangers should employ obscurants to conceal their approach. There are times when Rangers will want to employ obscurants to enhance existing cover and concealment. Members of the breach / entry team should be numbered for identification, communication, and control purposes.

(1) The # 1 Ranger should always be the most experienced and mature member of the team, usually the team leader. The # 1 Ranger is responsible for frontal and entry and breach point security. The team leader is responsible for initiating all voice and physical commands. The team leader must exercise situational awareness at all time with respect to the task, friendly force, and enemy activity.

(2) The # 2 Ranger is directly behind the # 1 Ranger in the order of movement, and he moves through the breach point in the opposite direction from the #1 Ranger.

(3) The # 3 Ranger will simply go opposite the # 2 Ranger inside the room at least 1 meter from the door.

(4) The # 4 Ranger moves the opposite the # 3 Ranger and is responsible for rear security (and is normally the last Ranger into the room). An additional duty of the # 4 Ranger is breaching.

***Note:*** Consider how much firepower each Ranger delivers. Where do you put the SAW gunner in the order? You must weigh the benefits of firepower against those of quick, accurate shots. If the # 4 Ranger has breaching responsibilities, it should not be the SAW gunner, because this would reduce your firepower.

d. **Actions Outside the Point of Entry**. Entry point position and individual weapon positions are important. The clearing team members should stand as close to the entry point as possible, ready to enter. Weapons are oriented in such a manner that the team provides itself with 360 degree security at all times. Team members must signal to one another that they are ready at the point of entry. This is best accomplished by sending up a "squeeze." If a tap method is used, an inadvertent bump may be misunderstood as a tap.

e. **Enter Building/Clear Room**. See Battle Drill 5, Chapter 6.

f. **Locking Down the Room**.

(1) Control the situation within the room.

(2) Use clear, concise arm and had signals. Voice commands should be kept to a minimum to reduce the amount of confusion and to prevent the enemy--who might be in the next room--from discerning what is going on. This enhances the opportunity for surprise and allows the assault force the opportunity to detect any approaching force.

(3) Physically and psychologically dominate.

(4) Establish security and report status.

(5) Cursory search of the room to include the ceiling (3 Dimensional Fight).

(6) Identify the dead using reflexive response techniques (Eye thump method).

(7) Search the room for PIR, precious cargo as per the mission and time available.

(8) Evacuate personnel.

(9) Mark room clear using chemical lights, engineer tape, chalk, paint, VS-17 panels, and so on.

14-6. **TTPS FOR MARKING BUILDINGS AND ROOMS**. Units have long identified a need to mark specific buildings and rooms during UO. Sometimes rooms need to be marked as having been cleared, or buildings need to be marked as containing friendly forces. The US Army Infantry School is currently testing a remote marking device that can be used to mark doors from as far away as across a wide street. In the past, units have tried several different field-expedient marking devices; some with more success than others. Chalk has been the most common. It is light and easily obtained but not as visible as other markings. Some of the other techniques have been to use spray paint, and paintball guns.

a. **NATO Standard Marking SOP**. The North Atlantic Treaty Organization (NATO) has developed a standard marking SOP for use during urban combat. It uses a combination of colors, shapes, and symbols. These markings can be fabricated from any material available. (Figure 14-1 shows examples.)

**Figure 14-1. EXAMPLE NATO STANDARD MARKINGS**

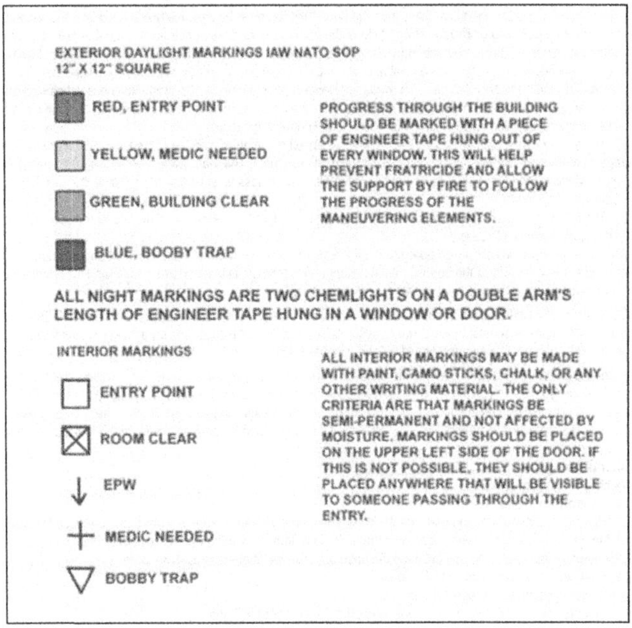


b. **Spray Paint**. Canned spray paint is easily obtained and comes in a wide assortment of colors including florescent shades that are highly visible in daylight. It cannot be removed once used. Cans of spray paint are bulky and hard to carry with other combat equipment. Paint is not visible during darkness nor does it show up well through thermal sights.

c. **Paintball Guns**. Commercial paintball guns have been purchased by some units and issued to small unit leaders. Some models can be carried in standard military holsters. They can mark a building or door from about 30 meters. The ammunition and propellant gas is not easily obtainable. The ammunition is fragile and often jams the gun if it gets wet. The available colors are not very bright, and just like spray paint, cannot be seen at night or through thermal sights.

d. **Wolf Tail.** A simple, effective, easy-to-make, lightweight device called a "wolf tail" can be fabricated to mark buildings, doorways, and windows (Figure 14-1). A unit has changed its tactical SOP to require that each Infantryman carry one of these devices in his ACU cargo pocket. Wolf tails, when used IAW a simple signaling plan understood by all members of the unit, can aid in command and control, reduce the chances of fratricide, and speed up casualty collection during urban combat.

(1) The wolf tail marking device is simple to make and versatile. It can be used together with the NATO marking scheme. Rolled up, it makes a small, easily accessible package that can be carried in the cargo pocket of the ACUs. It can be recovered easily and used again if the situation changes. All its components can be easily obtained through unit supply. It combines a variety of visual signals (colored strapping and one or more chemical lights of varying colors) with a distinctive heat signature that is easily identified through a thermal weapon sight. An infrared chemlight can be used either as a substitute for the colored chemlight(s) or in addition to them.

(2) Constructing the wolf-tail marking device requires the following material:

- A 2-foot length of nylon strap (the type used for cargo tie-downs) (engineer tape can be substituted).
- About 5 feet of 550 cord.
- A small weight such as a bolt or similar object.
- Duct tape.
- Chemical lights (colored or IR).
- Two 9-volt batteries.

(3) Assemble the items by tying or taping the cord to the small weight. Attach the other end of the cord to the nylon strapping, securing it with duct tape. Attach the 9-volt batteries in pairs to the lower end of the strapping with several wraps of duct tape, making sure that the negative terminals are opposite the positive, but not actually touching. Use more duct tape to attach the chemical lights, about 2 inches above the batteries, to the strapping.

(4) When you want to mark your position, push the batteries together firmly until the male and female plugs lock. This shorts out the battery, causing it to heat up rapidly. The hot battery is easily identified through the thermal sights of tanks or BFVs. The batteries stay visible for about 45 minutes. Activating the chemical lights provides an easily identified light source visible to the naked eye. You can use infrared chemical lights if you want them to be seen through night vision devices but not with the naked eye.

**Figure 14-2. EXAMPLE OF A WOLF-TAIL MARKING DEVICE**

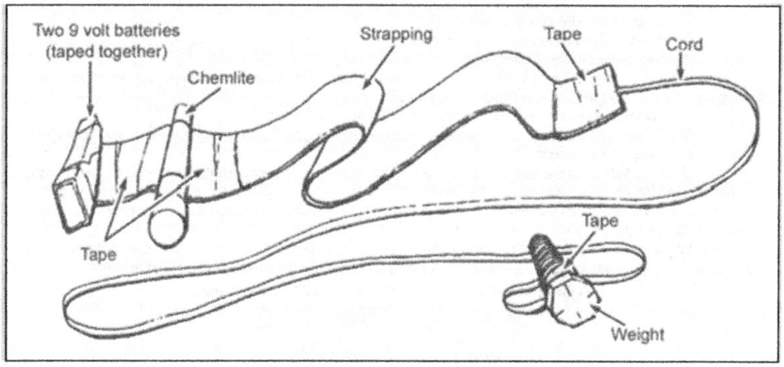


***Note:*** One option is to place chemical lights and batteries at both ends of the wolf tail to mark the inside and outside of a building or room.

(5) Use the cord and the small weight to hold the wolf tail in position by tying or draping it out a window or hanging it on a door, wherever it is best seen by other friendly troops. Squads or platoons can vary the numbers and colors of chemical lights, or use multiple battery sets to identify precisely what unit is in which building.

(6) Medics and combat lifesavers can carry a standardized variation that can be used to clearly identify a building as containing wounded personnel needing evacuation. This could be a white strap with multiple red chemical lights, or any other easily identified combination.

14-7. **URBAN ASSAULT BREACHES**. Understanding how to employ and incorporate breaching as part of a leader's planning process is an important part of urban operations. It is imperative that elements of your patrol be skilled in the art of breaching. Whether infiltrating or exfiltrating from an objective, leaders must plan for either option. One constant disadvantage with the employment of explosives is that placement requires Rangers to expose themselves to possible enemy fires. Breach teams need to be supported by fires or obscurants and breaching operations should be performed during hours of limited visibility whenever possible. Breaching classes follow:

a. **Mechanical**. Mechanical breaching should be an important part of a leader's breaching training program because it is usually an option. Mechanical breaching is best described as gaining access by the use of tools or saws. Although most tools and saws used are recognizable and self explanatory to the individual Ranger, one must practice on various techniques to increase speed an effectiveness. This reduces fatigue and expedites the actual assault.

(1) **Tools**.

(a) Hooligan tools..........................Doors/windows of all types.
(b) Sledge hammer.......................Heavy duty doors, locks, and window frames.
(c) Picket pounder.........................Doors of all types, light walls.
(d) Bolt Cutters.............................Chain link fence, locks, and wire obstacles.
(e) Pick Ax....................................Lightweight doors and locks.
(f) Saws........................................Fences, light doors, locks.

b. **Ballistic**. Ballistic breaching is defined as a forced entry or exit by the use of weapons. Whether using shotguns, M16A2/M4, M249 SAW, specific considerations must be addressed:

(a) Type of round and ricochet factor.
(b) Composition of the breaching point.
(c) Composition of the floor beyond the door.
(d) Personnel behind the door (friendly/enemy).
(e) Always shoot at a 45-degree angle.

c. **Explosive**. Explosive breaching is the most viable because it is the most effective. When employing explosives during breaching operations, leaders must consider three major factors.

(1) ***Overpressure***. The amount of PSI released from the concussion of the blast.

(2) ***Missile Hazard***. Fragmentation or projectiles sent at tremendous speed from the explosion area. This occurs from either the charge or target being breached.

(3) ***Minimum Safe Distance Requirements***. Use of explosives in the urban environment must consider the presence of noncombatants and friendly forces. Additionally, the are many hazardous materials located in the urban environment, these may include chemicals as well as construction materials. There is always a risk of secondary explosions and fires, when employing explosive breaching techniques.

(4) ***Charges***. Various charges can be used for explosive breaching. Leaders must conduct extensive training on the use of the charges to get proper target feedback. Figure 14-3 shows examples of charges used for explosive breaching.

**Figure 14-3. EXAMPLE CHARGES FOR EXPLOSIVE BREACHING**

| Charge | Target |
|---|---|
| Water Impulse................ | Steel/wood doors |
| Flexible linear................. | Wood doors |
| Ranger wall breach........ | Masonry/brick walls |
| Chain link ladder............ | Chain link fence |
| E-Type silhouette........... | Wood doors |
| Brashier breacher.......... | Concertina wire |

# Chapter 15
# VEHICLE CONVOY OPERATIONS

This chapter outlines a technique for conducting vehicle convoy operations. Convoy operations present a challenge to the Ranger leader. Trucks and other combat vehicles produce a large signature on the battlefield and increase your unit's value as a target. Vehicle movement is restricted to roads and terrain that they can traverse; therefore, a sound plan must be implemented to minimize the possibility of compromise.

15-1. **PLANNING CONSIDERATIONS**. This paragraph explains how to plan a vehicle convoy IAW the eight steps of the TLP.

a. **Receive the Mission**. (PL will extract the following information from Company OPORD).
- (1) Truck Support (number and type of Truck, ACL).
- (2) Weather. Road conditions.
- (3) Truck pickup and drop-off location / marking.
- (4) Truck movement Timeline (Pick-up time, roll time, H-hour).
- (5) Truck Routes (primary and alternate, Check points).
- (6) Drop-off locations (primary and alternate).
- (7) IED/Contact compromise and contingencies.

b. **Issue the Warning Order**. During the WARNO the PL gives a basic overview of the Truck Movement and gives specific Task to Maneuver Units.
- (1) Manifest Personnel (PSG).
- (2) Pick up point security and set-up (PSG).
- (3) Primary and alternate routes (Security SL).

c. **Make a tentative Plan**. PL considers METT-TC.
- (1) ***Mission***. Develop your Ground Tactical Plan and a then develop a truck plan that best supports it.
- (2) ***Enemy***. Consider enemy patrols and activity in AO.
- (3) ***Terrain***. Evaluate routes and dropoff and pickup points using OCOKA.
- (4) ***Troops***.
- (a) Truck ACL.
- (b) Chalks and chalk leaders identified.
- (c) Chalks planed with tactical integrity & self sufficiency.
- (d) Key leaders and weapons are cross-loaded.
- (e) Pickup point marking team identified.
- (f) Pickup point security teams identified.
- (5) ***Time***. PL considers the following when evaluating time available.
- (a) Movement to Pickup point.
- (b) Recon of Pickup point.
- (c) Emplacement of security.
- (d) Marking the Pickup point.
- (e) Pickup point Posture time.

d. **Initiate Necessary Movement**. PL decides whether to move the platoon to the pickup location or to conduct a map reconnaissance and complete the plan.

e. **Conduct Reconnaissance**. (Is not necessary if pickup point is secured and established by Higher HQ's).
- (1) Leader's Recon consists of PL, RTO, Chalk Leaders and Pickup point Security element.
- (2) PL determines suitability of Pickup point.
- (3) PL determines security plan (either overwatch or strong point).

f. **Complete the Plan**. Complete Truck Movement Annex on Page ??? the Ranger Handbook.

g. **Issue the OPORD**.

h. **Supervise**.
- (1) Back brief from chalk leaders.
- (2) Rock drill of truck movement and contingency plans.

15-2. **FIVE PHASES OF TRUCK MOVEMENT**. Each phase must support the ground tactical plan, which specifies actions in the objective area to accomplish the commander's intent for the assigned mission, be it a raid, ambush, recon or other follow-on missions.

a. **Staging Plan**.

(1) Secure the pickup point (strongpoint or overwatch).

(2) Mark pickup point (day/night).

b. **Loading Plan**. Each Ranger is assigned a truck.

c. **Movement Plan**.

(1) Troops awake and alert during movement.

(2) Platoon leader and chalk leaders tracking route progress.

(3) Compromise and contingency plan.

(a) React to IED.

(b) React to ambush.

(c) Truck breakdown.

d. **Dropoff Plan**.

(1) Establish security of dropoff point (overwatch or strongpoint).

(2) Dismount vehicles.

(3) Recon and secure assembly area.

(4) Adjust perimeter as chalks arrive.

(5) PSG clears the trucks when last chalk departs.

e. **Ground Tactical Plan**. Execute after platoon is reconsolidated or minimum force required is assembled.

15-3. **CONVOY TECHNIQUES**. The following convoy techniques have been included for your convenience:

- Convoy warning order
- Convoy brief
- React to Ambush Near/Far
- Convoy forced to stop (Methods 1 and 2)
- Break contact
- Recovery/CASEVAC Operations

## Convoy Warning Order

1. SITUATION: A brief statement of the enemy and friendly situation. Who, What, Where
2. MISSION: Who, What, When, Where, Why
3. TASK ORGANIZATION: Based on tasking from higher WARNO.
4. INITIAL TIME SCHEDULE:

| When | Who | What |
|---|---|---|
| | | |
| | | |
| | | |
| | | |

5. SPECIAL INSTRUCTIONS: PCC/PCI Guidance, Rehearsals, additional tasks to be accomplished.
6. SERVICE SUPPORT:

Class I: (Rations & Water)

| | | |
|---|---|---|
| | | |
| | | |
| | | |

Class III: (POL)

| | | |
|---|---|---|
| | | |
| | | |
| | | |

Class V: (Ammunition /Pyrotechnics)

| Weapon System | Rounds | Type |
|---|---|---|
| | | |
| | | |
| | | |

| Pyrotechnic Device | Number | Location |
|---|---|---|
| | | |
| | | |
| | | |

Class VIII: (Medical)

| | | |
|---|---|---|
| | | |
| | | |
| | | |

7. UNIFORM AND EQUIPMENT COMMON TO ALL.

**CONVOY BRIEF** (Modified from FM 55-30)

Movement Order No. ____

References: A. ______________________________ *(Maps, tables and relevant documents)*

B. ______________________________

TASK ORGANIZATION: (Internal organization for convoy – Manifest)

1. **SITUATION:**

a. Enemy Forces:

(1) Weather. General forecast.

(2) Light Data (EENT, % Illumination, MR, MS, BMNT)

(3) Discuss Enemy.

Identification of enemy (If known).

Composition / capabilities / strength / equipment

Location

(Hot Spots highlighted on map)

b. Friendly Forces:

(1) Operational support provided by higher headquarters.

Helicopter / Gunships

Quick Reaction Forces (QRFs)

MP Escorts / Rat Patrols

Fire Support elements

c. Attachments: (From outside the organization)

2. **MISSION:** (WHO, WHAT, WHEN, WHERE, & WHY)

3. **Employment Tactics, Techniques and Procedures**

a. Readiness Level i. Ride / Scanning (Observation)

ii. At the Ready

b. Scanning / Sector of Fire

i. Driver 1. Sector of Scan is 9 – 1 clock position

2. Observation with Mirrors

3. Sector of Fire is 9 – 11 o'clock

ii. TC 4. Sector of Scan is 11 – 3 clock position

5. Sector of Fire is 1 – 3 clock position

iii. Other Systems are based off vehicle type and load

b. Target ID i. Communicate / Signal

c. Body Positioning i. Engage as you Train (Right or Left handed Firing)

ii. Firing-side Shoulder Down

iii. Maintain your Body Position

iv. Weapon to Head

d. Rules of Engagement Concerns – Employ the appropriate systems based off the threat.

e. Point of Aim i. Moving Platform – Stationary Target requires aim to the rear (trail) and low dependant upon speed.

**CONVOY BRIEF** **(Modified from FM 55-30)** **(Continued)**

**3. Employment Tactics, Techniques and Procedures**

e. Point of Aim
   ii. Moving Platform – Moving Target requires aim directly on and low.
   iii. Stationary Platform – Moving Target requires aim to the front (lead) and low dependent upon speed.

f. Rate of Fire
   i. Burst Mode
   ii. Steady Suppression (ROE)

g. Magazine Awareness
   i. Serviceability / maintenance
   ii. Tracer Mix
   iii. Magazine Storage / Placement
   iv. Mounted, Reload when time is available

**Dismounted, Seek cover prior to the need to Reload**

## REACT TO AMBUSH (NEAR)

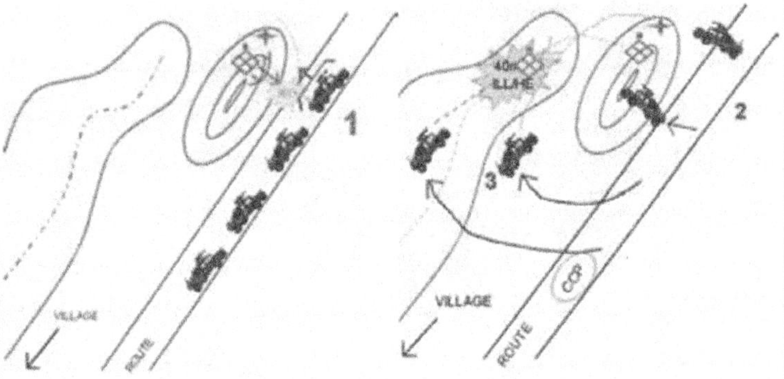


### (Enemy Initiates Weapons Fire / RPG / IED)

1. Lead element of convoy reacts to contact and immediately returns fire. Lead fire team fires 40mm HE (at night also white star if NVGs drown out by vehicle headlights – leader calls it. FYI, most travel at night will be with headlights on).
2. Lead elements assualt through ambush – fire team dismounts while gun trucks use M249, MK-19, and .50 cal in close support.
3. Trail fire team and trail gun truck move to block most likely enemy escape route. Fire team uses 40mm HE/Illumination if needed. (Use of 40mm becomes less likely as units get closer to urban areas due to collateral damage risk.) Establish CCP upon action completion or enemy neutralized.

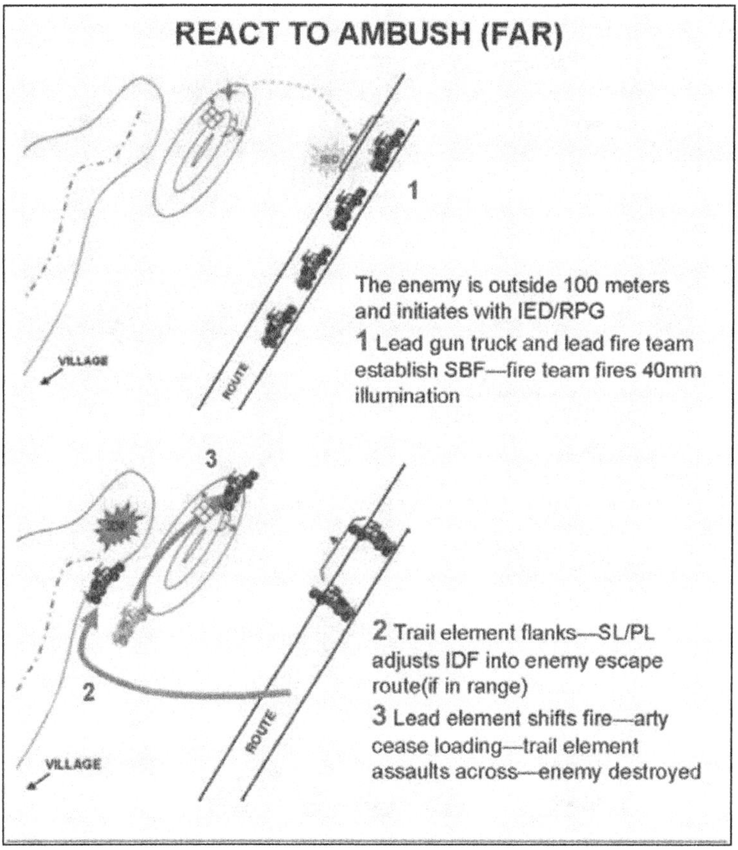
REACT TO AMBUSH (FAR)
1
The enemy is outside 100 meters and initiates with IED/RPG
1 Lead gun truck and lead fire team establish SBF—fire team fires 40mm illumination
VILLAGE
ROUTE
3
2
2 Trail element flanks—SL/PL adjusts IDF into enemy escape route(if in range)
3 Lead element shifts fire—arty cease loading—trail element assaults across—enemy destroyed
VILLAGE
ROUTE

# CONVOY FORCED TO STOP

## (Weapons Fire / RPG / IED / Indirect)

1. Vehicles forced to stop. Activate vehicle turn signal to indicate direction of contact.
2. Vehicle(s) / Personnel not in direct contact, report on internal communication, identifying truck number, type of contact, clock direction and eight digit grid if available.

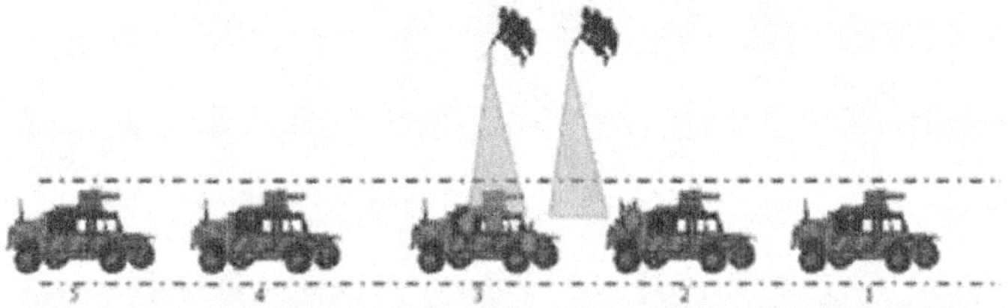


3. Personnel on vehicle(s) forced to stop dismount on the non-contact side, assume covered position(s) and provide initial base of fire.
4. The <u>entire</u> convoy halts; personnel will dismount vehicle(s) on the non-contact side and provide additional base of fire on the enemy if in range. Vehicle(s) not in contact will reposition to attack the attackers. M16's should now be on <u>Semi-Automatic</u> in order to conserve ammunition.
5. PL/CC/SC will move to better assess the situation and position the Gun Truck(s) in order to best suppress the enemy while maintaining <u>standoff</u>. Gun trucks will close with and engage the enemy.
6. Once the PL/CC/SC determines the convoy has either gained fire superiority or defeated the enemy contact, Recovery / CASEVAC operations will begin.
7. If the PL/CC/SC determines the convoy cannot gain fire superiority, leader will then conduct Break Contact procedures

# CONVOY FORCED TO STOP - METHOD 2

## (Weapons Fire /RPG / IED / Indirect)

1. Vehicle(s) forced to stop. Activate vehicle turn signal to indicate direction of contact.

*** ALL PERSONNEL STAY IN VEHICLES ***

2. All vehicles immediately drive forward out of the kill zone.

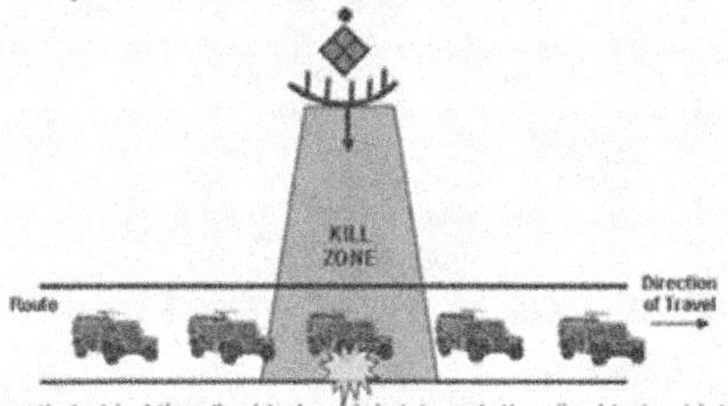


3. The vehicle(s) directly behind the disabled vehicle(s) push the disabled vehicle(s) out of the kill zone.
4. The vehicle(s) that are not disabled will establish a base of fire towards the suspected/known contact.
5. If fire superiority can be gained the PL will use the minimum force necessary to destroy the enemy.
6. If the PL determines the convoy cannot gain fire superiority, the leader will conduct break contact.

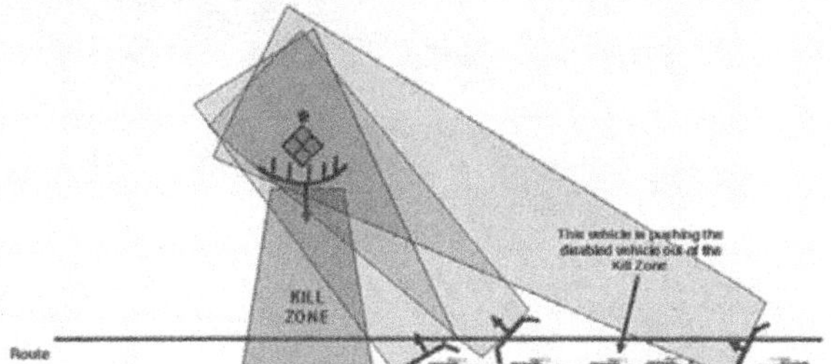

# BREAK CONTACT

**Always try to close with and destroy the enemy first. This way he cannot come back later on to attack you or another convoy again. However – if you must...**

1. The PL/CC/SC determines the convoy cannot gain fire superiority and the decision has been made to Break Contact.
2. The PL/CC/SC designates that either Rally Point "Rear" or "Forward" will be used. If necessary, both Rally Points may be used. Communication systems and appropriate pyrotechnic signals will be used to communicate Break Contact and Rally Point.
3. Personnel will deploy obscuration measures if available. Utilizing cover and concealment, Aid & Litter team(s) will evacuate all casualty(s) under support of Gun Truck and other protective fire(s).
4. Personnel will maintain position and suppression in contact zone and assist Aid & Litter team(s) as necessary.
5. Disabled vehicle(s) will be hauled back or destroyed as directed by leaders. (Thermite or explosives)

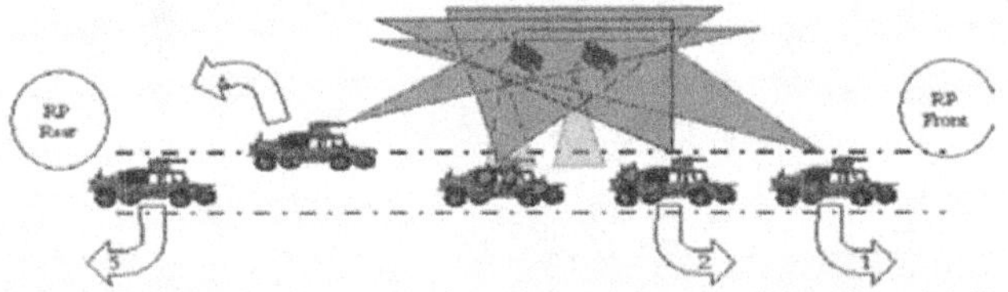


6. Vehicles will displace either backwards or forwards through the convoy lines under control of leaders. The most forward vehicle in the contact zone moves first, then the next most forward vehicle moves second. Vehicles will continue to displace. As vehicles displace, Gun Truck repositions as necessary until contact is broken.

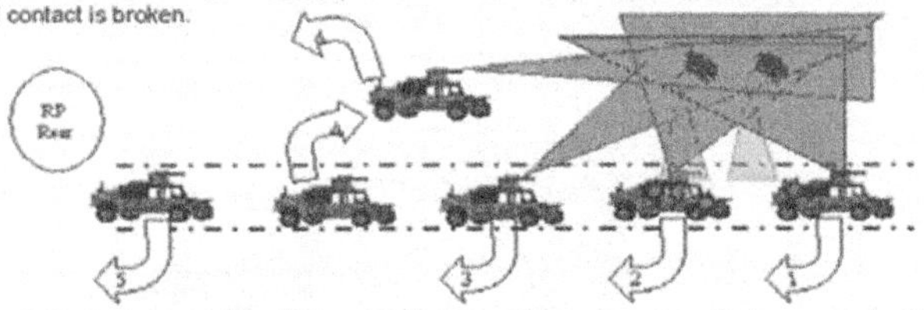


7. If Break Contact occurs with vehicles on both sides of the contact zone, displacement of vehicles will occur using an alternating displacement technique.
8. Upon occupation of the Rally Point, leaders will immediately position vehicles, setting 360 degree security, and conduct Consolidation and Reorganization.
9. If the convoy vehicles get separated when not in contact with the enemy, personnel and vehicles stay together and move to the closest Rally Point or Check Point.

# RECOVERY / CASEVAC OPERATIONS

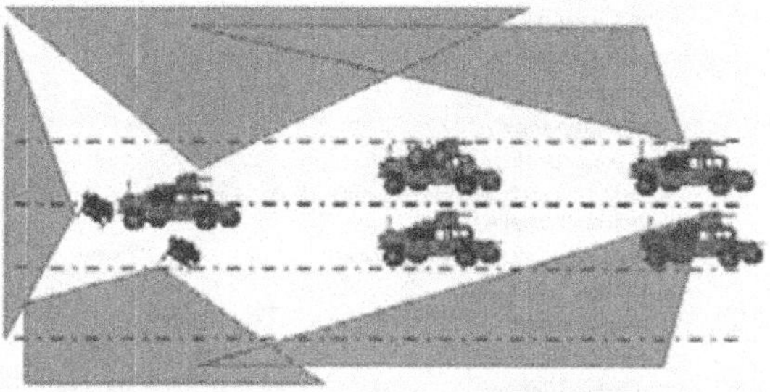

1. Once the leader assesses the enemy threat to be destroyed, neutralized, defeated and the area secured – Recovery / CASEVAC operations will begin. This keeps soldiers focused first on defeating and destroying the threat.
2. CASEVAC:

   a. Aid & Litter team will position on the safe side of the vehicle and extract casualty(s) and personnel.

   b. Treatment of casualty(s) will occur once they are safely removed from the contact area.
3. Vehicle Recovery Procedures:

   a. Recovery team will position on the safe side of the disabled vehicle.

   b. TC will dismount and assess the disabled vehicle.

   c. If determined the vehicle can be safely recovered, TC guide the recovery vehicle into position and conduct a hasty hook-up. TC will operate the disabled vehicle is necessary.

   d. Upon exiting the contact area, complete and correct hook-up procedures will occur.

   e. If assessment results in outside support necessary for recovery, leader will contact higher for guidance.
4. Once recovery operations are complete, the team will displace and conduct link-up with the convoy at the Rally Point.
5. Disabled vehicle(s) will be abandoned or destroyed as directed by leaders. (Thermite or explosives)

# GLOSSARY

| | |
|---|---|
| 1SG | first sergeant |
| 5 W's | Who, What, Where, When, and Why? |
| 5-S Rule | Search, Silence, Segregate, Safeguard, and Speed to rear |
| AA | avenue(s) of approach |
| AAR | after-action review |
| AATF | air assault task force |
| ABCDE | method of identification and response to life-threatening conditions: Airway, Breathing, Circulation, Disability, Exposure |
| ACE | ammunition, casualties, and equipment |
| ACL | allowable combat load |
| ACP | aerial checkpoint |
| ACU | Army combat uniform |
| ADA | air defense artillery |
| AG | assistant gunner |
| ATC | air traffic controller; a mechanical belay device that locks down on itself when tension is applied in opposite directions |
| ALT | alternate |
| AMC | air mission commander |
| ammo | ammunition |
| ANCD | automated net-control device |
| AO | area of operations |
| AOO | actions on the objective |
| AR | automatic rifleman |
| ABF | attack by fire |
| ATL | Alpha team leader |
| ATM | Alpha Team |
| bangalore torpedo | A manually emplaced, 1.5-meter long explosive-filled tube used to breach wire and detonate simple, pressure-activated antipersonnel mines. Ten tubes clear a 1- by 15-meter lane. |
| belay | any action taken to stop a climber's fall or to control the rate a load descends |
| binos | binoculars |
| BMNT | begin morning nautical twilight |
| BN | battalion |
| body belay | belay that uses the belayer's body to apply friction by routing the rope around the his body |
| bowline on a coil | knot used to secure a climber to the end of a climbing rope |
| BP | battle position |
| British junction knot | knot used to join the ends of detonation cords from multiple charges to one initiation system |
| BTC | bridge team commander |
| BTL | Bravo Team leader |
| BTM | Bravo Team |

| | |
|---|---|
| C2 | command and control |
| CAS | close air support |
| CASEVAC | casualty evacuation |
| CCA | close combat attack |
| CCIR | commander's critical information requirements |
| CCP | casualty collection point |
| CDR | commander |
| CDS | Camp Darby Special (map) |
| CLS | combat lifesaver |
| CO | company |
| COA | course of action |
| commo | communications |
| COMSEC | communications security |
| cordelette | short section of static rope or static cord. Also called "sling rope" |
| COTS | commercial off the shelf |
| CP | command post |
| CPR | cardiopulmonary resuscitation |
| CQC | close quarters combat |
| CS | combat support |
| CSS | *obsolete: now referred to as* sustainment |
| CTT | common task test |
| DAR | designated area of recovery |
| DOL | direction of landing |
| double figure eight knot | knot used to form a fixed loop in the end of the rope; loops are large enough to insert a carabiner |
| double overhand knot | knot used to secure the end of detonation cord |
| DP | duty position; decision point (depending on context) |
| DST | distance |
| DTG | date-time group |
| dynamic ropes | one of two classifications of kernmantle rope; used for climbing; *see also* static rope |
| DZ | drop zone |
| EA | engagement area |
| EEFI | essential elements of friendly information |
| EENT | end evening nautical twilight |
| end-of-the-rope clove hitch | intermediate anchor knot that requires constant tension |
| end-of-the-rope Prusik | knot used to attach a movable rope to a fixed rope; *see also* middle-of-the-rope Prusik |
| ENY | enemy |
| EPW | enemy prisoner of war |
| FA | field artillery |
| FDC | fire direction center |
| FFIR | friendly force information requirements |

| | |
|---|---|
| figure 8 slip knot | knot used to form an adjustable bight in the middle of a rope |
| FLIR | forward looking infrared |
| FLOT | forward line of own troops |
| FO | forward observer |
| FPF | final protective fires |
| FRAGO | fragmentary order |
| FSC | fire support coordinator |
| FSO | fire support officer |
| GOTWA | G - Where leader is GOING<br>O - Others he is taking with him<br>T - Time he plans to go<br>W - WHAT to do if the leader does not return in time<br>A - The unit's and the leader's ACTIONS on chance contact while the leader is gone |
| GPS | global positioning system |
| GRN | grenadier |
| HDG | heading |
| HE | high explosive |
| H-Hour | hit hour (the time the unit plans to accomplish the mission) |
| HI | high temperature (weather) |
| HPT | high-payoff target |
| HQ | headquarters |
| IAW | in accordance with |
| ICM | improved conventional munitions |
| ID | identification |
| IP | initial point |
| IR | intelligence requirements |
| ERRP | en route to release point |
| JAAT | joint air attack team |
| JD | Julian date |
| KIA | killed in action |
| LAW | light antiarmor weapon |
| LBV | load-bearing vest |
| LD | line of departure |
| LDA | linear danger area |
| LO | low temperature (weather) |
| LOGSTAT | logistical status |
| LP | listening post |
| LZ | landing zone |
| material factor | the strength, hardness, and mass of the material to be demolished |
| mb | millibar (a metric unit used to measure air pressure) |

| | |
|---|---|
| MDI | modernized demolition initiator |
| ME | main effort |
| mechanical belay | a belay that uses mechanical devices to help the belayer control the rope, as in rappelling |
| MEDEVAC | medical evacuation |
| METL | mission-essential task list |
| METT-TC | mission, enemy, terrain (and weather), troops (and support) available, time available, and civil considerations |
| middle of the rope clove hitch | knot that secures the middle of a rope to an anchor |
| middle-of-the-rope Prusik | knot that attaches a movable rope to a fixed rope, anywhere along the length of the fixed rope; *see also* end-of-the-rope Prusik |
| MG | machine gun/ner |
| MOPP | mission-oriented protective posture |
| MR | moonrise |
| MRE | meal, ready to eat |
| MS | moonset |
| MSD | minimum safe distance |
| MSL | mean sea level |
| munter hitch | commonly used belay that requires little equipment |
| NATO | North Atlantic Treaty Organization |
| NAV | navigation |
| NFA | no-fire area |
| NLT | no later than |
| NVD | night-vision device |
| NVG | night vision goggles |
| NVS | night vision system |
| OAKOC | observation and fields of fire, avenues of approach, key terrain, observation, and cover and concealment |
| OBJ | objective |
| OD | olive drab |
| OOM | order of movement |
| OP | observation post |
| OPORD | operations order |
| OPSKED | operational schedule |
| ORP | objective rally point |
| OT | observer-target |
| PB | patrol base |
| PCC | precombat checks |
| PCI | precombat inspection |
| PDF | principal direction of fire |

| | |
|---|---|
| PI | probability of incapacitation |
| PIR | priority intelligence requirements |
| PL | platoon leader |
| PLD | probable line of deployment |
| PLOT-CR | purpose, location, observer, trigger, communication method, resources (a format for planning fire support) |
| PLT | platoon |
| POL | petroleum, oils, and lubricants |
| PRI | primary |
| protection | a piece of equipment, natural or artificial, that is used to construct an anchor |
| PSG | platoon sergeant |
| PSI | pounds per square inch |
| PZ | pickup zone |
| R | rifleman |
| R&S | reconnaissance and surveillance |
| RACO | rear area combat operations |
| RAP | rocket-assisted projectile |
| rappel seat | a rope harness used in rappelling and climbing |
| RED | risk-estimate distance |
| REQ | required |
| rerouted figure 8 knot | anchor knot that also attaches a climber to a climbing rope |
| RFA | restrictive fire area |
| RFL | restrictive fire line |
| RFLM | rifleman |
| RHB | Ranger Handbook |
| ROE | rules of engagement |
| round-turn with two half hitches | a constant tension anchor knot |
| RP | release point |
| RTO | radio operator |
| S-2 | intelligence staff officer |
| S-3 | operations staff (and training) officer |
| SALUTE | Size, Activity, Location, Unit/Uniform, Time, and Equipment |
| SAW | squad automatic weapon |
| SBF | support by fire (position) |
| SDT | self-development test |
| SE | supporting effort |
| SEAD | suppression of enemy air defenses |
| SITREP | Situation Report |
| SITTEMP | situational template |

| | |
|---|---|
| SL | squad leader |
| sling rope | short section of static rope or static cord. Also called "cordelette" |
| SLLS | Stop, Look, Listen, Smell |
| SOC | succession of command |
| SOI | signal operating instructions |
| SOP | standing operating procedures |
| SP | start point |
| square knot | knot used to join two ropes of equal diameter; used to join the ends of the detonation cord to the explosive |
| SR | sunrise |
| SS | sunset |
| STANO | surveillance, target acquisition, and night observation |
| static ropes | one of two classifications of kernmantle rope; used where rope stretch is undesired, and when the rope is subjected to heavy static weight. *See also* dynamic rope |
| SURVIVAL | S Size up the situation, your surroundings, your physical condition, and your equipment.<br>U Undue haste makes waste; don't be too eager to move. Plan your moves.<br>R Remember where you are in relation to important friendly and locations and critical resources<br>V Vanquish fear and panic.<br>I Improvise. You can improve your situation. Use what you have. Use your Imagination.<br>V Value your life. Remember your goal: to get out alive. Remain stubborn. Refuse to give in to problems and obstacles that face you. This will give you the mental and physical strength to endure.<br>A Act like the natives; watch their daily routines. When, where, and how do they get food? Where they get water?<br>L Live by your wits. Learn basic skills. |
| suspension traverse | used to move personnel and equipment over rivers, ravines, chasms, and up or down a vertical obstacle |
| TAC | tactical air controller |
| tamping factor | depends on the location and tamping of the charge |
| technical climbing | using safe and proper equipment and techniques to climb on a rock formation in parties of two or more |
| tensionless anchor | used to anchor rope on high-load installations such as bridging |
| TL | team leader |
| TLP | troop-leading procedures |
| TL | team leader |
| TOC | tactical operations center |
| triple roll knot | knot used to join branches of detonation cord |
| TTP | tactics, techniques, and procedures |
| uli knot | knot used to securely fasten detonation cord to explosive |
| VIXL | video image crosslink |
| WARNO | warning order |
| WBGT | wet bulb globe temperature |

| | |
|---|---|
| WFFs | warfighting functions (fire support, movement and maneuver, protection, command and control, and sustainment) |
| WIA | wounded in action |
| XO | executive officer |
| WSL | weapons squad leader |

NOTES

# INDEX

Appendix A

# RESOURCES

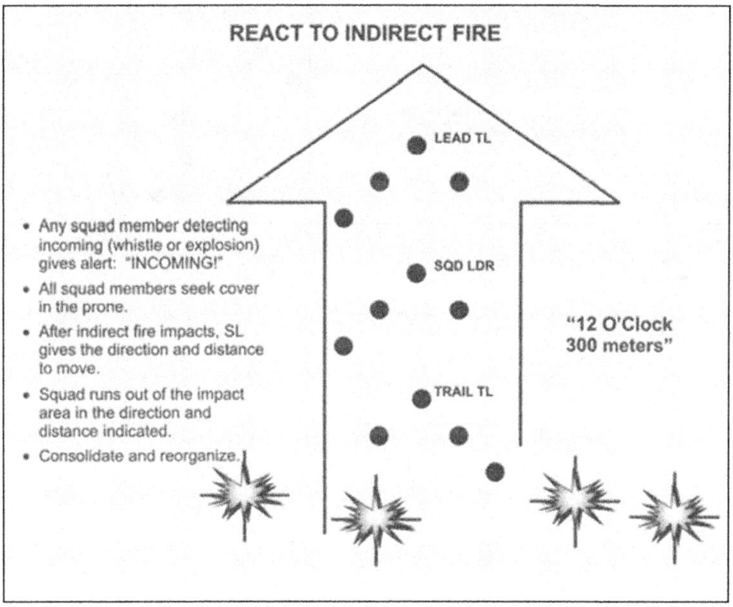

## REACT TO CONTACT

- Seek nearest cover.
- Return fire (known or suspected enemy position).
- TM LDRs control fire by using fire commands.
- Report enemy situation (number/position).
- Maintain contact (visual/oral) with team members.
- SQD LDR moves to team in contact and makes an assessment of the situation.
- Factors of his assessment:
  - Can squad move out to engagement area.
  - Can squad gain and maintain suppressive fire.
  - Location of enemy.
  - Size of enemy.
  - Vulnerable flanks.
  - Covered/concealed flanking routes.
- SQD LDR determines COA (break contact, squad attack, etc.)
- Report situation to PL.

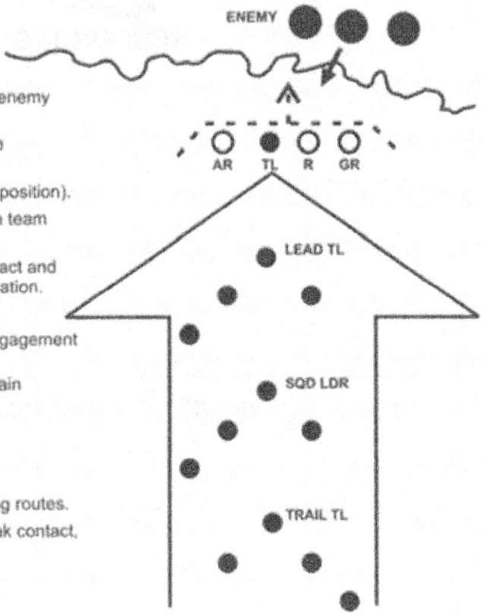

## REACT TO A NEAR AMBUSH

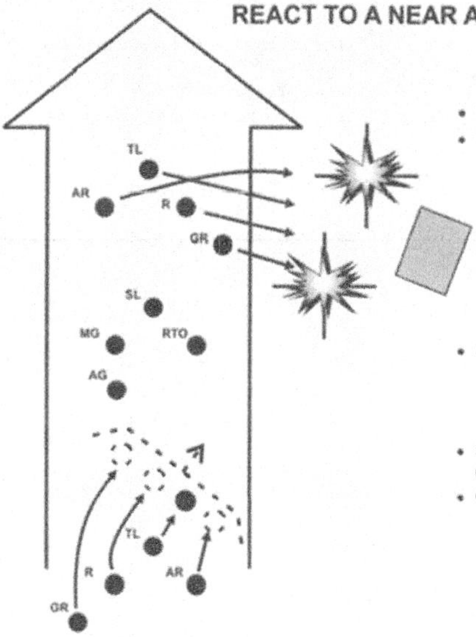


- Within hand grenade range - 35 meters.
- Soldiers in the kill zone: (without orders)
  - Return fire immediately.
  - Seek nearest available cover.
  - Assume prone position.
  - Throw concussion, frag, or smoke grenades.
  - After explosion of grenades, assault through ambush using fire and movement..
- Soldiers not in kill zone:
  - Identify enemy loaction.
  - Place accurate suppressive fire.
  - Shift fires as assault begins.
- Soldiers in kill zone continue to assault to eliminate ambush or until contact is broken.
- Consolidate and reorganize.

## REACT TO A FAR AMBUSH

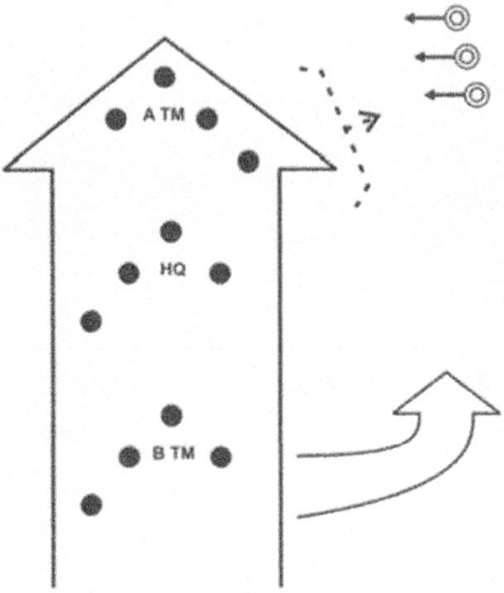


- More than 35 meters.
- TM in the kill zone: (without orders)
  - Return fire immediately.
  - Seek cover and concealment.
  - Suppress enemy (overwatch).
- SL assess situation.
  - Determines COA (flank).
- TM not in contact:
  - Moves along covered and concealed route.
  - Assaults enemy on weak flank.
  - Suppress enemy (overwatch).
- Overwatch TM continues to suppress shifts/cease fire as bounding team enters sector.
- Bounding team continues to assault through enemy.
- SL may request indirect fire.
- Consolidate and reorganize.

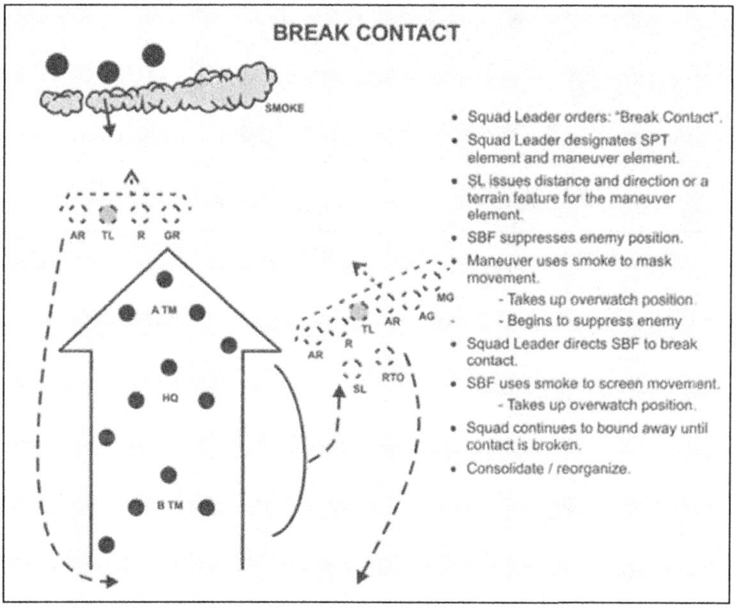
BREAK CONTACT
SMOKE
AR TL R GR
A TM
HQ
B TM
MG
AR AG
TL
R
AR
SL
RTO
• Squad Leader orders: "Break Contact".
• Squad Leader designates SPT element and maneuver element.
• SL issues distance and direction or a terrain feature for the maneuver element.
• SBF suppresses enemy position.
• Maneuver uses smoke to mask movement.
- Takes up overwatch position.
- Begins to suppress enemy
• Squad Leader directs SBF to break contact.
• SBF uses smoke to screen movement.
- Takes up overwatch position.
• Squad continues to bound away until contact is broken.
• Consolidate / reorganize.

## FORMATIONS AND ORDERS OF MOVEMENT

I. Movement Formation: Fire Team Wedge; MG Team attached.

II. Three Movement Techniques utilized:

   A. Traveling technique used behind FFL when contact is not likely.

   B. Traveling Overwatch forward of the FFL when enemy contact is possible.

   C. Bounding Overwatch used forward of the FFL when enemy contact is expected.

III. Distances are based on but not dictated by visibility, terrain, and vegetation.

IV. Actions at Night: Modified Wedge

V. Actions at the Halt: Short and Long Halt (GV/LV)

VI. Leader Location: Fixed/Unfixed

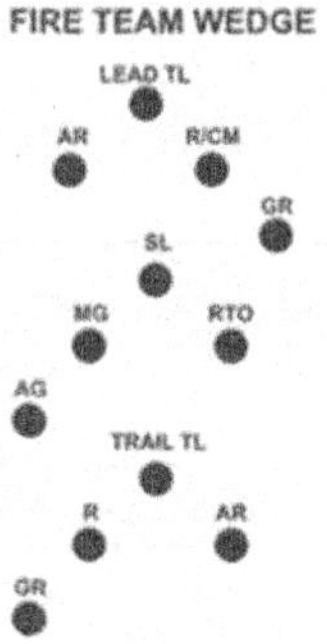


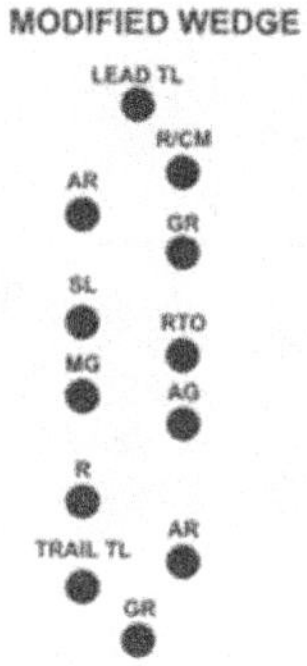

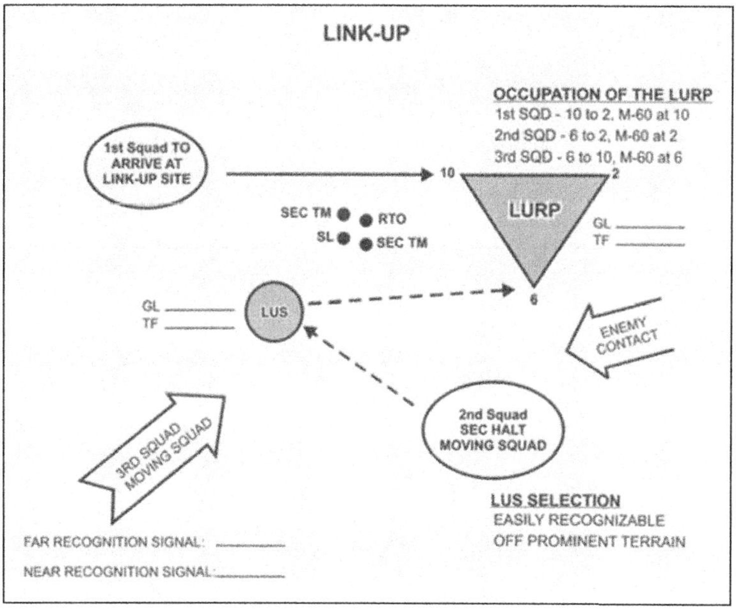
LINK-UP
OCCUPATION OF THE LURP
1st SQD - 10 to 2, M-60 at 10
2nd SQD - 6 to 2, M-60 at 2
3rd SQD - 6 to 10, M-60 at 6
1st Squad TO ARRIVE AT LINK-UP SITE
10
2
LURP
6
SEC TM
RTO
SL
SEC TM
GL
TF
GL
TF
LUS
ENEMY CONTACT
2nd Squad SEC HALT MOVING SQUAD
3RD SQUAD MOVING SQUAD
LUS SELECTION
EASILY RECOGNIZABLE
OFF PROMINENT TERRAIN
FAR RECOGNITION SIGNAL:
NEAR RECOGNITION SIGNAL:

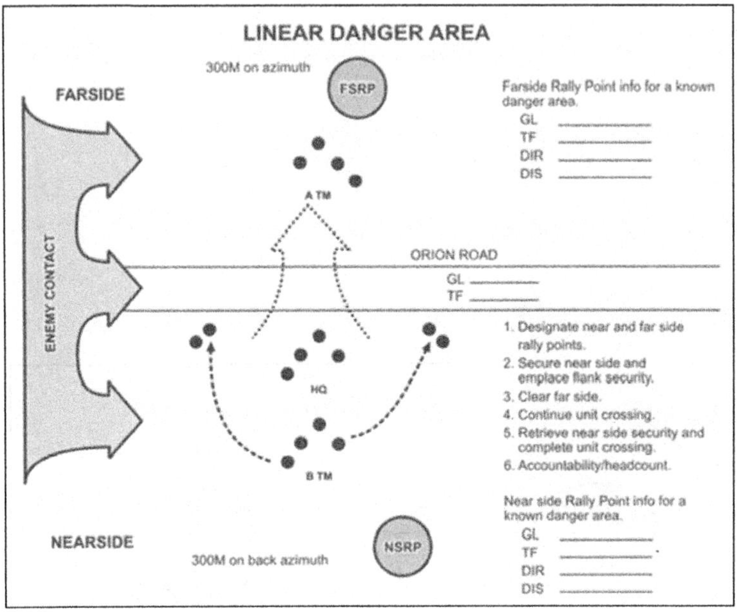
LINEAR DANGER AREA
300M on azimuth
FSRP
FARSIDE
Farside Rally Point info for a known danger area.
GL
TF
DIR
DIS
A TM
ENEMY CONTACT
ORION ROAD
GL
TF
1. Designate near and far side rally points.
2. Secure near side and emplace flank security.
3. Clear far side.
4. Continue unit crossing.
5. Retrieve near side security and complete unit crossing.
6. Accountability/headcount.
HQ
B TM
Near side Rally Point info for a known danger area.
GL
TF
DIR
DIS
NEARSIDE
300M on back azimuth
NSRP

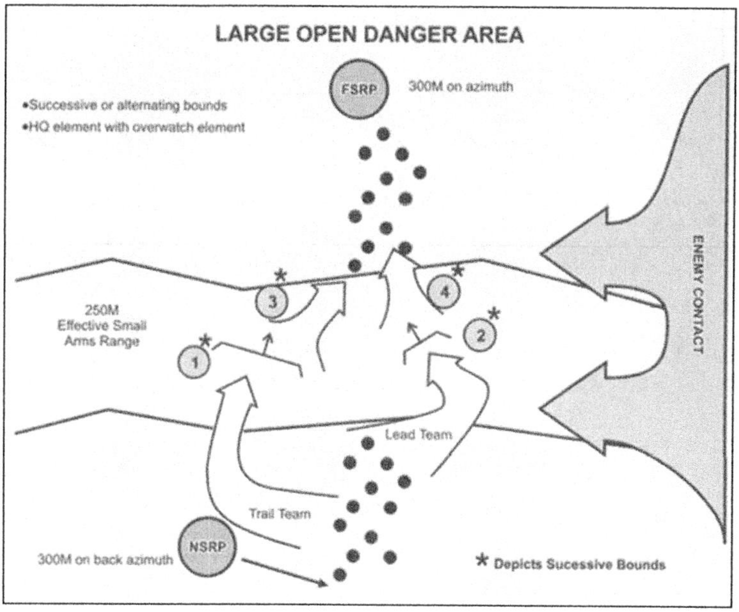
LARGE OPEN DANGER AREA
FSRP
300M on azimuth
•Successive or alternating bounds
•HQ element with overwatch element
ENEMY CONTACT
250M
Effective Small
Arms Range
1
2
3
4
Lead Team
Trail Team
NSRP
300M on back azimuth
* Depicts Sucessive Bounds

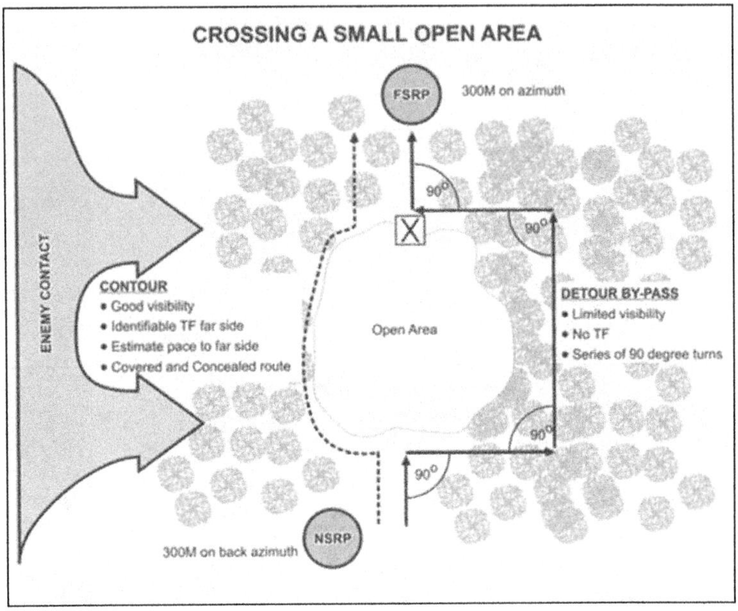

CROSSING A SMALL OPEN AREA
FSRP
300M on azimuth
ENEMY CONTACT
CONTOUR
• Good visibility
• Identifiable TF far side
• Estimate pace to far side
• Covered and Concealed route
Open Area
DETOUR BY-PASS
• Limited visibility
• No TF
• Series of 90 degree turns
90°
90°
90°
90°
NSRP
300M on back azimuth

## SQUAD ATTACK

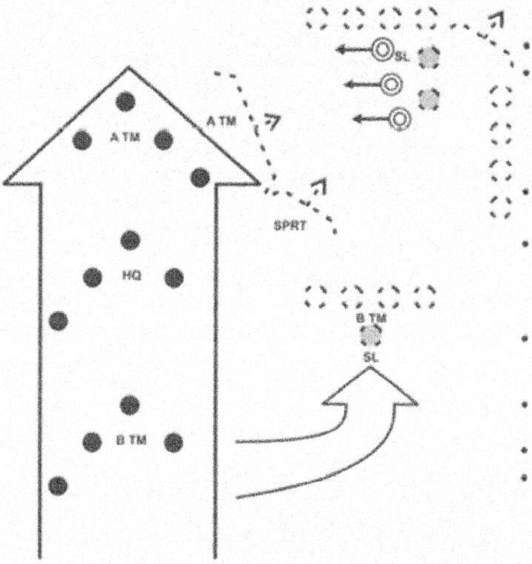


- React to contact.
- TM in the kill zone: (without orders)
  - Return fire.
  - Seek cover and concealment.
  - Suppress enemy calls out 3 D's becomes overwatch.
- SL assess situation.
  - Determines COA (flank/attack).
- TM not in contact: (with SL)
  - Moves along covered and concealed route.
  - Assaults enemy on weak flank.
- Overwatch TM continues to suppress shifts/cease fire as bounding team enters sector.
- Bounding team continues to assault through enemy.
- SL may request indirect fire.
- Consolidate and reorganize.

## RAID BOARDS (LEFT)
## GENERAL INFO

**Raid References:**
- RHB Chapter 5
- FM 3-21.8

**Purposes:**
1. Destroy
2. Liberate
3. Collect Intel

**Planning Considerations:**
- Mission
- Enemy
- Troops
- Terrain / OAKOC
- Time
- Civilians

STUDENT INPUT WRITTEN HERE

**Definition:**
A surprise attack on a fixed position or installation ending in a planned withdrawal.

**Characteristics:**
1. Surprise - Gain
2. Coordinated fires - Maintain
3. Violence of action - Retain
4. Planned withdrawal

INITIATIVE

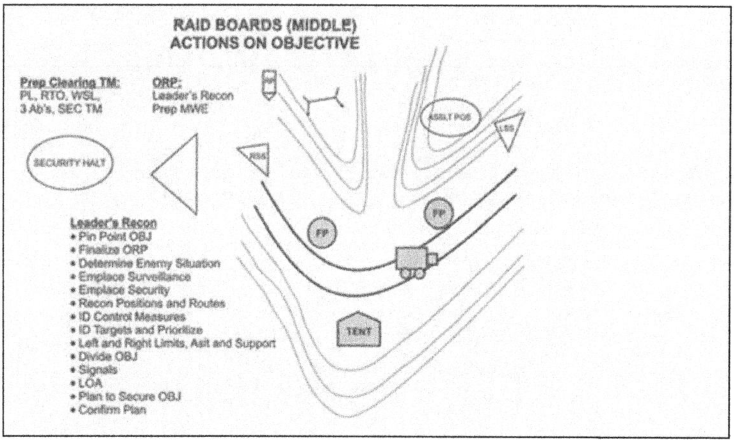
RAID BOARDS (MIDDLE)
ACTIONS ON OBJECTIVE
Prep Clearing TM:
PL, RTO, WSL,
3 Ab's, SEC TM
ORP:
Leader's Recon
Prep MWE
SECURITY HALT
ASSLT POS
FP
FP
TENT
Leader's Recon
• Pin Point OBJ
• Finalize ORP
• Determine Enemy Situation
• Emplace Surveillance
• Emplace Security
• Recon Positions and Routes
• ID Control Measures
• ID Targets and Prioritize
• Left and Right Limits, Aslt and Support
• Divide OBJ
• Signals
• LOA
• Plan to Secure OBJ
• Confirm Plan

## RAID BOARDS (MIDDLE)
## TASK ORGANIZATION

| ELEMENT | WHO | WHY | WHAT |
|---|---|---|---|
| HQ | PL, APL<br>RTO, FO, MEDIC | C2/FACILITATE<br>SIGNALS, PEQ-2, STANO | 1x119, ICOM, SKEDCO, PLUGGER |
| ASSLT 1 | SQUAD<br>(+)(-) | DESTROY | ICOM, STANO, M18A1, AT4,<br>SPECIAL TEAM KITS |
| ASSLT 2 | SQUAD<br>(+)(-) | DESTROY | ICOM, STANO, M18A1, AT4,<br>SPECIAL TEAM KITS |
| SUPPORT | WPN S SQD<br>(+)(-) | SUPPRESS | 3xM240B COMPLETE, ICOM, PEQ-2,<br>PVS-4, AT, STANO, SIGNALS, MAP |
| SECURITY | SQUAD<br>(+)(-) | DELAY | ICOMs, M18A1, AT, STANO, MAP<br>SIGNALS |

## RAID BOARDS (RIGHT)
## SOP'S

**CONTINGENCIES**

1. COMPROMISE
   a. ORP
   b. LEADERS RECON
   c. OCCUPATION
2. MASS CAL
3. COUNTERATTACK

**SOP's**

1. EPW SEARCH
2. AID LITTER
3. MED EVAC
4. CCP
5. WITHDRAW PLAN

## AMBUSH SOPS (LEFT)

**References:**
- RHB Chapter 5
- FM 3-21.8

**Ambush:**
A surprise attack from a covered and concealed position on a moving or temporarily halted target.

**Types:**
- Point
- Area

**Fundamentals:**
1. Surprise - Gain
2. Fire superiority - Maintain
3. Violence of action - Retain

**Purposes:**
1. Disrupt/Destroy
2. Collect Intelligence
3. Block or Deny Access
4. Canalize

**Categories:**
*Deliberate:* has specific target at a predetermined time and location.

*Hasty:* Platoon makes visual contact with the enemy and has time to establish an ambush without being detected.

**Planning Considerations:**
- Mission - Task/Purpose
- Enemy - MPCOA, MDCOA
- Terrain - Map Recon/Ldrs Recon
- Time - 1/3 2/3 Rule
- Troops - Task Org
- Civilians -

## AMBUSH BOARDS (MIDDLE)

| | | | |
|---|---|---|---|
| HQ | PL, PL RTO, PSG, MEDIC, FO | COMMAND AND CONTROL | PRC-119s, ICOM, NVGs, BINO's, SKEDCO, PEQ-2, SIGNALS, MAP |
| SECURITY | MINIMUM OF 3-2 MAN TEAMS | EARLY WARNING, SEALS OFF OBJ, LFS, RFS, ORP SECURITY | COMMO, NVGs, CLAYMORES, AT WPNS, BINO's, SIGNALS, MAP |
| SUPPORT | WSL, 3 GUN TEAMS COMPLETE +/- | SUPPRESSIVE FIRE ON KILL ZONE, SEC ON OBJ DURING RECON/REORG | MG's, COMMO, NVGs, AT WPNS, BINO's, SKEDCO, PEQ-2, SIGNALS., MAP |
| ASSAULT | 2 SQUADS +/- | MAIN EFFORT, BLOCK, DESTROY, CANALIZE, CAPTURE PIR | NVGs, BINO's, CLAYMORE, AT, FLEX CUFFS, POLELESS LITTER, MARKINGS, SIGNALS |

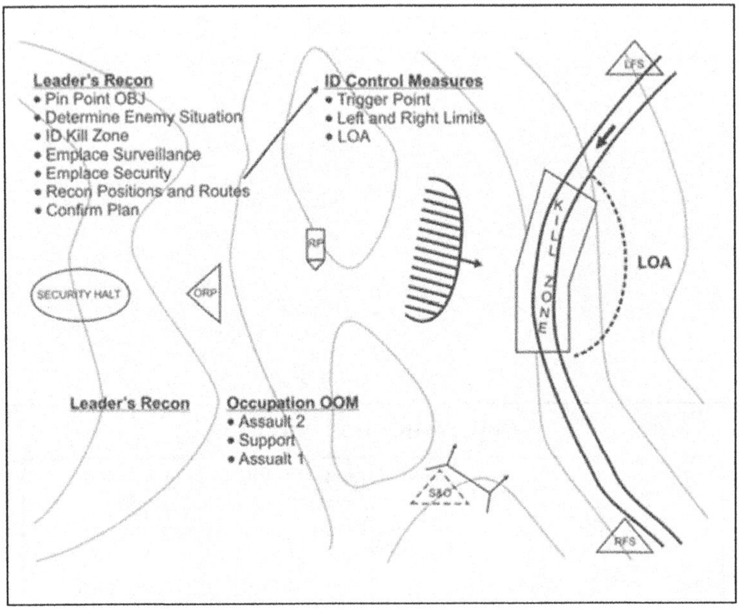

Leader's Recon
• Pin Point OBJ
• Determine Enemy Situation
• ID Kill Zone
• Emplace Surveillance
• Emplace Security
• Recon Positions and Routes
• Confirm Plan
ID Control Measures
• Trigger Point
• Left and Right Limits
• LOA
LFS
RP
KILL ZONE
LOA
SECURITY HALT
ORP
Leader's Recon
Occupation OOM
• Assault 2
• Support
• Assault 1
S&O
RFS

## AMBUSH BOARDS (RIGHT)
## SOP'S

**CONTINGENCIES**

1. COMPROMISE
   a. ORP
   b. LEADERS RECON
   c. OCCUPATION
2. MASS CAL
3. COUNTERATTACK

**SOP's**

1. EPW SEARCH
2. AID LITTER
3. MASS CAL
4. MED EVAC
5. CCP

NOTES

Appendix B

# QUICK REFERENCE CARDS

## CASUALTY FEEDER CARD

NAME __________________ ROS # __________

COMPANY ________ ALLEGERIES____________

LOCATION/EVENT ___________________________

TIME ____________ TIME ____________

TEMP ____________ TEMP ____________

PULSE ____________ PULSE ____________

RESP ____________ RESP ____________

B/P ____________ B/P ____________

MENTAL STATUS __________________________

A & O x 1 2 3

NPA OPA ET Tube

SAO$^2$ ____________ GLUC ______________

IV ____ AC FOREARM HAND OTHER

1000-1000-500-500-500- NS RL $D_5W$

NOTES: ________________________________

______________________________________

**TROOP-LEADING PROCEDURES**

1. Receive mission.
2. Issue warning order.
3. Make a tentative plan.
4. Start movement.
5. Reconnoiter.
6. Complete plan.
7. Issue plan.
8. Supervise.

**FIRE REQUEST (FM 6-30)**

1. Identification.
2. Warning order.
3. Target location.
4. Target description.
5. Method of engagement.
6. Method of control.

**ESTIMATE OF SITUATION (FM 7-10)**

1. Conduct a detailed mission analysis.
2. Analyze the situation and develop courses of action.
3. Analyze courses of action (wargame).
4. Compare courses of action.
5. Make a decision.

**SPOT REPORT**

1. Size.
2. Activity.
3. Location.
4. Unit/uniform.
5. Time.
6. Equipment.

**SHELL REPORT (FM 6-121)**

1. Observer identification.
2. Location (coded).
3. Azimuth to flash or sound.
4. Time (from and to).
5. Area shelled.
6. Nature of fire.
7. Type rounds received (artillery, mortars, etc.).
8. Damage (coded).

# GTA 07-01-038
# INFANTRY LEADERS' REFERENCE CARD
## January 1995

HEADQUARTERS, DEPARTMENT OF THE ARMY

References: FM 3-3, FM 6-30, FM 6-121, FM 7-8, FM 7-10, FM 7-90

(Supersedes GTA 7-1-31)

DISTRIBUTION RESTRICTION: Distribution authorized to US Government agencies only to protect technical or operational information from automatic dissemination under the International Exchange Program or by other means. Other requests for this document will be referred to Commander, US Army Training Support Center, ATTN: ATIC-DMF, Ft Eustis, VA 23604-6166.

DESTRUCTION NOTICE: Destroy by any method that will prevent disclosure of contents or reconstruction of the document.

DISTRIBUTION: US Army Training Support Centers (TSCs).

**OPERATION ORDER (FM 7-10)**

Task Organization

1. Situation.
   a. Enemy.
   b. Friendly.
   c. Attachments/detachments.
2. Mission.
   - Who, what, where, when, and why.
   - Task and purpose.
3. Execution.
   a. Concept of operation.
      (1) Maneuver.
      (2) Fire support.
      (3) Engineer.
   b. Tasks to maneuver units.
   c. Tasks to combat support units.
   d. Coordinating instructions.
4. Service support.
   a. General.
   b. Materiel and services.
   c. Casualty evacuation.
   d. Miscellaneous.
5. Command and signal.
   a. Command.
   b. Signal.

**WEAPONS (FM 7-8)**

| TYPE | MAX EFF RANGE (m) |
|---|---|
| M16A2 | 580 (pt) 800 (area) 200 (mov) |
| M203 | 150(pt) 350 (area) |
| M249 | 600 (pt) 800 (area) |
| M136 (AT4) | 300 |
| M47 (Dragon) | 1,000 (sta) 100 (mov) |
| MK19 | 1500 (pt) 2,212 (area) |
| M3 RAAWS | 700 (sta) 60 (mov) |
| M60 MG | 1,100 (800 grazing) |
| .50 Caliber MG | 1,800 (1,000 grazing) |
| TOW | 3,000 (ping purposes) |
| TOW 2 | 3,750 |
| 105-mm | 11,500 |
| 105-mm Tank | *2 to 2.5 km |
| 120-mm Tank | *2 to 2.5 km |
| 25-mm BFV | 2,200 |
| 155-mm M109A3 | 18,100 |
| M198 | 24,000 |
| 8-in Howitzer | 22,900 |

WEAPONS (MORTAR) HE ONLY (FM 7-90)

| | MIN | MAX |
|---|---|---|
| 60-mm | 70m | 3,500m |
| 81-mm (M252) | 80m | 5,800m |
| 81-mm (M29A1) | 73m | 4,790m |
| 4.2-inch | 770m | 6,840m |
| 120-mm | 200m | 7,200m |

FPFs (FM 7-90)

| GUNS | MORT | WIDTH | | DEPTH |
|---|---|---|---|---|
| 2 | 60-mm | 60m | x | 30m |
| 4 | 81 (M252) | 150m | x | 50m |
| 4 | 81 (M29A1) | 140m | x | 40m |
| 6 | 4.2-in | 240m | x | 40m |
| 6 | 120-mm | 360m | x | 60m |
| Plt | 155-mm | 200m | x | 50m |
| Btry | 155-mm | 400m | x | 50m |

*Optimum engagement ranges

**MEDEVAC REQUEST**

1. Requesting unit identification.
2. Location.
3. Number of patients by type (litter or ambulatory).
4. Type of injuries.
5. Special equipment needed.
8. Tactical situation.

**AIRCRAFT REQUEST**

1. Identification.
2. Precedence/priority.
3. Target description.
4. Target location.
5. Target time/date.
6. Desired ordinance/results.
7. Final control.

**DELIBERATE ATTACK CONSIDERATIONS**

1. Reconnoiter, pinpoint objective/enemy positions/obstacles.
2. Determine weak points; designate supporting positions.
3. Assign platoon/squad objectives—identify the decisive point.
4. Determine main attack, supporting attack, reserve.
5. Assign breach, support, assault missions.
6. Designate fire control measures.
7. Coordinate indirect/direct fires and CAS to time of attack.
8. Control measures during attack.
9. Secure (ground and air).
10. Consolidate and reorganize.

**NBC 1 OBSERVER'S INITIAL OR FOLLOW-UP REPORT**

**Instructions**

1. Line items D and H are mandatory for NBC 1 reports.
2. Line items A, E, G, I, K, L, M, S, Y, and ZA are optional for NBC 1 reports.
3. Line Items B, C, F, PAR, and PBR are reported if data is available.

**Section I. Chemical or Biological Only**

A. Strike serial number, if known (assigned by NBCE).
B. Position of observer.
C. Azimuth of attack from observer (state degrees or mils).
D. Date and time attack started (Zulu, local, or letter zone).
E. Time attack ended, if known.
F. Location of attack (UTM or place) (state actual or estimated).
G. Means of delivery, if known.
H. Type of agent and height of burst, if known.
I. Type and number of munitions or aircraft (state which).
K. Description of terrain (bare, scrubby vegetation, wooded, urban, or unknown).
S. Date and time contamination detected (Zulu, local, or letter zone).
Y. Representative downwind direction—4 digits (state degrees or mils), wind speed—3 digits (data kmph or knots).
ZA. Temperature (centigrade)—2 digits, cloud cover— 1 digit, significant weather phenomena—1 digit, air stability—1 digit
ZB. Remarks

**Section II. Nuclear Only**

A. Strike serial number, if known (assigned by NBCE).
B. Position of observer.
C. Azimuth of attack from observer (state degrees or mils and grid or magnetic).
D. Date and time attack started (Zulu, local, or letter zone).
F. Location of attack (UTM or place) (state actual or estimated).
G. Means of delivery, if known.
H. Type of burst (state air, surface, or unknown).
J. Flash-to-bang time (seconds).
K. Crater diameter (meters), if known.
L. Cloud width at H +5 minutes (degrees or mils).
M. Cloud angle(top or bottom) or cloud height(top or bottom) at H +10 minutes (state degrees, mils, meters, or feet).
PAR. Location of radioactive cloud outline (UTM).
PAB. Downwind direction of radioactive cloud (state degrees or mils).
ZB. Remarks.

**DEFENSE PLANNING CONSIDERATIONS**

1. Establish security (OP/patrols, PWs, M8),
2. Position key weapons:
   a. Coordinate with units on left and right
   b. Establish FPF or PDF for machine gun.
   c. Ensure mutual support between machine guns.
   d. Cover armor approaches with antiarmor systems.
   e. Establish fire control measures.
3. Prepare positions:
   a. Check sectors of fire
   b. Check overhead cover and view positions from enemy vantage.
   c. Position in depth and achieve mutual support between positions.
   d. Select/prepare alternate and supplementary positions.
4. Integrate indirect fires, CAS, and obstacles with direct and indirect fire.
5. Check communications and establish emergency signals.
6. Designate ammunition, supply, PW, and casualty points.

## IED / UXO

### Procedures when IEDs are found

**Security -** Maintain 360 degrees. Scan close in and far out, up high and down low.

**Always -** Scan you immediate surroundings for more IEDs.

**Move -** Move away. Vary safe distances, but plan for 300m minimum safe distance and adapt to your METT-TC.

**Attempt -** To confirm suspected IEDs using optics while staying back as far as possible.

**Cordon -** Off the area. Direct people out of danger area. Do not allow anyone to enter except for EOD. Question, search, and detain suspects as defined by your Existing ROE.

**Report -** Your situation using the 9 line IED / UXO spot report format.

---

This could be your hand if you try to dispose of UXOs or IEDs. The enemy has developed anti-handling to catch you when you try defusing. Leave it to experts!

**Call EOD - Don't be a Hero!**

### IED / UXO Report

**LINE 1.** DATE-TIME-GROUP: When the item was discovered.

**LINE 2.** REPORT ACTIVITY AND LOCATION: Unit and grid location of the IED/UXO.

**LINE 3.** CONTACT METHOD: Radio frequency, call sign, POC, and telephone number.

**LINE 4.** TYPE OF ORDNANCE: Dropped, projected, placed, or thrown. Give the number of items, if more than one.

**LINE 5.** NBC CONTAMINATIONS: Be as specific as possible.

**LINE 6.** RESOURCES THREATENED: Equipment, facilities, or other assets that are threatened.

**LINE 7.** IMPACT ON MISSION: Short description of current tactical situation and how the IED/UXO affects the status of the mission.

**LINE 8.** PROTECTIVE MEASURES: Any measures taken to protect personnel and equipment.

**LINE 9.** RECOMMENDED PRIORITY: Immediate, Indirect, Minor, No Threat.

---

### Priority

**Immediate:** Stops unit's maneuver and mission capability or threatens critical assets vital to the mission.

**Indirect:** Stops the unit's maneuver and mission capability or threatens critical assets important to the mission.

**Minor:** Reduces the unit's maneuver and mission capability or threatens non-critical assets of value.

**No Threat:** Has little or no effect on the unit's capabilities or assets.

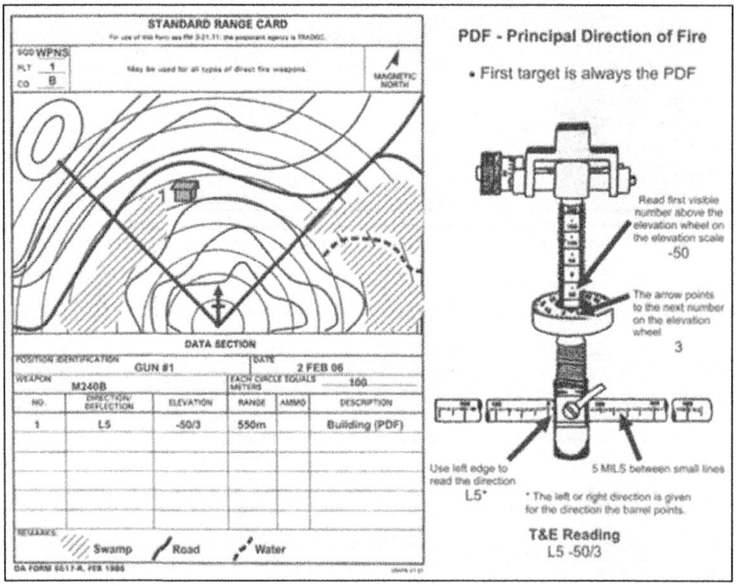
STANDARD RANGE CARD
SQD WPNS
PLT 1
CO B
May be used for all types of direct fire weapons.
MAGNETIC NORTH
1
DATA SECTION
POSITION IDENTIFICATION GUN #1
DATE 2 FEB 06
WEAPON M240B
EACH CIRCLE EQUALS 100 METERS
NO.
DIRECTION/DEFLECTION
ELEVATION
RANGE
AMMO
DESCRIPTION
1
L5
-50/3
550m
Building (PDF)
REMARKS:
Swamp
Road
Water
DA FORM 5517-R, FEB 1986
PDF - Principal Direction of Fire
• First target is always the PDF
Read first visible number above the elevation wheel on the elevation scale
-50
The arrow points to the next number on the elevation wheel
3
Use left edge to read the direction
L5*
5 MILS between small lines
* The left or right direction is given for the direction the barrel points.
T&E Reading
L5 -50/3

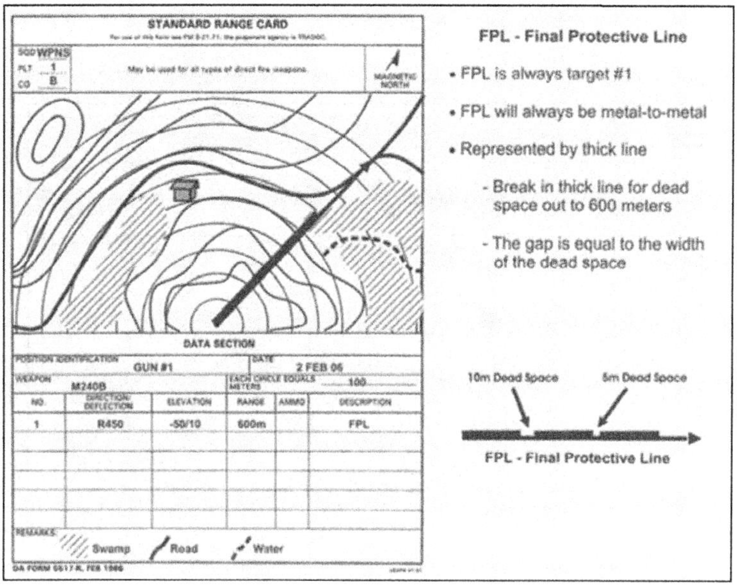
STANDARD RANGE CARD
SQD WPNS
PLT 1
CO B
May be used for all types of direct fire weapons.
MAGNETIC NORTH
DATA SECTION
POSITION IDENTIFICATION GUN #1
DATE 2 FEB 06
WEAPON M240B
EACH CIRCLE EQUALS METERS 100
NO.
DIRECTION/ DEFLECTION
ELEVATION
RANGE
AMMO
DESCRIPTION
1
R450
-50/10
600m
FPL
REMARKS:
Swamp
Road
Water
DA FORM 5517-R, FEB 1986
FPL - Final Protective Line
• FPL is always target #1
• FPL will always be metal-to-metal
• Represented by thick line
- Break in thick line for dead space out to 600 meters
- The gap is equal to the width of the dead space
10m Dead Space
5m Dead Space
FPL - Final Protective Line

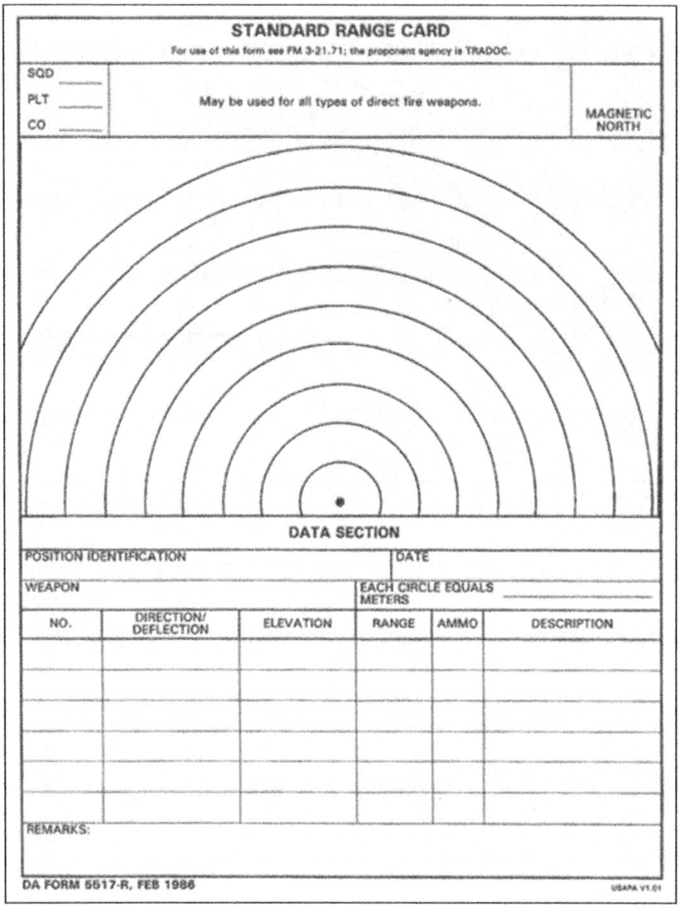

STANDARD RANGE CARD

For use of this form see FM 3-21.71; the proponent agency is TRADOC.

SQD ______
PLT ______
CO ______

May be used for all types of direct fire weapons.

MAGNETIC NORTH

DATA SECTION

POSITION IDENTIFICATION | DATE

WEAPON | EACH CIRCLE EQUALS ______ METERS

| NO. | DIRECTION/ DEFLECTION | ELEVATION | RANGE | AMMO | DESCRIPTION |
|---|---|---|---|---|---|
| | | | | | |
| | | | | | |
| | | | | | |
| | | | | | |
| | | | | | |
| | | | | | |

REMARKS:

DA FORM 5517-R, FEB 1986

USAPA V1.01

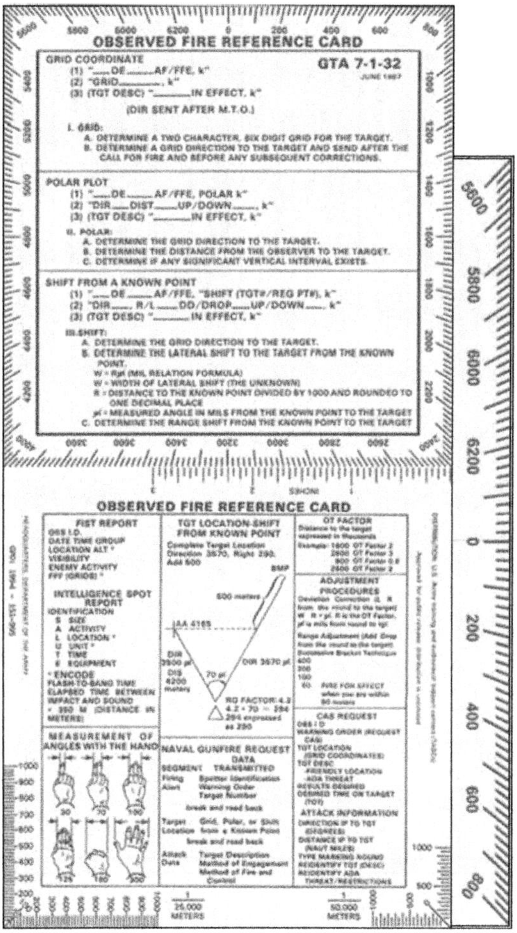

OBSERVED FIRE REFERENCE CARD

GTA 7-1-32

JUNE 1987

GRID COORDINATE
(1) "____DE____AF/FFE, k"
(2) "GRID________, k"
(3) (TGT DESC) "________IN EFFECT, k"

(DIR SENT AFTER M.T.O.)

I. GRID:
A. DETERMINE A TWO CHARACTER, SIX DIGIT GRID FOR THE TARGET.
B. DETERMINE A GRID DIRECTION TO THE TARGET AND SEND AFTER THE CALL FOR FIRE AND BEFORE ANY SUBSEQUENT CORRECTIONS.

POLAR PLOT
(1) "____DE____AF/FFE, POLAR k"
(2) "DIR____DIST____UP/DOWN____, k"
(3) (TGT DESC) "________IN EFFECT, k"

II. POLAR:
A. DETERMINE THE GRID DIRECTION TO THE TARGET.
B. DETERMINE THE DISTANCE FROM THE OBSERVER TO THE TARGET.
C. DETERMINE IF ANY SIGNIFICANT VERTICAL INTERVAL EXISTS.

SHIFT FROM A KNOWN POINT
(1) "____DE____AF/FFE, "SHIFT (TGT#/REG PT#), k"
(2) "DIR____, R/L____DD/DROP____UP/DOWN____, k"
(3) (TGT DESC) "________IN EFFECT, k"

III. SHIFT:
A. DETERMINE THE GRID DIRECTION TO THE TARGET.
B. DETERMINE THE LATERAL SHIFT TO THE TARGET FROM THE KNOWN POINT.
W = Rµ (MIL RELATION FORMULA)
W = WIDTH OF LATERAL SHIFT (THE UNKNOWN)
R = DISTANCE TO THE KNOWN POINT DIVIDED BY 1000 AND ROUNDED TO ONE DECIMAL PLACE
µ = MEASURED ANGLE IN MILS FROM THE KNOWN POINT TO THE TARGET
C. DETERMINE THE RANGE SHIFT FROM THE KNOWN POINT TO THE TARGET

INCHES

OBSERVED FIRE REFERENCE CARD

FIST REPORT
OBS I.D.
DATE TIME GROUP
LOCATION ALT *
VISIBILITY
ENEMY ACTIVITY
FPF (GRIDS) *

INTELLIGENCE SPOT REPORT
IDENTIFICATION
S SIZE
A ACTIVITY
L LOCATION *
U UNIT *
T TIME
E EQUIPMENT
* ENCODE
FLASH-TO-BANG TIME
ELAPSED TIME BETWEEN IMPACT AND SOUND × 350 M (DISTANCE IN METERS)

MEASUREMENT OF ANGLES WITH THE HAND
30 70 100
125 180 300

TGT LOCATION-SHIFT FROM KNOWN POINT
Complete Target Location
Direction 3570, Right 290, Add 500
BMP
500 meters
AA 4165
DIR 3500 µ
DIS 4200 meters
70 µ
DIR 3570 µ
RO FACTOR: 4.2
4.2 × 70 = 294
294 expressed as 290

NAVAL GUNFIRE REQUEST DATA
SEGMENT TRANSMITTED
Firing Alert — Spotter Identification, Warning Order, Target Number
break and read back
Target Location — Grid, Polar, or Shift from a Known Point
break and read back
Attack Data — Target Description, Method of Engagement, Method of Fire and Control

OT FACTOR
Distance to the target expressed in thousands
Example: 1800 OT Factor 2
2800 OT Factor 3
800 OT Factor 0.8
2500 OT Factor 2

ADJUSTMENT PROCEDURES
Deviation Correction (L, R from the round to the target)
W = R × µ. R is the OT Factor, µ is mils from round to tgt.
Range Adjustment (Add, Drop from the round to the target)
Successive Bracket Technique
400
200
100
50 FIRE FOR EFFECT when you are within 50 meters

CAS REQUEST
OBS I D
WARNING ORDER (REQUEST CAS)
TGT LOCATION (GRID COORDINATES)
TGT DESC
-FRIENDLY LOCATION
-ADA THREAT
RESULTS DESIRED
DESIRED TIME ON TARGET (TOT)

ATTACK INFORMATION
DIRECTION IP TO TGT (DEGREES)
DISTANCE IP TO TGT (NAUT MILES)
TYPE MARKING ROUND
REIDENTIFY TGT (DESC)
REIDENTIFY ADA THREAT/RESTRICTIONS

1/25,000 METERS

1/50,000 METERS

HEADQUARTERS, DEPARTMENT OF THE ARMY

GPO: 1994 — 333-995

DISTRIBUTION: U.S. Army training and evaluation support centers (TASCs)

Approved for public release; distribution is unlimited

## 9 Line MEDEVAC

| LINE ITEM | EXPLANATION |
|---|---|
| **1. Location of Pickup Site.** | Encrypt grid coordinates. When using *DRYAD Numeral Cipher, the same SET line* will be used to encrypt grid zone letters and coordinates. To preclude misunderstanding, a statement is made that grid zone letters are included in the message (unless unit SOP specifies its use at all times). |
| **2. Radio Frequency, Call Sign, Suffix.** | Encrypt the frequency of the radio at the pickup site, *not* a relay frequency. The call sign (and suffix if used) of person to be contacted at the pickup site may be transmitted in the clear. |
| **3. No. of Patients by Precedence.** | Report only applicable info & encrypt brevity codes. A = Urgent, B = Urgent-Surg, C = Priority, D = Routine, E = Convenience. (If 2 or more categories reported in same request, insert the word "*break*" between each category). |
| **4. Spec Equipment.** | Encrypt applicable brevity codes. A = None, B = Hoist, C = Extraction equipment, D = Ventilator. |
| **5. No. of Patients by Type.** | Report only applicable information and encrypt brevity code. If requesting MEDEVAC for both types, insert the word "*break*" between the litter entry and ambulatory entry: L + # of Pnt - Litter; A + # of Pnt - Ambul (sitting). |
| **6. Security Pickup Site (Wartime).** | N = No enemy troops in area, P = Possibly enemy troops in area (approach with caution), E = Enemy troops in area (approach with caution), X = Enemy troops in area (armed escort required). |
| **6. Number and type of Wound, Injury, Illness (Peacetime).** | Specific information regarding patient wounds by type (gunshot or shrapnel). Report serious bleeding, along with patient blood type, if known. |
| **7. Method of Marking Pickup Site.** | Encrypt the brevity codes. *A* = Panels, *B* = Pyrotechnic signal, *C* = Smoke Signal, *D* = None, *E* = Other. |
| **8. Patient Nationality and Status.** | Number of patients in each category need not be transmitted. Encrypt only applicable brevity codes. A = US military, B = US civilian, C = Non-US mil, D = Non-US civilian, E = EPW. |
| **9. NBC Contamination (Wartime).** | Include this line only when applicable. Encrypt the applicable brevity codes. N = nuclear, B = biological, C = chemical. |
| **9. Terrain Description (Peacetime).** | Include details of terrain features in and around proposed landing site. If possible, describe the relationship of site to a prominent terrain feature (lake, mountain, tower). |

Reference: FM 8-10-6, *Medical Evacuation in a Theater of Operations, pages 7-7 through 7-9.*

Made in the USA
Monee, IL
05 November 2026

67811405R00275